Reporting Experimental Data

Selected Reprints

Howard J. White, Jr., EDITOR
Guest Scientist, National Institute of Standards and Technology

Compiled under the auspices
of the American Chemical Society
Task Force on Scientific Numeric Data

American Chemical Society, Washington, DC 1993

Library of Congress Cataloging-in-Publication Data

Reporting experimental data: selected reprints / Howard J. White, Jr., editor; compiled under the auspices of the American Chemical Society, Task Force on Numeric Data.

p. cm.

Includes bibliographical references and index.

ISBN 0–8412–2529–X

1. Chemical literature—Standards. 2. Chemistry—Notation—Standards.

I. White, Howard J., Jr. II. American Chemical Society. Task Force on Numeric Data

QD8.5.R47 1993
808′.06654—dc20 93–14521
CIP

The paper used in this publication meets the minimum requirements of American National Standard for Information Sciences—Permanence of Paper for Printed Library Materials, ANSI Z39.48–1984. ∞

PRINTED IN THE UNITED STATES OF AMERICA

1993 Advisory Board

Contents

Preface

CHEMICAL RESEARCH FROM THE LABORATORY to the use of the resulting papers was studied as an entity in the 1950s and 1960s. The following observations emerged: The chemical literature was large and rapidly increasing, a good set of chemical measurements was expensive and costs were rising, and the need for new data was also quickly expanding as the increasing base of knowledge caused the number of new justifiable measurements to increase accordingly.

These observations, which might seem seem contradictory or divergent but which were amply verified, led to the conclusions that we would be unable to measure everything we wanted to measure and that we must make maximum use of the previously measured data. The abstract journals, which had begun as heavily subsidized ventures by individuals and professional societies, became highly specialized operations with substantial increases in costs. Programs to provide critically evaluated databases started in some countries, and a variety of evaluated and unevaluated databases began to appear, many of which were automated. Major companies in the chemical process industry began to prepare databases for use in design processes. In short, the importance of secondary products produced by selection and analysis from the primary literature was recognized, and methods of how to prepare such products were expanded, improved, and diversified.

This scrutiny of the primary chemical literature made it clear that in the process of preparing measurements for publication, authors were not including important information needed by subsequent users of the data. Because it is important to obtain as much value from each measurement as possible, ways were sought to strengthen this seemingly remediable weakness. Associations of experts began preparing recommendations for publication of measurements in their specialties. The International Union of Pure and Applied Chemistry (IUPAC) and the Committee on Data for Science and Technology of the International Council of Scientific Unions (CODATA) took the lead in providing such recommendations in a broader based, more systematic way. During the same period, the standardization of nomenclature, symbols, and units was being developed by many of the same organizations, particularly by IUPAC.

A Task Force on Scientific Data was established first under the American Chemical Society Committee in Science and later converted to a Society Task Force. The task force recognized that data published in the chemical literature often could not be fully used because of the authors' failure to provide the auxiliary information required by the user. Although many organizations had prepared guidelines on the presentation of data in the literature to alleviate this problem, the guidelines were widely scattered and generally unknown to most chemists. Therefore, the task force thought that ACS could provide a valuable service by reprinting these guidelines in one book readily available to authors.

This volume gathers the recommendations for publication of data for measurements of a physical chemical nature together with current recommendations for quantities, symbols, and units. The recommendations for quantities, symbols, and units are those made by IUPAC. The recommendations for the publication of data were gathered from a comprehensive literature search. In this latter endeavor, I thank the following individuals at the National Institute of Science and Technology (NIST) for finding relevant papers in their area of expertise: Robert Hampson, Frank J. Lovas, William C. Martin, Alan D. Mighell, and Wolfgang L. Wiese. My special thanks go to David R. Lide, Jr., who provided valuable assistance at every stage of the way, and who suggested that I might undertake this task in the first place. It would have been impossible to cover as wide a range without their assistance. Financial assistance from NIST is also appreciated.

HOWARD J. WHITE, JR.
National Institute of Standards and Technology
Building 221, Room A323
Gaithersburg, MD 20899

July 10, 1992

Section 1

Introduction: Describing and Presenting Measurements

Purpose

Chemists and engineers who are writing results of quantitative measurements of physical or chemical properties of substances for publication are the audience for this book. It gives guidelines on how to present the measured data.

Some researchers may feel that, because they understand theier own data better than anyone else, they do not need to read a book on presenting data. However, the purpose of publishing is making data available so other people can study and use them. The purpose of this book is to assist the researcher in providing those bits of information a reader will need to fully use the data. In other words, this book will help maximize the value of the data.

This book is similar to an airline pilot's checklist, which doesn't teach the pilot how to fly. It is meant to ensure that, in the process of getting the plane away from the gate, onto the runway, and into the air, something of great importance for a future step doesn't get overlooked. As is true of the pilot's checklist, the points in in this book are usually obvious once they are mentioned. As is also true of the pilot's checklist, a long history of obvious but overlooked points led to the conclusion that providing such a checklist is desirable.

What problems does this checklist help resolve and what items does it include? To answer these questions, we not only need to focus on the total publication process—including readers as well as an author, and a publisher—but also to consider the readers' situation in approaching an article. This step will be done first with some general comments and then with a specific example.

Consider any measurement of any quality. Clearly, some type of property of the sample has been measured. If replicate measurements seem to fluctuate randomly around some average value, a statistical limit can be placed upon the precision within which the measurement can be reproduced. The measurer is now entitled to say that some type of property of the sample has been measured within a certain specified limit of reproducibility.

The measurer is not interested in "some type of property," however, but in a specific, well-defined property that will be recognized and accepted by others as being of value in describing the sample's behavior. The data user is also usually not interested in the properties of the specific unique sample, but rather in a category of substances that the sample is intended to represent. In the measurer's mind, and in the paper that will be written, the data will be attributed to that well-defined property of that category of substance. In chemistry, the properties can usually be classed as thermodynamic, spectroscopic, and so forth, and the categories of substances can usually be specified in terms of atoms or molecules of certain kinds or mixtures of them in solid, liquid, or gaseous states.

Consider a potential reader, who is not just browsing or keeping current but who has a specific interest involving data. The reader has in mind a specific property for a specific category of substances and has been drawn to the measurer's paper, perhaps by its title, an abstract, or a periodic systematic scanning of the literature. Do the data in the paper meet the reader's requirements?

For example, the reader is interested in the thermal conductivity of copper. Figures 1.1 and 1.2 (*1*) show what some readers with an interest in the thermal conductivity of copper found when making a comprehensive search of the literature a few years ago. Both figures cover the same data set; the plot has been changed to enhance the low-temperature region in Figure 1.2. Each figure contains a large number of curves, all

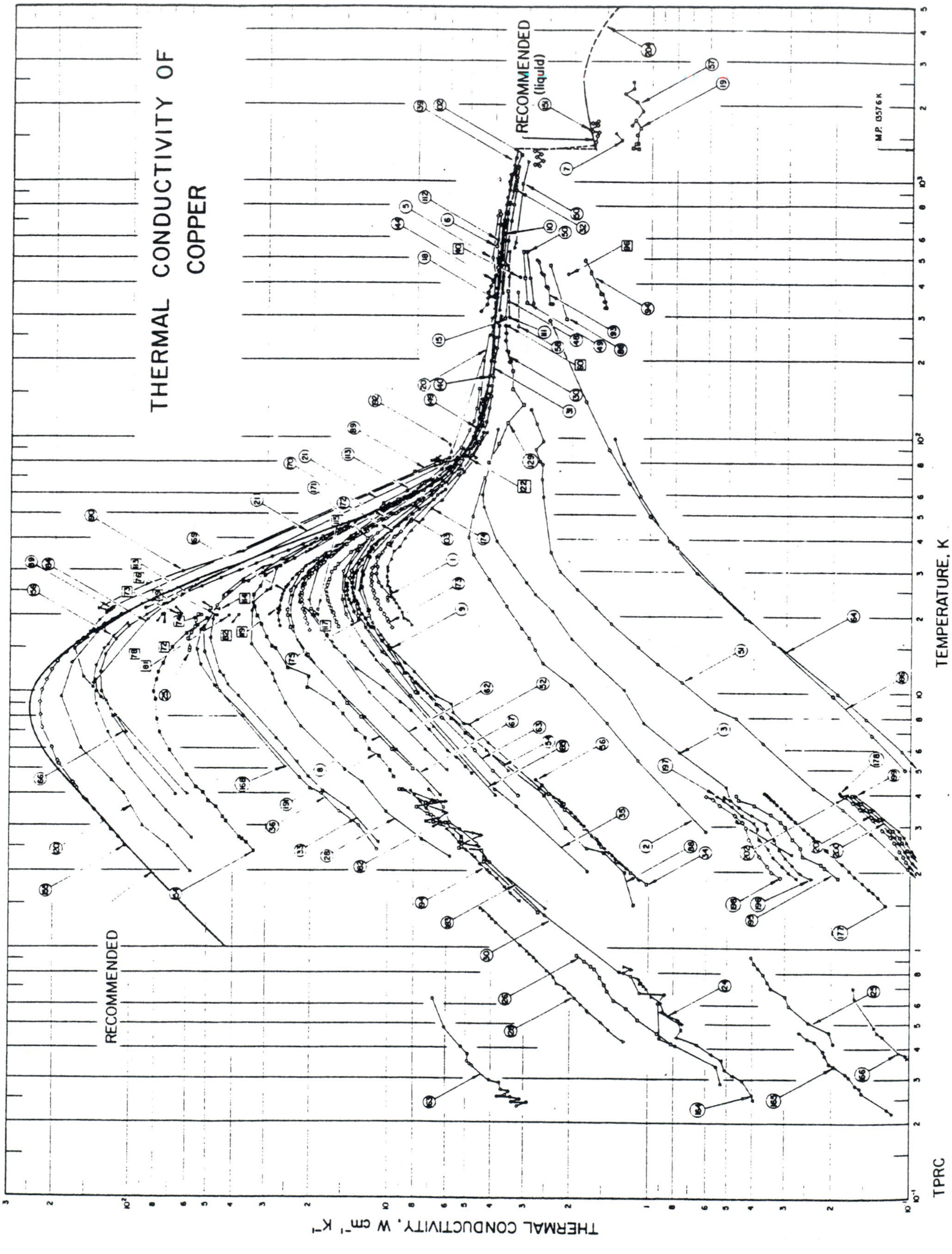

Figure 1.1. Thermal Conductivity of Copper. (*Reproduced from reference 1. Copyright 1974 American Chemical Society.*)

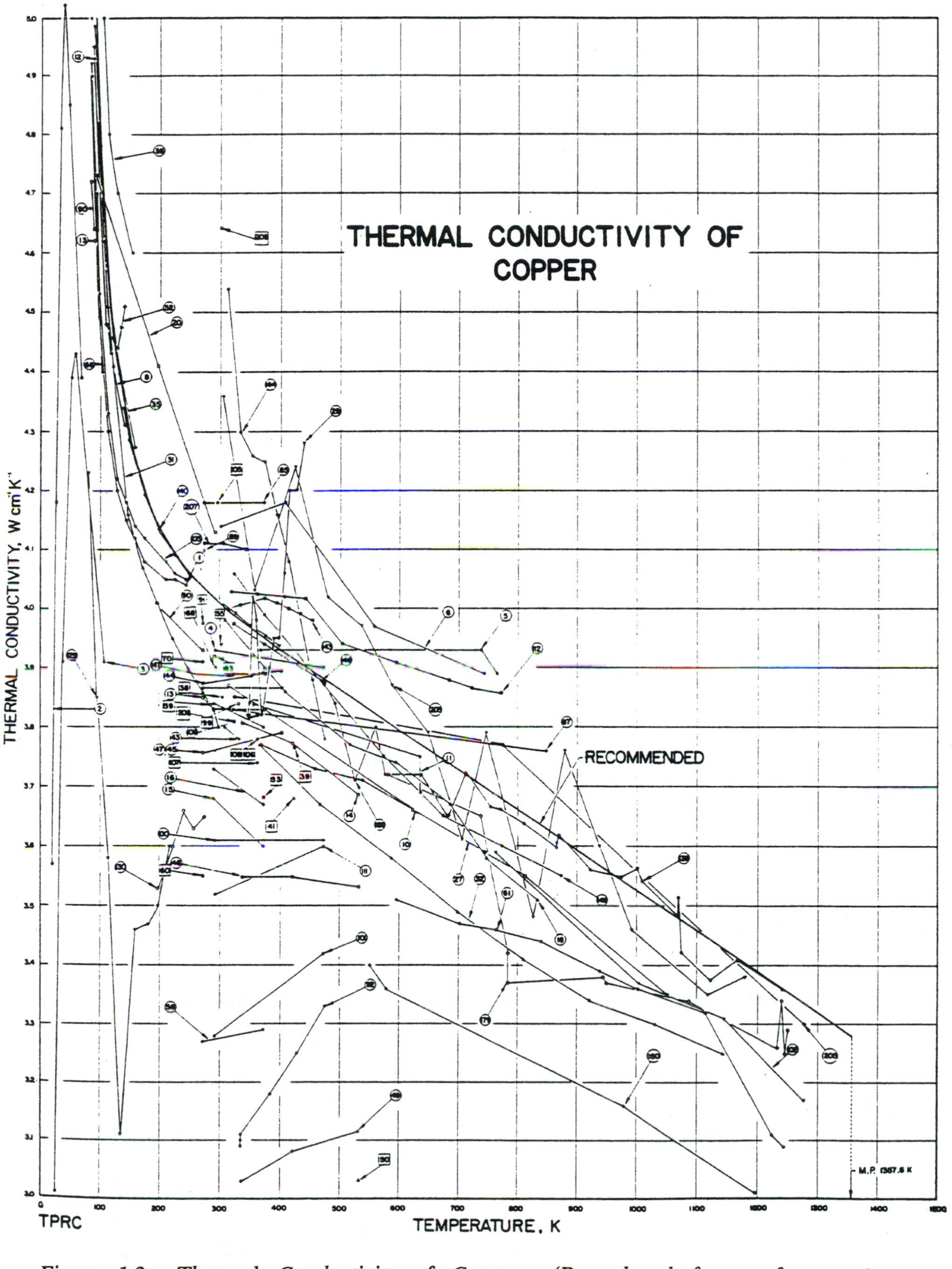

Figure 1.2. Thermal Conductivity of Copper. (Reproduced from reference 1. Copyright 1974 American Chemical Society.)

but one of which connect sets of measured data points. The curve labeled "recommended" represents a data evaluator's estimate of the locus of the most probable values. All data sets were found in a search of the literature for data on the thermal conductivity of copper, but, clearly, they don't all represent the same thermal conductivity for the same copper.

The designation "copper" implies that the properties refer to a material composed exclusively of atoms of copper, or if impurities are present, their contribution to the behavior of the sample is vanishingly small or falls within the stated uncertainties of the measurement. In practice, for convenience in categorizing and indexing, the designation may be extended to include copper containing small amounts of impurities whether or not they have a perceptible effect. Thus the study of the effects of small amounts of impurities on the properties of copper could be classified under copper in an index.

To some extent, this is the case for the thermal conductivity of copper below 150 K. The thermal conductivity of metals peaks at low temperatures, and very small amounts of impurities have an observable effect in depressing the peak. The purer the metal, the higher the peak. Because some, but not all, measurers were aware of the theoretically predicted peak and the possible influence of small amounts of impurities, they carefully described the composition of their samples and may have provided data respectably representative of the behavior of copper of that composition, although not of copper in the stricter sense.

Some evaluators have generated a family of recommended curves at low temperatures, each representative of a different level of purity. Note that the "recommended" curve in Figures 1.1 and 1.2 follows the highest curves rather closely. It is possible that measurements on still purer copper would be even higher than those shown, so that the recommended values for "copper" would need to be adjusted upward.

Because the sensitivity to small quantities of impurities sharply declines above 150 K, factors other than chemical composition must be considered to explain the differences between the data sets. Measurement of the thermal conductivity of solids is a difficult task. For example, the interface between the sample and the source of heat is difficult to define because neither presents a uniform plane surface to the other. The researcher must account for many kinds of heat losses. Factors affecting the physical composition of the sample, such as method of production and pretreatment, may be important. In short, many potential sources of systematic differences between measurements exist. Most data sets describe the same type of behavior with substantial offsets among them. A few data sets simply don't belong.

The thermal conductivity of copper makes a useful example because there are many sets of data and it is a difficult measurement. However, there is no reason to consider it unique. Experienced data users realize that every property of every substance presents a similar problem.

What are you, as the prospective author of a paper containing experimental data to think of this situation? One obvious conclusion is that making measurements is tricky, but you already knew this. A second conclusion is that the sample must be adequately described, and a third is that the measurement must be adequately described. A prospective user will face a situation similar to that shown in the example. The important thing is to provide enough information so the user can decide if your sample and measurement match his or her needs.

An adequate description of the sample requires a careful analysis of the degree to which the measurements are affected by the composition and structure of the sample. What is adequate changes from measurement to measurement and may even change over the range of an independent variable, as in the case of the thermal conductivity at low temperatures. Presumably, most of this analysis was performed when the measurements were planned, but it is possible that a greater appreciation of the influence of some variable will be developed during the course of the experiment. If providing adequate compensation for the effect of some variable that is expected to have an influence on the results is impossible, an estimate of the magnitude of the uncertainty thus introduced should be discussed in the process of estimating the overall uncertainty of the measurement.

An adequate description of the measurement should include a clear description of the actual measurement process and a discussion of the relationship of the measurements actually made to the desired and expressed data. An adequate description includes a discussion of any assumptions and approximations used in the conversion process and the influence of uncertainties in any collateral data needed. The measurement and control of the independent variables that define the ambient conditions are usually important. The reproducibility of replicate measurements can be treated statistically; the degree to which possible systematic errors have been avoided or compensated for usually requires analysis and some degree of judgment.

One additional point is worth making. You may be thinking, "I'm an engineer with neither the time nor the inclination to make a world-class measurement. What's written here doesn't apply to me." Or perhaps, "I'm an organic chemist. What's important to me is my newly discovered method for synthesizing a new class of compounds. The measured data on the compounds represent a collateral part of the paper and don't merit an extended description. My focus is on the synthetic method."

If you have had similar thoughts, consider the following. Your data are valuable in their own right. The example that has been discussed is not characteristic in one important respect. Such an array of data is rare; a large proportion of the sets in the literature contains measurements that have not been and probably will not be repeated. World-class measurements are very nice, but any measurement may be useful. The point is to describe the measurement sufficiently so that the reader can make an informed estimate of its reliability and therefore make maximum use of it. If a point is not covered, the reader or user has no recourse but to assume that it was not considered and to adjust any estimate of reliability downward accordingly.

Scope and Format

The subject considered in this book has already been characterized as the result of quantitative measurements of physical and chemical properties of substances, and reasons have been given for spending some effort on preparing adequate descriptions of the measured samples and the procedures used in making the measurement. Some general considerations of what "adequate" means have also been given.

Section 2 presents more details on what constitutes an adequate description. However, because there are limits beyond which it is difficult to go by taking a general approach, sections devoted to the following specific categories of measurements are included:

- thermodynamics including biothermodynamics
- chemical kinetics and transport properties
- electrochemistry
- colloid and surface chemistry
- photochemistry
- analytical chemistry
- crystallography and electron diffraction
- spectroscopies

A final section entitled "Automated Products" deals briefly with the impact automation is beginning to have on the subject covered in this book.

Dealing with each of these categories in some detail requires the help of experts in each field. Fortunately, such help is already available.

As the volume of technical literature increased and began to be used for a greater variety of purposes, the question of what constituted suitable publication of data began to be considered. An International Union of Pure and Applied Chemistry (IUPAC) report on thermochemistry from 1934 (*2*) contained some recommendations on the publication of data. This is the earliest such document we found. The Eighth (U.S.) Calorimetry Conference in 1953 adopted a resolution outlining minimum publication standards for the guidance of authors, editors, and referees of papers on calorimetry. This resolution was circulated and well received, and followed by a revised and expanded resolution that was adopted by the 15th IUPAC Conference in 1940, and then published (*3*). Similar recommendations in other areas began to appear at about the same time and have been appearing ever since.

These documents are prepared by experts, often by committees of experts assigned the task by appropriate scientific or technical societies. The reports also tend to have received extensive review and critiquing within the field. For these reasons, a prospective user of materials is reluctant to summarize or make editorial changes. Because the content is crucial to this book, the documents will be reproduced in their entirety or selected sections will be used without any textual changes except for possible deletion of redundant material or changes of reference numbers.

As is always true when arbitrary categories have been established, the placement of things into specific categories is also arbitrary. Readers are advised to check related sections to ensure that items of interest haven't been put there instead. This suggestion applies particularly to the area of analytical chemistry, the practitioners of which regularly use spectroscopic, thermodynamic, and other techniques, for example. Conversely, those interested in other categories may find items of interest in the analytical section.

The primary purpose of this book is to assemble much material related to the proper publication of experimental data in one volume to provide easy access to those who need it. An attempt has been made to provide comprehensive coverage in this area. Additional collateral material is cited by reference.

Quantities, Symbols, and Units

One of the first things that must be done in preparing measurements for publication is to decide how to express quantities, symbols, and units. These are large (and often contentious) subjects in their own rights.

The Commission on Physicochemical Symbols, Terminology, and Units of the Physical Chemistry Division of IUPAC emerged as the leader in establishing a set of quantities and symbols for chemistry. This set has been widely reviewed and agreed upon by chemists from many countries and has been coordinated with sets prepared by similar groups representing physicists, biologists, and so on, as well as by the International Organization for Standardization.

A comprehensive summary, entitled *Quantities, Units and Symbols in Physical Chemistry* (*4*) that is widely known as the Green Book, has recently been published. This book condenses three previous editions, five appendices, and a number of expanded sections some of which will be referred to later (*4a–4k*).

We strongly recommend that the Green Book be followed. Because our purpose is to provide a rather comprehensive set of guidelines, relevant segments from the Green Book are reproduced in the various specific sections that follow. Terminology should be defined at all times. If, for some reason, you must use terminology that differs from the usage recommended in the Green Book, you should define the terminology and relate it to that of the Green Book.

The Green Book recommends that the International System of Units (SI units) be used for scientific and technical purposes. The SI units were adopted by the 11th General Conference on Weights and Measures (reference 2 in the Green Book reference section) in 1960. SI units provide a coherent system of units built from seven SI base units, one for each of the seven dimensionally independent-base quantities: length, mass, time, electric current, thermodynamic temperature, amount of substance, and luminous intensity. The remaining quantities are expressed by derived units that can be expressed as products of the base units. Approved SI decimal prefixes may be used for convenience.

We recommend here that the SI units be used together with the SI prefixes as appropriate. A few non-SI units with strong traditional chemical usage are also accepted on a transitional basis. Appendix 1 provides a brief review of SI units. If special reasons for using non-SI units exist, you should define these units in terms of SI units and give appropriate conversion factors.

In preparing the Green Book, the IUPAC committee referred to most of the papers containing recommendations on units, symbols, or terminology. Papers in which such recommendations are the exclusive or primary concern, which are referred to in the Green Book, are considered to be superseded by it. However, some papers cover specific fields in more depth than the Green Book does, and may include recommendations beyond the Green Book's scope of coverage. These papers may even include recommendations for publication, for example, collected information that must be included to specify completely complex properties. Certainly, the papers expand and clarify the material in the Green Book, which has been highly condensed and very carefully worded.

We could make a case for including all of this supplementary material for completeness. However, what might be ideal from one point of view is often troublesome from another. Because such a procedure would substantially increase the size and price of this volume, we compromised.

All supplementary material is referenced and described, but only a few limited papers are reprinted in this book. Prospective authors of papers in certain areas are encouraged to read the appropriate supplementary material. This advice applies particularly to prospective authors in fields such as surface chemistry or electrochemistry, in which, by the nature of the field, the definitions of quantities tend to be subtle and complex. Any prospective author who is not thoroughly grounded in the field that includes the involved measurements is advised to read the relevant supplementary material in addition to the printed section from the Green Book.

Just as papers on quantities, symbols, and units may contain advice on what must be included in a suitable publication, papers recommending procedures for publication often make recommendations on quantities, symbols, and units. A more or less continuous range of papers, from those wholly or primarily concerned with nomenclature and terminology to those wholly or primarily concerned with recommendations about publication of data, is available.

Where discrepancies between the recommendations for quantities, symbols, or units in reprinted papers and the recommendations in the Green Book occur, those in the Green Book are to be taken as definitive.

Statistical Treatment of Data

Most data users are interested in the reliability or uncertainty of the data. For one user, an order-of-magnitude estimate may suffice because the value of that particular property exerts very little leverage on the calculation the user has in mind. For a second user, the calculation and the project that depends on the calculation may stand or fall on the degree of reliability of the value.

With appropriate experimental design and statistical treatment, the reproducibility of a measurement can be quite precisely determined. The degree of reliability for the desired property depends also upon possible systematic errors. Most often, uncertainties from recognized sources of possible systematic errors are larger than those associated with the reproducibility of the measurement. They tend to be more amenable to physical analysis, often involving collateral measurements, than to statistical treatment. Systematic errors from unrecognized sources can be discovered only by comparison with values obtained from many types of determinations or by comparison with related data taken under different conditions or on related substances. This is the field of the data evaluator and will not be discussed here. The preceding discussion refers implicitly to a single isolated data point. Where sets of data covering a range of conditions or interdependent properties are concerned, the treatment of the errors in reproducibility (or random errors) and of systematic errors becomes more complex.

We will not cover the field of statistics, which has a large and flourishing literature. Many of the reprinted papers have sections on the estimation of uncertainties. We have included in the reference list a few papers that cover estimation of uncertainties in specific fields that are intended as aids in preparing papers for publication.

Automated Products

Automation impinges on the publication of data in two primary ways: automated databases and automated procedures for publication. Some relevant aspects of each of these two ways are discussed in the final chapter of this book.

References

Because of the nature of this book, the organization of the references is somewhat more complicated than is usual. Each of the reprinted papers has its own references, which are an integral part of the paper and therefore are reproduced as part of it. The reprinted sections of the Green Book also contain their own reference numbers. The reference list for the Green Book is included as a separate item at the end of the text. Because the reprinted sections of the Green Book are always clearly indicated, it is easy for the reader to distinguish between sections taken from the Green Book and the original text. Finally, the text written for this volume has its own reference list.

Section 2

Items of General Use in Chemistry

Table 2.1, reproduced from the Green Book, provides information on quantities and symbols of general use in chemistry. These quantities and symbols refer to processes that underlie the information given in the other sections, and should be consulted by those primarily interested in the specialized sections.

Two papers are reprinted in this section. "Guide for the Presentation in the Primary Literature of Numerical Data Derived from Experiments" (*5*) (which supersedes reference 5a, "Guidelines for the Reporting of Numerical Data in Experimental Procedures") provides detailed general guidelines for preparing experimental data for publication. "Guide for the Presentation in the Primary Literature of Physical Property Correlations and Estimation Procedures" (*6*) is concerned with the preparation and presentation of correlations and estimation procedures. These two procedures are not only frequently used in the scientific literature, particularly in thermodynamics, transport properties, and electrochemistry, but also to some extent in most of the specialized areas mentioned in this volume. Some authors regress a correlating equation through their data and other data taken from the literature. The purpose is to extend their data's use over a wide range of temperature or concentration, for example, or perhaps to test the correlation procedure's validity. In any case, having some guidelines covering this widely used technique available is desirable.

Table 2.1. General chemistry. Continued on next page.

The symbols given by IUPAP [3] and by ISO [4.e,i] are in agreement with the recommendations given here.

Name	*Symbol*	*Definition*	*SI unit*	*Notes*
number of entities (e.g. molecules, atoms, ions, formula units)	N		1	
amount (of substance)	n	$n_B = N_B/L$	mol	1,2
Avogadro constant	L, N_A		mol^{-1}	
mass of atom, atomic mass	m_a, m		kg	
mass of entity (molecule, or formula unit)	m_f, m		kg	
atomic mass constant	m_u	$m_u = m_a(^{12}C)/12$	kg	3
molar mass	M	$M_B = m/n_B$	kg mol^{-1}	2, 4

(1) The words 'of substance' may be replaced by the specification of the entity.

Example When the amount of O_2 is equal to 3 moles, $n(O_2)=3$ mol, then the amount of $\frac{1}{2}O_2$ is equal to 6 moles, $n(\frac{1}{2}O_2)=6$ mol. Thus $n(\frac{1}{2}O_2)=2n(O_2)$. See also p.41 below.

(2) The definition applies to entities B which should always be indicated by a subscript or in parentheses, e.g. n_B or $n(B)$.

(3) m_u is equal to the unified atomic mass unit, with symbol u, i.e. $m_u = 1$ u (see section 3.7). In biochemistry this unit is called the dalton, with symbol Da, although the name and symbol have not been approved by CGPM.

(4) The definition applies to pure substance, where m is the total mass and V is the total volume. However, corresponding quantities may also be defined for a mixture as m/n and V/n, where $n=\sum_i n_i$. These quantities are called the mean molar mass and the mean molar volume respectively.

Table 2.1. Continued. Continued on next page.

Name	*Symbol*	*Definition*	*SI unit*	*Notes*
relative molecular mass (relative molar mass, molecular weight)	M_r	$M_{r,B}=m_B/m_u$	1	2, 5
molar volume	V_m	$V_{m,B}=V/n_B$	$m^3\ mol^{-1}$	2, 4
mass fraction	w	$w_B=m_B/\Sigma m_i$	1	2
volume fraction	ϕ	$\phi_B=V_B/\Sigma V_i$	1	2, 6
mole fraction, amount fraction, number fraction	x, y	$x_B=n_B/\Sigma n_i$	1	2, 7
(total) pressure	p, P		Pa	
partial pressure	p_B	$p_B=y_Bp$	Pa	8
mass concentration (mass density)	γ, ρ	$\gamma_B=m_B/V$	$kg\ m^{-3}$	2, 9, 10
number concentration, number density of entities	C, n	$C_B=N_B/V$	m^{-3}	2, 9, 11
amount concentration, concentration	c	$c_B=n_B/V$	$mol\ m^{-3}$	2, 9, 12
solubility	s	$s_B=c_B$ (saturated solution)	$mol\ m^{-3}$	2
molality (of a solute)	m, (b)	$m_B=n_B/m_A$	$mol\ kg^{-1}$	2, 13
surface concentration	Γ	$\Gamma_B=n_B/A$	$mol\ m^{-2}$	2
stoichiometric number	ν		1	14
extent of reaction, advancement	ξ	$\Delta\xi=\Delta n_B/\nu_B$	mol	2, 15
degree of dissociation	α		1	

(5) For molecules M_r is the relative molecular mass or molecular weight; for atoms M_r is the relative atomic mass or atomic weight and the symbol A_r may be used. M_r may also be called the relative molar mass, $M_{r,B}=M_B/(g\ mol^{-1})$.
(6) V_B and V_i are the volumes of appropriate components prior to mixing.
(7) For condensed phases x is used, and for gaseous mixtures y may be used.
(8) The symbol and the definition apply to entities B, which should be specified.
(9) V is the volume of the mixture.
(10) In polymer science the symbol c is often used for mass concentration.
(11) The term number concentration and symbol C is preferred for mixtures.
(12) 'Amount concentration' is an abbreviation for 'amount-of-substance concentration'. When there is no risk of confusion the word 'concentration' may be used alone. The symbol [B] is often used for amount concentration of entities B. This quantity is also sometimes called molarity. A solution of, for example, 1 $mol\ dm^{-3}$ is often called a 1 molar solution, denoted 1 M solution. However, M should not be treated as a symbol for the unit $mol\ dm^{-3}$, in the sense that it should not be used with SI prefixes or in conjunction with other units.
(13) In the definition m_B denotes the molality of solute B, and m_A denotes the mass of solvent A; thus the same symbol m is used with two different meanings. This confusion of notation may be avoided by using the symbol b for molality, but this is seldom done.

A solution of molality 1 mol/kg is occasionally called a 1 molal solution, denoted 1 m solution; however, the symbol m should not be treated as a symbol for the unit $mol\ kg^{-1}$.
(14) The stoichiometric number is defined through the reaction equation. It is negative for reactants and positive for products. The values of the stoichiometric numbers depend on how the reaction equation is written.

Example $(1/2)N_2+(3/2)H_2=NH_3$: $\nu(N_2)=-1/2$,
$\nu(H_2)=-3/2$,
$\nu(NH_3)=+1$.

(15) The extent of reaction also depends on how the reaction equation is written, but it is independent of which entity in the reaction equation is used in the definition.

Example For the reaction in footnote (14), when $\Delta\xi=2$ mol, $\Delta n(N_2)=-1$ mol, $\Delta n(H_2)=-3$ mol, and $\Delta n(NH_3)=+2$ mol.

This quantity was originally introduced as *degré d'avancement* by de Donder.

Table 2.1. Continued. Continued on next page.

Other symbols and conventions in chemistry

(i) *Symbols for particles and nuclear reactions*

neutron	n	helion	h
proton	p	alpha particle	α
deuteron	d	electron	e
triton	t	photon	γ
positive muon	μ^+	negative muon	μ^-

The electric charge of particles may be indicated by adding the superscript +, −, or 0; e.g. p^+, n^0, e^-, etc. If the symbols p and e are used without a charge, they refer to the positive proton and negative electron respectively.

The meaning of the symbolic expression indicating a nuclear reaction should be as follows:

$$\text{initial nuclide} \left(\text{incoming particles or quanta}, \text{outgoing particles or quanta}\right) \text{final nuclide}$$

Examples $^{14}N(\alpha,p)^{17}O$, $^{59}Co(n,\gamma)^{60}Co$,
$^{23}Na(\gamma,3n)^{20}Na$, $^{31}P(\gamma,pn)^{29}Si$

(ii) *Chemical symbols for the elements*

The chemical symbols of elements are (in most cases) derived from their Latin names and consist of one or two letters which should always be printed in roman (upright) type. Only for elements of atomic number greater than 103, the systematic symbols consist of three letters (see footnote U to table 6.2). A complete list is given in table 6.2. The symbol is not followed by a full stop except at the end of a sentence.

Examples I, U, Pa, C

The symbols can have different meanings:

(a) They can denote an atom of the element. For example, Cl can denote a chlorine atom having 17 protons and 18 or 20 neutrons (giving a mass number of 35 or 37), the difference being ignored. Its mass is on average 35.4527 u in terrestrial samples.

(b) The symbol may, as a kind of shorthand, denote a sample of the element. For example, Fe can denote a sample of iron, and He a sample of helium gas.

The term *nuclide* implies an atom of specified atomic number (proton number) and mass number (nucleon number). Nuclides having the same atomic number but different mass numbers are called isotopic nuclides or *isotopes*. Nuclides having the same mass number but different atomic numbers are called isobaric nuclides or *isobars*.

A nuclide may be specified by attaching the mass number as a left superscript to the symbol for the element.The atomic number may also be attached as a left subscript, if desired, although this is rarely done. If no left superscript is attached, the symbol is read as including all isotopes in natural abundance.

Examples ^{14}N, ^{12}C, ^{13}C, $^{16}_{8}O$, $n(Cl) = n(^{35}Cl) + n(^{37}Cl)$

The ionic charge number is denoted by a right superscript, the sign being given after its absolute value (which may be omitted when equal to 1).

Examples	Na^+	a sodium positive ion (cation)
	$^{79}Br^-$	a bromine-79 negative ion (anion, bromide ion)
	$3\ Al^{3+}$	three aluminium triply positive ions
	S^{2-}	a sulfur doubly negative ion (sulfide ion)

The right superscript position is also used to convey other information: Excited electronic states may be denoted by an asterisk,

Examples H^*, Cl^*

Table 2.1. Continued. Continued on next page.

Oxidation numbers are denoted by positive or negative roman numerals or by zero (see also (iv) below),

Examples Mn^{VII}, O^{-II}, Ni^{0}

The positions and meanings of indices around the symbol of the element are summarized as follows.

left superscript	mass number
left subscript	atomic number
right superscript	charge number, oxidation number, excitation symbol
right subscript	number of atoms per entity (see (iii) below)

(iii) *Chemical formulae*

Chemical formulae denote entities composed of more than one atom (molecules, complex ions, groups of atoms, etc.).

Examples N_2, P_4, C_6H_6, $CaSO_4$, $PtCl_4{}^{2-}$, $Fe_{0.91}S$

They may also be used as a shorthand to denote a sample of the corresponding chemical substance.

Examples	CH_3OH	methanol
	$\rho(H_2SO_4)$	mass density of sulfuric acid

The number of atoms in an entity is indicated by a right subscript (the numeral 1 being omitted). Groups of atoms may also be enclosed in parentheses. Entities may be specified by giving the corresponding formula, often multiplied by a factor. Charge numbers of complex ions, and excitation symbols, are added as right superscripts to the whole formula. The free radical nature of some entities may be stressed by adding a dot to the symbol.

Examples	H_2O	one water molecule, water
	$1/2\ O_2$	half an oxygen molecule
	$Zn_3(PO_4)_2$	one zinc phosphate formula unit, zinc phosphate
	$2\ MgSO_4$	two formula units of magnesium sulfate
	$1/5\ KMnO_4$	one-fifth of a potassium permanganate formula unit
	$1/2\ SO_4{}^{2-}$	half a sulfate ion
	$CH_3\cdot$	one methyl radical
	$CH_3\dot{C}HCH_3$	isopropyl radical
	$NO_2{}^*$	electronically excited nitrogen dioxide molecule

In the above examples, $\frac{1}{2}O_2$, $1/5\ KMnO_4$ and $\frac{1}{2}SO_4{}^{2-}$ are artificial in the sense that such fractions of a molecule cannot exist. However, it may often be convenient to specify entities this way when calculating amounts of substance; see (v) below.

Specific electronic states of entities (atoms, molecules, ions) can be denoted by giving the electronic term symbol (see section 2.6) in parentheses. Vibrational and rotational states can be specified by giving the corresponding quantum numbers.

Examples	$Hg(^3P_1)$	a mercury atom in the triplet-P-one state
	$HF(v=2, J=6)$	a hydrogen fluoride molecule in the vibrational state $v=2$ and the rotational state $J=6$
	$H_2O^+(^2A_1)$	a water molecule ion in the doublet-A-one state

Chemical formulae may be written in different ways according to the information that they convey, as follows:

Table 2.1. Continued. Continued on next page.

Formula	*Information conveyed*	*Example for lactic acid*
empirical	stoichiometric proportion only	CH_2O
molecular	in accord with molecular mass	$C_3H_6O_3$
structural	structural relationship of atoms	$CH_3CHOHCOOH$
displayed	projection of atoms and bonds	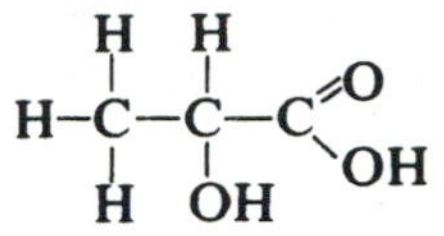
stereochemical	stereochemical relationship	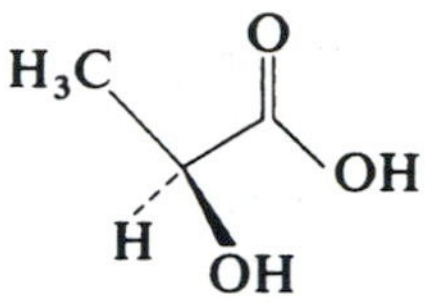

Further conventions for writing chemical formulae are described in [21, 22].

(iv) *Equations for chemical reactions*

Symbols connecting the reactants and products in a chemical reaction equation have the following meanings:

$H_2 + Br_2 \rightarrow 2HBr$	forward reaction
$H_2 + Br_2 \leftrightarrows 2HBr$	reaction, both directions
$H_2 + Br_2 \rightleftharpoons 2HBr$	equilibrium
$H_2 + Br_2 = 2HBr$	stoichiometric relation

If a reaction is considered elementary, for example $H + Br_2 \rightarrow HBr + Br$, it should be made clear in the text.

Redox equations are often written so that the absolute value of the stoichiometric coefficient for the electrons transferred (which are normally omitted from the overall equation) is equal to one.

Example $1/5\,KMn^{VII}O_4 + 8/5\,HCl = 1/5\,Mn^{II}Cl_2 + 1/2\,Cl_2 + 1/5\,KCl + 4/5\,H_2O$

Similarly a reaction in an electrochemical cell may be written so that the charge number of the cell reaction is equal to one:

Example $1/3\,In^0(s) + 1/2\,Hg^{I}{}_2SO_4(s) = 1/6\,In^{III}{}_2(SO_4)_3(aq) + Hg^0(l)$

(the symbols in parentheses denote the state, see (vi) below).

(v) *Amount of substance and the specification of entities*

The quantity 'amount of substance' has been used by chemists for a long time without a proper name. It was simply referred to as the 'number of moles'. This practice should be abandoned, because it is wrong to confuse the name of a physical quantity with the name of a unit (in a similar way it would be wrong to use 'number of metres' as a synonym for 'length'). The amount of substance is proportional to the number of specified elementary entities of that substance; the proportionality factor is the same for all substances and is the reciprocal of the Avogadro constant. The elementary entities may be chosen as convenient, not necessarily as physically real individual particles. Since the amount of substance and all physical quantities derived from it depend on this choice it is essential to specify the entities to avoid ambiguities.

Table 2.1. Continued.

Examples	n_{Cl}, $n(Cl)$	amount of Cl, amount of chlorine atoms
	$n(Cl_2)$	amount of Cl_2, amount of chlorine molecules
	$n(H_2SO_4)$	amount of (entities) H_2SO_4
	$n(1/5\,KMnO_4)$	amount of (entities) $1/5\,KMnO_4$
	$M(P_4)$	molar mass of (tetraphosphorus) P_4
	c_{HCl}, $c(HCl)$, $[HCl]$	amount concentration of HCl
	$\Lambda(MgSO_4)$	molar conductivity of (magnesium sulfate entities) $MgSO_4$
	$\Lambda(\frac{1}{2}MgSO_4)$	molar conductivity of (entities) $\frac{1}{2}MgSO_4$

$n(1/5\,KMnO_4) = 5n\,(KMnO_4)$
$\lambda(\frac{1}{2}Mg^{2+}) = \frac{1}{2}\lambda(Mg^{2+})$
$[\frac{1}{2}H_2SO_4] = 2[H_2SO_4]$
(see also examples in section 3.2, p.64.)

Note that 'amount of sulfur' is an ambiguous statement, because it might imply $n(S)$, $n(S_8)$, or $n(S_2)$, etc. In some cases analogous statements are less ambiguous. Thus for compounds the implied entity is usually the molecule or the common formula entity, and for solid metals it is the atom.

Examples '2 moles of water' implies $n(H_2O) = 2$ mol; '0.5 moles of sodium chloride' implies $n(NaCl) = 0.5$ mol; '3 millimoles of iron' implies $n(Fe) = 3$ mmol, but such statements should be avoided whenever there might be ambiguity.

(vi) *States of aggregation*

The following one-, two- or three-letter symbols are used to represent the states of aggregation of chemical species [1.j]. The letters are appended to the formula symbol in parentheses, and should be printed in roman (upright) type without a full stop (period).

g	gas or vapour	vit	vitreous substance
l	liquid	a, ads	species adsorbed on a substrate
s	solid	mon	monomeric form
cd	condensed phase (i.e. solid or liquid)	pol	polymeric form
		sln	solution
fl	fluid phase (i.e. gas or liquid)	aq	aqueous solution
		aq, ∞	aqueous solution at infinite dilution
cr	crystalline		
lc	liquid crystal	am	amorphous solid

Examples	HCl(g)	hydrogen chloride in the gaseous state
	C_V(fl)	heat capacity of a fluid at constant volume
	V_m(lc)	molar volume of a liquid crystal
	U(cr)	internal energy of a crystalline solid
	MnO_2(am)	manganese dioxide as an amorphous solid
	MnO_2(cr, I)	manganese dioxide as crystal form I
	NaOH(aq)	aqueous solution of sodium hydroxide
	NaOH(aq, ∞)	. . . as above, at infinite dilution
	$\Delta_f H^\circ(H_2O, l)$	standard enthalpy of formation of liquid water

The symbols g, l, to denote gas phase, liquid phase, etc., are also sometimes used as a right superscript, and the Greek letter symbols α, β, may be similarly used to denote phase α, phase β, etc., in a general notation.

Examples $V_m^{\ l}$, $V_m^{\ s}$ molar volume of the liquid phase, . . . of the solid phase
$S_m^{\ \alpha}$, $S_m^{\ \beta}$ molar entropy of phase α, . . . of phase β

GUIDE FOR THE PRESENTATION IN THE PRIMARY LITERATURE OF NUMERICAL DATA DERIVED FROM EXPERIMENTS*

Report of the CODATA Task Group on Publication of Data in the Primary Literature, September 1973

This Guide contains general recommendations on the reporting of numerical data for the guidance of authors, editors and referees. It is not a style manual. The recommendations are intended to facilitate the use, evaluation and comparison of data. They cover description of experiments, the treatment of data derived from them and presentation of the final, numerical results. The emphasis is on data that may be verified by repetition, e. g., data on physical and chemical properties and processes.

I. INTRODUCTION

This report is concerned with the presentation of numerical data that are capable of verification by repetition such as those on the physical and chemical properties and behaviour of material systems. The "properties" or "behaviour" may be microscopic (nuclear, atomic, or molecular) or macroscopic (e. g. density, crystal structure, transport properties, or energy transfer) or quantities associated with chemical transformations (equilibria, enthalpies of reaction, or rate constants). The "material systems" must be sufficiently well-defined to permit the measurements to be reproduced. Usually this means that the systems themselves must be of known purity and have a well-defined composition and state.

Recommendations are made concerning the information that should be reported so that the data may be compared and correlated with those obtained from other studies. These are general recommendations that apply to all branches of the physical sciences. Some existing specialized guides that expand upon these recommendations are listed in the bibliography; others are under preparation.

This is not a style manual for writing scientific papers. It is a statement of the minimum information that is needed to ensure that the reader can understand the quantitative data, can assess their precision and accuracy, and can recalculate the results when values for auxiliary data change.

The author of a paper has the primary responsibility for providing the reader with the type of information outlined in this Guide.

The Guide also provides journal editors and referees with a set of consistent, considered criteria for judging the completeness and acceptability of papers in so far as the reporting of numerical quantities are concerned. The recommendations reflect the experience of data evaluators; hence adherence to them will permit evaluators to consolidate the author's results with existing data and facilitate their incorporation in critical compilations.

The word "data" has been used loosely above. Terms used in the later sections of the Guide are defined here. *Measurements* are made on material systems using suitable instruments and procedures. The instrument readings are the initial numerical *data*. The *results*, usually stated in terms of physical and chemical properties, may be the initial data. More frequently, however, they are derived from these data by a process called *reduction of the data*. Associated with the results are two numbers: the *imprecision* which records the reproducibility and the *inaccuracy* which estimates the overall reliability of the measurements.

Recommendations relating to these matters are considered below under three headings: description of experimental procedures, reduction of data, and presentation of results.

---

* Reprinted from Unesco-UNISIST Guide, prepared with financial support of Unesco by the CODATA Task Group composed of:

Dr. David Garvin, National Bureau of Standards, Washington, D.C. 20234, U.S.A.;

Dr. Tangis Golashvili (until March 1973), CODATA Central Office, 19 Westendstr., Frankfurt/Main, Germany, Fed. Rep.;

Dr. Henry V. Kehiaian, Centre de Microcalorimetrie et Thermochimie du CNRS, 26 Rue du 141e RIA, 13003 Marseille, France;

Prof. Nicholas Kurti, F.R.S., The Clarendon Laboratory, Parks Road, Oxford OX1 3PU, United Kingdom;

Prof. Edgar F. Westrum, Jr., (Chairman), Department of Chemistry, University of Michigan, Ann Arbor, Michigan 48104, U.S.A.

II. THE DESCRIPTION OF EXPERIMENTAL PROCEDURES

Provide an adequate description of the experimental procedures used to obtain the numerical data. The quality of the information about *how* measurements were made often determines the acceptability of the results. When it becomes necessary for a reader to compare the results of several studies, or to reinterpret data, he may need to know whether the author paid attention to details that have since been realized to be important or whether his technique could have revealed more recently reported phenomena. If these questions cannot be answered, later workers and evaluators may assign very little weight to the results.

The major points to be considered are:

1) *Definition of the system studied.* This must be defined as precisely as necessary to ensure the reproducibility of the investigated properties within the limits of experimental error. All relevant details about the physical state and the constraints on the system must be given, as well as information about the origin, treatment, history and chemical composition of the samples.

2) *Description or identification of (a) physical or chemical methods used in the measurements for analysis of composition or purity and (b) reference substances or methods used to test the reliability of the results obtained.* The results of the tests should be given.

3) *Brief qualitative description of the type of measurements made and apparatus used.*

4) *Description of apparatus.* Novel apparatus should be described in detail and the results of its testing given. In other cases the details may be given by reference to another publication. Where relevant, the author should indicate dimensions, constructional material, electrical and other components, major modifications in equipment, etc. For commercial apparatus and components the manufacturer and model number should be specified.

5) *Description of experimental procedures.* The quantities actually measured should be stated clearly since these may differ considerably from the derived results. If the procedures are novel, how they were tested must be explained. Standard procedures may be identified by reference to another publication. The laboratory environment (temperature, pressure, humidity, geographical location, etc.), should be stated if relevant.

6) *Performance of measurement system.* The sensitivity or resolution achieved in the measurements should be stated and proved. Methods of calibration and reference substances used therefore should be identified. International or national standards of scales that were used in the calibration should be cited.

III. THE REDUCTION OF EXPERIMENTAL DATA

Explain the conversion of the measurements to the reported results. This reduction of data often is a long and complex process difficult for a reader to reconstruct. Inclusion of an example may be desirable. The procedures used for reduction of the data if adequately described in one publication may be given by reference in later publications.

Important components of this process are:

1) *Assumptions made about the experiments.* Usually some parameters are assumed to be unimportant, others are assumed constant, and corrections are made for the variations in still others. These assumptions and procedures should be stated; if they are based upon auxiliary experiments, those should be described.

2) *Complete description of physical models (including relevant mathematical expressions) used to convert the data to results.* Approximations should be explained. References to important computer programs employed should be given.

3) *Experimental results and physical constants taken from other sources.* These should be identified clearly, as should their sources, and their values stated.

4) *Identification of standards used to relate the measurements to the fundamental units of measure.* Examples are: a particular international temperature scale, a dated list of atomic weights, a national standard volt, etc.

IV. PRESENTATION OF NUMERICAL RESULTS

As a general principle, **report results in a form as free from interpretation as possible (i. e. as closely as is practical to experimentally observed quantities). These results should be reported in such a manner that the degree of experimental randomness can be assessed.** The reader should be able to recover enough of the experimental data so that he can reanalyze them in terms of different hypotheses.

A. **Citable Results. List important numerical results in explicitly titled tables.** These are the results that the author expects other workers to cite or use. Separate them from the discussion of the work. Results from other sources that are included in tables containing the new material should be clearly identified and referenced. Graphical and analytical representations of important results, although convenient for the reader, *are not acceptable substitutes* for tabular presentation of *accurate* experimental results. The author will, no doubt, interpret the primary results and present derived quantities. However, such secondary results should never be published at the cost of omitting the primary results upon which they are based.

B. **Compressed Presentation of Unsmoothed Data. An acceptable alternative to a complete table** (when there are many measurements) **is an easily used analytical expression supplemented by a deviation plot showing the individual points.** This procedure may save space and promote clarity, but must be sufficiently sensitive to permit full recovery of individual results.

C. **Presentation of Smoothed Data.** In addition to showing his work in the manner described above, an author may include tables of smoothed numerical results intended for use by the reader, for example, electrical resistivity at selected temperatures. In such cases, **arrange the tables with values of the argument so spaced that no serious loss of accuracy will result during interpolation (6) and give a sufficient number of digits to make such interpolation feasible.** Alternatively, such smoothed data may be provided by empirical equations which not only provide ready analytical interpolation, differentiation or integration, but can also save journal space. It is important that the deviation of the experimental values from the equation be within the imprecision of the experimental data.

D. **The "Imprecision" and "Inaccuracy" of the Results. Evaluate both in clearly defined terms.**

The various sources of uncertainty should be described rigorously, with clear separation between measurement imprecisions, numerical analysis limitations (or deviations from a model), and possible systematic biases.

Imprecision. The statistical or random uncertainty should be estimated using an appropriate standard statistical technique. It is only one component of the total error analysis and is not a sufficient statement of the reliability of the experiments.

Inaccuracy. Estimation of the other potential sources of error or limitations of the work is more difficult than is that for imprecision. There are no clear rules; subjective judgement is involved. However, these estimates of inaccuracies are more important than is that for imprecision because unexplained differences between two sets of measurements usually are larger than are the random errors.

Important components in the evaluation of inaccuracy (i. e., of possible systematic errors) for which estimates should be made are:

1) *Sensitivity of measurement or resolution* possible in the experiments. This provides a lower bound for the inaccuracy.
2) *Effects of assumptions* made in processing the data. In particular, include uncertainties or defects in the physical model used.
3) Possible *sources and magnitudes of errors* due to limitations of the measurement system. These should be discussed both for those for which corrections were made and those for which this could not be done. Those inherent in the calibration procedures or standards used should be included.
4) *Uncertainties in auxiliary data* taken from other sources.

These estimates for the components of the measurement should be combined to give the total estimated inaccuracy. Statistical procedures used in the reduction and estimation of the imprecision and inaccuracy of the data should be cited. The inaccuracy estimate may be important in some applications of the results and not in others. A discussion of this point should be included.

E. **Symbols, Units and Nomenclature. Use symbols, units and nomenclature recommended by the International Organization for Standardization and by the various international unions.**

In particular:

1) *Use SI units* and their accepted symbols (1, 2) as far as possible.
2) *Identify symbols* used for all physical quantities and, where available, use those recognized internationally (3—6).
3) *Use internationally accepted names* for chemical compounds. Commercial and common (trivial) names and abbreviations should be defined.
4) *Make figures and tables logically complete units*, independent of the main text. The caption should explain the table or figure. Particular care should be taken in labelling columns and rows of tables. The label must permit unambiguous recovery of the value for the physical quantity. Any label is acceptable as long as the meaning is clear.

Authors are reminded of the preferred convention:
physical quantity = number × power of ten × units
e. g., $\lambda = 5.896 \times 10^{-7}$ m.

It is the *number* that is tabulated.

The preferred label, obtained by rearrangement of the equation above, is:

physical quantity divided by units.

That is, in the equation $\lambda/\text{m} = 5.896 \times 10^{-7}$, the left hand side is the label and the right hand side is the number to be tabulated.

Other examples are: pressure as "P/MPa", temperature as "$T/10^5$K" and a first order rate constant as "log (k/s^{-1})".

An alternative form, e. g. entropy as "S in JK^{-1} mol^{-1}" is acceptable.

The use of non-SI units and, even more, the use of nonmetric units for reporting scientific data is discouraged. Occasionally there may be overriding reasons for using such units. In these cases the author should give factors for conversion to SI units. This may be done in a footnote: "Throughout this paper 1 torr = (101.325/760) kPa, 1 cal_{th} = 4.184 J, and 1 Å = 0.1 nm".

V. DISCUSSION

We have attempted to set forth the important aspects of numerical data presentation so as to save the data and to promote the usefulness of the quantitative results of scientific research.

There is an apparent conflict between these recommendations and the usual exhortations to authors for brevity as well as claritiy in their papers. The needs of the general reader often can be met by a brief article not containing the detailed information we prescribe, but those of the specialist and of the evaluator cannot. Although these recommendations call for more (but not much more) detail than is commonly provided, they do not exceed what appears in the better papers today. The needed statements may be terse and factual.

The ideal situation is to have all of the relevant information in the published article. However, if this is not practical then the supplementary material should be put in an auxiliary publication (submitted together with the shorter manuscript) and either placed in a suitable depository service or published as microform together with the article. In any event, the details must be available to the public from some source other than the author. The means of obtaining such auxiliary information must be clearly stated in the publication.

In order to provide the author with more guidance these general recommendations have been supplemented by a bibliography. It includes a non-exhaustive list of useful reports on generaltopics (e. g., symbols, units, nomenclature, physical constants, temperature scales, precision and accuracy) that should be consulted (1—13).

For application to particular fields there is a need for interpretation and implementation of our recommendations. Some specialized guides already exist. They are noted in the bibliography (14—33). Specialists in the fields for which guides exist believe that these have improved the quality of data reporting.

This Task Group is endeavouring to promote the preparation by appropriate international scientific bodies of discipline-wide guides for important areas as well as more specialized guides for sub-disciplines. We will welcome suggestions concerning what should be done and by whom.

VI. BIBLIOGRAPHY

SYMBOLS, UNITS, AND NOMENCLATURE

1. **ISO Recommendation R 1000, Feb. 1969. Rules for the use of Units of the International Systems of Units and a Selection of the Decimal Multiples and Sub-Multiples of the SI Units.** 1st. Edition. American National Standards Institute, New York, 1969.

2. **Le Système International d'Unités** (SI). International Bureau of Weights and Measures, Sèvres, France, 1970. Authorized English translations are available: (a) Her Majesty's Stationery Office, London, 1970, and (b) C. H. Page and P. Vigoreux, eds. National Bureau of Standards, Special Publication 330, U.S. Government Printing Office, Washington, D. C., 1970.

3. **Symbols, Units, and Nomenclature in Physics.** International Union of Pure and Applied Physics, Commission for Symbols, Units and Nomenclature, Document U.I.P. **11** (S.U.N. 65-3), (1965).

4. **Manual of Symbols and Terminology for Physicochemical Quantities and Units.** M. L. McGlashan, ed. *Pure Appl. Chem.* **21**, 3, (1970).

5. **Information Bulletin.** International Union of Pure and Applied Chemistry, Cowley Centre, Oxford, England (Nomenclature, etc.).

6. **Quantities, Units and Symbols.** The Symbols Committee of the Royal Society. The Royal Society, 6 Carlton House Terrace, London SWlY 5AG (1971).

PHYSICAL CONSTANTS AND TEMPERATURE SCALES

7. **Report of CODATA Task Group on Fundamental Constants.** *CODATA Bulletin No.* **11** (1973).

8. **International Practical Temperature Scale of 1968.** *Metrologia* **5,** 35 (1969). International Bureau of Weights and Measures (BIPM) Subcommittee on Thermometry.

9. **Relationships between the International Practical Temperature Scale of 1968 and the NBS-55, NPL-61, PRMI-54, and PSU-54 Temperature Scales in the Range from 13.81 to 90.18 K.** R. E. Bedford, M. Durieux, R. Muijlivijk, and C. R. Barber, *Metrologia* **5,** 47 (1969).

10. **Conversion of Existing Calorimetrically Determined Thermodynamic Properties to the Basis of the International Practical Temperature Scale of 1968.** T. B. Douglas, *J. Research National Bureau of Standards (U.S.)*, **73A,** 451 (1969).

11. **Ambiguities in the Use of Unit Names.** C. H. Page, *Science* **179,** 873 (1973).

12. **Atomic Weights of the Elements 1971.** *Pure Appl. Chem.* **30** (3—4), 637 (1972).

PRECISION AND ACCURACY

13. **Statistical Concepts and Procedures.** Vol. 1 of **Precision Measurement and Calibration.** H. H. Ku, ed., National Bureau of Standards (U.S.), Special Publication 300, 1969.

AVAILABLE GUIDES FOR REPORTING DATA IN VARIOUS DISCIPLINES

CHEMICAL KINETICS

14. [The CODATA Task Group on Data for Chemical Kinetics is now preparing (1973) guidelines on the reporting of rate data.]

CRYSTALLOGRAPHY

15. **Information and Suggestions on Presentation of the Results of Crystal Structure Studies.** J. A. Ibers, *Inorg. Chim. Acta* **3,** (9th prelim. page), 1969.

16. **Primary Crystallographic Data.** (mainly single crystal) O. Kennard, J. C. Speakman, and J. D. H. Donnay, *Acta Cryst.* **22,** 445 (1967).

17. **Powder Data.** O. Kennard, J. D. Hanawalt, A. J. Wilson, P. M. DeWolff and V. A. Frank-Kamenetsky. *J. Appl. Cryst.* **4,** Part 1, 81 (1971).

INFRARED SPECTRA

18. **Specifications for Evaluation of Infrared Reference Spectra.** Coblentz Society Board of Managers, *Anal. Chem.* **38,** 27A (1966).

19. **Recommendations for Infrared Spectra of New Compounds.** *Anal. Chem.* **40,** 2272 (1968).

MAGNETIC SPECTRA

20. **Manual on Mössbauer Data Presentation.** in **Mössbauer Effect Data Index.** 1971. J. G. Stevens, ed. Plenum Publishing Corp. New York (1972 — in press).

21. **Nomenclature and Conventions for Reporting Mössbauer Spectroscopic Data.** (Tentative Recommendations). Atomic and Molecular Properties ad hoc Panel on Mössbauer Data of the NDAB, National Research Council, Washington, 1971.

OPTICAL SPECTRA

22. **Notation for Atomic Spectra.** W. F. Meggers and C. F. Moore, Report of Subcommittee e, Joint Commission (IAU & IUPAP) for Spectroscopy, *J. Opt. Soc. Am.* **43,** 442 (1953).

23. **Notation for the Spectra of Diatomic Molecules.** F. A. Jenkins, Report of Subcommittee f, Joint Commission (IAU & IUPAP) for Spectroscopy, *J. Opt. Soc. Am.* **43,** 425 (1953).

24. **Report on Notation for the Spectra of Polyatomic Molecules.** Joint Commission (IAU & IUPAP) for Spectroscopy, *J. Chem. Phys.* **23,** 1997 (1955).

25. **Report of Subcommittee on Units and Terminology.** Appendix A to Minutes of the Meeting for the Triple Commission for Spectroscopy (IAU, IUPAP, IUPAC) at Tokyo, 9 Sept. 1962, *J. Opt. Soc. Am.* **53,** 883 (1963).

26. **Nomenclature, Symbols, Units and Their Usage in Spectro-Chemical Analysis — I.** *Appendix No. 1. (1969) to IUPAC Information Bulletins.*

27. **Recommendations for the Presentation of Raman Spectra for Cataloging and Documentation in Permanent Data Collections.** *Appendix No. 11 (1971) to IUPAC Information Bulletins.*

28. **Spectrometry Nomenclature.** *Anal. Chem.* **40,** 2271 (1968).

THERMAL CONDUCTIVITY

29. **Recommendations for Data Compilations and for the Reporting of Measurements of the Thermal Conductivity of Gases.** H. M. Hanley, et al., *J. Heat Transfer*, **93,** Series C, 479 (1971).

THERMOCHEMISTRY AND THERMODYNAMICS

30. **A Guide to Procedures for the Publication of Thermodynamic Data.** Commission I.2 on Thermodynamics and Thermochemistry of the International Union of Pure and Applied Chemistry. Definitive publication made in *Pure Appl. Chem.*, **29**, 397 (1972); verbatim publication in English:
    At. Energy Rev., **9**, 869 (1972),
    J. Chem. Thermodyn., **4**, 511 (1972),
    Indian J. Chem., **10**, 51 (1972),
    Indian J. Phys., **12**, 51 (1972),
    CODATA Newsletter, **8**, 4 (1972),
    J. Chem. Eng. Data., **18** (No. 1), 3 (1973),
    in French:
    J. Chim. Phys., **69** (No. 10), (Oct. 1972),
    Bull. Soc. Chim. France, Special number, (Mar. 1973),
    in Japanese:
    The Society of Calorimetry and Thermal Analysis, (Nov. 1971),
    and in Russian and German [In Press].

AVAILABLE GUIDES FOR REPORTING DATA IN VARIOUS SUB-DISCIPLINES

GAS CHROMATOGRAPHY

31. **Proposed Recommended Practice for Gas Chromatography Terms and Relationships.** ASTM Committee E-19, *J. Gas Chromatog.* **6**, 1 (1968).

THERMAL ANALYSIS

32. **Recommendations for the Presentation of Thermal Analysis Data.** H. G. McAdie, *Anal. Chem.* **39**, 543 (1967).

MASS SPECTROMETRY

33. **Suggestions for the Use of Symbols and Abbreviations in Papers Dealing with Topics in Organic Mass Spectrometry.** H. Budzikiewicz, *Organic Mass Spectrometry* **2**, 249—52 (1969).

MISCELLANEOUS GUIDES

34. **A.S.M.E. Resolution on Communication of Experimental Data.** *Mech. Eng.* **80**, 147 (Nov. 1958).
35. **Guide to Authors on the Preparation of Scientific Papers for Publication.** UNESCO Document SC/MD/5, Paris, Aug. 1968.

No. 7 (Aug. 1972) 4 pp, *Tentative Set of Key Values for Thermodynamics - Part III*
(Report of the CODATA Task Group on Key Values for Thermodynamics, June 1972), superseded by Bulletin No. 10

No. 8 (Dec. 1972), 32 pp, *Geological Data Files: Survey of International Activity*
(Report of COGEODATA, Committee on Storage, Automatic Processing and Retrieval of Geological Data of the International Union of Geological Sciences (IUGS)).

No. 9 (Dec. 1973), 6 pp, *Guide for the Presentation in the Primary Literature of Numerical Data Derived from Experiments*
(Report of the CODATA Task Group on Publication of Data in the Primary Literature, Sept. 1973).

No. 10 (Dec. 1973), 12 pp, *CODATA Recommended Key Values for Thermodynamics, 1973*
(Report of the CODATA Task Group on Key Values for Thermodynamics, Nov. 1973).

No. 11 (Dec. 1973), 8 pp, *Recommended Consistent Values of the Fundamental Physical Constants, 1973*
(Report of the CODATA Task Group on Fundamental Constants, August 1973).

GUIDE FOR THE PRESENTATION IN THE PRIMARY LITERATURE OF PHYSICAL PROPERTY CORRELATIONS AND ESTIMATION PROCEDURES

ABSTRACT

This Guide for the presentation of physical property correlations and estimation procedures sets forth the statistical quality criteria by which authors and journal editors should assess the practical utility of a proposed new method. Widely available standard regression program packages can be used for tests against available data. The minimum regression requirements of this Guide can be implemented on programmed hand-held calculators. References to the needed collections of data are included with this Guide.

I. INTRODUCTION

This report is concerned with the presentation of correlations and estimation procedures covering the physical properties of pure chemical substances. The "properties" will in general be macroscopic (equilibrium or transport properties) or quantities associated with chemical transformations, such as enthalpies of reaction. In the development of correlations or estimation procedures claiming a high degree of reliability, the experimental data upon which they are based should have been obtained on systems of known (and stated) purity and have a well defined composition and state. This quality control on data input and rigorous statistical significance standards described in this Guide set conditions for acceptability in the traditional professional journals.

On the other hand, the technical community should have access to the many ad hoc correlations of narrower scope originating in specific product and process development projects. They are generally based on results obtained experimentally from substances only "identified" but not highly purified. A logical place for the publication of such narrowly focused, yet useful correlation and estimation procedures would be the widely read technical magazines and trade journals dedicated to the dissemination of specific product and process R/D information. This Guide includes a proposal for their quality control [Section III, (j)-(l)].

Recommendations are made concerning the information that should be reported so that the validity of a proposed correlation or estimation procedure can be evaluated in comparison with other methods, as well as against experimental data.

This is not a style manual for writing scientific or engineering papers. It is a statement of the minimum information that is needed to ensure that the reader understands the proposed correlation or estimation procedure, can properly assess its correlating and predictive ability, especially with regard to the range of applicability and the precision and accuracy of the results to be expected, and can revise the coefficients in the light of new data.

The author of a paper has the primary responsibility

for providing the reader with the type of information outlined in this Guide.

The Guide also provides journal editors and referees with a set of consistent, considered criteria for judging the completeness and acceptability of papers in so far as the reporting of the mathematical relations are concerned. The recommendations reflect the experience of data evaluators; hence adherence to them will permit evaluators to consolidate the author's results with existing methods and facilitate their incorporation in critical compilations.

Recommendations relating to these matters are considered below under three headings: description of procedures, evaluation of statistical significance, and presentation of results.

II. THE DESCRIPTION OF ESTIMATING AND CORRELATING PROCEDURES

An adequate description should be provided of the computational procedures used to obtain the numerical coefficients in the equations as well as the data sources used in the process. If these sources are not the commonly accepted collections of critically evaluated data, the author should demonstrate the quality of the data by their internal and/or external consistency with other relevant data.* Failure to do so may result in users assigning little weight to the work.

The major points to be considered are:

1) Basic assumptions underlying the correlations or estimation procedures and definition of the physical state of substances to which they refer, as well as the limitations, if any, in ranges of T and P in which they can be applied.

2) Description of (or readily accessible reference to) the computation procedure(s) used. Either in the paper (or its Appendix) or in the references the computational procedure should be spelled out in sufficient detail for repetition and/or execution by any other reasonably skilled professional.

The procedure, as written, may not contravene the basic principles of physics, especially of thermodynamics, nor should it produce results which are inconsistent with Maxwell's thermodynamic relations, either relative to state variables or relative to other properties of the substance for which the calculation is recommended. If a particular formulation can produce such undersirable results even in the limit, this should either be stated explicitly or provisions be made in the algebra which will prevent such inadvertent misuse of the proposed procedure.

III. EVALUATION OF STATISTICAL SIGNIFICANCE†

The reliability of the proposed procedure should be demonstrated by suitable comparison of computed results with high quality experimental data. Thus a necessary but not sufficient set of conditions for acceptability of a correlation is that:

a) the coefficient of variation (σ/μ) where σ = standard deviation, μ = mean, be smaller than that for previous correlations, or that a small deterioration of σ/μ compared to previous schemes can be justified by the computational advantage or greater generality of the proposed new method, and

*See item III-j for the exceptions, when only "technical grade substances" are available.

† The definitions of the terms used can be found in references 13 through 16, especially 16.

b) there be no significant systematic trend of deviations between calculated and experimental values with the absolute value of the variables or with molecular structure (see also item (i) Autocorrelation), and

c) the change of the confidence interval of the dependent variables with the magnitude of the independent variable should be shown, if it is large.

The significance of the proposed procedure should be demonstrated by suitable statistical procedures such as:

d) The F-test.

e) Estimation of the adjusted coefficient of determination $\bar{R}^2$.

f) If the proposed equation has more than two terms, the significance of the individual terms should be demonstrated by the usual F-test procedure for the coefficients of each term. This is particularly important with purely empirical relations such as power series and other polynomial expansions.

Inclusion of the quality measures (a) through (f) should be considered the minimum demand for the publication of property correlations and estimation procedures in professional journals. Inclusion of the more sophisticated quality measures (g) through (i) would be desirable and authors would be well advised to use them. However, they are too unfamiliar at present in the technical community to whom this Guide is addressed to make their inclusion mandatory.

g) The significance of individual independent variables in a multiple regression should be demonstrated by at least one procedure, such as the F-test for each variable.

h) Multicollinearity, correlation between the variables in a multiple regression, or between the coefficients in a multi-term expression for a single independent variable distorts all reliability measures and should be avoided, if possible. A typical example for an unavoidable correlation is that which exists between ΔH°_V and ΔCp_V in the vapor pressure vs. temperature relation in series of polyatomic compounds.

i) Autocorrelation, the existence of correlation between successive error residuals—rather than random distribution of errors—indicates that there is something wrong with the proposed procedure. The phenomenon can be due to a missed non-linearity in the relation between the dependent and the independent variable(s), or it may point to the existence of a missing (ordering) variable. One consequence of the existence of autocorrelation is a spuriously high coefficient of determination.

The existence of autocorrelation is easily noticed by inspection of the plot of the proposed procedure's regression line through the experimental data. An objective measure of its presence is the Durbin-Watson test which is more often found in statistics textbooks used by economists (ref. 16) than in those used by engineers and scientists. Autocorrelation can occur equally in all areas where a natural phenomenon has to be described by a minimum number of variables.

The following set of quality measures is recommended for the "narrow scope" correlations and estimation procedures mentioned in the second paragraph of the Introduction to this Guide.

j) Following a description of the scope of chemical substances (including their characterization and purity) and of temperature and pressure ranges of the underlying experimental data, authors should demonstrate the internal consistency of those data if they involve thermodynamic properties. They should demonstrate the external consistency of the data, showing how closely their numerical value would have been approached by existing molecular structure correlation or estimation procedures if the substances in question fall within the scope of such procedures as described in "The Properties of Gases and Liquids" by R. C. Reid, J. M. Prausnitz, and T. K. Sherwood, McGraw-Hill, (1977).

k) They should regress the proposed procedure

against the experimental data of the investigation, and whatever literature data (preferably evaluated) of the investigation's compound class can be found, and present the standard error of estimate (σ) between experiment and "prediction" as well as the coefficient of determination ($\bar{R}^2$).*

1) The existence, and population of deviations in excess of 2σ should be reported.

IV. PRESENTATION OF FINAL RESULTS

A. If the proposed procedure of qualifying tests described in Section III is met, the proposed procedure should be outlined in useable summary form, and all the numerical coefficients be tabulated together with their error ranges and significance measures.

B. If the author recommends the use of different ranges of the independent variables, or for different families of chemical compounds, there should be a clearly marked separate table of coefficients (and their error ranges and significance measures) for each range, and/or for each family of chemical compounds for which they are applicable.

C. In addition to the error ranges and significance tests which are measures of imprecision of the estimates made by his procedure, the author should evaluate and state the inaccuracy to be expected for the estimates made by his procedure.

The author should also state the sensitivity of his procedure to uncertainties or errors in the independent variables that serve as input for his calculating procedure.

D. In principle, of course, any algebraic systematization of physical property data is "legitimate" which permits their precise compact representation and easy retrieval. However, preparers of correlations and estimation procedures should be encouraged to make all variables dimensionless either with the appropriate critical constants or with other properties that express intermolecular force, molecular dimensions, and appropriate intramolecular characteristics. Such procedures are usually more generally applicable than purely empirical correlations and estimations, and their limitations are more easily seen.

E. Mathematical formulations that are bound to produce indeterminate constants should not be used. For example, authors who use the poorly regressed form

$$y = a/(x + b)$$

should be advised to use a form that avoids statistical correlation between the fitted parameters and is more easily regressed, such as

$$1/y = c(x - x_o) + d'$$

where $d' = d + cx_o$, and where x_o is chosen such as to reduce the correlation coefficient between c and d' to zero. In least-squares fitting, x_o is just the simple mean x_o. (This example was suggested by Dr. E. R. Cohen).

F. Symbols, Units, and Nomenclature. Use symbols, units, and nomenclature recommended by the International Organization for Standardization (ISO) and by the various international unions.

In particular:

1) *Use SI units* and their accepted symbols (1,2) as far as possible.

2) *Identify symbols* used for all physical quantities and, where available, use those recognized internationally (3-6).

3) *Use internationally accepted names* for chemical compounds. Commercial and common (trivial) names and abbreviations should be defined.

---

*Keep in mind that for the user $\bar{R}^2$ is a crude measure of the correlation's ability to represent the trend with the independent variables, and that σ^2 is a crude measure of its reliability.

4) *Make figures and tables logically complete units,* independent of the main text. The caption should explain the table or figure. Particular care should be taken in labeling columns and rows of tables. The label must permit unambiguous recovery of the value for the physical quantity. Any label is acceptable as long as the meaning is clear.

Authors are reminded of the preferred convention: physical quantity = number × power of ten × units (e.g., $\lambda = 5.896 \times 10^{-7}$ m).

It is the *number* that is tabulated.

The preferred label, obtained by rearrangement of the equation above, is:

physical quantity divided by units.

That is, in the equation $\lambda/\mathrm{m} = 5.896 \times 10^{-7}$, the left-hand side is the label and the right-hand side is the number to be tabulated.

Other examples are: pressure as "P/MPa," temperature as "$T/(10^5$ K)" and a first order rate constant as "$\log(k/\mathrm{s}^{-1})$."

Alternative forms, e.g., entropy as "S in J K^{-1} mol^{-1}" or as the dimensionless entropy, S/R, are acceptable.

The use of non-SI units and, even more, the use of nonmetric units for reporting scientific data is discouraged. Occasionally there may be overriding reasons for using such units. In these cases, the author should give factors for conversion to SI units. This may be done in a footnote: "Throughout this paper 1 Torr = (101.325/760) kPa, 1 $\mathrm{cal_{th}}$ = 4.184 J, and 1 Å = 0.1 nm."

V. DISCUSSION

There is an apparent conflict between these recommendations and the usual exhortations to authors for brevity as well as clarity in their papers. The needs of the general reader often can be met by a brief article not containing the detailed information we prescribe, but those of the specialist and of the evaluator cannot. Although these recommendations call for more (but not much more) detail than is commonly provided, they do not exceed what appears in the better papers today. The needed statements may be terse and factual.

The ideal situation is to have all of the relevant information in the published article. However, if this is not practical then the supplementary material should be put in an auxiliary publication (submitted together with the shorter manuscript) and either placed in a suitable depository service or published as microform together with the article. In any event, the details must be available to the public from some source other than the author. The means of obtaining such auxiliary information must be clearly stated in the publication.

In order to provide the author with more guidance these general recommendations have been supplemented by a bibliography. It includes a non-exhaustive list of useful reports on general topics (e.g., symbols, units, nomenclature, physical constants, temperature scales, precision, and accuracy) that should be consulted (1-12).

VI. BIBLIOGRAPHY

SYMBOLS, UNITS, AND NOMENCLATURE

1. "International Standard ISO 1000 - 1973, SI Units and Recommendations for the use of Their Multiples and of Certain Other Units," 1st Edition, 1973-02-01. Available in any country adhering to ISO from the "Member Body," usually the national standardizing organization of this country.

2. "Le Système International d'Unités (SI)," third

Edition, 1977. Bureau International des Poids et Mesures (BIPM), 92310 Sèvres, France; available from OFFILIB, 48 rue Gay Lussac, 75005 Paris, France. English translation prepared jointly by the National Physical Laboratory, UK, and the National Bureau of Standards, USA, has been approved by BIPM: available from: (a) Her Majesty's Stationery Office, 49 High Holborn, London WCIV 6HB, UK: (b) US Government Printing Office, Washington, DC 20402, USA (National Bureau of Standards Special Publication 330).

3. "Symbols, Units, and Nomenclature in Physics." International Union of Pure and Applied Physics, Commission for Symbols, Units, and Nomenclature, Document U.I.P. 11 (S.U.N. 65-3), (1965).

4. "Manual of Symbols and Terminology for Physicochemical Quantities and Units," M. L. McGlashan, ed. 1973 Edition, IUPAC additional publication, Butterworths, London.

5. "Information Bulletin," International Union of Pure and Applied Chemistry, Cowley Centre, Oxford, England (Nomenclature, etc.).

6. "Quantities, Units, and Symbols," The Symbols Committee of the Royal Society. The Royal Society, 6 Carlton House Terrace, London SW1Y 5AG (1975).

PHYSICAL CONSTANTS AND TEMPERATURE SCALES

7. Report of CODATA Task Group on Fundamental Constants. ***CODATA Bulletin*** No. 11 (1973).

8. Echelle Internationale Pratique de Température de 1968. Edition amendée de 1975. Compte Rendus de la 15eme Conference Générale des Poids et Mesures (1975) et Comité Consultatif de Thermométrie, 10eme Session, Bureau International des Poids et Mesures, Pavillon de Breteuil, 92310 Sèvres, France. English version: H. Preston-Thomas, The International Practical Temperature Scale of 1968. Amended Edition of 1975, *Metrologia*, 12, 7-17 (1976).

9. R. E. Bedford, M. Durieux, R. Muijlivijk, and C. R. Barber, Relationships between the International Practical Temperature Scale of 1968 and the NBS-55, NPL-61, PRMI-54, and PSU-54 Temperature Scales in the Range from 13.81 to 90.18 K. *Metrologia* 5, 47 (1969).

10. T. B. Douglas, Conversion of Existing Calorimetrically Determined Thermodynamic Properties to the Basis of the International Practical Temperature Scale of 1968. *J. Research National Bureau of Standards* (U.S.), 73A, 451 (1969).

11. C. H. Page, Ambiguities in the Use of Unit Names. *Science* 179, 873 (1973)

12. Atomic Weight of the Elements 1973 (Table of Relative Atomic Masses of Selected Nuclides, pp 600-602), *Pure Appl. Chem.* 37, 591-603 (1974); Atomic Weights of the Elements 1975 (Table of isotopic composition of the elements as determined by mass-spectrometry, pp 86-91; table of relative atomic masses and half lives of selected radionuclides, p.94), *Pure Appl. Chem.* 47, 75-95 (1976).

STATISTICAL PROCEDURES

13. C. Daniel and F. S. Wood, "Fitting Equations to Data," Wiley-Interscience, New York 1971. (Gives a very conveniently used and widely available linear least-squaring curve-fitting program.)

14. O. J. Dunn and V. A. Clark, "Applied Statistics: Analysis of Variance and Regression," Wiley, New York, 1974.

15. P. D. Lark, R. B. Craven and R. C. L. Bosworth, "The Handling of Chemical Data," Pergamon, Oxford, 1968.

16. G. O. Wesolowsky, "Multiple Regression and Analysis of Variance," Wiley-Interscience, New York, 1976.

VII. SELECTED COMPILATIONS OF CRITICALLY EVALUATED DATA OF INTEREST TO ANTICIPATED USERS OF THIS GUIDE.

Prepared by Dr. D. R. Lide of the U.S. National Bureau of Standards. (These references will be brought up-to-date by the CODATA Secretariat whenever appropriate, but at least biennially.)

17. Reports of CODATA Task Group on Key Values for Thermodynamics, *CODATA Bulletin* No. 17 (1976); No. 22 (1977).

18. D. D. Wagman, et al., "Selected Values of Chemical Thermodynamic Properties," National Bureau of Standards Technical Note 270, Parts 3 (1968) to 7 (1973).

19. V. P. Glushko, Editor, "Thermal Constants of Substances," Vol. 1 (1965) to Vol.7 (1974), Akademiya Nauk, Moscow.

20. V. P. Glushko, Editor, "Thermodynamic and Thermophysical Properties of Combustion Products," Akademiya Nauk, Moscow. A handbook in ten parts (1971-1974).

21. E. S. Domalski, Selected Values of Heats of Formation of Organic Compounds, Containing the Elements C, H, N, O, P, and S, *J. Phys. Chem. Ref. Data* 1, 221-278 (1972).

22. J. D. Cox and G. Pilcher, "Thermochemistry of Organic and Organometallic Compounds," John Wiley & Sons, Inc., New York (1969).

23. D. R. Stull, E. F. Westrum, Jr., and G. C. Sinke, "The Chemical Thermodynamics of Organic Compounds," John Wiley & Sons, Inc., New York (1969).

24. R. C. Wilhoit and B. J. Zwolinski, Physical and Thermodynamic Properties of Aliphatic Alcohols, *J. Phys. Chem. Ref. Data* 2, Supplement 1 (1973).

25. D. R. Stull, and H. Prophet, "JANAF Thermochemical Tables, "Second Edition, Nat. Stand. Ref. Data Ser., Nat. Bur. Stand. (U.S.) 37 (1971); supplements have appeared in *J. Phys. Chem. Ref. Data* 3, 311-480 (1974); 4, 1-175 (1975); 7, 793-940 (1978).

26. J. J. Jasper, The Surface Tension of Pure Liquid Compounds, *J. Phys. Chem. Ref. Data* 1, 841-1009 (1972).

27. J. H. Dymond and E. B. Smith, "Virial Coefficients of Gases: A Critical Compilation," Clarendon Press, Oxford (1969), Oxford Science Research Paper No. 2.

28. J. H. Keenan, et al., "Steam Tables, Thermodynamic Properties of Water Including Vapor, Liquid and Solid Phases," John Wiley & Sons, Inc., New York (1966).

The following journals and continuing publication series also contain pertinent compilations of evaluated data:

29. *Journal of Physical & Chemical Reference Data*, American Chemical Society, Washington, DC.

30. V. A. Rabinovich, Editor, "Thermophysical Properties of Substances and Materials" Izdatel'stvo Standartov, Moscow, Nos. 1-6 (1966-72).

31. "International Thermodynamic Tables of the Fluid State" International Union of Pure and Applied Chemistry, Oxford, U.K.

*32. American Petroleum Institute Tables, "Selected Values of Properties of Hydrocarbons and Related Compounds" Thermodynamics Research Center, Texas A & M University, College Station, Texas, (1955 et seq.).

VIII. ACKNOWLEDGMENT

The Task Group gratefully acknowledges valuable criticism and suggestions received from Dr. Philip L. Altman, Dr. E. R. Cohen, Dr. G. A. Martin, Prof. Maurice Menache, Prof. G. A. Wilkins, and Prof. A. J. C. Wilson.

---

*These data compilations do not always differentiate between experimental data and estimated properties. This, of course, limits their usefulness for the test of new correlations and estimation procedures.

MEMBERS OF TASK GROUP

Dr. Arnold A. Bondi, Chairman

Shell Development Co.

Houston, Texas, 77001 U.S.A.

Dr. Malcolm Chase

Thermal Research, 1707 Building

Dow Chemical Co.

Midland, Michigan, 48640 U.S.A.

Prof. Ronald P. Danner

Department of Chemical Engineering

Pennsylvania State University

University Park, Pennsylvania, 16802 U.S.A.

Dr. L. J. Lawrenson

Division of Chemical Standards

National Physical Laboratory

Teddington, England, U.K.

Prof. Eiji O' shima

Chemical Engineering Department

University of Tokyo

Tokyo, Japan

Section 3

Thermodynamics Including Biothermodynamics

Table 3.1, reprinted from the Green Book, contains the section on terminology pertinent to thermodynamics.

The first reprinted paper, "A Guide to Procedures for the Publication of Thermodynamic Data" (*7*), is a general document that is basic to the other more specialized papers that follow. It is the current successor to the earlier documents (*2, 3, 7m*). A reprinted paper, "Guidelines for Reporting Experimental Data on Vapor–Liquid Equilibrium of Mixtures at Low and Moderate Pressures" (*8*), contains recommendations intended to increase the usefulness of publications containing measurements on vapor–liquid equilibria at low and moderate pressures for systems in which all components are subcritical (i.e., below their critical temperatures and pressures).

Useful information on the assessment and presentation of uncertainties of the numerical results of thermodynamic measurements is found in reference 9.

A set of reprinted papers covers thermodynamic measurements in biology. The first paper, "Recommendations for Measurement and Presentation of Biochemical Equilibrium Data" (*10*), has been widely published and serves as a basis for the others. Recommendations for the Presentation of Thermodynamic and Related Data in Biology" (*11*) discusses biothermodynamic data. The second paper is "Calorimetric Measurements on Cellular Systems: Recommendations for Measurements and Presentation of Results" (*12*).

Reference 13 (not reprinted here) is concerned with the preparation of correlation procedures and tables of values for the thermodynamic properties of fluids and with the simultaneous fitting of large matrices of multiproperty data to produce equations that provide high-quality property values over a wide range of temperatures and pressure including two-phase lines, and the publication of useful tables of values from these correlations. As such, it is concerned with the use of experimental property data rather than with their publication. However, because measurements are sometimes made to improve such correlations, use of new data to improve such correlations is then discussed along with the data. Reference 6 (reprinted in Section 2) would also be relevant to such authors. Reference 14 (not reprinted here) presents guidelines for compilers, evaluators, and editors for contributions to the series published by the International Union for Pure and Applied Chemistry, Commission on Solubility Data. Reference 14 is addressed more to the users of experimental data than to the providers. However, by implication, the paper (*14*) provides a general discussion of the topics of importance in solubility data, and therefore should be of interest to authors of solubility papers.

Table 3.1. Chemical Thermodynamics. Continued on next page.

The names and symbols of the more generally used quantities given here are also recommended by IUPAP [3] and by ISO [4.e, i]. Additional information can be found in [1.d, j].

Name	*Symbol*	*Definition*	*SI unit*	*Notes*
heat	q, Q		J	1
work	w, W		J	1
internal energy	U	$\Delta U = q + w$	J	1
enthalpy	H	$H = U + pV$	J	
thermodynamic temperature	T		K	
Celsius temperature	θ, t	$\theta/°C = T/K - 273.15$	°C	2
entropy	S	$dS \geqslant dq/T$	$J\,K^{-1}$	
Helmholtz energy, (Helmholtz function)	A	$A = U - TS$	J	3
Gibbs energy, (Gibbs function)	G	$G = H - TS$	J	
Massieu function	J	$J = -A/T$	$J\,K^{-1}$	
Planck function	Y	$Y = -G/T$	$J\,K^{-1}$	
surface tension	γ, σ	$\gamma = (\partial G/\partial A_s)_{T,p}$	$J\,m^{-2}$, $N\,m^{-1}$	
molar quantity X	X_m	$X_m = X/n$	(varies)	4, 5
specific quantity X	x	$x = X/m$	(varies)	4, 5
pressure coefficient	β	$\beta = (\partial p/\partial T)_V$	$Pa\,K^{-1}$	
relative pressure coefficient	α_p	$\alpha_p = (1/p)(\partial p/\partial T)_V$	K^{-1}	
compressibility,				
isothermal	κ_T	$\kappa_T = -(1/V)(\partial V/\partial p)_T$	Pa^{-1}	
isentropic	κ_S	$\kappa_S = -(1/V)(\partial V/\partial p)_S$	Pa^{-1}	
linear expansion coefficient	α_l	$\alpha_l = (1/l)(\partial l/\partial T)$	K^{-1}	
cubic expansion coefficient	α, α_V, γ	$\alpha = (1/V)(\partial V/\partial T)_p$	K^{-1}	6
heat capacity,				
at constant pressure	C_p	$C_p = (\partial H/\partial T)_p$	$J\,K^{-1}$	
at constant volume	C_V	$C_V = (\partial U/\partial T)_V$	$J\,K^{-1}$	
ratio of heat capacities	γ, (κ)	$\gamma = C_p/C_V$	1	
Joule–Thomson coefficient	μ, μ_{JT}	$\mu = (\partial T/\partial p)_H$	$K\,Pa^{-1}$	
second virial coefficient	B	$pV_m = RT(1 + B/V_m + \cdots)$	$m^3\,mol^{-1}$	
compression factor (compressibility factor)	Z	$Z = pV_m/RT$	1	
partial molar quantity X	X_B, (X'_B)	$X_B = (\partial X/\partial n_B)_{T,p,n_{j\neq B}}$	(varies)	7

(1) Both $q > 0$ and $w > 0$ indicate an increase in the energy of the system; $\Delta U = q + w$.

(2) This quantity is sometimes misnamed 'centigrade temperature'.

(3) It is sometimes convenient to use the symbol F for Helmholtz energy in the context of surface chemistry, to avoid confusion with A for area.

(4) The definition applies to pure substance. However, the concept of molar and specific quantities may also be applied to mixtures.

(5) X is an extensive quantity. The unit depends on the quantity. In the case of molar quantities the entities should be specified.

Example molar volume of B, $V_m(B) = V/n_B$

(6) This quantity is also called the coefficient of thermal expansion, or the expansivity coefficient.

(7) The symbol applies to entities B which should be specified. The prime may be used to distinguish partial molar X from X when necessary.

Example The partial molar volume of Na_2SO_4 in aqueous solution may be denoted $V'(Na_2SO_4, aq)$, in order to distinguish it from the volume of the solution $V(Na_2SO_4, aq)$.

Table 3.1. Continued. Continued on next page.

Name	*Symbol*	*Definition*	*SI unit*	*Notes*
chemical potential (partial molar Gibbs energy)	μ	$\mu_B=(\partial G/\partial n_B)_{T,p,n_{j\neq B}}$	$J\,mol^{-1}$	8
absolute activity	λ	$\lambda_B=\exp(\mu_B/RT)$	1	8
standard chemical potential	$\mu^\ominus$, μ°		$J\,mol^{-1}$	9
standard partial molar enthalpy	$H_B^\ominus$	$H_B^\ominus=\mu_B^\ominus+TS_B^\ominus$	$J\,mol^{-1}$	8, 9
standard partial molar entropy	$S_B^\ominus$	$S_B^\ominus=-(\partial\mu_B^\ominus/\partial T)_p$	$J\,mol^{-1}\,K^{-1}$	8, 9
standard reaction Gibbs energy (function)	$\Delta_r G^\ominus$	$\Delta_r G^\ominus=\sum_B \nu_B\mu_B^\ominus$	$J\,mol^{-1}$	9, 10, 11
affinity of reaction	A, ($\mathscr{A}$)	$A=-(\partial G/\partial\xi)_{p,T}$ $=-\sum_B \nu_B\mu_B$	$J\,mol^{-1}$	11
standard reaction enthalpy	$\Delta_r H^\ominus$	$\Delta_r H^\ominus=\sum_B \nu_B H_B^\ominus$	$J\,mol^{-1}$	9, 10, 11
standard reaction entropy	$\Delta_r S^\ominus$	$\Delta_r S^\ominus=\sum_B \nu_B S_B^\ominus$	$J\,mol^{-1}\,K^{-1}$	9, 10, 11
equilibrium constant	$K^\ominus$, K	$K^\ominus=\exp(-\Delta_r G^\ominus/RT)$	1	9, 11, 12
equilibrium constant, pressure basis	K_p	$K_p=\prod_B p_B^{\nu_B}$	$Pa^{\Sigma\nu}$	11, 13
equilibrium constant concentration basis	K_c	$K_c=\prod_B c_B^{\nu_B}$	$(mol\,m^{-3})^{\Sigma\nu}$	11, 13
molality basis	K_m	$K_m=\prod_B m_B^{\nu_B}$	$(mol\,kg^{-1})^{\Sigma\nu}$	11, 13
fugacity	f, $\tilde{p}$	$f_B=\lambda_B \lim_{p\to 0}(p_B/\lambda_B)_T$	Pa	8
fugacity coefficient	ϕ	$\phi_B=f_B/p_B$	1	
activity and activity coefficient referenced to Raoult's law, (relative) activity	a	$a_B=\exp\left[\frac{\mu_B-\mu_B^*}{RT}\right]$	1	8, 14, 15

(8) The definition applies to entities B which should be specified.
(9) The symbol $^\ominus$ or $^\circ$ is used to indicate standard. They are equally acceptable. Definitions of standard states are discussed below (p.47). Whenever a standard chemical potential $\mu^\ominus$ or a standard equilibrium constant $K^\ominus$ or other standard quantity is used, the standard state must be specified.
(10) The symbol r indicates reaction in general. In particular cases r can be replaced by another appropriate subscript, e.g. $\Delta_f H^\ominus$ denotes the standard molar enthalpy of formation; see p.45 below for a list of subscripts.
(11) The reaction must be specified for which this quantity applies.
(12) This quantity is dimensionless and its value depends on the choice of standard state, which must be specified. ISO [4.i] recommend the symbol $K^\ominus$ and the name 'standard equilibrium constant'. The IUPAC Thermodynamics Commission [1.j] recommend the symbol K and the name 'thermodynamic equilibrium constant'.
(13) These quantities are not in general dimensionless. One can define in an analogous way an equilibrium constant in terms of fugacity K_f, etc. At low pressures K_p is approximately related to $K^\ominus$ by the equation $K^\ominus\approx K_p/(p^\ominus)^{\Sigma\nu}$, and similarly in dilute solutions K_c is approximately related to $K^\ominus$ by $K^\ominus\approx K_c/(c^\ominus)^{\Sigma\nu}$; however, the exact relations involve fugacity coefficients or activity coefficients [1.j].

The equilibrium constant of dissolution of an electrolyte (describing the equilibrium between excess solid phase and solvated ions) is often called a solubility product, denoted K_{sol} or K_s (or $K_{sol}^\ominus$ or $K_s^\ominus$ as appropriate). In a similar way the equilibrium constant for an acid dissociation is often written K_a, for base hydrolysis K_b, and for water dissociation K_w.
(14) An equivalent definition is $a_B=\lambda_B/\lambda_B^*$. The symbol * denotes pure substance. The symbol $^\circ$ should not be used with this meaning, although it has been so used in the past.
(15) In the defining equations given here the pressure dependence of the activity has been neglected, as is often done for condensed phases at atmospheric pressure.

Table 3.1. Continued. Continued on next page.

Name	*Symbol*	*Definition*	*SI Unit*	*Notes*
activity coefficient	f	$f_B = a_B / x_B$	1	8
activities and activity coefficients referenced to Henry's law, (relative) activity,				
molality basis	a_m	$a_{m,B} = \exp\left[\frac{\mu_B - \mu_B^\ominus}{RT}\right]$	1	8, 15, 16
concentration basis	a_c	$a_{c,B} = \exp\left[\frac{\mu_B - \mu_B^\ominus}{RT}\right]$	1	8, 15, 16
mole fraction basis	a_x	$a_{x,B} = \exp\left[\frac{\mu_B - \mu_B^\ominus}{RT}\right]$	1	8, 15, 16
activity coefficient,				
molality basis	γ_m	$a_{m,B} = \gamma_{m,B} m_B / m^\ominus$	1	8
concentration basis	γ_c	$a_{c,B} = \gamma_{c,B} c_B / c^\ominus$	1	8
mole fraction basis	γ_x	$a_{x,B} = \gamma_{x,B} x_B$	1	8
ionic strength,				
molality basis	I_m, I	$I_m = \frac{1}{2}\Sigma m_B z_B^2$	mol kg^{-1}	
concentration basis	I_c, I	$I_c = \frac{1}{2}\Sigma c_B z_B^2$	mol m^{-3}	
osmotic coefficient,				
molality basis	ϕ_m	$\phi_m = (\mu_A^* - \mu_A)/(RTM_A \Sigma m_B)$	1	
mole fraction basis	ϕ_x	$\phi_x = (\mu_A - \mu_A^*)/(RT \ln x_A)$	1	
osmotic pressure	Π	$\Pi = c_B RT$ (ideal dilute solution)	Pa	

(16) An equivalent definition is $a_B = \lambda_B / \lambda_B^\ominus$, where $\lambda_B^\ominus = \exp(\mu_B^\ominus / RT)$. The definition of $\mu^\ominus$ is different for a molality basis, a concentration basis, or a mole fraction basis; see p.47 below.

Other symbols and conventions in chemical thermodynamics

A more extensive description of this subject can be found in [1.j].

(i) *Symbols used as subscripts to denote a chemical process or reaction*
These symbols should be printed in roman (upright) type, without a full stop (period).

vaporization, evaporation (liquid→gas)	vap
sublimation (solid→gas)	sub
melting, fusion (solid→liquid)	fus
transition (between two phases)	trs
mixing of fluids	mix
solution (of solute in solvent)	sol
dilution (of a solution)	dil
adsorption	ads
displacement	dpl
immersion	imm
reaction in general	r
atomization	at
combustion reaction	c
formation reaction	f

(ii) *Recommended superscripts*

standard	$\ominus$, o
pure substance	*
infinite dilution	∞
ideal	id
activated complex, transition state	‡
excess quantity	E

Table 3.1. Continued. Continued on next page.

(iii) *Examples of the use of these symbols*

The subscripts used to denote a chemical process, listed under (i) above, should be used as subscripts to the Δ symbol to denote the change in an extensive thermodynamic quantity associated with the process.

Example $\Delta_{vap}H = H(g) - H(l)$, for the enthalpy of vaporization, an extensive quantity proportional to the amount of substance vaporized.

The more useful quantity is usually the change divided by the amount of substance transferred, which should be denoted with an additional subscript m.

Example $\Delta_{vap}H_m$ for the molar enthalpy of vapourization.

However the subscript m is frequently omitted, particularly when the reader may tell from the units that a molar quantity is implied.

Example $\Delta_{vap}H = 40.7\ \text{kJ mol}^{-1}$ for H_2O at 373.15 K and 1 atm.

The subscript specifying the change is also sometimes attached to the symbol for the quantity rather than the Δ, so that the above quantity is denoted $\Delta H_{vap,m}$ or simply ΔH_{vap}, but this is not recommended.

The subscript r is used to denote changes associated with a *chemical reaction*. Although symbols such as $\Delta_r H$ should denote the integral enthalpy of reaction, $\Delta_r H = H(\xi_2) - H(\xi_1)$, in practice this symbol is usually used to denote the change divided by the amount transferred, i.e. the change per extent of reaction, defined by the equation

$$\Delta_r H = \sum_B \nu_B H_B = (\partial H/\partial \xi)_{T,p}$$

It is thus essential to specify the stoichiometric reaction equation when giving numerical values for such quantities in order to define the extent of reaction ξ and the values of the stoichiometric numbers ν_B.

Example $N_2(g) + 3H_2(g) = 2NH_3(g)$, $\Delta_r H^\circ = -92.4\ \text{kJ mol}^{-1}$
$\Delta_r S^\circ = -199\ \text{J mol}^{-1}\ \text{K}^{-1}$

The mol^{-1} in the units identifies the quantities in this example as the change per extent of reaction. They may be called the molar enthalpy and entropy of reaction, and a subscript m may be added to the symbol, to emphasize the difference from the integral quantities if required.

The *standard reaction quantities* are particulary important. They are defined by the equations

$$\Delta_r H^\circ \quad (= \Delta_r H_m{}^\circ = \Delta H_m{}^\circ) = \sum_B \nu_B H_B{}^\circ$$

$$\Delta_r S^\circ \quad (= \Delta_r S_m{}^\circ = \Delta S_m{}^\circ) = \sum_B \nu_B S_B{}^\circ$$

$$\Delta_r G^\circ \quad (= \Delta_r G_m{}^\circ = \Delta G_m{}^\circ) = \sum_B \nu_B \mu_B{}^\circ$$

The symbols in parentheses are alternatives. In view of the variety of styles in current use it is important to specify notation with care for these symbols. The relation to the affinity of the reaction is

$$-A = \Delta_r G = \Delta_r G^\circ + RT \ln\left(\prod_B a_B{}^{\nu_B}\right),$$

and the relation to the standard equilibrium constant is $\Delta_r G^\circ = -RT \ln K^\circ$.

The term *combustion* and symbol c denotes the complete oxidation of a substance. For the definition of complete oxidation of substances containing elements other than C, H and O see [45]. The corresponding reaction equation is written so that the stoichiometric number ν of the substance is -1.

Example The standard enthalpy of combustion of gaseous methane is $\Delta_c H^\circ(CH_4, g, 298.15\ K) = -890.3\ \text{kJ mol}^{-1}$, implying the reaction $CH_4(g) + 2O_2(g) \rightarrow CO_2(g) + 2H_2O(l)$.

The term *formation* and symbol f denotes the formation of the substance from elements in their reference state (usually the most stable state of each element at the chosen temperature and pressure). The corresponding reaction equation is written so that the stoichiometric number ν of the substance is $+1$.

Example The standard entropy of formation of crystalline mercury II chloride is $\Delta_f S^\circ(HgCl_2, cr, 298.15\ K) = -154.3\ \text{J mol}^{-1}\ \text{K}^{-1}$, implying the reaction $Hg(l) + Cl_2(g) \rightarrow HgCl_2(cr)$.

Table 3.1. Continued.

The term *atomization*, symbol at, denotes a process in which a substance is separated into its constituent atoms in the ground state in the gas phase. The corresponding reaction equation is written so that the stoichiometric number ν of the substance is -1.

Example The standard (internal) energy of atomization of liquid water is $\Delta_{at}U^{\circ}(H_2O, l) = 625\ kJ\ mol^{-1}$ implying the reaction $H_2O(l) \rightarrow 2H(g) + O(g)$.

(iv) *Standard states* [1.j]
The standard chemical potential of substance B at temperature T, $\mu_B^{\circ}(T)$, is the value of the chemical potential under standard conditions, specified as follows. Three differently defined standard states are recognized.

For a gas phase. The standard state for a gaseous substance, whether pure or in a gaseous mixture, is the (hypothetical) state of the pure substance B in the gaseous phase at the standard pressure $p = p^{\circ}$ and exhibiting ideal gas behaviour. The standard chemical potential is defined as

$$\mu_B^{\circ}(T) = \lim_{p \to 0} [\mu_B(T, p, y_B, \ldots) - RT \ln(y_B p/p^{\circ})]$$

For a pure phase, or a mixture, or a solvent, in the liquid or solid state. The standard state for a liquid or solid substance, whether pure or in a mixture, or for a solvent, is the state of the pure substance B in the liquid or solid phase at the standard pressure $p = p^{\circ}$. The standard chemical potential is defined as

$$\mu_B^{\circ}(T) = \mu_B^{*}(T, p^{\circ})$$

For a solute in solution. For a solute in a liquid or solid solution the standard state is referenced to the ideal dilute behaviour of the solute. It is the (hypothetical) state of solute B at the standard molality m°, standard pressure p°, and exhibiting infinitely diluted solution behaviour. The standard chemical potential is defined as

$$\mu_B^{\circ}(T) = [\mu_B(T, p^{\circ}, m_B, \ldots) - RT \ln(m_B/m^{\circ})]^{\infty}.$$

The chemical potential of the solute B as a function of the molality m_B at constant pressure $p = p^{\circ}$ is then given by the expression

$$\mu_B(m_B) = \mu_B^{\circ} + RT \ln(m_B \gamma_{m,B}/m^{\circ}).$$

Sometimes (amount) concentration c is used as a variable in place of molality m; both of the above equations then have c in place of m throughout. Occasionally mole fraction x is used in place of m; both of the above equations then have x in place of m throughout, and $x^{\circ} = 1$. Although the standard state of a solute is always referenced to ideal dilute behaviour, the definition of the standard state and the value of the standard chemical potential μ° are different depending on whether molality m, concentration c, or mole fraction x is used as a variable.

(v) *Standard pressures, molality, and concentration*
The standard pressure recommended by IUPAC since 1982 [1.j] is

$$p^{\circ} = 10^5\ Pa \quad (= 1\ bar),$$

and this is known as the standard state pressure.

Up to 1982 the standard pressure was usually taken to be

$$p^{\circ} = 101\ 325\ Pa\ (= 1\ atm, \text{ called the 'standard atmosphere'}).$$

It may also sometimes be desirable to use a standard pressure substantially different from 10^5 Pa, for example in tabulating data appropriate to high-pressure chemistry. It is therefore important always to specify the standard pressure adopted. The conversion of values corresponding to different p° is described in [46].

The standard molality is always taken as $m^{\circ} = 1\ mol\ kg^{-1}$.
The standard concentration is always taken as $c^{\circ} = 1\ mol\ dm^{-3}$.

In principle other m° and c° values could be used, but they would have to be specified.

(vi) *Values of thermodynamic quantities*
Values of many thermodynamic quantities represent basic chemical properties of substances and serve for further calculations. Extensive tabulations exist, e.g. [47, 48, 49]. Special care has to be taken in reporting the data and their uncertainties [23, 24].

A GUIDE TO PROCEDURES FOR THE PUBLICATION OF THERMODYNAMIC DATA

PHYSICAL CHEMISTRY DIVISION

COMMISSION ON THERMODYNAMICS AND THERMOCHEMISTRY†

At the 1971 Washington. DC meeting of the IUPAC Commission on Thermodynamics and Thermochemistry it was resolved that the following guide should be given the widest publicity to aid in the resolution of some current problems in scientific communication and in the hope that it might stimulate similar action by other organizations concerned with specialized fields of science.

This guide was previously ratified by the All-Union Calorimetry Conference (USSR). the Calorimetry Conference (USA). the Experimental Thermodynamics Conference (UK). the Society of Calorimetry and Thermal Analysis (Japan). and the Société Française des Termiciens (France).

In 1953. the Eighth (US) Calorimetry Conference adopted a resolution providing guidance on minimum publication standards in calorimetry. This resulted in an improvement in the quality of publication. and so led the Calorimetry Conference to revise. extend and publish its recommendations in 1960[1]. A second revision was authorized in 1970. and it was intended that this new version should be submitted to various calorimetry and related conferences[2]. and to the International Union of Pure and Applied Chemistry for suggestions and ratification.

This has been done with the approval of Commission I.2 of IUPAC and the agreed text is hereby presented in accordance with the terms of the resolution cited at the head of this document. The Commission itself has also prescribed recommendations on the publication of thermochemical studies as early as 1934[3]. as has the scientific council on chemical thermodynamics of IONKh of the USSR Academy of Sciences[4].

INTRODUCTION

This guide is addressed not only to specialists in calorimetry or in the various aspects of thermodynamics but also to all those who measure and publish thermodynamic quantities as adjuvant aspects of their research endeavours. We would urge all who ever publish thermodynamic values for whatever purpose determined to follow these suggestions so that maximum benefit from their studies will be realized. Journal editors and referees should note that this document embodies a set of consistent. carefully considered criteria for judging the completeness and acceptability of papers reporting thermodynamic quantities. The ultimate needs of the compiler and correlator of such data have been considered also.

---

† *Chairman*: S. Sunner (Sweden); *Secretary*: E. F. Westrum Jr (USA); *Titular Members*: J. D. Cox (UK); E. U. Franck (Germany); L. V. Gurvich (USSR); F. D. Rossini (USA); S. Seki (Japan); B. Vodar (France). *Associate Members*: P. Heydemann (USA); R. J. Irving (UK); H. Kehiaian (Poland); M. Laffitte (France); G. C. Sinke (USA); N. J. Trappeniers (Netherlands); R. Vîlcu (Romania); I. Wadsö (Sweden).

ESSENTIAL INFORMATION

The highly interdependent nature of thermodynamic data imposes special obligations upon the author of papers reporting the results of thermodynamic investigation. He must give enough information about his experiment to allow readers to appraise the precision and accuracy of his results so they may be properly consolidated within the existing body of data in the literature. Further, as accepted values of physical constants change or as new thermodynamic data for related systems become available, subsequent investigators often can recalculate results if it is clear that they are based on good experiments for which adequate information is presented, however old they may be. For these reasons, an author's prime responsibility is to report his results in a form related as closely to experimentally observed quantities as is practical, with enough experimental details and auxiliary information to characterize the results adequately and to allow critical assessment of the accuracy claimed. For the convenience of the reader, the author may interpret and correlate the primary results as appropriate and present derived results in a form easy to utilize. However, such derived (or secondary) results *never* should be published at the cost of omitting the primary results on which they are based. Reference may be made to accessible earlier publications for some details.

In addition to the presentation of the data themselves, estimates of the precision indices and probable accuracy of the data should be given by the authors. The various sources of uncertainty should be rigorously described with clear separation of measurement imprecisions, numerical analysis deviations and possible systematic biases. The methods and assumptions for the statistical analyses should be indicated. Possible sources and magnitudes of systematic errors should be identified and enumerated.

Because temperature scales are of such great significance for all thermodynamic measurements the considerations and conversions cited by Rossini[5] and others should be observed. For accurately measured temperatures depending on an International Practical Temperature Scale, the scale used in calibration of the temperature-measuring instrument(s) (e.g. 'IPTS-48 as amended in 1960') and the scale to which the specified temperature values refer (sometimes these two scales are different, e.g. the latter might be IPTS-68) should be stated.

In instances where requisite primary results are too extensive or for other reasons do not merit journal publication in full, the use of auxiliary publication services may be appropriate[6]. A footnote in the publication indicates how the reader may obtain the adjuvant data.

Apparatus and procedures—A description of the apparatus including details of the reaction container or calorimeter vessel, the controlled environment, and measuring systems such as those for time, temperature, and pressure; the design of calibration heaters and heater lead placement; precautions as to shielding or isolation of calibration circuits; the method of calibration and the sensitivity of the instruments used in these measuring systems such as thermometers, bridges, potentiometers, flowmeters and weighing devices should be given. The history of a particular apparatus which is used in an on-going series of researches should be maintained and documented as to modifications, improvements etc., to the end that should corrections be made necessary by subsequent recalibrations or by revelation of systematic errors or bias, such corrections can be applied to all affected data either by the author or by compilers or reviewers. Information establishing the heat capacity of the calorimeter or the energy equivalent (preferably with traceability to a calibrating or standardizing laboratory), together with details of the observational procedures, the methods of evaluating the corrected temperature increment, methods of analysis of results, and the precision of the measurements should be given. The reliability of the results should be established by the use of recognized reference substances such as the samples likely to be recommended by the IUPAC Sub-commission on Standard Calibration Materials[7], those provided by the (US) National Bureau of Standards[8] or by VNIIM (USSR)[9], those authorized by the Calorimetry Conference for thermophysical or thermochemical measurements[10], or those systems generally recognized as standard for mixtures, e.g. the solubility of oxygen in water at 1 atm and 298.15 K for gas solubility[11], hexane + cyclohexane for enthalpy increments on mixing[12], and benzene + cyclohexane for volume increments on mixing[13]. Determination of the same quantity by two or more independent methods is often an advisable alternative method for detection of systematic error.

Materials—The source of and/or method of preparation for all materials used, including calibration, reference and auxiliary substances; experimental values for analyses and pertinent physical properties of materials, the criteria of characterization and purity, as well as the method, temperature, time interval etc. of storing samples and preparing them for measurements should be stated whenever this is important. The density used in reduction of weighing to mass, and special procedures such as for dealing with partially filled ampoules, should be specified. For studies made on solutions, the source, preparation and quality of the *solvent* should always be included, as should information as to dissolved gases (carbon dioxide, air etc.) whenever these impurities may influence the results. For all thermochemical studies, the methods and results of all analytical investigations on the initial and/or final system should be presented, including tests for incomplete reaction, side reactions, evaporation losses of components from mixtures, corrosion of apparatus etc. These considerations as well as possible dialysis of compounds prior to actual reaction, buffers employed etc. may be especially important in biothermodynamic studies.

Symbols, terminology, units—Authors are encouraged to follow as closely as possible the recommendations of the International System (SI) of units[14] and the symbols and nomenclature approved by IUPAC[15]. In particular, authors are urged (a) to use the recommended name for each physical quantity and the preferred symbol, (b) to use the internationally accepted symbols for units, (c) to use SI units, and (d) to adhere, particularly in the labels of columns or rows in tables and in the labels of the axes of graphs, to the implications of the convention: physical quantity = number × unit; for example, by the use of $S/(\mathrm{JK^{-1}\ mol^{-1}})$ as a label for a series of numbers rather than 'S' (with or without a statement somewhere as to what the units are), 'S, {J/K mol)}', 'S J/(K mol)', and (e) to functional expressions such as S° (H_2O, g, 298 K) rather than expressions like S°_{298,H_2O}. (Those authors who elect to use non-SI units should define them in terms of SI units. This might be done, for example, by means of such a footnote as: 'Throughout this paper $\mathrm{cal_{th}}$ = 4.184 J, Torr = (101.325/760) kPa and Å = 0.1 nm'.)

Mode of presentation of results—Although a table giving the appropriate independent and dependent variables is often the best form of presentation in that it permits recovery of the primary results, circumstances arise where especially for a great many determinations the same end may be achieved, for example, by the presentation of an equation representing the temperature dependence of the measured quantity and a *deviation* plot showing the individual points. This procedure saves space, promotes clarity and in many instances can be sensitive enough to permit full recovery of individual results without a table.

Occasionally, as in combustion calorimetry, each set of replicate measurements can be acceptably recorded as only the mean value, the magnitude and definition of its precision index, and the number of individual measurements made. For the precision indices on thermochemical data, the conventions suggested by Rossini[16] are recommended.

Occasionally, tabulation of high-temperature thermodynamic functions is given by empirical equations which, instead of being in predetermined form and number of disposable parameters, are derived to fit the enthalpy, vapour-pressure, or other data within their precision at every temperature of measurement. This not only provides ready analytical interpolation, differentiation or integration (as by a computer), but can save the journal space of longer-than-abbreviated tables when this is an issue. It is obviously not expedient, of course, when the given property function shows major lack of monotonic behaviour. Graphical and analytical representation of the primary results or secondary results are occasionally worthwhile for the convenience of the reader, but are not generally a satisfactory substitute for tabular presentation of accurate experimental results. Extensive tabulations of secondary values (e.g. smoothed values at rounded temperatures) should be so designed with values reported at temperatures (or pressures etc.) so spaced that no serious loss of accuracy will result by using an interpolation formula equivalent to five-place Lagrangian interpolation[17]. In particular, a sufficient number of digits should be retained to make such interpolation feasible.

It is not practicable to give detailed recommendations for presenting the results of all types of thermodynamic investigations. However, the following paragraphs do provide recommendations for some important kinds of thermochemical, thermophysical and equilibrium studies and will serve as guides for related areas.

PRESENTATION OF THERMOCHEMICAL CALORIMETRIC DATA

Reaction calorimetry—The following experimental quantities should be included if applicable: energy equivalent of the calorimetric system, mass of sample and/or mass of product used in determining the amount of reaction, masses of auxiliary substances, corrected temperature increment (or if appropriate initial and final temperatures plus heat exchange adjustment), total observed energy change, ignition energy, chemical and physical specification of the initial and final states of the reaction, conversion to 'standard' concentrations, correction for side reactions, reduction to standard states (e.g. the 'Washburn corrections'), temperature of experiments, final experimental energy (or enthalpy) of reaction and uncertainty interval (with the chemical reaction to which the result applies precisely specified). Derived values such as standard enthalpy and standard Gibbs energy of formation may be provided for the convenience of the reader. Detailed discussion of procedures is availablc[18, 19].

Solution calorimetry—Although most of the considerations and literature of the previous paragraph apply, additional material to be included would involve the specification of the temperature of the measurements, appropriately specified concentrations of all reactants, solvent, supporting electrolyte, ionic strength, the precise reaction occurring etc. Thermochemical studies on solutions should present primary and derived results in sufficient detail that the actual calorimetric process is clearly discernible. Dilution corrections to standard states should describe any approximations for unmixing, e.g. Young's rule[20] or approximations as to the relative apparent enthalpy of uncharged solutes such as NH_3, CO_2 etc. in the solution. Investigators of thermochemical properties of solutions are urged to ascertain densities of their solutions to facilitate theoretical comparison.

PRESENTATION OF THERMOPHYSICAL CALORIMETRIC DATA

The following information is important for delineating the temperature dependence of the thermodynamic (thermophysical) properties of non-reacting systems and ascertaining the influence of thermal history on measured properties and evaluating the reliability of the results: A table of experimental values of heat capacity or enthalpy increment values including the actual temperature increments used in the measurements (if important as e.g. in transition, pre-melting or anomalous regions), the chronological sequence of data where the thermal history may be significant (either implicitly by chronological presentation or by a general statement), values of the energy (or enthalpy) increments and temperatures of essentially isothermal phase changes, and the reliability of the data indicated by an estimated uncertainty. These primary experimental results may be supplemented, but never supplanted, by a tabulation of smoothed values of thermodynamic properties at selected temperatures for the convenience of the reader. Where applicable, such tabulations should include values of the appropriate heat capacity, the standard entropy (S°), the standard enthalpy increment $\{H^\circ(T) - H^\circ(0)\}$, the standard Gibbs energy functions $[\{G^\circ(T) - H^\circ(0)\}/T]$[21]. Frequently these have been tabulated at 5K intervals from 0 to 50 K, at 10 K intervals from 50 to 300 K or slightly higher temperatures, at 100 K intervals at higher temperatures, and at the temperatures of phase transitions. Such details need not be given for all substances but are desiderata for important compounds. Use of auxiliary publication services as mentioned above may be more appropriate as repositories for such tabulations. Certainly values at the two important reference temperatures, 273.15 and 298.15 K, should be included in the tabulations. Definitive information on procedures is available[22].

PRESENTATION OF CALCULATED THERMOPHYSICAL FUNCTIONS

Because the usefulness of calorimetric data is extended if calculated thermodynamic functions based upon them are provided, recommendations for the presentation of such calculated values are incorporated here. The following information with appropriate sources is needed to characterize the results of statistical thermodynamic calculations: details of the molecular model used, including bond distances and angles, specification of the exact conformation, moments of inertia or rotational constants, symmetry number, complete vibrational assignments, parameters used for calculating contributions of internal rotation, rotational isomerism, anharmonicity, centrifugal distortion etc., the citation (usually by reference) of formulae and special tabulations used, comparison with experimental thermodynamic data. The functions tabulated should include those indicated in previous sections except that the heat capacity should be C_v° or C_p° and in addition, values of the standard enthalpy of formation $\Delta_f H^\circ$, standard Gibbs energy of formation, $\Delta_f G^\circ$, and the common logarithm of the equilibrium constant of formation, $\log K_f$, may be published if warranted. The criterion suggested above for the spacing of values is recommended; this corresponds, for example, to 50 K intervals to 300 K, 100 K intervals to 2000 K, 200 K intervals to 3000 K, and 500 K intervals at higher temperatures. Again the values should be given also at the reference temperatures, 273.15 and 298.15 K.

For purposes of smoothness, interpolation and internal consistency, one (but seldom more than one) more digit may be retained than is justified by the absolute accuracy. For calculations on small, rigid molecules, involving anharmonicity, centrifugal stretching, rotation–vibration interaction, Fermi resonance, isotopic composition and any other significant effect, functions may be reported to three decimal digits, e.g.

$$12.345 \text{ cal K}^{-1} \text{ mol}^{-1} \text{ or } 51.651 \text{ JK}^{-1} \text{ mol}^{-1}$$

For more approximate calculations, but with agreement with calorimetric values of C_p^0 and S^0 obtained by empirical anharmonicity corrections or other semi-empirical procedure, two decimal digits are appropriate. For calculations to the rigid-rotator, harmonic-oscillator, independent-internal-rotator approximation, only a single decimal digit should be retained. For still more approximate calculations (based on non-definitive calorimetric data, rough incremental calculations etc.) the single decimal digit should be subscripted, e.g.

$$12._3 \text{ cal K}^{-1}\text{ mol}^{-1} \text{ or } 51._7 \text{ JK}^{-1}\text{ mol}^{-1}.$$

PRESENTATION OF *p*, *V*, *T* DATA

Full disclosure of the use and calibration of the pressure gauge including buoyancy, local gravitational acceleration, fluid heads, pressure reference standards and pressure point in the sample. Volumetric measurements should be referred to a calibration standard as, for example, the density of mercury or water, and the effects from combined pressure and temperature should be described. Also, volumetric data presented in mass or molar units should include the values assumed for atomic weights as well as a statement of the corrections applied for impurity in the sample.

Here also, the importance of the presentation of primary results should be observed. Primary results are not the instrument readings but the values of physical properties derived from these instrument readings after the application of all calibration corrections and evaluating equations etc. These experimental values should be reported in full, together with a brief discussion of the application of various corrections and adjustments. The steps necessary to transform experimental values into final results should be made clear. Particularly, methods of smoothing results and of numerical differentiation or integration should be given as well as the sequence in which they are used. The final results should be accompanied by a statement regarding the magnitude of the differences between the smooth and the unsmooth results. Care should be taken to indicate clearly any results of other workers which are included in the preparation of a composite table. The base for tabulating entropy or Gibbs energy of a compressed fluid should be defined in one of the following ways:

$$\{S(p, T, \text{real}) - S(p, T, \text{ideal})\} \text{ or } \{S(V, T, \text{real}) - S(V, T, \text{ideal})\}.$$

When results are presented in tabular form, the spacing of data should be sufficiently close to permit accurate mapping of the *p*, *V*, *T* surface. Consistency tests for anomalous trends should be applied whenever possible. A related guideline[23] for the specialized communication of results may be of interest.

The results of thermophysical measurements leading to Joule–Thomson coefficients or $(\partial H/\partial p)_T$ should include adequate descriptions of apparatus, procedures, calibrations and comparison measurements, as well as procedures for data reduction.

Since more accurate characterization of the pressure scale to increasingly high values is now possible[24], authors should be scrupulous in indicating the pressure scale involved.

PRESENTATION OF OTHER EQUILIBRIUM MEASUREMENTS

Included in this category are such diverse measurements as vapour pressures, either by equilibrium or effusion techniques; studies of solubility, distribution, dissociation, adsorption or other chemical equilibria, by whatever techniques are applicable; electrochemical measurements etc. Such measurements provide information leading to the standard Gibbs energy increment for a reaction ($\Delta_r G^\circ$), but care is needed before associating this value with a *particular* reaction. Here again, details of apparatus, calibration and experimental procedures should be clearly presented, together with preparations of materials, analytical procedures and any special procedures applicable to the problem. Evidence as to attainment of equilibrium should be given. Primary results, as defined in connection with *p, V, T* data, should be reported in full, together with results and methods for reduction of results to standard state conditions, including any auxiliary information or equations, such as those from the Debye–Hückel theory of electrolytes. The chosen standard state should be clearly defined, and the final results, such as $\Delta_r G^\circ$, log K or log p, should be accompanied by an explicit description or statement of the related reaction, process or change in state. The presentation should include a discussion of errors, estimates of other factors, including instrumental and analytical limitations, effects of non-attainment of equilibrium, side reactions, non-ideality corrections etc., to the end that the precision indices for $\Delta_r G^\circ$ of reaction include such contributions.

Definitive studies on activity or osmotic coefficients should likewise present clear descriptions of apparatus, materials, experimental and computational procedures, together with sufficient primary results to permit verification of derived results. A forthcoming book will discuss procedures in the areas of equilibrium measurement[25]. In reporting adsorption studies, the mass of adsorbent used as well as the duration of the experiment are important.

NON-DEFINITIVE DATA

Although this resolution is concerned primarily with precise and accurate data taken by definitive techniques on well-characterized samples it is recognized that survey measurements are often made for technical and/or analytical purposes and these results are submitted for publication. Such values may be of doubtful significance as a basis for theoretical deductions or for incorporation in critical tables of scientific data. Therefore, the foregoing recommendations do not fully apply and presentation of a brief note summarizing the most important values and heralding a more detailed documentary presentation may be appropriate. However, in differential scanning calorimetric measurements, for example, the characterization of the solid phases before and after the experiment, the composition and pressure of the gas phase, the heating rate, the arrangement of the sample in the cell, the extent of baseline drift during the thermal effect and the way it is taken into account should be noted. Additional recommendations for presentation of thermal analysis data have been cited[26].

This document represents the endeavours of the Project Group of Commission I.2 on Thermodynamics and Thermochemistry of IUPAC composed of:

V. P. Kolesov (USSR)
M. L. McGlashan (UK)
Jean Rouquerol (France)
Syûzô Seki (Japan)
C. E. Vanderzee (USA)
E. F. Westrum Jr (USA, *Chairman*)

REFERENCES

1 *Science*, **132**, 1658 (1960); *Physics Today*, **14**, No. 2, 47 (1961).

2 The All-Union Calorimetry Conference (USSR), The Calorimetry Conference (USA), The Experimental Thermodynamics Conference (UK), The Society of Calorimetry and Thermal Analysis (Japan), La Section de Calorimetrie de la Société Française des Termiciens, The International Conference on Calorimetry and Thermodynamics (meeting in Maine), and the Colloque International de Thermochimie (meeting in Marseilles).

[3] W. Swietoslawski and L. Keffler. 'First Report from the Standing Commission for Thermochemistry'. IUPAC General Secretariat. Paris (1934).

[4] Institute of General and Inorganic Chemistry (IONKh) cf. *Zhur. Fiz. Khim.* **39**. 1298 (1965).

[5] F. D. Rossini. *J. Chem. Thermodynamics*. **2**. 447 (1970).

[6] For example. National Auxiliary Publications Service of the American Society for Information Science. c/o CCM Information Corp.: 909 3rd Ave., New York. New York 10022.

[7] IUPAC Subcommission on Standard Calibration Materials; personal communications.

[8] US National Bureau of Standards. Office of Standard Reference Materials. Washington. DC. 20234: e.g. NBS Standard Reference Material 720 (Sapphire heat capacity standard). benzoic acid combustion standard 39i. No. 736 trishydroxymethylaminomethane (THAM) for aqueous solution calorimetry.

[9] Vsesoyuzny Nauchno-Issledovatel'skiĭ Institutt Metrologiĭ (All-Union Scientific Research Institute for Metrology).

[10] Calorimetry Conference standards such as *n*-heptane. benzoic acid. synthetic sapphire (aluminium oxide). and copper for heat capacity and/or enthalpy measurements.

[11] cf. R. Battino and H. L. Clever. *Chem. Rev.* **66**. 395 (1966).

[12] cf. M. L. McGlashan and H. F. Stoeckli. *J. Chem. Thermodynamics*. **1**. 589 (1969).

[13] cf. R. Battino. *Chem. Rev.* **71**. 5 (1971).

[14] 'Le système International d'Unités (SI)' International Bureau of Weights and Measures. Sevres. France (1970). Authorized English translations are available: (a) H.M.S.O.. London (1970) and (b) C. H. Page and P. Vigoreux. eds. *NBS Spec. Publ. No. 330*. US Government Printing Office: Washington. DC (1970).

[15] A convenient source for this is: *Manual of Symbols and Terminology for Physicochemical Quantities and Units*. M. L. McGlashan. ed. *Pure and Applied Chemistry*. **21**. 3 (1970).

[16] F. D. Rossini. Chapter 14. in *Experimental Thermochemistry*. Vol. I. F. D. Rossini. ed.. Interscience: New York (1956).

[17] *Tables of Lagrangian Interpolation Coefficients*. Prepared by the Mathematical Tables Project. Works Projects Administration of the Federal Works Agency. Columbia University Press: New York (1944).

[18] *Experimental Thermochemistry*. Vol. I. F. D. Rossini. ed.. Interscience: New York (1956).

[19] *Experimental Thermochemistry*. Vol. II. H. A. Skinner. ed.. Interscience: New York (1962).

[20] T. F. Young. Y. C. Yu and A. A. Krametz. *Disc. Faraday Soc.* **24**. 37 (1957).

[21] The nomenclature 'Gibbs energy function' (and 'enthalpy function') for $(\{H^{\circ}(T) - H^{\circ}(0)\}/T)$ have been widely used. They are presently being reconsidered by IUPAC and other international scientific unions.

[22] *Experimental Thermodynamics*. Vol. I. J. P. McCullough and D. W. Scott. eds. Butterworths: London (1969).

[23] 'ASME Resolution on Communication of Experimental Data'. *Mechanical Engineering*. **80**. 147 (November 1958).

[24] *Accurate Characterization of the High Pressure Environment*. *NBS Spec. Publ. No. 326*. US Government Printing Office: Washington. DC (1971). cf. especially pp 1–3. 313–340.

[25] *Experimental Thermodynamics*. Vol. II. B. Vodar and B. LeNeindre. eds. Butterworths: London (to be published).

[26] H. G. McAdie. *Analyt. Chem.* **39**. 543 (1967).

GUIDE FOR REPORTING EXPERIMENTAL DATA ON VAPOR-LIQUID EQUILIBRIA OF MIXTURES AT LOW AND MODERATE PRESSURES

SUMMARY

This Guide contains recommendations intended to increase the usefulness of publications containing measurements on vapor-liquid equilibria (VLE) at low and moderate pressures for systems in which all components are subcritical, i.e. below their critical temperatures and pressures.

It highlights the needs for:

- *clear statements on the purity of the substances used,*
- *describing clearly the experimental procedures, apparatus, and appraising the precision and accuracy of the results,*
- *measuring a consistent set of pure component vapor pressures, VLE, and volumetric properties of phases,*
- *making an adequate number of measurements over a wide composition range,*
- *reporting original experimental results and clearly distinguishing them from derived values, etc.,*
- *measurements of VLE or H_m^E over a wide temperature range.*

1 INTRODUCTION

This guide is addressed to researchers planning to publish results of measurements of phase equilibrium, to reviewers, and to the editors of journals. The purpose is to document the types of information which are considered to be most important and particularly to encourage the publication of the essential information required for future utilization of the data by their users. This document reflects the opinion and experience of those involved in the design, optimization, and economic evaluation of industrial plants as well as those involved in the measurement and correlation of phase equilibrium data. Data on vapor-liquid equilibria of mixtures are of vital importance in two broad areas: design and optimization of industrial processes and as a source of information on the thermodynamic properties of the liquid phase. Unfortunately, contemporary papers reporting results on *VLE* measurements frequently omit information important for the diverse set of users.

Neglect of the requirements needed for the assessment of the precision and accuracy of the measurements and/or lack of essential information by the experimenter not only prevents readers from taking advantage of the published data but makes it difficult to incorporate the data into data banks.

Consequently there is a need for a guidance document to assist experimentalists in presenting the information required for the user to make a proper assessment of the reliability of the measurements.

The need for standard procedures and requirements for reporting experimental data in thermodynamics has been a concern of various international organizations. The International Union of Pure and Applied Chemistry (IUPAC) has published a "Guide to Procedures for the Publication of Thermodynamic Data" [1] and a "Report on Assignment and Presentation of Uncertainties of the Numerical Results of Thermodynamic Measurements" [2]. These documents contain valuable discussions and general recommendations which should be observed in all physicochemical measurements as well as detailed information concerning presentation of calorimetric data. The International Council of Scientific Unions (ICSU) Committee on Data for Science and Technology (CODATA) has published the "Guide for Presentation of Numerical Data Derived from Experiments in the Primary Literature" [3], a document of considerable generality and has initiated a series of more specialized guides [5], [6], and [7]. In the Preface to the CODATA Guide [6], Professor Edgar Westrum, acting as CODATA Secretary General, encouraged scientists in international scientific bodies to extend the coverage of the general CODATA Guides with disciplinary and subdisciplinary guides.

This Guide for reporting *VLE* measurements data is an example of such an endeavor.

2 ESSENTIAL INFORMATION

2.1 Properties of Pure Components

Phase equilibrium measurements should be made on well defined mixtures. Therefore, pertinent information should be reported on the pure compounds.
To prevent wrong identification of chemicals, the IUPAC recommended names should be accompanied by the "Chemical Abstracts Registry numbers".

The purity of the compounds should be established. The sources and methods of purification as well as the method of characterizing the overall impurity content should be given. As far as possible all the impurities present should be identified, and their content determined.

High purity substances should be used for the *VLE* measurements. If all other experimental conditions ensure that the results could serve as reference data, the purity of the components should be no lower than 99.95 mol percent.

The substances used must be as anhydrous (free of solvent) as possible before introducing them to the measuring device; evidence for the absence of water (or other solvents) should be stated.

If, in addition, some physicochemical properties of the substance such as density, refractive index, boiling or freezing point are determined and compared with literature values the accuracy of the methods of determination used and the reliability of the reference data should be estimated.

Vapor pressures of all pure components should be measured over the range of pressures and temperatures at which the vapor-liquid equilibrium measurements are made. Preferably, these measurements should be made with the same apparatus, and with the same or higher accuracy than that achieved in *VLE* measurements on the mixture.

The value obtained should be compared with evaluated literature data if available.

Of particular concern is the stability of the components. If the components are subject to reactions, decomposition, polymerization etc. at the conditions under which the measurements are carried out, pertinent information on the kinetics of this process should be obtained and taken into consideration when establishing the experimental procedure and in appraising the accuracy of the *VLE* results.

2.2 Apparatus and Experimental Procedures

The experimental procedures and the apparatus used should be described clearly. If the study reported is a continuation of earlier work and the procedure and equipment have already been described, reference thereto should be given and the text should include short descriptions of modifications, if any. The description should be made in sufficient detail to enable the reader to perform his own evaluation of the apparatus, procedures, and modifications as well as to ascertain the effects on measured and/or derived values.

The procedure for introducing the substances into the measuring device should be described.

This procedure should ensure prevention of contamination of the substances by the surroundings--e.g. by absorption of moisture. If a static method is used the procedure of degassing the substances should be presented.

2.3 Stability and Control

Only rarely do researchers discuss the mechanism by which the stability of an experimental procedure is controlled. The key variables that influence the accuracy of data are: pressure, temperature, feed composition, and the nature of the components. The order of importance of these variables varies with the method used. Experience has shown that the mechanisms by which the pressure or temperature are controlled and monitored greatly influence the level of reliability of the results.

For example in an ebulliometer arrangement in which the system pressure is controlled the temperature is assumed to be constant. However, the system pressure may not be constant. For example, the pressure may be controlled by a manostat, which allows on-off connection between the cell and the surge tank. It is incumbent on the researcher to study and report on the characteristics of the manostat set-up and its rate and level of response. Moreover, if the heat input into the ebulliometer is manually controlled, the system's response to variations in pressure gives an automatic change in the rate of vaporization.

In static cells when the temperature is controlled by a thermostat, the rate of response of the pressure to changes in temperature should be studied and reported. Adequate degassing of samples is crucial in this method.

A very important matter infrequently mentioned in publications is a discussion of the laboratory barometer, the accuracy and the calibration of its scale and how the readings are used to establish the absolute pressure inside a cell. Authors are urged to discuss all aspects of the data-taking process and should report on the logic of decision on how the system is judged to be stable. In those cases where the temperature, pressure, and heat input are recorded continuously, the variations in the chart's curves must be reviewed in the manuscripts. Discussions of the results of manual recording of the visual observation of instruments' readings are considered to be very important.

When experiments are carried out in a be circulating still, sample-withdrawal can cause instantaneous changes in the system pressure. With experience and care, researchers can avoid this upset. However, they are urged to report the procedure by which stability is assured while a sample is taken.

In *VLE* measurements--as in other phase equilibria investigations--special attention should be paid that equilibrium within the entire measuring system has been attained. This includes uniformity of the composition of each of the coexisting phases and uniformity of temperature in the measuring cell. For example, when circulating stills are used, the heat input as well as the rate of circulation can affect the uniformity of sample compositions.

Uncertainties connected with the presence of heat flow due to temperature differences in the measuring system are often hard to establish. The experimenter should pay attention to this problem and inform the reader in what manner the attainment of the equilibrium state was ascertained.

2.4 Reporting on the Accuracy of Measurements

It is mandatory that researchers report the accuracy and the precision of all measured values. Most important is the need to report the precision and accuracy in the determination of pressure, temperature, and composition of the coexisting phases. Moreover, distinction should be made between the accuracy achieved in measuring differences of these properties. Detailed description or traceable references to the procedures used for calibration of each of the measuring devices should be given.

For the assignment of uncertainties, the IUPAC Commission I.2 Report [2] should be followed. Three clearly defined levels are distinguished:

- uncertainties which can be expressed as standard deviations which represent the variations of the individual primary results,
- uncertainties arising as attributes of measuring system such as those associated with conversion factors in transformation of initial readings into final results, precision limits of measuring devices, calibration errors, and errors connected with the variations of external conditions,
- systematic errors - which can seldom be clearly recognized - but which should as far as possible be identified and eliminated through calibration of the measuring instruments and/or by application of comparative measurements with recognized reference substances.

Particularly if the measuring system is a new apparatus, it should be tested with a reference system. The experimenter should be aware that absolute measurements may always be affected by systematic errors which could be much larger than that of all uncertainties stated and he should discuss possible sources of these errors beyond his control.

Systematic errors may be often detected and evaluated by tests of thermodynamic consistency.

2.5 Original Experimental Results Must be Reported

The original numerical results obtained in measurements of the pressure, temperature, and composition of the vapor and liquid phases should be reported. Description of (or references to) the measuring techniques used, accompanied by the discussion of possible sources of inaccuracies and imprecisions as stated in sections 3.1 to 3.4 should be provided.

As the physical properties measured (primary results) are often not the instrument readings but the values derived from these readings after application of all calibration corrections and evaluating equations according to the IUPAC Guide [1], the experimental values should be reported in full together with the brief discussion of the application of various corrections and adjustments. The steps necessary to transform experimental values into final results should be made clear.

When a full description of the method for conversion of the instrument readings into primary data is too expensive or for other reasons does not merit journal publication, a footnote in the publication should indicate how the reader may obtain the additional data.
To avoid misunderstanding, misuse, or erroneous citations; directly measured values should be identified and clearly distinguished from computed values. This will eliminate the influence of spurious information generated or based on various assumptions. For example, if a sample is prepared by weighing, equilibrated, and the composition of the liquid phase is computed on the assumptions that the mass of the evaporated part and its compositions are known, the composition of the original sample should be reported, accompanied by the description of the computational procedure together with an estimate of uncertainties.

If--in addition to primary values--smoothed values are presented, the methods of smoothing should be given.

2.6 Derived Results Must be Reported as Such

All derived values should be distinguished clearly from the experimental values. The authors can mislead their audience if they report the derived results as if they were experimental values. Thus, computational procedure for calculating vapor compositions from the liquid composition and total pressure data at constant temperature should be described clearly.

In such cases all auxiliary data--e.g., liquid and vapor molar volumes data of pure components--for the equation of state and other equations used in computation should be reported in full.

If due to the size of the manuscript, its publication is split into a main paper and supplementary material, it is recommended that the primary results be always in tabulated form in the main paper. It is also advisable to present the correlating equations and parameters in the main paper.

If a correlating equation is reported, its performance should be presented as a table or a plot giving the differences between experimental and computed values as a function of the independent variable. Plots of calculated and experimental values are not recommended. When numerical values of differences are reported, the original values of the measured independent and dependent variables also should be reported in the same table and the method used for computation of residuals clearly described.

3 MODE OF PRESENTATION AND SCOPE OF EXPERIMENTAL DATA

3.1 Symbols and Units

All numerical values should be accompanied with units.

For units and symbols, the IUPAC recommendations given in the "Manual for Symbols and Terminology for Physico-Chemical Quantities and Units" [4] should be followed.

3.2 Type and Number of Data Points

For each isothermal or isobaric set of measurements at least seven concentrations - and preferably more than ten concentrations - evenly spaced over the composition range are needed. This number should be sufficient to allow consistency tests and the use of correlating equations with three, four, or even more parameters.

It is better to make isothermal rather than isobaric measurements. If isothermal *VLE* measurements are made at two or more temperatures, an enthalpy of mixing should be estimated (at least for the equimolar mixture) and compared with directly measured values if they are available. This procedure is strongly recommended for checking the reliability of measurements.

Whenever possible, it is advisable to combine *VLE* studies with measurements of excess volumes and excess enthalpies.

The experimenter should realize that the usefulness of his work will be greatly enhanced if isothermal *VLE* measurements are performed at several temperatures preferably in the range 30-80 K. Alternatively *VLE* measurements at one temperature can be combined with excess enthalpy measurements over a temperature range. It may be convenient to perform the measurements at selected temperatures with the same sample of liquid mixture in the measuring cell.

3.3 Presentation of Data

The recommendations given in the CODATA Guide on Presentation in Primary Literature [6] should be followed in *VLE* data publication.

A table giving numerical values of the appropriate independent and dependent variables obtained as a set of primary results derived from experiment should be recommended as the most appropriate mode of presentation of both *VLE* mixture data and vapor pressures pure components. All experimental data should be included in this set and if the data were submitted to critical evaluation and some of them were rejected, this should be clearly stated.

Tables with smoothed numerical values may be included only as an additional mode of presentation.

Analytical expression used for smoothing along with the standard deviations or other information enabling the appraisal of the accuracy of the data should be reported.

If the vapor pressure data of pure substances are represented in analytical form--e.g., with constants off the Antoine equation--the temperature range for which these constants are valid should be indicated.

Graphical representation of the *VLE* data or derived values (as activity coefficient, excess enthalpy of mixing, etc.) may be worthwhile as an illustration of the shape of the functional relationship, but they do not replace the publication of the numerical values.

4 REFERENCES

1. (IUPAC) "Guide to Procedures for Publication of Thermodynamic Data." Pure and Appl. Chem., 29, 397 (1972).

2. (IUPAC) "Report on Assignment and Presentation of Uncertainties of the Numerical Results of Thermodynamic Measurements." J. Chem. Thermodynamics, 13, 603 (1981).
3. (CODATA) "Presentation in the Primary Literature of Numerical Data Derived from Experiments." CODATA Bull. No. 9 (1973).
4. (IUPAC) "Quantities, Symbols and Units in Physical Chemistry." Blackwells (1988).
5. (CODATA) "Biologists' Guide for the Presentation of Numerical Data in the Primary Literature." CODATA Bull. No. 25 (1977).
6. (CODATA) "Guide for the Presentation in the Primary Literature of Physical Property Correlation and Estimation Procedures." CODATA Bull. No. 30 (1978).
7. (CODATA) "Guide for the Presentation of Astronomical Data." CODATA Bull. No. 46 (1982).

COMPOSITION OF THE CODATA TASK GROUP ON CRITICALLY EVALUATED PHASE EQUILIBRIUM DATA

Members

Dr. A. Maczynski, Academy of Sciences Institute of Physical Chemistry, Warsaw, Poland (Chairman)

Prof. A. Fredenslund, Denmarks Tekniske Hojskole, Lyngby, Denmark

Dr. J. Gmehling, Universität Dortmund, Abteiling Chemietechnik, Dortmund, F.R.G.

Dr. K. Kojima, Nihon University, College of Science and Technology, Tokyo, Japan

Dr. S. Kemeny, University of Technical Sciences, Budapest, Hungary

Prof. D. Lempe, Technische Hochschule "Carl Schorlemmer", Merseburg, G.D.R.

Dr. K.N. Marsh, Texas A&M University System, College Station, Texas, U.S.A.

Prof. J. Vidal, Institut Français du Pétrole, Rueil Malmaison, France

Prof. D. Zudkevitch, Allied Chemical Corporation, Morristown, New Jersey, U.S.A.

Corresponding Members

Prof. P. Alessi, Università degli Studi di Trieste, Trieste, Italy

Prof. H. Knapp, Technische Universität Berlin, Berlin, F.R.G.

Dr. S. Malanowski, Polish Academy of Sciences, Warsaw, Poland

Recommendations for Measurement and Presentation of Biochemical Equilibrium Data*

PREPARED BY THE INTERUNION COMMISSION ON BIOTHERMODYNAMICS‡

Equilibrium data are of importance in describing and understanding biochemical systems. At present there is marked variation among different investigators in the choice of experimental conditions for the study of similar or identical reactions and in the manner of reporting the data. In many cases, the description of the system does not provide all the essential information that would permit reproduction of the experiments. This can lead to confusion and difficulties in correlating the results of different workers.

Equilibrium studies in biochemistry involve special problems that are not encountered in general chemistry. The attainment of biochemical equilibria commonly involves the addition of a specific enzyme to the system to catalyze the reaction studied; sometimes two or more enzymes must be added. In addition, the enzyme may require the presence of certain cofactors, such as metal ions. The reactants, or products, or both, may bind or release protons or other ions during the reaction under study. Thus, the experimental system must be described with particular care.

The acquisition and reporting of meaningful thermodynamic data relating to biochemical systems, as well as to other complex reacting systems, involves two fundamental problems.

1. It is not always possible to define the reacting species precisely.

2. Even in cases where the reacting species can be defined, their thermodynamic activities frequently cannot be determined.

In this report, we offer several recommendations with the aim of increasing the usefulness of biochemical equilibrium data and coordinating the results of different investigators. These recommendations include a set of standard conditions which would facilitate the attainment of a common body of knowledge of a wide range of biochemical equilibria. This does not preclude the choice of special experimental conditions that may be appropriate for certain reactions; but whenever possible these reactions should also be studied under the recommended standard conditions. To avoid confusion in interpretation we also recommend standardization of terminology, symbols, and units in the presentation of such data.

For other discussions of the presentation of numerical data and of thermodynamic data derived from experiments we call the attention of the reader to guides prepared by CODATA (1) and IUPAC (2).

Part I. Standard Conditions for Equilibrium Measurements

True thermodynamic equilibrium constants are defined in terms of activities of the reactants and products. In many systems of biochemical interest it is not possible to evaluate the activities of all components. It is therefore frequently necessary to calculate equilibrium constants in terms of concentrations. The proper quotient of equilibrium concentrations is acceptably constant for many purposes, and will be referred to in this document as the concentration equilibrium constant, with the symbol K_c. However, it should be recognized that values for such equilibrium constants, K_c, and corresponding Gibbs energy changes, ΔG_c°, may not be truly constant as the composition of the system is changed.[1] It is essential that reported values for such quantities be accompanied by a complete and accurate description of the composition of the reacting system, and the methods by which the composition was established. References to prior work will in many cases provide adequate descriptions of these methods.

Experimental Conditions

It is recommended that measurements be made at 25 °C (or 37 °C) and an ionic strength brought to 0.10 mol·dm^{-3} with potassium chloride,[2] using the lowest effective buffer concentration. If these conditions are not practicable, conditions should be chosen that are well defined and can be maintained constant throughout a series of experiments. Reagents should be as pure as practicable, and the method of purification, estimated purity, and source of each reagent should be given. The existence of a true equilibrium condition should be demonstrated by approaching the equilibrium state from both directions and by making determinations at several different concentrations of the reactants. It is desirable also to study the effect of variation of the ionic strength and the presence of specific salt effects. Complete or representative experimental data from which equilibrium constants are derived should be reported.

* This document is published with the approval of IUPAC, IUB, and IUPAB. Comments on and suggestions for future revisions of the Recommendations may be sent to any member of the Interunion Commission on Biothermodynamics (see below).

Reprints of this publication may be obtained from W. E. Cohn, Director, NRC Office of Biochemical Nomenclature, Biology Division, Oak Ridge National Laboratory, Box Y, Oak Ridge, Tenn., U. S. A. 37830.

‡ The members of the Commission are H. Gutfreund (Bristol, U. K.) and P. Privalov (Poustchino, U. S. S. R.) representing International Union of Biochemistry (IUB), J. T. Edsall (Cambridge, Mass., U. S. A.) and W. P. Jencks (Waltham, Mass., U. S. A.) representing International Union of Pure and Applied Biophysics (IUPAB), G. T. Armstrong (Washington, D. C., U. S. A.) and I. Wadsö (chairman; Lund, Sweden) representing International Union of Pure and Applied Chemistry (IUPAC), and, R. L. Biltonen (associated member; Charlottesville, Va., U. S. A.).

[1] The standard Gibbs energy change, ΔG°, is now the correct term for what is frequently called the standard Gibbs free energy change or simply the free energy change and often given the symbol ΔF°.

[2] The recommended unit of concentration is mol·dm^{-3}, commonly denoted by M in the biochemical literature.

Temperature. A standard temperature of 25 °C (298.15 K) is preferred to permit comparison with most available chemical data. The numerical values of the equilibrium constant at both 25 °C and 37 °C should be reported when possible.

It is desirable to measure equilibrium constants at several temperatures. The standard enthalpy and entropy changes for a specific reaction can then be calculated from the temperature dependence of the equilibrium constant. However, when such experiments are performed it is essential to take account of changes in the dissociation constants of reactants and buffers with changing temperature and the resulting changes in pH and in the concentrations of dissociating species. It may be noted that calorimetry is generally a more accurate technique for determination of the enthalpy change of a reaction. The calorimetrically determined enthalpy change must be corrected for the enthalpy change from any reaction of a buffer involving protons that are taken up or given off in the reaction.

Buffer and pH. If only a limited number of measurements are to be made, they should be carried out at pH = 7.0 and, if possible, also at a pH value at which the apparent equilibrium constant K_c', has little or no dependence on pH. (K_c' is defined in a later section.) If direct measurements at pH = 7.0 are not practicable, the calculated values for this pH should be reported. The procedure used in making these calculations must be carefully described. Care should be taken that the solution is adequately buffered so that the pH is well defined throughout the experiment. It is desirable to determine the effect of varying the nature and concentration of the buffer in order to identify buffer effects. Buffers that are known to interact with reactants (including macromolecules) or salts, such as phosphate or pyrophosphate in the presence of divalent metal ions, should be avoided.

Complexations with ions. The equilibrium constants for complexation of ions with reactants, such as magnesium ion with phosphate compounds, should be determined under the conditions of the experiments and the concentrations of reactants should be corrected for complexation. If this is not practicable, there are three alternatives: (*a*) Metal ions that are necessary for activation of enzymes may be added in trace amounts that are sufficient to permit attainment of equilibrium but are too small to change the concentrations of reactants significantly by complexation. (*b*) Measurements may be carried out at a concentration of metal ions such that essentially all of one (or more) of the reactants exists in the complexed form and the degree of complexation of other reactants is not significant; *i.e.* the equilibrium constant and standard Gibbs energy change are determined for the reaction involving the complexed species. (*c*) The measurements may be carried out under physiological conditions, to obtain equilibria and standard Gibbs energy changes that are directly applicable to biological systems.

In any case, the concentrations of metal ions should be specified and, when possible, equilibrium constants should be reported corresponding to the three situations above, i.e. uncomplexed reactants, fully complexed reactants (*e.g.* Mg·ATP), and physiological conditions. In view of the limited availability and occasional unreliability of equilibrium constants for ion complexation, it is highly desirable to determine directly the effect of variation of the concentration of metal ions under the experimental conditions employed. In some cases it may be necessary to correct for complexation of reactants by appropriate extrapolation procedures or to use special experimental conditions to avoid complexation of the reactants with components of the reaction medium. For example, tetramethylammonium chloride may be used instead of potassium chloride for measurements with polyphosphates or other polyanions that complex with alkali ions.

General Considerations in Reporting Equilibrium Data

Consider the following reaction

$$aA + bB \rightleftharpoons cC + dD \quad (1)$$

where *a*, *b*, *c*, and *d*, respectively, are the stoichiometric coefficients of components A, B, C, and D. In this case the concentration equilibrium constant is

$$K_c = \frac{[C]^c \cdot [D]^d}{[A]^a \cdot [B]^b} \quad (2)$$

and the corresponding standard Gibbs energy[3] change is

$$\Delta G_c^\circ = -RT \ln K_c \quad (3)$$

where [A], [B], etc. are the concentrations of distinct molecular components involved in the chemical reaction, R is the gas constant and T is the thermodynamic temperature.

Water as a reactant in *dilute aqueous* solutions provides a special problem. In order to conform to the general principles of dilute solutions, the value for the factor $[H_2O]$ in equilibrium expressions should be taken as unity in such cases (see the last paragraph of Part I).

In some cases it is known that the designated concentration of a component includes several distinct molecular species (*e.g.* [C] is $[CH_3COOH] + [CH_3COO^-]$) but insufficient information exists to allow a further description of the equilibrium system. For example, in a buffered system it may not be possible to evaluate the participation of the H^+ ion in all of the possible equilibria. However, if the $[H^+]$ may be maintained constant it is possible to write an apparent equilibrium constant, K_c', in which the $[H^+]$ does not appear explicitly. Such an apparent equilibrium constant will be pH-dependent and will generally have a different value from K_c (Equation 2) which will be pH-independent. The K_c' is written in terms of the total concentrations of the measurable components

$$K_c' \text{ (pH} = x\text{, etc.)} = \frac{[C_1 + C_2 + \ldots]^c \, [D_1 + D_2 + \ldots]^d}{[A_1 + A_2 + \ldots]^a \, [B_1 + B_2 + \ldots]^b} \quad (4)$$

and an apparent standard Gibbs energy change

$$\Delta G_c^{\circ\prime} \text{ (pH} = x\text{, etc.)} = -RT \ln K_c' \quad (5)$$

The value of K_c' is subject to certain constraints. The value of pH, and perhaps that of the concentration of certain metal ions, or other factors must be fixed and specified in order to obtain a definite K_c' value. The values of these constraints

[3] Equation 3 implies that the equilibrium in question is for a "dilute-ideal" solution, in which the activity coefficients of all reactants and products are independent of concentration over the concentration range under consideration. Although this condition cannot be expected to hold rigorously in any actual system, in practice ΔG_c° is frequently found to be constant within the experimental error of the measurements in biochemical systems. Therefore, we shall refer to ΔG_c° as a "standard Gibbs energy change," although it should be recognized that this usage is an approximation. It is also to be noted that the argument of a logarithm (*e.g.*, K_c in Equation 3) must be dimensionless. For concentration equilibrium constants this can be achieved by using "relative concentrations," that is, concentration divided by a standard concentration = 1. This operation is of importance in principle, but can be ignored in practice. It is necessary, however, that the standard state concentration units be explicitly described when reporting such values.

must be stated, as indicated above, but these constraining factors do not explicitly appear among the concentrations of components on the right hand side of Equation 4. Usually in biochemistry subscript c in K_c can be deleted. However, it is recommended that the notation "apparent" should never be left out in connection with K_c' and $\Delta G_c^{0\prime}$.

A simple example should help clarify the situation. Consider the reaction:

$$R_1COOR_2 + H_2O \overset{K_I}{\rightleftharpoons} R_1COOH + HOR_2 \overset{K_{II}}{\rightleftharpoons} R_1COO^- + HOR_2 + H^+ \quad (6)$$

The concentration equilibrium constants and standard Gibbs energy changes for steps I and II are

$$K_I = \frac{[R_1COOH]\,[HOR_2]}{[R_1COOR_2]\,[H_2O]} \quad (7)$$

Since the factor for water in Equation 7 is taken as unity in *dilute aqueous solutions*, it may be omitted in writing the equations for equilibria in such systems.

$$\Delta G_I^\circ = -RT \ln K_I \quad (8)$$

$$K_{II} = \frac{[R_1COO^-]\,[HOR_2]\,[H^+]}{[R_1COOH]\,[HOR_2]} = \frac{[R_1COO^-]\,[H^+]}{[R_1COOH]} \quad (9)$$

$$\Delta G_{II}^\circ = -RT \ln K_{II} \quad (10)$$

The concentration equilibrium constant and standard Gibbs energy change for the overall reaction are

$$K = \frac{[R_1COO^-]\,[HOR_2]\,[H^+]}{[R_1COOR_2]} = K_I K_{II} \quad (11)$$

$$\Delta G^\circ = -RT \ln K \quad (12)$$

However, an apparent equilibrium constant at some pH = x can be written in which the concentrations of R_1COO^- and R_1COOH are summed and do not have to be differentiated.

$$K'(\text{pH} = x) = \frac{([R_1COO^-] + [R_1COOH])\,[HOR_2]}{[R_1COOR_2]} \quad (13)$$

The corresponding standard Gibbs energy change is

$$\Delta G^{\circ\prime}(\text{pH} = x) = -RT \ln K' \quad (14)$$

In cases where the details of the chemical equilibria are known the interconversion between K and K' is generally straightforward. For the example just described

$$[R_1COOH] = \frac{[R_1COO^-]\,[H^+]}{K_{II}} \quad (15)$$

and

$$K'(\text{pH} = x) = \frac{[R_1COO^-]\,(1 + [H^+]/K_{II})\,[HOR_2]}{[R_1COOR_2]} = K\left(1 + \frac{[H^+]}{K_{II}}\right)\cdot\frac{1}{[H^+]} \quad (16)$$

The well known problem of the standard Gibbs energy change of ATP hydrolysis to ADP and inorganic phosphate (P_i) furnishes a more complex example of concentration and apparent equilibrium constants. The overall reaction may be written

$$\text{total ATP} + H_2O \rightleftharpoons \text{total ADP} + \text{total } P_i$$

Thus, the apparent equilibrium constant and $\Delta G^{\circ\prime}$ values at some specified pH are:

$$K' = \frac{[\text{total ADP}]\,[\text{total } P_i]}{[\text{total ATP}]} \quad (17)$$

$$\Delta G^{\circ\prime} = -RT \ln \frac{[\text{total ADP}]\,[\text{total } P_i]}{[\text{total ATP}]} \quad (18)$$

Data are available to convert these apparent constants into concentration constants. In order to deal with these phosphate compounds in a medium related to their biological environment, it is necessary to consider their interactions with protons and with magnesium ions. At pH values near 7 and at physiological $[Mg^{2+}]$ values, it is found that each of the three phosphates can be described by one acidic dissociation, and each acid and its conjugate base may bind 1 Mg^{2+} ion. Thus the total concentrations in K' and $\Delta G^{0\prime}$ may be resolved as follows:

$$[\text{total ATP}] = [ATP^{4-}] + [ATP^{3-}] + [ATPMg^{2-}] + [ATPMg^-] \quad (19)$$

$$[\text{total ADP}] = [ADP^{3-}] + [ADP^{2-}] + [ADPMg^-] + [ADPMg] \quad (20)$$

$$[\text{total } P_i] = [P_i^{2-}] + [P_i^-] + [P_iMg] + [P_iMg^+] \quad (21)$$

In fact $[P_iMg^+]$ is always so small as to be negligible, and $[P_iMg]$ is of minor importance.

The three pK_a values involved—one for each of the compounds ATP, ADP, and P_i—are known, as are the association constants of the various species for Mg^{2+} ion. We can then formulate the reaction in terms of the relations between particular species of reactants and products. For instance in the limiting case of high pH (>8) and $[Mg^{2+}] = 0$, the predominant reaction becomes:

$$ATP^{4-} + H_2O \rightleftharpoons ADP^{3-} + P_i^{2-} + H^+ \quad (22)$$

It would be equally valid to formulate the reaction in terms of the equation:

$$ATP^{3-} + H_2O \rightleftharpoons ADP^{2-} + P_i^- \quad (23)$$

and other choices are obviously possible, provided that the equations balance with respect to total charge and stoichiometry. We note that the equilibrium constants of all such reactions must be independent of pH, whether or not $[H^+]$ enters explicitly into the reaction. Formulation of such equilibrium constants implies nothing as to the mechanism of the reaction in this or any other case.

If we choose to formulate the reaction in terms of Equation 22, we have:

$$K_{ATP^{4-}} = \frac{[ADP^{3-}]\,[P_i^{2-}]\,[H^+]}{[ATP^{4-}]} \quad (24)$$

In the limiting case where $[Mg^{2+}] = 0$, we can then write for the relation between K' (of Equation 17) and $K_{ATP^{4-}}$ (Equation 24):

$$K'(\text{pH} = x, [Mg^{2+}] = 0) = \frac{([ADP^{3-}] + [ADP^{2-}])\,([P_i^{2-}] + [P_i^-])}{([ATP^{4-}] + [ATP^{3-}])} = K_{ATP^{4-}}\frac{\left(1 + \frac{[H^+]}{K_\beta}\right)\left(1 + \frac{[H^+]}{K_\gamma}\right)}{\left(1 + \frac{[H^+]}{K_\alpha}\right)[H^+]} \quad (25)$$

where K_α, K_β, and K_γ are, respectively, the acidic dissociation constants of ATP^{3-}, ADP^{2-}, and P_i^-.

Since all the constants in Equation 25 are known, the apparent equilibrium constant Equation 17 is thus explicitly related to the pH-independent constant $K_{ATP^{4-}}$. In the presence of Mg^{2+} ion the analysis is in principle exactly similar but is much more complicated in detail. A detailed discussion is given for instance by Alberty (3).

We note that the usefulness of Equation 17 depends upon the fact that it is possible to determine analytically the *total* concentrations of ATP, ADP, and P_i, without distinguishing between the different forms listed in Equations 19, 20, and 21.

We recommend that concentration equilibrium constants and corresponding standard Gibbs energy changes be reported when the reaction can be completely and accurately defined. In cases where this is impossible corresponding apparent quantities can be reported, if possible over a range of pH values and other constraining variables. In any case the values must be accompanied by a precise definition of the reaction scheme assumed.

Quantities. It is recommended that values of equilibrium constants and standard Gibbs energy changes should normally be reported based on concentrations (amount of substance divided by volume, with units of moles per cubic decimetre) except for the proton factor, for water in aqueous systems in which water is involved as a reactant or product, for gases, or for solids. For special purposes it may be desirable to report additional values that are based on molalities, mole fractions, or partial pressures; in any case the unit employed should be clearly identified. The use of equilibrium constants based on concentrations expressed in such units as millimoles or micromoles per cubic decimetre (litre) should be avoided.

It is recommended that for reactions in dilute aqueous solutions the factor for H_2O be taken as unity, in the absence of a measurement of its relative activity. An actual concentration such as the conventional 55.5 $mol \cdot dm^{-3}$ should only be used for liquid water for well defined reasons. It is recommended that 10^{-pH} be used for the proton factor. The conventions chosen for the liquid water and the proton factor must always be clearly and explicitly stated. For gases the partial pressure in atmospheres or (if the partial pressure is expressed in kilopascals) the ratio of the partial pressure to 101.325 kPa should be used. Gibbs energies should be reported in joules per mole ($J \cdot mol^{-1}$). Until this unit comes into general use, results may in addition be reported in thermochemical calories per mole ($1\ cal_{th} = 4.184$ J).

Part II. Symbols, Units, and Terminology

General

The precision of statement, the universal intelligibility, and the enduring quality of a thermodynamic document will be enhanced if the symbols, units, and terminology used in it are those recommended by an international standardizing body. The following information summarizes recommendations of international scientific bodies in the physical and measurement sciences, selected for particular relevance to the thermodynamics of biological processes. The author is referred to recommendations of the IUPAC (4)[a] for more detailed lists of recommended symbols and terminology for physicochemical quantities, and for a discussion of the International System of Units (SI). The SI is also recommended by the International Organization for Standardization (ISO), which provides rules for their use (5). Other descriptions of the SI are provided by the International Bureau of Weights and Measures (6).

In adhering to these recommendations for the presentation of scientific data, authors are urged:

1. To use the recommended name of each physical quantity and the preferred symbol, as far as possible.
2. To use SI units and their internationally accepted symbols (4–6).
3. To use functional expressions where feasible, rather than subscripts or superscripts, for specifying substance, physical state, and temperature to which the quantity refers. *e.g.* for a process such as Reaction 1

$$\Delta G_1^{0\prime}\ (\mathrm{pH} = 7.0,\ I_c = 0.10\ \mathrm{mol \cdot dm^{-3}},\ \mathrm{at}\ 25\ ^\circ\mathrm{C})$$

or for a dissolved substance

$$C_p(\mathrm{C_2H_5OH,\ in\ H_2O,\ 25\ ^\circ C})$$

4. To indicate both numerical values and units in referring to physical quantities in the text, and in labels for axes of graphs and rows and columns of tables.

The preferred convention for statements in the text is:

$$\text{physical quantity} = \text{numerical value} \times \text{unit}$$

e.g.

$$\Delta G^{0\prime}(\mathrm{pH} = 7.0,\ I_c = 0.10\ \mathrm{mol \cdot dm^{-3}},\ \mathrm{at}\ 25\ ^\circ\mathrm{C}) = -8.7\ \mathrm{kJ \cdot mol^{-1}}.$$

The preferred label for tables and graphs, obtained by rearrangement of the equation above is:

$$\text{physical quantity/unit} = \text{numerical value}$$

that is, in the equation $\Delta G/\mathrm{kJ \cdot mol^{-1}} = -8.7$, the left hand side is the label and the right hand side is the numerical quantity to be tabulated. If a power of 10 is used in addition to or in place of a prefix on the units the power of 10 should be shown correctly with the units (thus $\Delta G/\mathrm{J \cdot mol^{-1}} = -0.087 \times 10^5$, or $\Delta G/10^5\ \mathrm{J \cdot mol^{-1}} = -0.087$, or any other algebraically correct form).

Values other than SI units may be included in parentheses when it is felt this will improve communication between the author and his readers. This may be considered desirable when there is still widespread current usage of a non-SI unit for a certain quantity; such as the calorie or the atmosphere. When a non-SI unit is used, the author should define it in terms of SI units.

The symbols for physical quantities should be printed in italic (sloping) type or underlined in typescript and the symbols for units should be printed in roman (upright) type.

The usage of these recommendations is illustrated below for certain specific applications in thermodynamic measurements.

No attempt will be made here to discuss all of the names and symbols for physical quantities, which are adequately described in the references given. Our attention will be directed

[a] Ref. 4 gives a summary of the International System of Units (SI) and recommendations for the usage of these units including reference to usage of other commonly found non-SI units. It also lists recommended names and symbols for quantities relating to space and time, mechanical properties, molecular masses and concentrations, thermodynamics, electricity and magnetism, electrochemistry, light and electromagnetic radiation, and transport properties. Recommendations are also given in Ref. 4 for subscripts and superscripts and for format of presenting formulas, symbols, units, numerical values, mathematical symbols, and relationships. Special chapters discuss notation and conventions for electromotive force and electrode processes, pH measurements, and reaction rates and related quantities. Also of interest to the biochemist and biothermodynamicist is an appendix which gives recommendations concerning the definitions of activities and related properties in aqueous solutions as well as other phases.

toward those for which the International recommendations form a distinct variation from much established practice.

Temperature

Thermodynamic temperatures and temperature differences are expressed in kelvins, symbol K, not degrees Kelvin or °K. Celsius temperatures and temperature differences are expressed in degrees Celsius, symbol °C. The degree Celsius, which is identical to the kelvin, is sometimes improperly called the degree "centigrade."

Pressure

The unit of pressure is the pascal, which is one newton per square metre ($1\ Pa = 1\ N \cdot m^{-2}$). A convenient unit for many pressure measurements is the kilopascal (symbol kPa). The biological thermodynamicist should recognize that the common thermodynamic term, pV, is an energy term, and if p is in pascals and V is in cubic metres, then the $p \cdot V$ product is obtained directly in joules. No combination of customary non-SI units avoids the use of a conversion factor; and such awkward energy units as litre-atmosphere are avoided by the recommended usage.

The commonly found units mmHg or Torr, and their submultiples, should be avoided.

The standard atmosphere is defined as 101 325 pascals (1 atm = 101.325 kPa). This is a non-SI unit. In calculating equilibrium constants and standard thermodynamic functions based on pressure measurements it should be recognized that the accepted standard pressure is 101.325 kPa, which is often called one atmosphere in this context.

Energy

Energy measurements, including all thermal measurements, should be reported in joules (J), kilojoules (kJ), or millijoules (mJ) as appropriate. In cases where energies are also expressed in thermochemical calories (cal_{th}) or a multiple or submultiple, the author should state the conversion factor used ($1\ cal_{th} = 4.184\ J$). The "nutritional" or "large" calorie (sometimes abbreviated "Cal" and equal to one kilocalorie) has little use outside of its special area, and its use should be avoided. Other definitions of calorie have little or no use in biochemistry or in other thermochemical work.

The author should realize that the use of calories is on the decline because of the acceptance of the SI units, so that data reported in calories will probably require conversion relatively soon.

For other units of measure, the author is referred to IUPAC and BIPM recommendations (4–6).

Composition of Solutions

For thermodynamic applications the composition of a solution is commonly described in terms of concentration, molality, or mole fraction. Other descriptors, such as mass fraction or volume fraction, may be found convenient for special applications. All of these should be clearly distinguished, and the solutions used should be unambiguously described using proper units.

The concentration of a solute substance B is the amount of B (moles) divided by the volume of the solution. Accepted symbols are c_B and [B]. A convenient unit of concentration is $mol \cdot dm^{-3}$. Concentration is sometimes called "molarity." A solution with a concentration of $0.1\ mol \cdot dm^{-3}$ is often called a 0.1 molar solution or a 0.1 M solution. Because the term molarity and the unit M are liable to be confused with molality, the term concentration and the unit $mol \cdot dm^{-3}$ are preferred.

The mass concentration of a solute substance B is the mass of B divided by the volume of the solution. An accepted symbol is ρ_B and an appropriate unit is $kg \cdot m^{-3}$.

The molality of solute substance B is the amount of B (moles) divided by the mass of solvent. An accepted symbol is m_B and an appropriate unit is $mol \cdot kg^{-1}$. A solution having a molality equal to $0.1\ mol \cdot kg^{-1}$ is sometimes called a 0.1 molal solution or a 0.1 m solution. Because of possible confusion of "molal" with "molar" and because m is the symbol for the metre, the unit $mol \cdot kg^{-1}$ is preferred.

The mole fraction of substance B is the amount of B (moles) in a solution divided by the total amount (moles) of all substances in the solution. Accepted symbols are x_B and y_B, where x_B or $y_B = n_B/\Sigma_i n_i$. The mole fraction is a dimensionless number.

State Functions

In stating energy quantities associated with the state functions of a substance, care should be taken to use the precisely defined functions and the recommended symbols, internal energy (U), enthalpy (H), Helmholtz energy (A), Gibbs energy (G), entropy (S), or heat capacity (C).

The term heat should be avoided in reference to state thermodynamic functions. Thus enthalpy is preferable to heat content and enthalpy of formation is preferable to heat of formation. An exception to this restriction is heat capacity for which no alternative has been accepted. The term specific heat capacity should refer only to heat capacity per unit mass. The term specific heat should be avoided. The corresponding molar quantity is molar heat capacity.

The term free energy is ambiguous and should not be used. Proper terms are Helmholtz energy or Helmholtz function and Gibbs energy or Gibbs function, symbols A and G, respectively, and referring to $U - TS$ or $H - TS$, respectively. The extensive quantities have the unit joule (J). The specific quantities have the unit joule per kilogram ($J \cdot kg^{-1}$). The molar quantities have the unit joule per mole ($J \cdot mol^{-1}$). Entropy, symbol S, an extensive quantity, has the unit joule per kelvin ($J \cdot K^{-1}$); the specific entropy has the unit joule per kilogram kelvin ($J \cdot kg^{-1} \cdot K^{-1}$); the molar entropy has the unit joule per mole kelvin ($J \cdot mol^{-1} \cdot K^{-1}$).

Equilibrium Constants

Ideally the thermodynamic equilibrium constant, based on activities of the reacting components, should be reported. In the case of most biochemical studies, however, this is not practicable. Therefore, the recommendations made in the preceding section of this document regarding the calculation and reporting of equilibrium constants and Gibbs energy changes should be observed. In accordance with the general recommendation given above, that symbols for physical quantities be differentiated from symbols for units, the symbol K should be in italic (sloping) type in printed documents, or underlined in typewritten documents, to differentiate it from the symbol for kelvin, K, which should be in roman (upright) type.

Electrode Measurements

Conventions concerning the signs of electric potential differences, electromotive forces, and electrode potentials, are given in Ref. 4 (Chapter 9) and are sometimes called the "Stockholm

Convention" of 1953. The author should strictly adhere to them.

By these conventions the direction of a written chemical reaction, the order of representation of the corresponding electrolytic cell elements, and the assignments of sign to the electromotive force, E, of the cell obey the following consistency relationships.

1. The cell reactions and the cell diagram are written from left to right in such a way that *when* reaction occurs in this direction positive electric charge flows in the cell from left to right. The electric potential difference is equal in sign and magnitude to the electric potential of the electrode on the right minus that of the electrode on the left.

2. The electrode potential of an electrode (half-cell) is the electromotive force of a cell in which the electrode on the right is the electrode in question and the electrode on the left is a standard hydrogen electrode. A more positive electromotive force indicates a greater oxidizing potential.

TABLE I(a)

Thermodynamic terms used in these recommendations

[A]	concentration of A
y_B	activity coefficient of substance B (concentration basis)
a_B	relative activity of solute B (Note: $a_B = y_B \cdot [B]$)
I	ionic strength ($I_c = \frac{1}{2}\sum_{1}^{S} c_i z_i^2$; $I_m = \frac{1}{2}\sum_{1}^{S} m_i \cdot z_i^2$) where c_i is the concentration of the ith ion, m_i is the molality of the ith ion, and z_i is the charge number of the ith ion. S is the number of ion types present.
K_c	pH-independent proper product of equilibrium concentrations (concentration equilibrium constant)
K	thermodynamic equilibrium constant
K' (pH = x; etc.)	pH-dependent apparent proper product of summed equilibrium concentrations, constrained with respect to concentrations of stated species (apparent equilibrium constant)
K_{exp}	see K'
K_{app}	see K'
K_{obsd}	see K'
ΔG_c^0	Standard Gibbs energy change corresponding to the pH-independent product of equilibrium concentrations ($\Delta G_c^0 = -RT \ln K_c$)
$\Delta G_c^{0\prime}$ (pH = x)	Apparent standard Gibbs energy change corresponding to the apparent product of equilibrium concentrations at fixed pH in a buffered solution. ($\Delta G_c^{0\prime} = -RT \ln K_c$)
ΔG_{exp}^0	see $\Delta G^{0\prime}$
ΔG_{app}^0	see $\Delta G^{0\prime}$
ΔF	formerly common symbol for free energy change. Now more properly called Gibbs energy change and given the symbol ΔG.
ΔF^0	see ΔF, ΔG^0, $\Delta G^{0\prime}$
ΔS	entropy change
ΔH	enthalpy change
C_p	heat capacity at constant pressure

Summary Tables

Table I lists the thermodynamic functions found in the recommendations of the main text, gives their units and a few relationships involving them. It also summarizes some of the principal recommended conditions and practices. Table II gives some SI units and their symbols.

The units of molar enthalpy change and molar Gibbs energy change are generally expressed in joules per mole ($J \cdot mol^{-1}$) or kilojoules per mole ($kJ \cdot mol^{-1}$) of a reaction shown, but the units should be stated in every case. For particular purposes specific enthalpy change or specific Gibbs energy change may be given per unit mass of a given reactant or product (joules per kilogram) ($J \cdot kg^{-1}$).

The units of entropy change are generally expressed in joules per mole kelvin ($J \cdot mol^{-1} \cdot K^{-1}$) for a particular process, but the units should be stated in every case. For particular purposes specific entropy change may be given per unit mass of a given reactant or product, joule per kilogram kelvin ($J \cdot kg^{-1} \cdot K^{-1}$), or the total entropy change for a stated process in which case the units are joules per kelvin ($J \cdot K^{-1}$).

The units of absolute entropy and heat capacity are the same as those for entropy change ($J \cdot mol^{-1} \cdot K^{-1}$, $J \cdot kg^{-1} \cdot K^{-1}$, or $J \cdot K^{-1}$) but refer to a particular substance or aggregate of substances rather than to a process.

TABLE I(b)

Summary of thermodynamic symbols and other recommendations

Quantity	Apparent value at pH = constant(x)	pH-independent value
Equilibrium constant[a]	K_c'	K_c
Standard molar Gibbs energy change[b]	$\Delta G^{0\prime}$ (pH = x)	ΔG^0
Concentration of reactant A	$[A]_{total}$	[A]

[a] Approximately constant proper quotient of equilibrium concentrations.

[b] Formally calculated from the equilibrium constant.

Recommended conventions concerning factors in expressions for equilibrium constants.

In cases where water occurs as a reactant or product, state whether the factor for water is taken as unity or as 55.5 or other number.

State whether 10^{-pH} or some other measure is used for the hydrogen ion factor. The pH is not uncontested as an accurate measure of concentration or activity of the hydrogen ion.

Recommended measurement conditions

Temperature t (or T)	Primary conditions 25 °C (298.15 K) (also, vary t)	Secondary conditions 37 °C (310.15 K) (also, vary t)
Ionic strength, $I/mol \cdot dm^{-3}$	0.1 (made up with KCl)	0.1 (made up with KCl)
Hydrogen ion concentration	pH = 7	
Buffer concentration	lowest effective	

Some thermodynamic relations: the units for these quantities are joules, J, for the extensive quantity total energy, joules per mole, ($J \cdot mol^{-1}$) for molar energy, and joule per kilogram ($J \cdot kg^{-1}$) for specific energy.

$\Delta U = Q + W$	1st law of thermodynamics. The increase of internal energy of a system is the sum of heat supplied to the system and work done on the system.
$\Delta H = \Delta U + p\Delta V$	For a constant pressure system an increase in enthalpy of the system (ΔH) results in an increase in internal energy (ΔU) and work done by the system ($p\Delta V$).
$\Delta H = H(T_2) - H(T_1) = \int C_p dT$	For a system at constant pressure to which energy is supplied the increase of enthalpy is the integral of C_p over the range of the temperature change it causes.
$\Delta G = \Delta H - T\Delta S$	The increase in Gibbs energy of a system (ΔG) at constant temperature is the increase in enthalpy (ΔH) minus the increase in $T\Delta S$.

TABLE II

Physical quantities, SI units, and their symbols

	Name of unit	Symbol for unit	
	(a) Base units		
Length	metre	m	
Mass	kilogram	kg	
Time	second	s	
Electrical current	ampere	A	
Temperature	kelvin	K	
Amount of substance	mole	mol	
Luminous intensity	candela	cd	
	(b) Derived units (examples)		
Force	newton	N	$(kg \cdot m \cdot s^{-2})$
Pressure	pascal	Pa	$(N \cdot m^{-2})$
Energy	joule	J	$(kg \cdot m^2 \cdot s^{-2})$
Power	watt	W	$(J \cdot s^{-1})$
Electric charge	coulomb	C	$(A \cdot s)$
Electric potential difference	volt	V	$(J \cdot A^{-1} \cdot s^{-1})$
Electric resistance	ohm	Ω	$(V \cdot A^{-1})$
Frequency	hertz	Hz	(s^{-1})
Area	square metre	m^2	
Volume	cubic metre	m^3	
Density	kilogram per cubic metre	$kg \cdot m^{-3}$	

(c) Prefixes

Multiplier	Prefix	Symbol	Multiplier	Prefix	Symbol
10^{-1}	deci	d	10	deca	da
10^{-2}	centi	c	10^2	hecto	h
10^{-3}	milli	m	10^3	kilo	k
10^{-6}	micro	μ	10^6	mega	M
10^{-9}	nano	n	10^9	giga	G
10^{-12}	pico	p	10^{12}	tera	T
10^{-15}	femto	f	10^{15}	peta	P
10^{-18}	atto	a	10^{18}	exa	E

REFERENCES

1. Guide for the Presentation in the Primary Literature of Numerical Data Derived from Experiments. *CODATA Bulletin No. 9, December 1973.*
2. A Guide to Procedures for the Publication of Thermodynamic Data. (1972); *J. Chem. Thermodynamics* **4,** 511–520; (1972) *Pure Appl. Chem.* **29,** 395–407.
3. Alberty, R. A. (1969) *J. Biol. Chem.* **244,** 3290.
4. International Union of Pure and Applied Chemistry (IUPAC). McGlashan, M. L. (1970). *Manual of Symbols and Terminology for Physiochemical Quantities and Units,* Butterworth and Co., London, Toronto, etc. (SBN-408 89350 8); (1970) *Pure Appl. Chem.* **21,** No. 1, 3–44; Revision, M. A. Paul (1975) Butterworth and Co., London (ISBN 0 408 70671 6).
5. International Organization for Standardization (ISO). SI units and recommendations for the use of their multiples and of certain other units. International Standard ISO-1000, First edition-1973-02-01, American National Standards Institute, New York. See also, for more detail: International Standard ISO-31/0-1974 and ISO 31/I-XII (1965–1975) which deal with quantities, units, symbols, conversion factors, and conversion tables for various branches of science and technology. Copies are obtained through the ISO-member national standards organizations of various countries.
6. Bureau International des Poids et Mesures (BIPM): Le Système International d'Unités, 1970, OFFILIB, 49 Rue Gay-Lussac, F75 Paris 5, France. Authorized English translations are available from: (*a*) The International System of Units (SI), National Bureau of Standards (USA) Publication 330, 1974, United States Government Printing Office, Washington, DC. or (*b*) The International System of Units (SI), Her Majesty's Stationery Office, London.

RECOMMENDATIONS FOR THE PRESENTATION OF THERMODYNAMIC AND RELATED DATA IN BIOLOGY

(Recommendations 1985)

INTERUNION COMMISSION ON BIOTHERMODYNAMICS*

Prepared for publication by

INGEMAR WADSÖ

University of Lund, Sweden

Abstract

Thermodynamic data are important in describing and in developing an understanding of biological systems. However, at present there is a marked variation in the terminology and symbols used in this connection. A guide has been prepared in which SI units and symbols of special importance for biothermodynamics are summarized together with examples and with comments concerning their use. It is recognized that in some cases recommendations made by IUPAC may have to be adjusted slightly or developed further in order to suit the practical needs in the biological sciences. In the present report suggestions are given with this purpose in mind.

The overall aim of the report has been to prepare a document which will serve as a practically useful guide for those who are involved in scientific writing and in teaching in the field of biothermodynamics. The document is therefore intended to be largely self-contained.

Physicochemical data are of considerable importance in describing and in developing an understanding of biological systems. At present, there is marked variation in the terminology and symbols used for physicochemical quantities and in the units of measurement in which they are reported in the biological literature. This situation leads to confusion and many difficulties in comparing and correlating the results from different laboratories. Recommendations concerning units, symbols and terminology have been made by the international standardizing bodies in order to facilitate communication and to remove ambiguities. These recommendations emphasize the use of the modern metric system of units called the Système International (SI) d'Unités (International System of Units) [1], which differs in several important ways from previous scientific usage.

Such recommendations cannot always be expected to be immediately accepted and used in all fields. It is hoped that the present transition period can be made as short as possible and that future generations of scientists will not be unnecessarily burdened by the variety of units and symbols currently found in textbooks, research reports and data compilations. We therefore encourage scientists studying biological systems to accept and to use the SI whenever suitable.

In this document, SI units and symbols of special importance for biological sciences and their applications to particular quantities are summarized. A more complete discussion of the SI, together with recommendations for symbols and terminology used to describe quantities of importance in physical chemistry is given in IUPAC's 'Manual of symbols and terminology for physicochemical quantities and units' [2] and also [3]. For other discussions of the presentation of numerical data and of thermodynamic data derived from experiments, we call attention to guides prepared by CODATA [4] and IUPAC [5].

Although we believe it is important to avoid differences in terminology and symbols in physical chemistry and in the biological sciences, we realize that some recommendations made by IUPAC may have to be adjusted or further developed in order to suit the practical needs in the biological sciences. In the present document, some recommendations are given with this purpose in mind.

The Interunion Commission on Biothermodynamics has previously made recommendations for terminology in two special areas of biothermodynamics: 'Recommendations for presentation of biochemical equilibrium data' [6] and 'Calorimetric measurements on cellular systems. Recommendations for measurements and presentation of results' [7]. In order to facilitate the practical use of the recommendations, they have been prepared to be largely self-contained. There is thus some

Reproduced from *Eur.J.Biochem.*, Vol. 153, pp.429-434 (1985) by courtesy of Springer-Verlag, Heidelberg, FRG.

*Membership of the Commission during the period (1977–85) in which the report was prepared was as follows:

Chairmen: I. Wadsö (1977–80, 1983–85) and R. L. Biltonen (1980–83). *Regular members*: G. T. Armstrong (USA, 1977–82), and I. Wadsö (Sweden, 1977–85) representing IUPAC; H. Gutfreund (UK, 1977–81), P. L. Privalov (USSR, 1977–82), M. Klingenberg (FRG, 1982–84), and W. Pfeil (GDR, 1982–85) representing IUB; R. L. Biltonen (USA, 1977–83), J. T. Edsall (USA, 1977–78), H. Eisenberg (Israel, 1978–85) representing IUPAB. *Associate members*: A. E. Beezer (UK, 1977–85), J. P. Belaich (France, 1979–84), H. Eisenberg (Israel, 1977–78), S. J. Gill (USA, 1979–85), L. G. Hepler (Canada, 1977–84), H.-J. Hinz (FRG, 1982–85), K. Hiromi (Japan, 1977–79), G. Rialdi (Italy, 1982–85), H. Suga (Japan, 1979–85), and D. F. Wilson (USA, 1977–79).

unavoidable overlap between the present and previous documents.

PHYSICAL QUANTITIES, SI UNITS AND THEIR SYMBOLS

A physical quantity is the product of a numerical value (a pure number) and a unit. By international agreement a set of seven dimensionally independent units form the so-called SI base units [1 – 3]. The base physical quantities and units and their recommended symbols are summarized in Table 1.

The symbols for physical quantities should be printed in italic (sloping) type of the Latin or Greek alphabets or underlined in typescript, and the symbols for units should be printed in roman (upright) type.

Table 1. *SI base quantities and units*

Base quantities		Base units	
Name	symbol	name	symbol
Length	l	metre	m
Mass	m	kilogram	kg
Time	t	second	s
Electric current	I	ampere	A
Temperature*	T	kelvin	K
Amount of substance	n	mole	mol
Luminous intensity	I_v	candela	cd

* Thermodynamic ('absolute') temperature. Recommended symbol for Celsius temperature is t or θ. Where symbols are needed to represent both time and Celsius temperature, t is the preferred symbol for time and θ for Celsius temperature [2].

All other physical quantities and units are regarded as being derived from the base quantities and units. Certain SI-derived units have been given special names and symbols. In Table 2, symbols and units for some thermodynamic quantities (functions) are given. For more complete tabulations of quantities and units recommended for chemistry and physics, see [2, 3].

The SI base units are often cumbersome to use in practical work and therefore the use of certain prefixes denoting multiples or submultiples is convenient. Recommended prefixes are listed in Table 3.

Table 3. *SI prefixes*

Multiplier	Prefix	Symbol	Multiplier	Prefix	Symbol
10^{-1}	deci	d	10	deca	da
10^{-2}	centi	c	10^{2}	hecto	h
10^{-3}	milli	m	10^{3}	kilo	k
10^{-6}	micro	µ	10^{6}	mega	M
10^{-9}	nano	n	10^{9}	giga	G
10^{-12}	pico	p	10^{12}	tera	T
10^{-15}	femto	f	10^{15}	peta	P
10^{-18}	atto	a	10^{18}	exa	E

SOME COMMENTS CONCERNING CHOICE OF UNITS IN BIOTHERMODYNAMICS

Mass: specific and molar quantities

The SI base unit of mass is the kilogram. However, its multiples and submultiples are named and symbolized as multiples and submultiples of the gram. *Example:* µg, not nkg for 10^{-9} kg.

The term 'specific' preceding the name of an extensive quantity means 'divided by mass' and the specific quantity should preferably be designated by a lower case letter. (An 'extensive' physical quantity for a system depends on the amount of material it contains.) *Example:* c_p is the specific

Table 2. *Symbols and units for some thermodynamic quantities*

Quantity		SI unit	
Name	symbol	name	symbol
Volume	V	cubic metre	m^3
Force	F	newton	$N = m\,kg\,s^{-2}$
Density	ϱ	kilogram per cubic metre	$kg\,m^{-3}$
Pressure	p	pascal	$Pa = N\,m^{-2}$
Viscosity	η	pascal second	Pa s
Energy	E	joule	J = N m
heat	q, Q*	joule	J = N m
work	w, W*	joule	J = N m
internal energy	U, (E)	joule	J = N m
enthalpy: $U + pV$	H	joule	J = N m
Gibbs energy: $H - TS$	G	joule	J = N m
Helmholtz energy: $U - TS$	A	joule	J = N m
Entropy	S	joule per kelvin	$J\,K^{-1}$
Power	P	watt	$W = J\,s^{-1}$
Heat capacity			
at constant pressure: $(\delta H/\delta T)_p$	C_p	joule per kelvin	$J\,K^{-1}$
at constant volume: $(\delta U/\delta T)_V$	C_V	joule per kelvin	$J\,K^{-1}$
Cubic expansion coefficient: $V^{-1}(\delta V/\delta T)_p$	α	per kelvin	K^{-1}
Isothermal compressibility: $-V^{-1}(\delta V/\delta p)_T$	κ	per pascal	Pa^{-1}
Osmotic pressure	Π	pascal	Pa
Chemical potential of substance B	μ_B	joule per mole	$J\,mol^{-1}$
Absolute activity of substance B	λ_B	dimensionless	
Relative activity of substance B	a_B	dimensionless	
Activity coefficient, mole fraction basis	f_B	dimensionless	
Activity coefficient, molality basis	γ_B	dimensionless	
Activity coefficient, concentration basis	y_B	dimensionless	
Osmotic coefficient	ϕ	dimensionless	

* It is recommended that $q > 0$ and $w > 0$ both indicate increases in the energy of the system under discussion [2]. Thus $\Delta U = q + w$.

heat capacity at constant pressure. A suitable unit is $J\ K^{-1}\ g^{-1}$. (The term 'specific heat' is an inappropriate term for specific heat capacity.)

The term 'molar' preceding the name of an extensive quantity means 'divided by amount of substance', giving, in effect, the quantity per mole. The quantity should be represented by a capital letter and the attached lower case subscript, m. *Example:* $C_{p,m}$ is the molar heat capacity ($J\ K^{-1}\ mol^{-1}$) at constant pressure. The subscript, m, can be omitted when there is no risk of ambiguity.

Thermodynamic quantities should, where possible, be reported in terms of molar quantities. For biochemical macromolecules the molecular mass (dalton, a non-SI unit of mass, symbol Da) or, the numerically identical, molar mass ($g\ mol^{-1}$) used in the calculations should always be reported (cf. [8]). If the quantity of a substance is determined by weighing, the water content and other known or estimated impurities should be reported and corrections applied. In cases where the molecular mass is not known, the amount of compound should, where possible, be reported in an SI mass unit.

Volume

The SI unit for volume is the cubic metre (m^3). More convenient units in biothermodynamics are usually its submultiples: cubic decimetre (dm^3), cubic centimetre (cm^3) and cubic millimetre (mm^3). The symbol 'cc' (to denote cm^3) should not be used.

The cubic decimetre is identical to the litre (l or L). This non-SI unit and its submultiples, the millilitre ($1\ ml \equiv 1\ cm^3$), the microlitre ($1\ \mu l \equiv 1\ mm^3$), and the nanolitre ($1\ nl \equiv 10^{-12}\ m^3$) are in biology often considered to be more convenient and more readily understood than corresponding SI units. One can therefore expect the litre-based volume units to be used alongside the SI units (cf., however, comments under ***Pressure***).

Time

The SI base unit for time is the second (s). Other units for time which are exactly defined in terms of the second are the minute (min, not mn), hour (h, not hr), and day (d). However, when reporting measured values of properties involving time, the use of min, h and d is discouraged. It should be realized that the second, or its multiples or submultiples, is employed as the basic unit of time in most electronic instruments used for determination of time-dependent physical quantities. Furthermore, power in watts is obtained directly by dividing energy in joules by time in seconds.

Temperature

Thermodynamic temperatures and temperature differences are expressed in the SI base unit, the kelvin, symbol K (not degrees Kelvin or °K). Temperature and temperature differences may also be expressed in degrees Celsius, symbol °C. The degree Celsius was formerly called the degree 'centigrade'. To many biologists the Celsius scale is more convenient to use, e.g. to express an experimental temperature. However, it is recommended that whenever values for temperatures are used in connection with thermodynamic calculations, they are reported in kelvins.

Pressure

The SI unit of pressure is the pascal, which is one newton per square metre ($Pa = N\ m^{-2}$). A convenient unit for many pressure measurements is the kilopascal (symbol kPa). The biological thermodynamicist should recognize that the common thermodynamic term, pV, is an energy term, and if p is in pascals and V is in cubic metres, then the pV product is obtained directly in joules. Note that combinations of customary non-SI units all require the use of a conversion factor.

The commonly found units mmHg or Torr are exactly defined in terms of Pa but should be avoided (cf. [2]).

One atmosphere is defined as 101 325 pascals (1 atm = 101.325 kPa). However, the atmosphere is a non-SI unit and its use should be avoided. The IUPAC Commission on Thermodynamics has recently recommended that 10^5 Pa (1 bar) be adopted as the standard state pressure in chemical thermodynamics [9]. In calculating equilibrium constants and standard thermodynamic functions based on pressure measurements, it should be recognized, however, that the accepted standard state pressure for many years has been 1 atmosphere.

Viscosity

The SI unit for viscosity (η) is the pascal-second ($Pa\ s = kg\ s^{-1}\ m^{-1}$). Traditionally used non-SI units are the poise (P) or the centipoise (cP), 1 Pa s being equal to 10 P.

Energy

Energy measurements, including all thermal measurements, should be reported in joules (J), kilojoules (kJ) or millijoules (mJ) as appropriate. The use of 'calorie' is discouraged (cf. [7]). The 'thermochemical calorie' (cal_{th}) is defined as $1\ cal_{th} \equiv 4.184\ J$.

Entropy

The recommended SI unit is joule per kelvin ($J\ K^{-1}$). The use of 'entropy units' ('e.u.') is not acceptable.

Power

The SI unit of power is watt (W). More convenient units in biothermodynamics are usually milliwatt (mW), microwatt (μW), picowatt (pW) or femtowatt (fW). The use of units such as calories per hour, which is currently common in biology, is discouraged (cf. [6]).

Density

The SI for density (ϱ) is $kg\ m^{-3}$. A more convenient unit is usually $g\ cm^{-3}$.

NOTATIONS FOR VARIABLES, STATES AND PROCESSES

Variables

Variable parameters for the thermodynamic functions can be indicated in parentheses after the symbols, e.g. $C_p(T,p)$.

Table 4. *Some recommended superscripts*

Superscript	Meaning
$^\circ$ (or $\ominus$)	standard
*	pure substance
∞	infinite dilution
id	ideal
'	apparent (cf. the text)
E	excess
‡	activated complex

Table 5. *Symbols for states of aggregation*

State	Symbol
Gas	g
Liquid	l
Solid	s
Fluid	fl
Liquid crystal	lc
Crystalline solid	cr
Amorphous solid	am
Vitreous substance	vit
Solution	sln
Aqueous solution	aq

Similarly, values for quantities can be represented by an appropriate symbol with values for the established conditions given in parentheses. *Example:* C_p (298.2 K, 0.1 MPa, pH = 7.0) or C_p (25.0°C, 1 bar, pH = 7.0).

States

Superscripts to symbols for properties (thermodynamic functions) are frequently used to denote a particular state (Table 4). *Example:* $C^*_{p,B}$ is the (molar) heat capacity for the pure substance B at constant pressure.

The term 'apparent' indicated by a superscript is used to mean that a process is not well known or that its value carries uncertainties which are not known. *Example:* $\Delta G^{0\prime}$ is the apparent standard Gibbs energy change. For a detailed discussion of K′ (apparent equilibrium constant) and $\Delta G^{0\prime}$, see [6].

The word 'apparent' is used in a different sense in the context of partial molar quantities. As a symbol for 'apparent' in this connection, the use of subscript ϕ, as in Y_ϕ, is recommended. Other notations employed for this property include $^{\phi}X$ and ϕ_X. *Example:* The apparent molar volume V_B of a solute B is defined by $V_{B,\phi} = (V - n_A V^*_A)/n_B$, where n_A and n_B are the amounts of solvent and solute, respectively. V is the total volume of the solution and V^*_A is the molar volume of the pure substance A.

To denote a state of aggregation it is recommended that an appropriate symbol be given in parentheses after the symbol for the thermodynamic quantity. The symbols given in Table 5 have recently been recommended by the IUPAC Commission on Thermodynamics [9]. *Example:* V^*_B(cr) is the (molar) volume of substance B in its pure crystalline state (cf. Table 5).

It is not practical to give definite recommendations for symbols for all states which are of relevance in biological systems. In many cases the terms used, for example 'helix' and 'coil', imply structural knowledge which is only presumed. In other cases the terms used, for example 'native' and 'denatured', are strictly operational and their meanings vary from situation to situation. Therefore it is urged that when such terms are used they be well defined within the context of their use and that their associated symbolic representation be judiciously selected. Any such notation should be used with caution and in all cases the text should clearly describe the state to which a value or function refers.

Processes

A thermodynamic change is denoted by the symbol Δ before the corresponding quantity. The nature of the change is signified by annotation of the Δ. Two methods of annotation are now recommended by the IUPAC Commission on Thermodynamics [9]:

i) use of regular symbols as superscripts and subscripts: the notation $\Delta_\alpha^\beta X$ is recommended to denote the change in property X for the process $\alpha \rightarrow \beta$ where α and β are symbols for states or species. *Example:* $\Delta_s^l H$ meaning the change in enthalpy when a substance changes from the solid to the liquid state.

ii) use of special subscripts to denote the process. The subscript symbols given in Table 6 have been recommended [9]. *Example:* $\Delta_c S_B$, meaning the entropy of combustion of substance B.

Currently the symbol for a process is usually placed as subscript to the symbol for the property, e.g. ΔS_c. This practice is likely to prevail for some time in biothermodynamics.

For many processes of biological relevance, the symbols listed in Table 6 are insufficient. In those cases, for example ionization, protonation and oxidation, the process should be well defined in the text and the symbol used should be judiciously selected.

Table 6. *Symbols for processes*

Process	Symbol (subscript)
Vaporization (evaporation)	vap
Sublimation (evaporation)	sub
Melting (fusion)	fus
Transition of one solid phase to another	trs
Mixing of fluids	mix
Solution (dissolution)	sol
Reaction (except for combustion)	r
Combustion	c
Formation (of a component from its elements)	f

It should be realized that in biochemical thermodynamics processes are not always well defined. Rather than using special symbols it is therefore often advisable to use 'neutral' symbols like a, b, . . . or 1, 2, . . ., which should be clearly identified in the text. In cases where one is dealing with complex reaction schemes, these should always be represented

Table 7. *Notations for composition of solutions*

Quantity Name	symbol	Practical unit
Amount of substance B	n_B	mol
Concentration of solute substance B (the amount of substance of B divided by the volume of the solution)	c_B, [B]	mol dm^{-3}
Mass concentration of substance B (mass of B divided by the volume of the solution)	ϱ_B	g dm^{-3}
Molality of solute substance B (the amount of substance of B divided by mass of principal solvent)	m_B	mol kg^{-1}
Mole fraction of substance B ($n_B/\Sigma_j n_j$)	x_B	dimensionless
Mass fraction of substance B ($m_B/\Sigma_j m_j$)	w_B	dimensionless
Volume fraction of substance B ($V_B/\Sigma_j V_j$)	ϕ_B	dimensionless

by a figure in the text. It is then recommended that each reaction step be given a 'neutral' symbol. *Example:*

$$
\begin{array}{c}
\quad 1 \quad\; 2 \\
A \rightleftharpoons B \rightleftharpoons C \\
{}_3\searrow \quad \nearrow{}_4 \\
D
\end{array}
$$

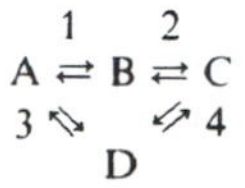

SOLUTIONS

The word 'solution' is used to describe a liquid or solid phase containing more than one component. For convenience one of these substances is referred to as 'solvent' (normally the one present in the largest concentration) and all others as 'solutes'. In biological solutions, which generally contain a number of components, the distinction is not always clear and often a fixed mixture of components is referred to as 'solvent' (e.g. H_2O plus buffer components), with all other substances being defined 'solutes'.

Recommended notations for composition of solutions are summarized in Table 7.

Subscript A (or 1) is sometimes used to denote the solvent and B (or 2) the solute, but the designation of subscripts should always be defined.

A convenient unit of concentration, sometimes called 'molarity', is mol dm^{-3}. A solution with a concentration of 0.1 mol dm^{-3} is often called a 0.1 molar solution or a 0.1 M solution. Because the term molarity and the symbol M are liable to be confused with molality (Table 7), the term 'concentration' and the symbol 'mol dm^{-3} (mol l^{-1}, mol L^{-1})' are preferred.

In the particular case of aqueous solutions, the solvent can be denoted by 'aq', but usually not by 'H_2O'. The latter notation should only be used for pure water or, possibly, in cases where presence of solutes is insignificant for the problem considered.

Authors should be aware of the fact that concentrations of low-molecular-weight components diffusible through semi-permeable membranes in solutions of biological macromolecules prepared by equilibrium dialysis may differ from the concentrations of these solutes in the dialysis solvent. The chemical potentials, μ, of the diffusible solutes, though, are identical in both the macromolecular solution and in the dialysis solvent. The composition of the dialysis solvent should therefore be stated precisely (cf. [10, 11]).

A partial molar quantity of substance B is defined as $Y_B = (\delta Y/\delta n_B)_{T,p,n_C\ldots}$ where Y is an extensive quantity of a system. *Examples:* $V_B = (\delta V/\delta n_B)_{T,p,n_C\ldots}$, the partial molar volume of substance B; $v_B = (\delta v/\delta m_B)_{T,p,m_C\ldots}$, the partial specific volume of substrate B.

If the partial quantity refers to infinite dilution, it should be so designated by the superscript symbol ^{x} (cf. Table 4).

The symbol $\bar{Y}_B$ is often used instead of the recommended Y_B. The bar over Y resolves no ambiguity; in fact, it is misleading, as the bar is often used to indicate an average (cf. [3]).

The condition to which an apparent molar quantity, Y_ϕ, refers should be clearly stated in the text or indicated in the symbol. *Example:* $V_{B,\phi}(\text{aq}, c = 0.1 \text{ mol dm}^{-3})$, meaning the apparent molar volume of substance B in aqueous solution where the concentration of B is 0.1 mol dm^{-3}.

PRESENTATION OF RESULTS

Whenever values for quantities are reported — in a text, a table, a graph or in a projected slide — the units should also be given. In addition, necessary information concerning vital experimental parameters (temperature, concentration, pH, solvent composition, etc.) should be stated. It is not always practicable to include adequate experimental parameters in the table or graph itself, or even in the accompanying legend. The necessary additional information must then be given in the text.

When thermodynamic results are given in a table, it is usually most convenient to list them as pure numbers with the unit given in the table legend or preferably in the columnar heading. It is then correct to give the symbol for the quantity divided by the symbol for the unit. *Example:* $\Delta H/(\text{kJ mol}^{-1})$

or $\dfrac{\Delta H}{\text{kJ mol}^{-1}}$, meaning that the numbers reported in the table

are enthalpy changes expressed in the unit kilojoule per mole.

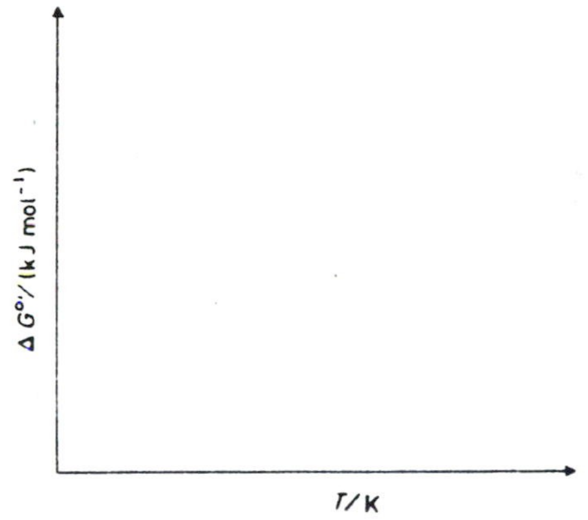

Fig. 1. *Apparent standard Gibbs energy change expressed in kilojoules per mole, as a function of temperature, given in kelvins*

The axes of graphs and the headings in tables should be labeled with both the name (or symbol) of the quantity and the unit employed so as to give a dimensionless ratio for plotting or tabulation. An example of this is given in Fig. 1.

REFERENCES

1. International Organization for the Standardization (1973), *SI units and recommendations for the use of their multiples and of certain other units*, American National Standards Institute, New York. For more details, see also: International Standard ISO-31/0-1974 and ISO-31/I — XII (1965—1975), which deal with quantities, units, symbols, conversion factors, and conversion tables for various branches of science and technology. Copies are obtained through the ISO-member national standards organizations of various countries.
2. IUPAC's Physical Chemistry Division, Commission on Physicochemical Symbols (1979) Manual of symbols and terminology for physicochemical quantities and units', 1979 edition, *Pure Appl. Chem.* 51, 1. Also available from Pergamon Press, Oxford.
3. McGlashan, M. L. (1971) *Physicochemical quantities and units. The grammar and spelling of physical chemistry*, 2nd edn, The Royal Institute of Chemistry, London.
4. Codata (1973) Guide for the presentation in the primary literature of numerical data derived from experiments, *Codata Bull.* no. 9.
5. IUPAC's Physical Chemistry Division (1972) A guide to procedures for the presentation of thermodynamic data. *J. Chem. Thermodyn. 4*, 551 (1972); and *Pure Appl. Chem. 29*, 395 (1972).

6. IUPAC-IUPAB-IUB Interunion Commission on Biothermodynamics: Recommendations for presentation of biochemical equilibrium data. *J. Biol. Chem.* *251*, 6879 (1976); *Quart. Rev. Biophys.* *9*, 439 (1976); *Codata Bull.* no. 20 (1976); *Handbook of Biochemistry and Molecular Biology*, 3rd edn, Physical and Chemical Data Section (G. D. Fasman, ed.) vol. 1, pp. 93–106, CRC Press, Cleveland, Ohio; *Eur. J. Biochem.* *72*, 1 (1977); *Biochim. Biophys. Acta* *461*, 1 (1976); *Biofizika (USSR)* *23*, 739 (1978); *Netsusokutei (Japan)* *4*, 172 (1977) and *5*, 77 (1978); and *Biochemical nomenclature and related documents*, pp. 45–51, The Biochemical Society, London (1978).

7. IUPAC-IUPAB-IUB Interunion Commission on Biothermodynamics: Calorimetric measurements on cellular systems. Recommendations for measurements and presentation of results. *Codata Bull.* No. 44 (1981); and *Pure Appl. Chem.* *54*, 671 (1982).

8. Nomenclature Committee of IUB and IUPAC-IUB Joint Commission on Biochemical Nomenclature (1981) Newsletter 1981; *Eur. J. Biochem.* *114*, 1 (1981).

9. IUPAC's Physical Chemistry Division, Commission on Thermodynamics (1982) Manual of symbols and terminology for physicochemical quantities and units. Appendix IV: Notation for states and processes, significance of the word standard in chemical thermodynamics, and remarks on commonly tabulated forms of thermodynamic functions. *Pure Appl. Chem.* *54*, 1239.

10. Casassa, E. F. & Eisenberg, H. (1964) Thermodynamic analysis of multicomponent solutions, *Adv. Prot. Chem.* *19*, 287.

11. Eisenberg, H. (1976) *Biological macromolecules and polyelectrolytes in solution*, Clarendon Press, Oxford.

CALORIMETRIC MEASUREMENTS ON CELLULAR SYSTEMS: RECOMMENDATIONS FOR MEASUREMENTS AND PRESENTATION OF RESULTS

INTERUNION COMMISSION ON BIOTHERMODYNAMICS*

INTRODUCTION

Calorimetric measurements are of basic importance in studies of the thermodynamic properties of living systems. However, calorimetric measurements on such systems are presently undertaken, to a large extent, as parts of diagnostic investigations without the objective of acquiring thermodynamic information. Examples of experiments of this kind are calorimetric monitoring of metabolic activity or identification of microorganisms by their calorimetric growth patterns. In such experiments heat quantities are determined, but, from a molecular thermodynamic point of view, the results are usually of marginal significance. This is because the systems investigated in general are poorly defined and the conditions under which the calorimetric measurements are made are frequently not reported in sufficient detail.

The Commission offers in this report some recommendations concerning experimental details and procedures for the reporting of results from calorimetric measurements on cellular systems. One aim has been to increase the usefulness of the thermodynamic quantities determined for such systems. In addition, we hope that the recommendations will assist in orienting some 'diagnostic' calorimetric studies towards procedures that will lead to results of thermodynamic significance. We believe that adherence to the recommendations should also be of value in purely 'diagnostic' calorimetric studies. It should facilitate reproduction of experiments within a laboratory, comparison of results obtained by different workers and comparison of results obtained for cells from different preparations.

Many types of calorimeters suitable for work on cellular systems are now available. The measured heats are often small and 'microcalorimeters' are normally employed. The term 'microcalorimeter' is not strictly defined but in current practice it usually implies (for reaction calorimeters) that the smallest measurable thermal power is of the order of 1 μW and that the reaction vessel is small, of the order of a few cm^3 or less.

There is a general and lasting value in well documented energy data for biological systems. Thus, the use of arbitrary units like 'mm recorder deflection' should be avoided. Instead the calorimeter should be

*Membership of the Commission during the period (1977–80) in which the report was prepared was as follows:

Chairman: I. WADSÖ (Sweden); *Vice-Chairman:* R. L. BILTONEN (USA); *Members:* G. T. ARMSTRONG (USA); H. EISENBERG (Israel); J. T. EDSALL (USA); H. GUTFREUND (UK); P. L. PRIVALOV (USSR); *Associate Members:* A. E. BEEZER (UK); J. P. BELAICH (France); S. J. GILL (USA); L. G. HEPLER (Canada); K. HIROMI (Japan); H. SUGA (Japan); D. F. WILSON (USA)

Members of the Task Group for the formulation of these recommendations:
J. P. BELAICH (France); A. E. BEEZER (UK); E. PROSEN (USA); I. WADSÖ (Sweden).

Reprinted with permission from *Pure and Applied Chemistry,* Volume 54, Number 3, 1982, pages 671–679.

0033–4545/82/030671–09$03.00

calibrated and the results reported in SI units, see e.g. (1). Calorimeters are normally calibrated electrically. The measurement of electrical energy is, today, a trivial procedure that can easily be made with an accuracy exceeding the needs of biological experiments. The problem is rather to make certain that the electrical energy is released in a manner closely comparable with that for the process studied. Microcalorimeters used in biological work are not always well suited for such strict comparisons. In particular for flow-through vessels* calibration procedures are often far from ideal, and it is currently rare that the results are checked by a suitable test reaction.

The response time of a calorimeter can be significant compared to the duration of the process being studied. Therefore, the calorimetric signal at any point in time may not have a simple relationship with the thermal power generated at that time within the reaction vessel (2). This must be taken into account when kinetic information is deduced from the results of calorimetric experiments.

It is necessary to specify carefully the physiological conditions for the cellular material during a calorimetric measurement. Special attention must be given to the practical design of the reaction vessel, the performance of the experiment and to the methods by which the physiological conditions in the reaction vessel are verified.

PHYSICAL QUANTITIES. SI UNITS AND THEIR SYMBOLS

By international agreement a set of seven dimensionally independent units form the so-called SI * base units. The base physical quantities and units and their recommended symbols are summarized in Table 1.

Table 1

Base quantities		Base units	
Name	Symbol	Name	Symbol
length	l	metre	m
mass	m	kilogram	kg
time	t	second	s
electrical current	I	ampere	A
temperature †	T	kelvin	K
amount of substance	n	mole	mol
luminous intensity	I_v	candela	cd

* We recommend the use of the term 'calorimetric vessel' in place of the often used term 'cell' to avoid confusion with 'biological cells' which may be the subject of the investigation.

Reproduced from *CODATA Bulletin*, No.44 (1981), pp.1-8.

* Le Système International d'Unités or The International System of Units; cf. (1).

† Thermodynamic ('absolute') temperature. Recommended symbol for Celsius temperature is t or θ. Where symbols are needed to represent both time and Celsius temperature, t is the preferred symbol for time and θ for Celsius temperature (1).

All other physical quantities and units are regarded as being derived from the base quantities and units. Certain SI derived units have been given special names and symbols. Some of these are summarized in Table 2.

Table 2

Physical quantity	SI unit		Definition of SI unit
	Name	Symbol	
force	newton	N	$m \cdot kg \cdot s^{-2}$
pressure, stress	pascal	Pa	$(= N \cdot m^{-2})$ $m^{-1} \cdot kg \cdot s^{-2}$
energy	joule	J	$(= N \cdot m)$ $m^{2} \cdot kg \cdot s^{-2}$
power	watt	W	$(= J \cdot s^{-1})$ $m^{2} \cdot kg \cdot s^{-3}$

In Table 3, SI symbols and units for some thermodynamic quantities (functions) are given. For an extensive tabulation of quantities and units recommended for chemistry and physics see (1).

Table 3

Quantity		Unit	
Name	Symbol	Name	Symbol
energy:			
heat	q, Q *	joule	J
work	w, W *	joule	J
internal energy	U, (E)	joule	J
enthalpy: $U + pV$	H	joule	J
Gibbs energy: $H\text{-}TS$	G	joule	J
Helmholtz energy: $U\text{-}TS$	A	joule	J
entropy	S	joule per kelvin	$J \cdot K^{-1}$
power	P	watt	W
electrical potential difference	U, ΔV, $\Delta \phi$	volt	V
volume	V	cubic metre	m^3
density	ρ	kilogram per cubic metre	$kg \cdot m^{-3}$

* It is recommended that $q > 0$ and $w > 0$ both indicate increase of energy of the system under discussion (1). Thus $\Delta U = q + w$.

The SI base units are sometimes cumbersome to use in practical work and therefore the use of certain prefixes denoting multiples or submultiples are often found convenient. Recommended SI prefixes are summarized in Table 4.

Table 4

Multiplier	Prefix	Symbol	Multiplier	Prefix	Symbol	Multiplier	Prefix	Symbol
10^{-18}	atto	a	10^{-2}	centi	c	10^{6}	mega	M
10^{-15}	femto	f	10^{-1}	deci	d	10^{9}	giga	G
10^{-12}	pico	p	10	deca	da	10^{12}	tera	T
10^{-9}	nano	n	10^{2}	hecto	h	10^{15}	peta	P
10^{-6}	micro	μ	10^{3}	kilo	k	10^{18}	exa	E
10^{-3}	milli	m						

Presentation of experimental and derived results

The output of many calorimetric experiments is a plot of power (energy evolution per unit of time) as a function of time. The recommended unit for power (P) is watt (W) and for time the second (s) or some multiple thereof [e.g. kilosecond (ks)]. With the use of these units the time integral of the power-curve is given directly in joules (J) or some multiple thereof.

In graphs the axes should be labeled with both the name (or the symbol) of the quantity and the unit employed. Preferably the label should show the quantity divided by the unit in order to emphasize that the scale markers represent pure (dimensionless) numbers.

Some comments on terminology, symbols and units

To avoid confusion and to simplify communication we recommend standardization of terminology, symbols and units in accordance with the SI and with other rules accepted by the International Unions. Some physical quantities, SI units and their symbols are summarized in Tables 1-3.

In a report, values other than SI units may be included in parentheses when it is felt this will improve communication between the author and the readers. This may be considered desirable when there is still widespread current usage of a non SI unit for a certain quantity such as the calorie or the atmosphere.

The symbols for physical quantities should be printed in italic (sloping) type or underlined in typescript and the symbols for units should be printed in roman (upright) type.

No attempt will be made here to discuss all the names and symbols for physical quantities. Our attention will be directed toward those for which the international recommendations form a distinct variation from much established practice.

Energy

Results of energy measurements should be reported in joules (J, kJ, mJ, etc. as appropriate). In cases where energies are also expressed in calories (more properly called thermochemical calories, cal_{th}) or a multiple or submultiple thereof, the author should state the conversion factor used ($1\ cal_{th} = 4.184\ J$). The 'nutritional' or 'large' calorie (sometimes abbreviated 'Cal' and equal to one kilocalorie) should not be used.

The author should realize that the use of calories is in decline because of the acceptance of the SI units, so that data reported in calories will probably require conversion relatively soon.

Temperature

Thermodynamic temperatures and temperature differences are expressed in kelvins, symbol K (e.g., not degree kelvin or ^{o}K). Celsius temperatures and temperature differences are expressed in degrees Celsius, symbol ^{o}C, (often improperly called the degree 'centigrade'). One degree Celsius is exactly equal to one kelvin.

Pressure

The unit of pressure is the pascal, which is one newton per square metre ($1\ Pa = 1\ N \cdot m^{-2}$). A convenient unit for many pressure measurements is the kilopascal (symbol kPa). The pressure unit commonly used until now, the atmosphere (atm), approximately equals 100 kPa.

One atmosphere, defined as 101 325 Pa, is presently the accepted standard pressure used in calculating equilibrium constants (cf. 3) and standard thermodynamic functions. *

The biological thermodynamicist should recognize that the common thermodynamic term pV is an energy term, and if p is in pascals and V is in cubic metres then the $p \cdot V$ product is obtained directly in joules. No combination of customary non SI units avoids the use of a conversion factor; and such awkward energy units as litre-atmosphere are avoided by the usage recommended. The commonly found units—mm Hg, Torr, or their multiples—should be avoided.

Volume

The basic SI unit for volume, cubic metre (m^3) is often cumbersome for use in biology. More convenient units are its submultiples: cubic decimetre (dm^3), cubic centimetre (cm^3) and cubic millimetre (mm^3). The cubic decimetre is identical to the litre (l or L). The use of the name litre and its submultiples millilitre ($ml \equiv cm^3$) and microlitre ($\mu l \equiv mm^3$) is not encouraged, but is likely to prevail for some time.

Composition of solutions

For thermodynamic applications the composition of a solution is commonly described in terms of concentration, molality, or mole fraction.

The molar concentration of a solute B is the amount of substance of B (expressed in moles) divided by the volume of the solution. Accepted symbols are c_B and [B]. Concentration is sometimes called 'molarity.' A solution with a concentration of $0.1\ mol \cdot dm^{-3}$ is often called a 0.1 molar solution or a 0.1 M solution. Because the term molarity and the symbol M are liable to be confused with molality, the term concentration and the symbol $mol \cdot dm^{-3}$ are preferred.

The mass concentration of a solute substance B is the mass of B divided by the volume of the solution. An accepted symbol is ρ_B and an appropriate unit is $kg \cdot m^{-3}$ ($g \cdot dm^{-3}$).

The molality of solute B is the amount of substance B (moles) divided by the mass of solvent. A solution having a molality equal to $0.1\ mol \cdot kg^{-1}$ is sometimes called a 0.1 molal or a 0.1 m solution. An accepted symbol is m_B and the appropriate unit is $mol \cdot kg^{-1}$.

---

* The IUPAC Commission on Thermodynamics has recently (4) proposed that the standard pressure shall be equal to 10^5 Pa (0.1 MPa) which is identical to 1 bar.

SOME SPECIFIC RECOMMENDATIONS

The cell preparation

Where possible, experiments should employ material obtained from type collections with well defined organisms or cells. In other cases detailed information about the origin and method of preparation (isolation) should be reported. For cells which have been stored, maintenance conditions such as time, temperature and media should be reported. Details concerning passages of the organisms from the state of storage to experimental use should also be described (number of subcultures and details of the growth conditions). Counts of contaminating cells, e.g., for blood cell preparations, should be reported.

The calorimeter

Full details of the design of the calorimeter and its performance should be reported. Alternatively, reference can be given to other reports provided these are readily available. However, even minor changes in apparatus design and experimental procedure should be indicated. In each report the following details require attention:

(1) The design of the calorimetric vessel, its volume and its construction material, gas phase, method for initiation of a process, method used for stirring or agitation of the liquid medium.

(2) Time constants of the apparatus.

(3) Results of calibration and test experiments including accuracy assignment.

(4) Base line stability (where relevant) over a period of time which normally should be greater than the measurement period.

(5) Amplification and recording equipment; pumps and other ancillary equipment.

(6) Method employed for cleaning and sterilization of the calorimetric vessel and flow lines where appropriate.

The calorimetric experiments

Experimentalists are reminded that all processes, both physical and chemical, will contribute to the observed heats. Thus, processes such as dilution, protonization, mixing, vaporization, etc. may give rise to significant systematic errors if not taken into account in the experimental design and in the evaluation of the results. In the special case of twin calorimeters, where two processes are compared, it is frequently useful to observe the separate output from each vessel to aid in the interpretation of the difference signal.

In each study the following details require attention:

(1) The measurement temperature must be reported.

(2) It is suggested that, where possible, chemically defined media should be used. Such media can be accurately reproduced and a detailed chemical analysis of reaction mixtures will be facilitated. In any case full details about the medium composition must be reported.

(3) If a gas or a liquid is perfused into, or through, the reaction vessel this must be reported in detail.

(4) Full details concerning inoculation of the medium must be reported (size, age and storage condition of inoculum, medium used and, where relevant, the growth phase). For preparations of, for instance, blood cells, counts of contaminating cells should be made and reported.

For growth experiments (microorganisms, cultured tissue cells) information about changes in the cell count is often essential. In any case initial and final cell concentrations (dry mass, etc.) should normally be reported, together with results of viability tests.

Details of the enumeration method followed, together with an estimate of the accuracy of the count, should be given. Relevant details of the extent of damaged cells should be given. (Note: intact enzyme substrate systems may contribute to the observed thermal power.)

(5) For the evaluation of results of calorimetric experiments, and to ensure their lasting value, heat measurements should ideally be accompanied by detailed analysis (pH, concentrations of O_2, CO_2, energy source(s), metabolites, etc.) performed continuously, or at intervals, on the reaction medium.

For almost all calorimetric experiments on cellular material it is essential that initial and final pH values be reported and that information be given concerning the concentration of oxygen. The problem with concentration gradients in experiments performed under static conditions should be noted, cf. effects of cell sedimentation.

Sometimes the necessary analytical information (including cell counts) must be obtained from parallel non-calorimetric experiments. It is important to perform such experiments under conditions which are as close as possible to those in the calorimetric experiment. They should be performed simultaneously, there should be no scale effect, vessels should be of the same material, medium and gassing rates, etc. should be identical. Even if great care is taken in performing such parallel experiments the cellular processes may not proceed exactly as expected. This must be taken into account when uncertainty assignments are made.

(6) Experiments with cells are often carried out in suspensions. Depending on the type of organism, and on the calorimetric method, the cells may stay in a uniform suspension or they may partly or completely sediment before, or during, the calorimetric experiment. Cells originally in suspension may adhere to the walls of the calorimetric vessels. In flow calorimetric experiments, therefore, it is sometimes difficult to assess accurately the number of cells contained in the reaction vessel (during the experiment or at the end of the experiment) as they may have been partially trapped in the flow lines or they may have accumulated in the reaction vessel. Effects of this nature can be of crucial importance for the evaluation of the calorimetric results. It is, therefore, most important to consider these effects in the design of the calorimetric experiments and in the reporting of the results.

(7) When using differential scanning calorimeters for the study of intact cellular systems attention must be paid to the scanning rates employed. The scanning rates used should attempt to ensure equilibrium conditions throughout the temperature range of interest. For comparative purposes it is advisable to extrapolate the recorded data to zero heating velocities. Thus, experiments should be conducted using a minimum of three different heating rates.

The use of the term power-time-curve is recommended in place of the commonly used term 'thermogram' to avoid confusion with, for example, its use in differential thermal analysis.

It is often useful to derive values for the thermal power produced per cell, for a defined mass (dry weight) of cells or for a defined volume (wet, packed) of cells. Example: For freshly prepared human erythrocytes in plasma an approximate power per cell at 37 $^{\circ}C$ and pH 7.40 is $P = 8$ fW. For wet packed cells this value corresponds to about 80 $mW \cdot dm^{-3}$. Alternatively it may be of interest to report heat quantities associated with the consumption of specified substrates. When such values are reported they must be accompanied by relevant information about the experimental conditions since the power or heat quantities evolved can be very sensitive to variations in experimental conditions such as pH, medium composition (including concentrations of oxygen and carbon dioxide), temperature, storage conditions for the cells (time, temperature), cell concentration, etc.

The power-time-curve reported should take into account the dynamic characteristics of both the calorimeter and the process under study, see e.g. (5). Two situations can arise

(1) where the dynamic parameters of the calorimeter do not affect the experimental power-time-curve (e.g. where steady state processes are observed or where the time constant for the process studied is long compared to that of the instrument),

(2) where the experimental power-time-curve requires correction to take account of these instrumental parameters.

The corrections under (2) should be expressed in sufficient detail to permit regeneration, if required, or the experimental power-time-curve.

Where instrumental data handling systems take into account these dynamic characteristics prior to presentation of results, the appropriate references or descriptions of the method of correction employed and the values of the defined parameters should be reported.

Interpretation of observed heat quantities associated with growth processes

If possible, the enthalpy change associated with growth phenomena should be expressed in thermodynamic terms (6). It is recommended that the enthalpy change per unit of substrate metabolized be reported, ΔH_{met}.* The recommended unit for ΔH_{met} is $J \cdot mol^{-1}$. When the catabolic process is well known it is possible to calculate the enthalpy change corresponding to catabolism, ΔH_{cat}. The enthalpy change corresponding to anabolism is similarly expressed as ΔH_{an}. Thus, in well defined systems when an energy substrate is metabolized, the enthalpy balance can be written:

$$\Delta H_{met} = (1-\alpha)\Delta H_{cat} + \alpha \Delta H_{an}$$

α represents the fraction of the carbon source which is incorporated into the cellular material and can be calculated from the molecular growth yield.

When the nitrogen source is simple (e.g. NH_3, NO_3^-, N_2) it is possible to calculate the enthalpy change corresponding to anabolism from values for the elemental composition of the cellular material, enthalpies of combustion of the dried cells and the molecular growth yields.† For this calculation it is simpler to express the results as the specific enthalpy change,‡ i.e., the enthalpy change per unit mass of dried cells, Δh_{an}.

$$\Delta h_{an} = \frac{\alpha \Delta H_{an}}{Y_s}$$

where Y_s is the molecular growth yield. Recommended units are for Δh_{an} $J \cdot g^{-1}$ and for Y_s $g \cdot mol^{-1}$. Equation (1) then becomes

$$\Delta H_{met} = (1-\alpha)\Delta H_{cat} + Y_s \Delta h_{an}.$$

---

* Reference (7) proposes that such symbols be written $\Delta_{met}H$ rather than ΔH_{met}. At present we leave the choice open.

† There is in the literature some confusion over the definition of the physical state of microorganism subjected to combustion calorimetry. Attention has to be directed in the future to obtaining agreement on the conditions under which such experiments should be conducted.

‡ When an extensive quantity is represented by a capital letter, the corresponding specific quantity may be denoted by the corresponding lower case letter (1).

References

1. International Union of Pure and Applied Chemistry: McGlashan, M.L. "Manual of Symbols and Terminology for Physicochemical Quantities and Units." Second revision prepared for publication by Whiffen, D.H. *Pure and Appl. Chem.* **51**, 1-41 (1979).

2. Randzio, S. and Suurkuusk, J. in *Biological Calorimetry* (A.E. Beezer, ed.), 311-341, Academic Press, London (1980).

3. Interunion Commission on Biothermodynamics (IUPAC, IUB, IUPAB): Recommendations for Measurement and Presentation of Biochemical Equilibrium Data, see e.g., *J. Biol. Chem.* **251**, 6879- (1976).

4. IUPAC Commission I.2 on Chemical Thermodynamics, Merseburg, August 1980.

5. Belaich, A. and Belaich, J.P. *J. Bact.* **125**, 19-24 (1976).

6. Belaich, J.P. in *Biological Calorimetry* (A.E. Beezer, ed.), 1-42, Academic Press, London (1980).

7. International Union of Pure and Applied Chemistry. Phys. Chem. Div., Commission on Thermodynamics: Cox, J.D. "Manual of Symbols and Terminology for Physicochemical Quantities and Units. Appendix IV: Notation for States and Processes, Significance of the Word *Standard* in Chemical Thermodynamics, and Remarks on Functions Used in Thermodynamic Tables." Provisional document. *Pure and Appl. Chem.* **51**, 393-403 (1979).

Section 4

Chemical Kinetics and Transport Properties

Chemical kinetics and transport properties were combined because they are time-dependent properties and because sometimes the transport properties influence chemical kinetics. Transport properties are concerned with mass flow (diffusion), heat flow (thermal conductivity), and resistance to flow (viscosity).

Tables 4.1 and 4.2 contain recommendations from the Green Book for chemical kinetics and transport properties, respectively. An additional paper reference 15 (not reprinted here) provides information on quantities, units, and symbols for chemical thermodynamics. This paper basically covers the same source material as Table 4.1, but in much more detail. Because Table 4.1 is rather condensed, we recommend that readers who have any uncertainties about the content of Table 4.1 also study reference 15. Reference 15 also briefly covers photochemical and radiation chemical reactions.

The presentation of chemical kinetics data in the primary literature is discussed in reference 16 (reprinted here).

Heterogeneous catalysis is an important area of catalysis that Table 4.1 does not cover. This is understandable because heterogeneous catalysis is inextricably related to surface chemistry. Although colloid and surface chemistry are covered in Table 6.1, only surface adsorption aspects of heterogeneous catalysis are covered. "Manual of Symbols and Terminology for Physiochemical Quantities and Units: Appendix II. Definitions, Terminology, and Symbols in Colloid and Chemistry, Part II: Heterogeneous Catalysis *(17)*, reprinted in this section, provides some recommendations relating to the publication of data. Enzyme catalysis, a specialized category of catalysis, is not covered in this book, but is covered in reference 18.

"Recommendations for Data Compilations and for the Reporting of Measurements of the Thermal Conductivity of Gases *(19)* is reprinted here. Transport properties in electrolytic solutions are covered in Section 5. (*See* references 26 and 27 and Tables 5.1 and 5.2.)

Table 4.1. Chemical kinetics. Continued on next page.

The recommendations given here are based on previous IUPAC recommendations [1.c,k and 25], which are not in complete agreement. For recommendations on reporting of chemical kinetic data see also [50].

Name	*Symbol*	*Definition*	*SI unit*	*Notes*
rate of change of quantity X	$\dot{X}$	$\dot{X} = \mathrm{d}X/\mathrm{d}t$	(varies)	1
rate of conversion	$\dot{\xi}$	$\dot{\xi} = \mathrm{d}\xi/\mathrm{d}t$	mol s^{-1}	2
rate of concentration change (due to chemical reaction)	r_B, v_B	$r_B = \mathrm{d}c_B/\mathrm{d}t$	mol m^{-3} s^{-1}	3,4
rate of reaction (based on amount concentration)	v	$v = \dot{\xi}/V$ $= \nu_B^{-1}\mathrm{d}c_B/\mathrm{d}t$	mol m^{-3} s^{-1}	2,4

(1) e.g. rate of change of pressure $\dot{p} = \mathrm{d}p/\mathrm{d}t$, for which the SI unit is Pa s^{-1}.
(2) The reaction must be specified for which this quantity applies.
(3) The symbol and the definition apply to entities B.
(4) Note that r_B, v_B and v can also be defined on the basis of partial pressure, number concentration, surface concentration, etc., with analogous definitions. If necessary differently defined rates of reaction can be distinguished by a subscript, e.g. $v_p = \nu_B^{-1}\,\mathrm{d}p_B/\mathrm{d}t$, etc. Note that the rate of reaction can only be defined for a reaction of known and time-independent stoichiometry, in terms of a specified reaction equation; also the second equation for the rate of reaction follows from the first only if the volume V is constant. The derivatives must be those due to the chemical reaction considered; in open systems, such as flow systems, effects due to input and output processes must also be taken into account.

Table 4.1. Continued.

Name	*Symbol*	*Definition*	*SI unit*	*Notes*
partial order of reaction	n_B	$v = k\Pi c_B{}^{n_B}$	1	5
overall order of reaction	n	$n = \Sigma n_B$	1	
rate constant, rate coefficient	k	$v = k\Pi c_B{}^{n_B}$	$(\text{mol}^{-1}\,\text{m}^3)^{n-1}\,\text{s}^{-1}$	
Boltzmann constant	k, k_B		J K^{-1}	
half life	$t_{\frac{1}{2}}$	$c(t_{\frac{1}{2}}) = c_0/2$	s	
relaxation time	τ	$\tau = 1/(k_1 + k_{-1})$	s	6
energy of activation, activation energy	E_a, E	$E_a = RT^2\,\text{d}\ln k/\text{d}T$	J mol^{-1}	
pre-exponential factor	A	$k = A\exp(-E_a/RT)$	$(\text{mol}^{-1}\,\text{m}^3)^{n-1}\,\text{s}^{-1}$	
volume of activation	$\Delta^{\ddagger}V$	$\Delta^{\ddagger}V = -RT \times (\partial \ln k/\partial p)_T$	$\text{m}^3\,\text{mol}^{-1}$	
collision diameter	d	$d_{AB} = r_A + r_B$	m	
collision cross-section	σ	$\sigma_{AB} = \pi d_{AB}{}^2$	m^2	
collision frequency	Z_A		s^{-1}	7
collision number	Z_{AB}, Z_{AA}		$\text{m}^{-3}\,\text{s}^{-1}$	7
collision frequency factor	z_{AB}, z_{AA}	$z_{AB} = Z_{AB}/Lc_Ac_B$	$\text{m}^3\,\text{mol}^{-1}\,\text{s}^{-1}$	7
standard enthalpy of activation	$\Delta^{\ddagger}H^{\circ}, \Delta H^{\ddagger}$		J mol^{-1}	8
standard entropy of activation	$\Delta^{\ddagger}S^{\circ}, \Delta S^{\ddagger}$		$\text{J mol}^{-1}\,\text{K}^{-1}$	8
standard Gibbs energy of activation	$\Delta^{\ddagger}G^{\circ}, \Delta G^{\ddagger}$		J mol^{-1}	8
quantum yield, photochemical yield	ϕ		1	9

(5) The symbol applies to reactant B.

(6) The definition applies to reactions of first order in forward and backward directions with rate constants k_1 and k_{-1} respectively.

(7) Z_A is the number of collisions per time experienced by a single particle with particles of type A. Z_{AA} or Z_{AB} is the total number of AA or AB collisions per time and volume in a system containing only A molecules, or containing two types of molecules A and B. Three-body collisions can be treated in a similar way.

(8) The quantities $\Delta^{\ddagger}H^{\circ}$, $\Delta^{\ddagger}S^{\circ}$ and $\Delta^{\ddagger}G^{\circ}$ are used in the transition state theory of chemical reaction. They are normally used only in connection with elementary reactions. The relation between the rate constant k and these quantities is

$$k = \kappa(k_B T/h)\exp(-\Delta^{\ddagger}G^{\circ}/RT),$$

where k has the dimensions of a first-order rate constant and is obtained by multiplication of an nth-order rate constant by $(c^{\circ})^{n-1}$. κ is a transmission coefficient, and $\Delta^{\ddagger}G^{\circ} = \Delta^{\ddagger}H^{\circ} - T\Delta^{\ddagger}S^{\circ}$. Unfortunately the standard symbol $^{\circ}$ is usually omitted, and these quantities are usually written $\Delta H^{\ddagger}$, $\Delta S^{\ddagger}$ and $\Delta G^{\ddagger}$.

(9) The quantum yield ϕ is defined by the equation

$$\phi = \frac{\text{number of product molecules formed}}{\text{number of quanta absorbed}}$$

Table 4.2. Transport Properties. Continued on next page.

The names and symbols recommended here are in agreement with those recommended by IUPAP [3] and ISO [4.m]. Further information on transport phenomena in electrochemical systems can also be found in [27].

Name	*Symbol*	*Definition*	*SI units*	*Notes*
flux (of a quantity X)	J_X, J	$J_X = A^{-1}\, dX/dt$	(varies)	
volume flow rate	q_V, $\dot{V}$	$q_v = dV/dt$	$m^3\, s^{-1}$	
mass flow rate	q_m, $\dot{m}$	$q_m = dm/dt$	$kg\, s^{-1}$	
mass transfer coefficient	k_d		$m\, s^{-1}$	
heat flow rate	ϕ	$\phi = dq/dt$	W	
heat flux	J_q	$J_q = \phi/A$	$W\, m^{-2}$	
thermal conductance	G	$G = \phi/\Delta T$	$W\, K^{-1}$	
thermal resistance	R	$R = 1/G$	$K\, W^{-1}$	
thermal conductivity	λ, k	$\lambda = J_q/(dT/dl)$	$W\, m^{-1}\, K^{-1}$	
coefficient of heat transfer	h, (k, K, α)	$h = J_q/\Delta T$	$W\, m^{-2}\, K^{-1}$	
thermal diffusivity	a	$a = \lambda/\rho c_p$	$m^2\, s^{-1}$	
diffusion coefficient	D	$D = J_n/(dc/dl)$	$m^2\, s^{-1}$	

The following symbols are used in the definitions of the dimensionless quantities: mass (m), time (t), volume (V), area (A), density (ρ), speed (v), length (l), viscosity (η), pressure (p), acceleration of free fall (g), cubic expansion coefficient (α), temperature (T), surface tension (γ), speed of sound (c), mean free path (λ), frequency (f), thermal diffusivity (a), coefficient of heat transfer (h), thermal conductivity (k), specific heat capacity at constant pressure (c_p), diffusion coefficient (D), mole fraction (x), mass transfer coefficient (k_d), permeability (μ), electric conductivity (κ), and magnetic flux density (B).

Name	*Symbol*	*Definition*	*SI unit*	*Notes*
Reynolds number	Re	$Re = \rho vl/\eta$	1	
Euler number	Eu	$Eu = \Delta p/\rho v^2$	1	
Froude number	Fr	$Fr = v/(lg)^{\frac{1}{2}}$	1	
Grashof number	Gr	$Gr = l^3 g\alpha\Delta T\rho^2/\eta^2$	1	
Weber number	We	$We = \rho v^2 l/\gamma$	1	
Mach number	Ma	$Ma = v/c$	1	
Knudsen number	Kn	$Kn = \lambda/l$	1	
Strouhal number	Sr	$Sr = lf/v$	1	
Fourier number	Fo	$Fo = at/l^2$	1	
Péclet number	Pe	$Pe = vl/a$	1	
Rayleigh number	Ra	$Ra = l^3 g\alpha\Delta T\rho/\eta a$	1	
Nusselt number	Nu	$Nu = hl/k$	1	
Stanton number	St	$St = h/\rho vc_p$	1	
Fourier number for mass transfer	Fo^*	$Fo^* = Dt/l^2$	1	1
Péclet number for mass transfer	Pe^*	$Pe^* = vl/D$	1	1
Grashof number for mass transfer	Gr^*	$Gr^* = l^3 g\left(\frac{\partial\rho}{\partial x}\right)_{T,p}\left(\frac{\Delta x\rho}{\eta}\right)$	1	1

(1) This quantity applies to the transport of matter in binary mixtures.

Table 4.2. Continued.

Name	*Symbol*	*Definition*	*SI unit*	*Notes*
Nusselt number for mass transfer	Nu^*	$Nu^* = k_d l/D$	1	1, 2
Stanton number for mass transfer	St^*	$St^* = k_d/v$	1	1
Prandtl number	Pr	$Pr = \eta/\rho a$	1	
Schmidt number	Sc	$Sc = \eta/\rho D$	1	
Lewis number	Le	$Le = a/D$	1	
magnetic Reynolds number	Rm, Re_m	$Rm = v\mu\kappa l$	1	
Alfvén number	Al	$Al = v(\rho\mu)^{\frac{1}{2}}/B$	1	
Hartmann number	Ha	$Ha = Bl(\kappa/\eta)^{\frac{1}{2}}$	1	
Cowling number	Co	$Co = B^2/\mu\rho v^2$	1	

(2) The name Sherwood number and symbol *Sh* has been widely used for this quantity.

Table 4.2 is reprinted with permission from *Quantities, Units and Symbols in Physical Chemistry,* Section 2.15, pages 58–59.

THE PRESENTATION OF CHEMICAL KINETICS DATA IN THE PRIMARY LITERATURE

REPORT OF THE CODATA TASK GROUP ON DATA FOR CHEMICAL KINETICS

A set of recommendations entitled "Reporting Experimental Data for Chemical Kinetics" is here submitted to CODATA for consideration, approval and publication.

The purpose of these recommendations is to ensure that accurate rate measurements are preserved and not lost through inadequate reporting. To this end the recommendations tell what types of data and auxiliary information should be reported and also point out the significant items of documentation that should be included. The Task Group believes that adherence by authors to these relatively few elementary recommendations when reporting experimental work can increase the durability of the results and improve the possibility of evaluating and comparing related studies.

These are recommendations for the discipline of chemical kinetics as a whole. They are part of a larger CODATA effort to provide guidelines for various branches of science. A more general document, "Guide for the Presentation in the Primary Literature of Numerical Data Derived from Experiments", has been prepared by the CODATA Task Group on Publication of Data in the Primary Literature and published as *CODATA bulletin* No. 9, December 1973. It too should be studied by kineticists interested in improving the quality of the published literature. We have made use of ideas in it and, at times, have included some of its statements verbatim in our recommendations.

General guidelines applicable to an entire discipline, however, are not sufficient; detailed recommendations are needed for various specialized types of rate measurements. The Task Group has made an interim attempt to do this for some sub-disciplines. These recommendations are given in Appendix I of the report. They should be replaced by more comprehensive recommendations prepared by panels of experts in each specialty.

The Task Group undertook this work at the request of CODATA. Our starting point was a good set of guidelines prepared by Academician V.N. Kondratiev. An expanded version was prepared and circulated to about 1400 kineticists. More than 200 replies were received, most of them containing substantive suggestions. The Task Group has reviewed these comments and has incorporated many of them in its present report. We are very grateful to our colleagues throughout the world for their careful and constructive review of our work.

It is our hope that, through continued cooperation, kineticists will be able to raise the accepted standard of reporting to a point where guidelines on the subject become unnecessary.

S.W. BENSON, Chairman
Stanford Research Institute
Menlo Park, California 94025, U.S.A.

D.L. BAULCH
School of Chemistry
The University of Leeds
Leeds LS2 9JT, U.K.

E.T. DENISOV
Institute of Chemical Physics
U.S.S.R. Academy of Sciences
Leninskiy Prospekt 14
Moscow B-21, U.S.S.R.

J.E. DUBOIS
Laboratory of Physical Organic Chemistry
University of Paris
1, rue Guy de la Brosse
75005 Paris, France

J. DURUP
Laboratory of Physical Chemistry
Orsay, France

D. GARVIN
Institute for Materials Research
National Bureau of Standards
Washington, D.C. 20234, U.S.A.

H. HARTMANN
Institute of Physical Chemistry
University of Frankfurt
6000 Frankfurt/Main 1,
Federal Republic of Germany

K.U. INGOLD
Division of Chemistry
National Research Council of Canada
Ottawa 7, Canada

J.A. KERR
Department of Chemistry
The University of Birmingham
Birmingham 15, U.K.

M. KOIZUMI
Tohoku University
Sendai, Japan

T.W. NEWTON
Los Alamos Scientific Laboratory
University of California
Los Alamos, New Mexico 87544, U.S.A.

G.A. SCHUIT
Technische Hogeschool
Eindhoven, Netherlands

R. TAMAMUSHI
Laboratory of Inorganic Chemistry
Institute of Physical and Chemical Research
Rikagaku Kenkyusho, 2-1 Kirosawa, Wako-Shi
Saitama, Japan

V.I. VEDENEEV
Institute of Chemical Physics
U.S.S.R. Academy of Sciences
Leninskiy Prospekt 14
Moscow B-71, U.S.S.R.

H.G. WAGNER
Institute of Physical Chemistry
University of Göttingen
34 Göttingen, Federal Republic of Germany

REPORTING EXPERIMENTAL DATA FOR CHEMICAL KINETICS

Because of the widespread and growing use of the numerical results of chemical kinetics research in other fields of science and technology, it is important that those results can be properly compared, evaluated and compiled for use by non-specialists. Because of the variety of techniques employed in chemical kinetics and the wide range of systems studied, the quality of this reevaluation depends critically on the standards of reporting in the primary literature. In these recommendations we give criteria for the presentation of experimentally determined numerical rate data.

Our criteria apply most specifically to kinetics papers in which numerical data are reported. Our aim is to ensure that accurate rate measurements are preserved and not lost through inadequate reporting. In general we confine ourselves to measurements of macroscopic rate constants; these guidelines are not concerned with the interpretation of rates in terms of mechanisms.

The recommendations are not necessaril applicable to all papers in chemical kinetics. There are valuable studies of a general or semiquantitative nature to which many of our suggestions may be irrelevant. We do not discourage the publication of these papers. We do want sufficient information included in them so that the work can be assessed. Nor are these recommendations a guide to literary style or the organization of a paper. Our entire concern is with the factual content.

1. GENERAL

We base our recommendations on the following principles:

(*i*) The work should be described in sufficient detail to allow the equipment, conditions, procedure and, logically, the results to be reproduced, and to allow the quality of the work to be appraised.

(*ii*) The results should be presented in a form that will permit them to be re-analyzed and reinterpreted by others.

(*iii*) There should be an assessment of the accuracy and reproducibility of the numerical results.

In expanding these points in the following paragraphs we cannot anticipate the consequences for all the many possible chemical systems. We confine ourselves to points of general importance.

2. EXPERIMENTAL

2.1 MATERIALS

The names of reactants, catalysts, ionic-strength adjusting chemicals, third body molecules (gas phase reactions), etc. should be given explicitly. Common names are often ambiguous and should be clarified; abbreviations should be specified, particularly for ligands in coordination compounds. The use of systematic nomenclature, chemical formulae and chemical structure diagrams is encouraged. Wherever possible, precise composition and structure and, where appropriate, stereo-descriptions should be given.

The origins, methods of purification and storage of reagents, solvents and other relevant materials should be stated. Criteria for purity, analytical results and known impurities should be given.

Where solid phases are used, other information may be necessary, such as the chemical composition, the structure (pellet, powder, film); the microstructure (where possible), and a description of the solid's pre-treatment and history.

2.2 APPARATUS

The report should include, either directly or by reference, descriptions of:

(*i*) Materials of construction, particularly those in contact with the reactants or products.

(*ii*) The reaction vessel including, where appropriate, dimensions, surface-to-volume ratio, materials of construction and pre-treatment. Where the reaction does not occur homogeneously throughout the vessel (e.g., in some high temperature processes or photolyses), the "effective reaction volume", spatial distribution of the reaction or both should be given.

(*iii*). All relevant ancillary equipment and measuring devices. Performance capabilities should be stated for detection devices (noise, baseline stabilities), thermostatting equipment (stability, temperature profile), temperature-measuring instruments, etc.

In the case of novel or complex apparatus it may be necessary to supplement the description with a realistic diagram of important features of the apparatus, giving dimensions and relevant geometry. Construction details should be made available on demand.

Modifications to standard or previously used apparatus should always be described.

2.3 EXPERIMENTAL METHODS

The account of the experimental method should include, either directly or by reference:

(*i*) Description of the preparation of the reaction mixture, any methods used for initiating the reaction and those used to follow the course of the reaction.

(*ii*) An indication of the sequence of operations, in mixing or generating reactants, in analyzing the change and in separating products.

(*iii*) An explanation of the way in which any method is used, with quantitative details, e.g., reactant concentrations generated, detection conditions (signal-to-noise ratio), or wavelength and extinction coefficient used in spectrophotometric measurements. For photochemistry and radiation chemistry the spectrum (energy distribution of the incident radiation), beam geometry and rate of energy input are of particular importance.

(*iv*) The applicability of novel procedures should be justified.

(*v*) It must be made clear what physical properties are measured and how they are related to the concentrations of the chemical species or to their rate of change.

3. PRESENTATION OF RESULTS

In this section we consider general points common to most kinetics studies. Problems peculiar to various sub-disciplines are considered in Appendix I.

By "results" we do not mean meter readings but those typical outputs of kinetics studies: the measured rate of reaction, the stoichiometric change in the system, the dependence of rate upon concentration, the rate constants and their variation with temperature, bond strengths, and so on.

3.1 GENERAL

(*i*) Whenever a rate is reported, care should be taken to state clearly to what process it refers. When possible the process should be stated as a chemical equation.

Ambiguities in the meaning of the rate may occur for reactions such as (1)

$$\nu_A A + \nu_B B + \cdots \rightarrow \nu_u U + \nu_v V + \cdots \qquad (1)$$

if stoichiometric numbers, ν_x, other than unity occur. A clear definition of the rate of reaction, R_1, is required. We recommend use of the convention

$$R_1 = -\frac{1}{\nu_A}\frac{d[A]}{dt} = -\frac{1}{\nu_B}\frac{d[B]}{dt} = \frac{1}{\nu_u}\frac{d[U]}{dt} = \cdots$$

$$= \text{some function of concentrations} \qquad (2)$$

(*ii*) The results should be reported in a form that does not depend on interpretations made in the paper, still showing the experimental scatter and with as little dependence as possible on theoretical assumptions. Some processing of instrument readings is usually necessary before a convenient reporting stage is reached. This is acceptable, but the relationship between the reported results and the initial data should not be lost.

(*iii*) The procedures and expressions used to derive quantities from direct experimental measurements, or from already processed results, should be explained. At times the procedure may be relatively trivial, but at least it should be outlined; in other instances a very complete explanation or a worked example may be necessary.

(*iv*) The use of graphs to show trends, to compare data and as a basis for conclusion and generalization is valid and often extremely valuable. Our experience suggests, however, that a graph is not appropriate for recording accurately determined rate data. Inaccuracies occur in transferring results to a graph and, unless the graph is unusually large, the results cannot be recovered with sufficient accuracy for evaluation and reprocessing. For these reasons it is vital that accurate numerical data be presented in tabular form.

When there is a large amount of data, an acceptable alternative may be an easily used analytical expression together with a graph showing the deviation of the individual points from the expression [e.g., see Blades, *Int. J. Chem. Kin. 3,* 187 (1971)]. However, in experiments where several conditions change over the range of experiments (e.g., initial concentration or degree of conversion), tabulation is much preferred.

As far as possible all tables and graphs should be self-explanatory.

(*v*) Care should be taken always to include a comprehensive statement, preferably in the tables, of the variables (pressure, temperatures, concentrations, etc.) relevant to a particular rate or rate constant measurement. All relevant variables should be given. For example, in flow experiments, the flow velocity and distance should be quoted in preference to (or as well as) the reaction time. The number of determinations on which a measured rate constant is based should be included.

(*vi*) The amount of a product and/or reactant being analyzed or the magnitude of the physical change being monitored should be given. Sufficient data should be quoted to allow the degree of conversion to be calculated. Possible perturbation of the reaction by the analytical procedures should be discussed.

(*vii*) Negative results and unsuccessful experiments are often important. Wherever relevant they should be reported.

3.2 RATE CONSTANTS AND RESULTS DERIVED FROM THEM

(*i*) All reported rate constants should be defined by giving an explicit mathematical statement of the pertinent rate equation, i.e., an expression equivalent to equation (2) above.

(*ii*) The formulation of the actual rate-determining step in a complex reaction is often difficult and subject to conflicting interpretation. Such formulations are not discouraged, but for the purposes of consistent data compilation, it is desirable that the rate laws and rate constants based on principal species be given in addition to other more speculative possibilities and that at some stage a rate constant defined in terms of the net, or observed, reaction be reported.

(*iii*) Rate constants are often derived by using numerical analysis to fit the observed change in a system to a hypothetical reaction scheme involving a number of variables, usually rate constants, some of which are known, some unknown. In such procedures:

(a) The reaction scheme should be justified, preferably on the basis of independent experiments or theory.

(b) An indication should be given of the sensitivity of the fit to the values assumed for the variables.

(c) The sensitivity of the fit to other possible models should be indicated.

These conditions should be met whether analytical or numerical analysis is being used, but they are particularly important for computer modeling.

Where steady-state expressions are used to analyze results, the validity of the steady-state approximation should be justified and any uncertainty in the procedure estimated.

(*iv*) Where auxiliary data are used in deriving rate constants and the like, the literature source should be given. If the auxiliary data are measured or calculated, the methods used should be described. In either case the accuracy of the data over the range of conditions in which they are applied should be quoted.

In the case of thermodynamic data it is important to define the standard state used.

(*v*) A variety of functions are used to express the rate constant, κ, as a function of temperature, T. We recommend the use of one of the following forms:

$$\kappa = A \exp(-B/T) \qquad (3)$$

$$\kappa = A'T^n \exp(-B'/T) \qquad (4)$$

$$\kappa = A''T^n \qquad (5)$$

where the A's, B's and n's are constants. Commonly, n in equation (4) is preset on the basis of theory.

Unless there are overriding reasons, equations of these types should be used to determine only two independent constants. When energies, enthalpies and entropies of activation are derived, it must be made clear which of these equations has been used.

3.3 ERRORS: ACCURACY AND PRECISION

(*i*) At some stage in the reporting of experimentally determined numerical data there should be a statement of the precision and accuracy.

The statistical or random uncertainty (precision) should be estimated using an appropriate standard statistical technique. However, it is only one component of the total error analysis and is not a sufficient statement of the reliability of the experiments.

Estimation of the other potential sources of error or limitations of the work is more difficult. There are no clear rules; subjective judgment is involved. However, these estimates of accuracy are more important than estimates of precision because the difference between values reported by two laboratories is often larger than the random errors.

(*ii*) Wherever possible the various sources of error, such as measurement precision, numerical analysis limitations, assumptions made in processing the data or uncertainties in auxiliary data, should be identified. The methods for estimating these errors should be indicated.

(*iii*) Precision. The quoted precision should be clearly defined, e.g., is the standard deviation of the value being given or a multiple thereof? It is important to specify the number of data being processed; techniques appropriate to the sample size should be applied.

When least-squares methods are used in processing the data, it is important to specify what quantity is minimized and/or what weighting scheme is used. For example, is the sum of squares of absolute, or relative, errors minimized?

It is important to take account of the fact that in experiments carried out over a range of a particular variable, say temperature, results obtained at the extremes of the range are usually much less accurate than those near the center. In these cases a weighted least-squares procedure should be used.

(*iv*) When data are presented on graphs, realistic error bars should be shown.

3.4 UNITS

It is extremely important that all physical quantities reported should have their units clearly specified. Table headings and axes of graphs should be labelled in dimensionless form, e.g., $k/m^3\ mol^{-1}\ s^{-1}$, and the units of quantities used in analytical expressions, e.g., Arrhenius equation, should be stated.

Units used in reporting the final results should be compatible with the International System of Units (SI).

The unit of time should be the second. For concentration, the SI unit is $mol\ m^{-3}$, but permissible alternatives are $mol\ dm^{-3}$, $mol\ cm^{-3}$, which are all commonly used. Rate constants may be given in any appropriate combination of these units.

The SI unit of energy is the joule.

The use of non-SI units, and especially the use of nonmetric units, for reporting rate data is discouraged.

Occasionally there may be overriding reasons for using such units. In these cases the author should give factors for conversion to SI units. This may be done in a footnote: "Throughout this paper 1 torr = (101.325/760) kPa, and 1 kcal = 4.184 kJ".

3.5 REFERENCES

Pertinent references must be given for all methods, results and auxiliary quantities quoted. In every case the reference cited should be *the actual reference consulted by the author*.

Readily accessible references are preferred; authors quoting less accessible references should indicate how they may be obtained.

References to long articles and books should cite specific pages wherever possible.

4. ACCESSIBILITY OF DATA

Our recommendations request authors to be more explicit and more careful in their reporting. We do not believe that their implementation will require a substantial increase in the length of papers, except possibly for 3.1 (*iv*), which recommends the use of tabulation of results. In most cases we expect that it will be sufficient to tabulate rate constants and the associated experimental conditions [section 3.1 (*v*)]. Only in exceptional cases will more be required.

The ideal situation is to have all the relevant information in the published article. However, if this is not practical, the supplementary material should be put in an auxiliary publication (submitted together with a shorter manuscript) and either placed in a suitable depository service or published as microform together with the article. In any event, the details must be available to the public from some source other than the author. The means of obtaining such auxiliary information must be clearly stated in the publication.

5. REFEREE'S CHECK LIST

One of the problems encountered by referees is that of keeping in mind all the criteria outlined here. To assist in this we recommend that referees be supplied with a check list on which our criteria are suitably itemized. A suggested form is given in Appendix II. It is, of course, the prerogative of the referee to decide whether the items on the list are appropriate or relevant to the paper under consideration.

APPENDIX I

INTERIM RECOMMENDATIONS APPLICABLE TO SPECIFIC FIELDS OF CHEMICAL KINETICS

In this appendix we supplement the general recommendations with discussions of problems frequently encountered in interpreting reports of research in the areas of homogeneous gas phase, liquid phase, solid phase and heterogeneous reaction kinetics.

These are interim recommendations. They are not comprehensive. We believe that all these areas require further scrutiny by experts before more specific guidelines are prepared.

1. HOMOGENEOUS GAS PHASE REACTIONS

(*i*) It is usual in defining a rate constant to assume constant volume conditions. In systems where the volume is not constant, appropriate corrections should be made. This is particularly important for shock tube and flow systems if the reaction is appreciably exothermic or endothermic.

(*ii*) Pressure is often an important variable in gas phase systems, particularly when addition and dissociation reactions are involved. Thus dissociation reactions, first order at high pressures, become second order at low pressures, and analogous changes from second to third order occur for addition reactions. Formulations of the type (1) and (2) are often used to denote the reaction (in this example, addition) in these two kinetic regions.

$$A + B \rightarrow AB \quad (1)$$

$$A + B + M \rightarrow AB + M \quad (2)$$

Although this practice is permissible, it is essential that reports on such reactions include the pressure at which the measurements were made and, if known, the order of the reaction under those conditions. Also, the composition of the system, often just denoted by M, should be given explicitly. In studies at very high pressures where the gas behavior is far from "ideal", it is preferable to express results in terms of concentrations rather than pressures.

(*iii*) Any effects due to reaction vessel surface should be reported directly or by reference. If no tests were made, this should be stated.

(*iv*) Modern kinetics techniques often involve reactants and products in states far removed from thermal equilibrium. Beam studies (neutral and ionic), flash photolysis, shock tube experiments and any strongly exothermic or endothermic reaction may produce nonequilibrium conditions. In such cases every effort must be made to specify completely the energy states or energy distribution of the species involved. This may require extensive tabulation of measured cross-sections, conditions of measurement (e.g., strength of electric fields, pressure) and internal and kinetic energy of reactants and products.

2. LIQUID PHASE REACTIONS

(*i*) The properties of the reaction medium often play a more important role in reactions in liquids than in the gas phase. It is important that the composition of the medium be specified in detail. This may include a statement of the solvent used, its origin and history (section 2.1), the composition of solvent mixture used (vague statements such as "60% acetone-water" are not acceptable without clarification), ionic strength, reactant concentrations and concentrations of other species present that may affect the reaction, e.g., hydrogen ion. Also, wherever possible, impurity concentrations, including the presence of dissolved air, should be given and their possible role considered.

(*ii*) It should be clearly stated whether the reported rate constants are based on concentrations or activities.

(*iii*) When rate constants are defined in terms of minor, but probably kinetically active species, the values used for the necessary ionization or complexing constants should be quoted. If required, information on the temperature coefficients of these constants should also be included. As for all such auxiliary data, literature source or measurement techniques should be given.

(*iv*) In liquid phase studies, results are often reported in terms of the transition state theory parameters $\Delta H^{\ddagger}$ and $\Delta S^{\ddagger}$. When this is done:

(a) It is extremely important to state clearly the process and the standard state to which these "thermodynamic" quantities refer. Ambiguity may arise when overall rate parameters are interpreted in terms of an assumed mechanism. It should be made quite clear whether the $\Delta H^{\ddagger}$ and $\Delta S^{\ddagger}$ refer to the overall process, to an assumed rate determining step, or to some combination of rate constants and possibly equilibrium constants for reactions involved in the mechanism.

(b) It is helpful for compilation and evaluation and for users of the data to have the Arrhenius parameters given as well as the transition state quantities.

3. SOLID PHASE REACTIONS

(*i*) Because rate processes in the bulk of a solid may be very sensitive to the solid's structure and to the impurities present, these should be specified (section 2.1). The "reactivity" of the solid may be affected by such physical properties as specific surface area, pore size and pore size distribution, and crystallite size and distribution. Defects, impurities and potentially active centers are similarly important. All these properties should be characterized wherever possible.

(*ii*) It is necessary to specify the stage of the reaction to which any rate parameters pertain since changes in the solid structure may affect the reaction rate.

(*iii*) In the early stages of reaction, appearance and growth of reaction centers will occur. It is desirable to give rate parameters for these processes and the distribution of these centers through the solid as well as parameters for the net reaction.

(*iv*) Kinetics of solid state reactions are affected by heat evolution and absorption and hence by physical properties such as thermal conductivity. In such cases the effects of self-heating or cooling should be considered in deriving kinetic data.

(*v*) The extent of a solid state reaction is often expressed in terms of the fractional reaction: the meaning of this term should be clearly explained.

(*vi*) Any equation used to relate the variation of reaction rate to extent of reaction should be stated and the terms defined. The range of conditions over which it is applicable should also be stated.

(*vii*) The simple Arrhenius relation is often not applicable to reactions occurring in solids. Where Arrhenius parameters are quoted, the method used for calculating them should be explained.

4. REACTIONS AT THE SURFACES OF SOLIDS

(*i*) The general criteria for the description of the solid (section 2.1) should include the method of preparation. The pre-treatment of the solid should be described.

(*ii*) Some attempt should be made to characterize the surface. The surface area should be determined and, if possible, the concentration of active sites on the surface. With supported metal catalysts, the metallic surface area and thence the dispersion of the metal should also be determined. The methods used in these determinations should be specified.

(*iii*) Information as to the texture of a catalyst or adsorbent should be given, in particular, the particle size, the pore volume and the pore size distribution. Methods used in determining the last two quantities should be specified.

(*iv*) The kinetic equations used to calculate the rate constants should be specified. Rates of reaction and rate constants may be given per unit mass, per mole, or per unit area of the catalyst or adsorbent, but the units chosen should be clearly indicated. Where possible, it may be convenient to express the rates of catalytic reactions as turnover numbers, the number of molecules reacting per site per second. The method used for determining the number of sites should be described.

(*v*) Since concentration gradients in the pores of a catalyst or adsorbent often influence rates, the data needed to evaluate the degree of severity of this effect should always be given. Diffusion limitations should be considered.

(*vi*) As for solid state reactions, the effects of self-heating or self-cooling should be considered.

(*vii*) In systems involving gas flow over a surface, both total flow and flow velocity may be important and should be specified.

APPENDIX II

REFEREE'S CHECK LIST

This check list attempts to cover only the most important, standard features to be considered in refereeing papers on chemical kinetics. In many instances further unlisted points will require comment.

Referees are requested to indicate whether they consider the particular point listed to be treated satisfactorily or whether a suggestion for improvement is appended.

	Satisfactory	Comment Appended
1. APPARATUS AND PROCEDURES		
Is the author's description of the following items satisfactory?		
(i) Reaction vessel		
(ii) Ancillary equipment and measuring device		
(iii) Procedure		
(iv) Calibration of apparatus		
2. MATERIALS		
Has the author provided sufficient information on:		
(i) Origins and/or preparation, and purity of reagents used?		
(ii) Chemical formula, structure, history, storage and pre-treatment of reagents and/or materials used?		
3. RESULTS		
(i) Are sufficient primary data presented to allow reworking of the results independent of any assumption or interpretations of the author?		
(ii) Are <u>all</u> of the relevant variables recorded for each set of rate measurements presented?		
(iii) Are the reported rates specified unambiguously in terms of the stoichiometric change in the reaction?		
(iv) Are all reported rate constants unambiguously defined in terms of rate equations?		
(v) Are the procedures and expressions used to derive numerical results from the data adequately explained and their use justified?		
(vi) Are error limits quoted and justified for measured, derived, and auxiliary quantities?		
4. GENERAL		
Is the referencing adequate? Is the nomenclature clear? Are all symbols defined?		

Definitions, Terminology, and Symbols in Colloid and Surface Chemistry

Part II: Heterogeneous Catalysis

Adopted by the IUPAC Council at Madrid, Spain, on 9 September 1975

Prepared for publication by
ROBERT L. BURWELL, JR.

PREFACE

This Part II of Appendix II† to the *Manual of Symbols and Terminology for Physicochemical Quantities and Units*‡ (hereinafter referred to as the *Manual*) has been prepared by the Commission on Colloid and Surface Chemistry of the Division of Physical Chemistry of the International Union of Pure and Applied Chemistry. It is the outcome of extensive discussions within the Commission§ and its Task Force headed by Professor Burwell, with other IUPAC Commissions, and with persons outside IUPAC during the period 1970–1975. Among the latter, special mention must be made to Professors M. Boudart (USA), J. B. Butt (USA), and F. S. Stone (UK). A tentative version of these proposals was issued as Appendix 39 (August 1974) on Tentative Nomenclature, Symbols, Units and Standards to IUPAC *Information Bulletin.* The text has been revised in the light of the criticisms, comments, and suggestions which were received, and the present version was prepared by the Commission and formally adopted by the IUPAC Council at its meeting in Madrid, Spain, in September 1975.

It was felt that the use of unambiguous terminology would promote communication and avoid misunderstandings among workers in heterogeneous catalysis and that a list of preferred symbols would be useful in many respects. Heterogeneous catalysis is primarily a branch of physical chemistry but it has substantial overlap with organic and inorganic chemistry and with chemical engineering. The Commission agreed that no term or symbol should be used in heterogeneous catalysis in a sense different from that in physical chemistry in general or, as far as possible, in a sense different from that in other branches of chemistry.

The present proposals are based on the same principles as those used in the *Manual*‡ and in Part I of this Appendix and are consistent with them. The most pertinent definitions of Part I are summarized and quoted in sections 1.2.1 and 1.2.2.

Historical and common usage of terms has been retained as far as is compatible with the above principles.

Since the present proposals should be considered as one of the sub-sets of the set of terms and symbols of physical chemistry, the general principles are not repeated here. Attention must be called, however to one point, namely the restriction of the term "specific" to the meaning, divided by mass. This necessitates either the repetitive use of "per unit area" or the introduction of a new term having this meaning. After careful consideration the Commission recommends that the term *areal,* meaning divided by area, be used. This is, however, at this time, a provisional recommendation subject to a decision on this and related terms by ICSU, the International Council of Scientific Unions.

La Jolla, California
29 December 1975

KAROL J. MYSELS
Chairman
Commission on Colloid and Surface Chemistry

†Part I of Appendix II, *Definitions, Terminology and Symbols in Colloid and Surface Chemistry*, prepared for publication by D. H. Everett, *Pure Appl. Chem.*, **31**, 579–638 (1972).

‡*Manual of Symbols and Terminology for Physicochemical Quantities and Units* (1973 Edn.), prepared for publication by M. L. McGlashan and M. A. Paul, Butterworths, London (1975).

§The membership of the Commission during this period was as follows:

Chairman: –1973 D. H. Everett (UK); 1973– K. J. Mysels (USA)
Secretary: H. van Olphen (USA)
Titular Members: S. Brunauer (USA); R. L. Burwell, Jr. (USA); R. Haul (Germany); V. B. Kazansky (USSR); 1971– C. Kemball (UK); –1973 K. J. Mysels (USA); –1971 M. Pretre (France); G. Schay (Hungary).
Associate Members: R. M. Barrer (UK); –1973 G. K. Boreskov (USSR); A. V. Kiselev (USSR); –1973 H. Lange (Germany); 1973– J. Lyklema (Netherlands); A. Scheludko (Bulgaria); G. A. Schuit (Netherlands); 1971– K. Tamaru (Japan).
Observer: –1971 Sir Eric Rideal (UK).
National Representatives: 1972– K. Morikawa (Japan); 1971–1974 Sir Eric Rideal (UK) (deceased); 1975 W. Schirmer (DDR).

CONTENTS

SECTION 1. DEFINITIONS AND TERMINOLOGY

1.1 *Catalysis and catalysts*

Catalysis is the phenomenon in which a relatively small amount of a foreign material, called a *catalyst*, augments the rate of a chemical reaction without itself being consumed. Cases occur with certain reactants in which the addition of a substance reduces the rate of a particular reaction, for example, the addition of an inhibitor in a chain reaction or a poison in a catalytic reaction. The term "negative catalysis" has been used for these phenomena but this usage is not recommended; terms such as inhibition or poisoning are preferred.

A catalyst provides for sets of *elementary processes* (often called *elementary steps*) which link reactants and products and which do not occur in the absence of the catalyst. For example, suppose the reaction

$$A = C$$

to proceed at some rate which might be measurable but might be essentially zero. The addition of X might now provide a new pathway involving the intermediate B,

$$A + X \longrightarrow B$$
$$B \longrightarrow C + X.$$

If reaction by this pathway proceeds at a rate significant with respect to the uncatalysed rate such that the total rate is increased, X is a catalyst. In this sense, a *catalytic reaction* is a closed sequence of elementary steps similar to the propagation steps of a gas-phase chain reaction.

The catalyst enters into reaction but is regenerated at the end of each reaction cycle. Thus, one unit of catalyst results in the conversion of many units of reactants (but see §1.7).

A catalyst, of course, may catalyse only one or some of several thermodynamically possible reactions.

It is difficult to separate Nature into water-tight compartments and probably no operational definition of catalysis can be entirely satisfactory. Thus, water might facilitate the reaction between two solids by dissolving them. This phenomenon might appear to constitute an example of catalysis but such solvent effects are not, in general, considered to fall within the scope of catalysis. The kinetic salt effect in solution is also usually excluded. Further, a catalyst must be material and, although an input of heat into a system usually augments the rate of a reaction, heat is not called a catalyst, nor is light a catalyst in leading to reaction between chlorine and hydrogen.

A catalyst should be distinguished from an *initiator*. An initiator starts a chain reaction, for example, di-t-butylperoxide in the polymerization of styrene, but the initiator is consumed in the reaction. It is not a catalyst.

In *homogeneous catalysis*, all reactants and the catalyst are molecularly dispersed in one phase.

In *heterogeneous catalysis*, the catalyst constitutes a separate phase. In the usual case, the catalyst is a crystalline or amorphous solid, the reactants and products being in one or more fluid phases. The catalytic reaction occurs at the surface of the solid and, ideally, its rate is proportional to the area of the catalyst. However, in practical cases, transport processes may restrict the rate (see §1.6).

Most examples of catalysis can be readily characterized as homogeneous or heterogeneous but there are examples of catalysis which overlap the two types. Consider a system in which intermediates are formed at the surface and then are desorbed into the gas phase and react there. Such intermediates might generate a chain reaction in the gas phase, i.e. chain initiation and chain termination occur at the surface but chain propagation occurs in the gas phase.

Enzyme catalysis may share some of the characteristics of homogeneous and heterogeneous catalysis, as when the catalyst is a macromolecule small enough to be molecularly dispersed in one phase with all reactants but large enough so that one may speak of active sites on its surface.

This manual deals with heterogeneous catalysis. Other types of catalysis will receive no further attention.

1.2 *Adsorption*

1.2.1 *General terms*

Although adsorption exists as a subject of scientific investigation independent of its role in heterogeneous catalysis, it requires particular attention here because of its central role in heterogeneous catalysis. Most or all catalytic reactions involve the adsorption of at least one of the reactants. Many terms related to adsorption have already been defined in Appendix II, Part I, §1.1. These include *surface, interface, area of surface or interface,* and *specific surface area.* Appendix II, Part I, recommends A or S and a or s as symbols for area and specific area, respectively. A_s and a_s may be used to avoid confusion with Helmholtz energy A or *entropy* S where necessary.

Other terms are *sorption, sorptive, sorbate* [a distinction being made between a species in its sorbed state (sorbate) and a substance in the fluid phase which is capable of being sorbed (sorptive)], *absorption, absorptive, absorbate, absorbent*; and *adsorption, adsorptive, adsorbate, adsorbent.*† The term *adsorption complex* is used to denote the entity constituted by the adsorbate and the part of the adsorbent to which it is bound.

Appendix II, Part I, § 1.1.5, treats the adsorbent/fluid‡ interface as follows.

†The use of *substrate* for adsorbent or support is to be discouraged because of its general use in enzyme chemistry to designate a reactant.

‡Appendix II, Part I, recommends: The use of a solidus to separate the names of bulk phases is preferred to the use of a hyphen which can lead to ambiguities.

"It is often useful to consider the adsorbent/fluid interface as comprising two regions. The region of the fluid phase (i.e. liquid or gas) forming part of the adsorbent/fluid interface may be called the *adsorption space*, while the portion of the adsorbent included in the interface is called the *surface layer of the adsorbent*."

When used to denote the process in which molecules† or dissociated molecules accumulate in the adsorption space or in the surface layer of the absorbent, adsorption has as its counterpart the term *desorption* which denotes the converse process (see Appendix II, Part I, §1.1.4). Adsorption is also used to denote the result of the process of adsorption, i.e. the presence of adsorbate on an adsorbent. The adsorbed state may or may not be in equilibrium with the adsorptive (see §1.2.2(c)).

Adsorption and desorption may also be used to indicate the direction from which equilibrium has been approached, e.g. adsorption curve (point), desorption curve (point).

1.2.2 *Chemisorption and physisorption*

For convenience, the relevant portions of §§1.1.6 and 1.1.7 of Appendix II, Part I, are reproduced here.

Chemisorption and physisorption

Chemisorption (or *Chemical Adsorption*) is adsorption in which the forces involved are valence forces of the same kind as those operating in the formation of chemical compounds. The problem of distinguishing between chemisorption and physisorption (see below) is basically the same as that of distinguishing between chemical and physical interaction in general. No absolutely sharp distinction can be made and intermediate cases exist, for example, adsorption involving strong hydrogen bonds or weak charge-transfer.

Some features which are useful in recognising chemisorption include:

(a) the phenomenon is characterised by chemical specificity;
(b) changes in the electronic state may be detectable by suitable physical means (e.g. u.v., infrared or microwave spectroscopy, electrical conductivity, magnetic susceptibility);
(c) the chemical nature of the adsorptive(s) may be altered by surface dissociation or reaction in such a way that on desorption the original species cannot be recovered; in this sense chemisorption may not be reversible;
(d) the energy of chemisorption is of the same order of magnitude as the energy change in a chemical reaction between a solid and a fluid: thus chemisorption, like chemical reactions in general, may be exothermic or endothermic and the magnitudes of the energy changes may range from very small to very large;
(e) the elementary step in chemisorption often involves an activation energy;
(f) where the activation energy for adsorption is large (*activated adsorption*), true equilibrium may be achieved slowly or in practice not at all. For example, in the adsorption of gases by solids the observed extent of adsorption, at a constant gas pressure after a fixed time, may in certain ranges of temperature increase with rise in temperature. In addition, where the activation energy for desorption is large, removal of the chemisorbed species from the surface may be possible only under extreme conditions of temperature or high vacuum, or by some suitable chemical treatment of the surface;
(g) since the adsorbed molecules are linked to the surface by valence bonds, they will usually occupy certain *adsorption sites* on the surface and only one layer of chemisorbed molecules is formed (monolayer adsorption).

Physisorption (or *Physical Adsorption*) is adsorption in which the forces involved are intermolecular forces (van der Waals forces) of the same kind as those responsible for the imperfection of real gases and the condensation of vapours, and which do not involve a significant change in the electronic orbital patterns of the species involved. The term *van der Waals adsorption* is synonymous with physical adsorption, but its use is not recommended.

Some features which are useful in recognising physisorption include:

(a′) the phenomenon is a general one and occurs in any solid/fluid system, although certain specific molecular interactions may occur, arising from particular geometrical or electronic properties of the adsorbent and/or adsorptive;
(b′) evidence for the perturbation of the electronic states of adsorbent and adsorbate is minimal;
(c′) the adsorbed species are chemically identical with those in the fluid phase, so that the chemical nature of the fluid is not altered by adsorption and subsequent desorption;
(d′) the energy of interaction between the molecules of adsorbate and the adsorbent is of the same order of magnitude as, but is usually greater than, the energy of condensation of the adsorptive;
(e′) the elementary step in physical adsorption does not involve an activation energy. Slow, temperature dependent, equilibration may however result from rate-determining transport processes;
(f′) in physical adsorption, equilibrium is established between the adsorbate and the fluid phase. In solid/gas systems at not too high pressures the extent of physical adsorption increases with increase in gas pressure and usually decreases with increasing temperature. In the case of systems showing hysteresis the equilibrium may be metastable.
(g′) under appropriate conditions of pressure and temperature, molecules from the gas phase can be adsorbed in excess of those in direct contact with the surface (multilayer adsorption or filling of micropores).

Monolayer and multilayer adsorption, micropore filling and capillary condensation

In *monolayer adsorption* all the adsorbed molecules are in contact with the surface layer of the adsorbent.

In *multilayer adsorption* the adsorption space accommodates more than one layer of molecules and not all adsorbed molecules are in contact with the surface layer of the adsorbent.

The *monolayer capacity* is defined, for chemisorption, as the amount of adsorbate which is needed to occupy all adsorption sites as determined by the structure of the adsorbent and by the chemical nature of the adsorptive; and, for physisorption, as the amount needed to cover the

†The term molecules is used in the general sense to denote any molecular species: atom, ion, neutral molecule or radical.

surface with a complete monolayer of molecules in close-packed array, the kind of close-packing having to be stated explicitly when necessary. Quantities relating to monolayer capacity may be denoted by subscript m.

The *surface coverage* (θ) for both monolayer and multilayer adsorption is defined as the ratio of the amount of adsorbed substance to the monolayer capacity.

The *area occupied by a molecule in a complete monolayer* is denoted by a_m; for example, for nitrogen molecules $a_m(N_2)$.

Micropore filling is the process in which molecules are adsorbed in the adsorption space within micropores.

The *micropore volume* is conventionally measured by the volume of the adsorbed material which completely fills the micropores, expressed in terms of bulk liquid at atmospheric pressure and at the temperature of measurement.

In certain cases (e.g. porous crystals) the micropore volume can be determined from structural data.

Capillary condensation is said to occur when, in porous solids, multilayer adsorption from a vapour proceeds to the point at which pore spaces are filled with liquid separated from the gas phase by menisci.

The concept of capillary condensation loses its sense when the dimensions of the pores are so small that the term meniscus ceases to have a physical significance. Capillary condensation is often accompanied by hysteresis."

1.2.3 *Types of chemisorption*

Non-dissociative, dissociative. If a molecules is adsorbed without fragmentation, the adsorption process is *non-dissociative.* Adsorption of carbon monoxide is frequently of this type. If a molecule is adsorbed with dissociation into two or more fragments both or all of which are bound to the surface of the adsorbent, the process is *dissociative.* Chemisorption of hydrogen is commonly of this type.

$$H_2(g) \longrightarrow 2H(ads) \quad \text{or}$$

$$H_2(g) + 2* \longrightarrow 2H*.$$

The asterisk represents a surface site.

Homolytic and *heterolytic* relate in the usual sense to the formal nature of the cleavage of a single bond. If the electron pair in the bond of the adsorptive A:B is divided in the course of its dissociative adsorption, the adsorption is *homolytic dissociative adsorption.* If A or B retains the electron pair, the adsorption is *heterolytic dissociative adsorption.* Examples follow.

(a) Homolytic dissociative adsorption of hydrogen on the surface of a metal:

$$H_2 + 2* \longrightarrow 2H*.$$

(b) Heterolytic dissociative adsorption of hydrogen at the surface of an oxide where the surface sites M^{n+} and O^{2-} are surface sites in which the ions are of lower coordination than the ions in the bulk phase:

$$H_2 + M^{n+} + O^{2-} \longrightarrow H^-M^{n+} + HO^-$$

Where clarity requires it, the equation may be written

$$H_2(g) + M_s^{n+} + O_s^{2-} \longrightarrow H^-M_s^{n+} + HO_s^-$$

where the subscript s indicates that the species indicated are part of the surface.

The notation H^-M^{n+} is used, as in conventional inorganic terminology, to indicate that the oxidation number of M has not changed.

(c) Heterolytic dissociative adsorption of water at the same pair of sites as in (b):

$$H_2O + M^{n+} + O^{2-} \longrightarrow HO^-M^{n+} + HO^-$$

Reductive and oxidative dissociative adsorption involve usage analogous to that in coordination chemistry in which one speaks of the following reaction as an oxidative addition

$$L_4M(I) + H_2 \longrightarrow L_4M(III)H_2.$$

Here, M represents a transition metal atom and L a ligand. H as a ligand is given an oxidation number of −1. If reductive, the electron pair which constitutes the bond in the sorptive, A:B, is transferred to surface species; if oxidative, a pair of electrons is removed from surface species. One would say that dissociative adsorption of Cl_2 on a metal is oxidative if chlorine forms Cl^- ions on the surface of the adsorbent. A dissociative adsorption would be reductive if, for example, it occurred thus (note that $H_2 \rightarrow 2H^+ + 2e$ here),

$$H_2(g) + 2[M(III)O^{2-}]_s \longrightarrow 2[M(II)(OH)^-]_s$$

Charge transfer adsorption represents oxidative or reductive chemisorption where reductive and oxidative refer to electron gain or loss on species in the solid. In simple cases it is non-dissociative, i.e. there is a mere transfer of charge between adsorptive and adsorbent in forming the adsorbate. Two examples follow.

$$\text{Reductive} \quad X + * \longrightarrow X^+*^-,$$

where X represents an aromatic molecule of low ionization potential such as anthracene or triphenylamine and * a site on silica-alumina.

$$\text{Oxidative} \quad O_2 + * \longrightarrow O_2^-*^+.$$

The term, charge transfer adsorption, has also been applied to adsorption which resembles the charge transfer complexes of Mulliken.

Immobile, mobile. These terms are used to describe the freedom of the molecules of adsorbate to move about the surface. Adsorption is immobile when kT is small compared to ΔE, the energy barrier separating adjacent sites. The adsorbate has little chance of migrating to neighbouring sites and such adsorption is necessarily *localized.* Mobility of the adsorbate will increase with temperature and mobile adsorption may be either localized or *non-localized.* In localized mobile adsorption, the adsorbate spends most of the time on the adsorption sites but can migrate or be desorbed and readsorbed elsewhere. In non-localized adsorption the mobility is so great that a small fraction of the adsorbed species are on the adsorption sites and a large fraction at other positions on the surface.

In some cases of localized adsorption the adsorbate is

ordered into a two-dimensional lattice or *net* in a particular range of surface coverage and temperature. If the net of the ordered adsorbed phase is in registry with the lattice of the adsorbent the structure is called *coherent*, if not it is called *incoherent* (see also §1.2.4).

Each of the various processes of adsorption may have desorptions of the reverse forms, for example, dissociative adsorption may have as its reverse, *associative desorption*. However, the process of chemisorption may not be reversible (§1.2.2(c)). Desorption may lead to species other than that adsorbed, for example, ethane dissociatively adsorbed on clean nickel gives little or no ethane upon desorption, 1-butene dissociatively adsorbed to methylallyl and H on zinc oxide gives mainly 2-butenes upon desorption, and some WO_3 may evaporate from tungsten covered with adsorbed oxygen.

Photoadsorption, photodesorption. Irradiation by light (usually visible or ultraviolet) may affect adsorption. In a system containing adsorptive and adsorbent exposure to light may lead to increased adsorption (*photoadsorption*) or it may lead to desorption of an adsorbate (*photodesorption*).

1.2.4 *Sites for chemisorption*

Sites may be classified according to their chemical nature in usual chemical terminology. The following terms are simple extensions of ordinary chemical usage: *basic sites, acidic sites, Lewis acid sites, proton* or Brønsted *acid sites, electron accepting sites* and *electron donating sites* (possible examples of the last two appear under *charge transfer adsorption*).

It is often useful to consider that sites for chemisorption result from *surface coordinative unsaturation*, i.e. that atoms at the surface have a lower coordination number than those in bulk. Thus, for example a chromium ion at the surface of chromium oxide has a coordination number less than that of a chromium ion in the bulk. The chromium ion will tend to bind a suitable adsorptive so as to restore its coordination number. An atom in the (100) surface of a face-centered cubic metal has a coordination number of 8 vs 12 for an atom in bulk; this, too, represents surface coordinative unsaturation. However, of course, there are sites to which the concept of surface coordinative unsaturation does not apply, for example, *Brønsted* acid sites.

One is rarely sure as to the exact identity and structure of sites in adsorption and heterogeneous catalysis. However, some symbolism is needed for theoretical discussion of possible sites. On the one hand one may wish to use a description which is general and non-specific. For this * and (ads) are recommended as, for example, H* and H(ads). Or, one may wish to use a symbolism which is as specific as possible. General chemical symbols may be useful in this case. A symbolism useful for metals involves the specification of C_j and B_n where C_j denotes a surface atom with j nearest neighbours and B_n denotes an ensemble of n surface atoms which together constitute an adsorption site, for example, the adsorption site lying above the centre of three surface atoms constituting the corners of an equilateral triangle is a B_3 site (for details see van Hardeveld and Hartog, *Surface Sci.* **15**, 189 (1969)).

Cases of chemisorption are known in which at high coverages the net (two-dimensional lattice) of the adsorbate is not in registry with the lattice of the adsorbent. In such situations, the concept of sites of precise location and fixed number may not be applicable. Similar difficulties about the definition of sites will occur if surface reconstruction takes place upon interaction of adsorbate and adsorbent.

Because of various difficulties which often appear in knowing the identity of surface sites, it is frequently convenient, particularly for metals, to define the surface coverage θ as the ratio of the number of adsorbed atoms or groups to the number of surface atoms (c.f. §1.2.2).

1.2.5 *Uniformity of sites*

Variations in the nature of the sites for adsorption or catalysis can occur even with pure metals where there is no question of differences in chemical composition between one part of the surface and another. These variations arise not only because of defects in the metal surfaces but also because the nature of a site depends on the structure of the surface. *Uniform sites* are more likely to be encountered when adsorption or catalysis is studied on an individual face of a single crystal, but even individual faces may present more than one kind of site. *Non-uniform sites* will normally occur with specimens of metal exposing more than one type of crystal face. There are two main kinds of non-uniformities. *Intrinsic non-uniformity* is a variation due solely to the nature of the adsorbent. *Induced non-uniformity* arises when the presence of an adsorbate molecule on one site leads to a variation in the strength of adsorption at a neighbouring site. Thus, a set of uniform sites on an individual crystal face may become non-uniform if the surface is partially covered with a chemisorbed species.

When the catalytic properties of metals are examined, the importance of the non-uniformity of sites depends on the reaction under study. For some reactions, the activity of the metal catalyst depends only on the total number of sites available and these are termed *structure-insensitive reactions*. For other reactions, classified as *structure-sensitive reactions*, activity may be much greater on sites associated with a particular crystal face or even with some type of defect structure. The alternative names of *facile* or *demanding* have been used to describe structure–insensitive or structure–sensitive reactions respectively.

The terms of §1.2.5 have been discussed with reference to metallic surfaces but they can be applied to other adsorbents and catalysts and, in particular, to the pair-sites involved in heterolytic dissociative adsorption.

1.2.6 *Active site, active centre*

The term *active sites* is often applied to those sites for adsorption which are the effective sites for a particular heterogeneous catalytic reaction. The terms active site and *active centre* are often used as synonyms, but active centre may also be used to describe an ensemble of sites at which a catalytic reaction takes place.

1.2.7 *Adsorption isotherms*

An *adsorption isotherm* for a single gaseous adsorptive on a solid is the function which relates at constant temperature the amount of substance adsorbed at equilibrium to the pressure (or concentration) of the adsorptive in the gas phase. The surface excess amount rather than the amount adsorbed is the quantity accessible to experimental measurement, but, at lower pressures, the difference between the two quantities becomes negligible (see Appendix II, Part I, §1.1.11).

Similarly, when two or more adsorptives adsorb competitively on a surface, the adsorption isotherm for adsorptive i at a given temperature is a function of the equilibrium partial pressures of all of the adsorptives. In the case of adsorption from a liquid solution, an adsorption isotherm for any preferentially adsorbed solute may be similarly defined in terms of the equilibrium concentration of the respective solution component, but the isotherm usually depends on the nature of the solvent and on the concentrations (mole fractions) of other solute components if present. Individual solute isotherms cannot be derived from surface excesses except on the basis of an appropriate model of the adsorption layer; when chemisorption occurs it is generally adequate to assume monolayer adsorption. Amounts adsorbed are often expressed in terms of coverages θ_i. In chemisorption, θ_i is the fraction of sites for adsorption covered by species i. Types of adsorption isotherms of interest to heterogeneous catalysis follow.

The linear adsorption isotherm. The simplest adsorption isotherm is the analogue of Henry's law. For a single adsorptive, it takes the form

$$\theta = Kp \quad \text{or} \quad \theta = Kc,$$

where p and c are the pressure and concentration of the adsorptive, θ is the coverage by adsorbate and K the *linear adsorption isotherm equilibrium constant*, or *Henry's law constant.* Most adsorption isotherms reduce to Henry's law when p or c becomes small enough provided that simple adsorption occurs, i.e. adsorption is neither dissociative nor associative. That is, at low enough coverages Henry's law usually applies to the first of the following equations but not the second and third.

$$A + * \rightleftharpoons A*;$$
$$A_2 + 2* \rightleftharpoons 2A*;$$
$$2A + * \rightleftharpoons A_2*.$$

The *Langmuir adsorption isotherm*,

$$\theta = \frac{Kp}{1+Kp} \quad \text{or} \quad \frac{\theta}{p(1-\theta)} = K,$$

or the equivalents in terms of concentrations, is commonly taken to result from simple (non-dissociative) adsorption from an ideal gas on a surface with a fixed number of uniform sites which can hold one and only one adsorbate species. K is called the *Langmuir adsorption equilibrium constant.* Further, the enthalpy of the adsorbed form must be independent of whether or not adjacent sites are occupied and consequently the enthalpy of adsorption is independent of θ. The second form of Langmuir's isotherm given above, emphasizes that the constant K is the equilibrium constant for $A + * \rightleftharpoons A*$. Since the constancy of enthalpy with coverage is analogous to the constancy of enthalpy with pressure in an ideal gas, the adsorbed state in a system following Langmuir's isotherm is sometimes called an *ideal adsorbed state.*

If chemisorption is dissociative,

$$A_2 + 2* \rightleftharpoons 2A*,$$

Langmuir's equation takes the form

$$\theta = \frac{K^{1/2}p^{1/2}}{1+K^{1/2}p^{1/2}} \quad \text{or} \quad K = \frac{\theta^2}{p(1-\theta)^2}.$$

For simple adsorption of two adsorptives A and B competing for the same sites, Langmuir's isotherm takes the form

$$\theta_A = \frac{K_A p_A}{1 + K_A p_A + K_B p_B},$$

where K_A and K_B are the equilibrium constants for the separate adsorption of A and B respectively. This equation can be generalized to cover adsorption of several adsorptives and to allow for dissociative adsorption of one or more adsorptives.

In the *Freundlich adsorption isotherm*, the amount adsorbed is proportional to a fractional power of the pressure of the adsorptive. For a particular system, the fractional power and the constant of proportionality are functions of temperature. In terms of coverage the isotherm assumes the form

$$\theta = ap^{1/n},$$

where n is a number greater than unity and a a constant. In the region of validity of the isotherm the (differential) enthalpy of adsorption is a linear function of $\ln \theta$.

In the *Temkin adsorption isotherm*, the amount adsorbed is related to the logarithm of the pressure of the adsorptive

$$\theta = A \ln p + B,$$

where A and B are constants. In the region of validity of the isotherm the (differential) enthalpy of adsorption is a linear function of θ.

The *Brunauer–Emmett–Teller* (or *BET*) *adsorption isotherm* applies only to the physisorption of vapours but it is important to heterogeneous catalysis because of its use for the determination of the surface areas of solids. The isotherm is given by the following equation,

$$\frac{n}{n_m} = \frac{c(p/p^0)}{(1-p/p^0)(1+(c-1)(p/p^0))} = \theta,$$

where c is a constant which depends upon the temperature, the adsorptive and the adsorbent, n is the amount adsorbed, n_m is the monolayer capacity and p^0 is the saturated vapour pressure of the pure, liquid adsorptive at the temperature in question. According to this equation, which is based on a model of multilayer adsorption, θ exceeds unity when p/p^0 is sufficiently large.

1.2.8 *Bifunctional catalysis*

Some heterogeneous catalytic reactions proceed by a sequence of elementary processes certain of which occur at one set of sites while others occur at sites which are of a completely different nature. For example, some of the processes in the reforming reactions of hydrocarbons on platinum/alumina occur at the surface of platinum, others at acidic sites on the alumina. Such catalytic reactions are said to represent *bifunctional catalysis*. The two types of sites are ordinarily intermixed on the same primary particles (§1.3.2) but similar reactions may result even when the catalyst is a mixture of particles each containing but one type of site. These ideas could, of course, be extended to create the concept of *polyfunctional catalysis*.

1.2.9 *Rates of adsorption and desorption*

Sticking coefficient is the ratio of the rate of adsorption to the rate at which the adsorptive strikes the total surface, i.e. covered and uncovered. It is usually a function of surface coverage, of temperature and of the details of the surface structure of the adsorbent.

Sticking probability is often used with the same meaning but in principle it is a microscopic quantity concerned with the individual collision process. Thus the sticking coefficient can be considered as a mean sticking probability averaged over all angles and energies of the impinging molecules and over the whole surface.

The *mean residence time* of adsorbed molecules is the mean time during which the molecules remain on the surface of the adsorbent, i.e. the mean time interval between impact and desorption. While residing on the surface the molecules may migrate between adsorption sites before desorption. If the residence time of an adsorbed species refers to specified adsorption sites it would be called the mean *life time* of the particular adsorption complex. When the rate of desorption is first order in coverage the residence time is independent of surface coverage and equal to the reciprocal of the rate constant of the desorption process. In this case it can be characterized unambiguously also by a half-life or by some other specified fractional-life of the desorption process. If the desorption process is not first order, e.g. due to mutual interactions of the adsorbed molecules and/or energetic heterogeneity of the surface, the residence time depends upon surface coverage and the operational definition of "residence time" needs to be specified precisely.

Unactivated and *activated adsorption*. If the temperature coefficient of the rate of adsorption is very small, the adsorption process is said to be *unactivated* (i.e. to have a negligible activation energy). In this case the sticking coefficient at low coverages may be near unity particularly for smaller molecules. If the temperature coefficient of the rate of adsorption is substantial, the adsorption process is said to be *activated* (i.e. to have a significant activation energy). In this case, the sticking coefficient is small. In general, the activation energy of activated adsorption is a function of coverage and it usually increases with increasing coverage.

A number of relations between rate of activated adsorption and coverage have been proposed. Of these, one has been particularly frequently used, the *Roginskii–Zeldovich equation* sometimes called the *Elovich equation*,

$$\frac{d\theta}{dt} = a\,e^{-b\theta},$$

where θ is the coverage, and a and b are constants characteristic of the system.

1.3 *Composition, structure and texture of catalysts*

1.3.1 *General terms*

Catalysts may be one-phase or multiphase. In the first case, they may be composed of one substance (for example, alumina or platinum black) or they may be a one phase solution of two or more substances. In this case, the components of the solution should be given and joined by a hyphen (for example, silica-alumina).

Support. In multiphase catalysts, the active catalytic material is often present as the minor component dispersed upon a *support* sometimes called a *carrier*. The support may be catalytically inert but it may contribute to the overall catalytic activity. Certain bifunctional catalysts (§1.2.8) constitute an extreme example of this. In naming such a catalyst, the active component should be listed first, the support second and the two words or phrases should be separated by a solidus, for example, platinum/silica or platinum/silica-alumina. The solidus is sometimes replaced by the word "on", for example, platinum on alumina.

Promoter. In some cases, a relatively small quantity of one or more substances, the *promoter* or *promoters*, when added to a catalyst improves the activity, the selectivity, or the useful lifetime of the catalyst. In general, a promoter may either augment a desired reaction or suppress an undesired one. There is no formal system of nomenclature for designating promoted catalysts. One may, however, for example, employ the phrase "iron promoted with alumina and potassium oxide".

A promoter which works by reducing the tendency for sintering and loss of area may be called a textural promoter (see §1.7.3).

Doping. In the case of semiconducting catalysts, a small amount of foreign material dissolved in the original catalyst may modify the rate of a particular reaction. This phenomenon is sometimes called *doping* by analogy with the effect of similar materials upon semiconductivity.

1.3.2 *Porosity and texture*

Many but not all catalysts are porous materials in which most of the surface area is internal. It is sometimes convenient to speak of the *structure* and *texture* of such materials. The *structure* is defined by the distribution in space of the atoms or ions in the material part of the catalyst and, in particular, by the distribution at the surface. The *texture* is defined by the detailed geometry of the void space in the particles of catalyst. *Porosity* is a concept related to texture and refers to the pore space in a material. With zeolites, however, much of the porosity is determined by the crystal structure.

An exact description of the texture of a porous catalyst would require the specification of a very large number of parameters. The following averaged properties are often used.

With respect to porous solids, the surface associated with pores may be called the *internal surface*. Because the accessibility of pores may depend on the size of the fluid molecules, the extent of the accessible internal surface may depend on the size of the molecules comprising the fluid, and may be different for the various components of a fluid mixture (*molecular sieve effect*).

When a porous solid consists of discrete particles, it is convenient to describe the outer boundary of the particles as *external surface*.

It is expedient to classify pores according to their sizes†

(i) pores with widths exceeding about 0.05 μm or 50 nm (500 Å) are called *macropores*;

†See Appendix II, Part I, §1.1.5.

(ii) pores with widths not exceeding about 2.0 nm (20 Å) are called *micropores*;
(iii) pores of intermediate size are called *mesopores*.

The terms *intermediate* or *transitional* pores, which have been used in the past, are not recommended.

In the case of micropores, the whole of their accessible volume may be regarded as adsorption space.

The above limits are to some extent arbitrary. In some circumstances it may prove convenient to choose somewhat different values.

Pore size distribution is the distribution of pore volume with respect to pore size; alternatively, it may be defined by the related distribution of pore area with respect to pore size. It is an important factor for the kinetic behaviour of a porous catalyst and thus an essential property for its characterization (see §1.6).

The computation of such a distribution involves arbitrary assumptions and a pore-size distribution should always be accompanied by an indication as to the method used in its determination. The methods usually involve either or both of the following (i) adsorption–desorption isotherms of nitrogen or other adsorptives in conjunction with a particular model for conversion of the isotherm into a pore-size distribution, (ii) data obtained by the mercury porosimeter. The isotherm gives a pore-size distribution for mesopores. The mercury porosimeter gives a distribution covering macropores and larger mesopores. In both cases what is measured is, strictly speaking, not the exact volume of pores having a given pore size, but the volume of pores accessible through pores of a given size. The relationship between these two functions depends on the geometrical nature of the pore system.

The *specific pore volume* is the total internal void volume per unit mass of adsorbent. Some of the pore volume may be completely enclosed, and thus inaccessible to molecules participating in a catalytic reaction.

The total accessible pore volume may be measured by the amount of adsorbate at the saturation pressure of the adsorptive, calculated as liquid volume, provided the adsorption on the external surface can be neglected or can be evaluated. The accessible pore volume may be different for molecules of different sizes. A method which is not subject to the effect of the external surface is the determination of the dead space by means of a non-sorbable gas (normally helium) in conjunction with the determination of the bulk volume of the adsorbent by means of a non-wetting liquid or by geometrical measurements.

Primary particles. Certain materials widely used as catalysts or supports consist of spheroids of about 10 nm (100 Å) in diameter loosely cemented into granules or pellets. The texture of these resembles that of a cemented, loose gravel bed. The 10 nm (100 Å) particles may be called *primary* particles.

Percentage exposed in metallic catalysts. The accessibility of the atoms of metal in metallic catalysts, supported or unsupported, depends upon the percentage of the total atoms of metal which are surface atoms. It is recommended that the term *percentage exposed* be employed for this quantity rather than the term *dispersion* which has been frequently employed.

Pretreatment and activation. Following the preparation of a catalyst or following its insertion into a catalytic reactor, a catalyst is often subjected to various treatments before the start of a catalytic run. The term *pretreatment* may, in general, be applied to this set of treatments. In some cases the word *activation* is used. It implies that the material is converted into a catalyst or into a very much more effective one by the pretreatment. *Outgassing* is a form of pretreatment in which a catalyst is heated *in vacuo* to remove adsorbed or dissolved gas. *Calcination* is a term which means heating in air or oxygen and is most likely to be applied to a step in the preparation of a catalyst.

1.4 *Catalytic reactors*

The vessel in which a catalytic reaction is carried out is called a *reactor*. Many different arrangements can be adopted for introducing the reactants and removing the products.

In a *batch* reactor the reactants and the catalyst are placed in the reactor which is then closed to transport of matter and the reaction is allowed to proceed for a given time whereupon the mixture of unreacted material together with the products is withdrawn. Provision for mixing may be required.

In a *flow* reactor, the reactants pass through the reactor while the catalysis is in progress. Many variations are possible.

The catalyst may be held in a *packed bed* and the reactants passed over the catalyst. A packed bed flow reactor is commonly called a *fixed bed reactor* and the term *plug-flow* is also used to indicate that no attempt is made to back-mix the reaction mixture as it passes through the catalyst bed. The main modes of operation of a flow reactor are *differential* involving a small amount of reaction so that the composition of the mixture is approximately constant throughout the catalyst bed, or *integral* involving a more substantial amount of reaction such that the composition of material in contact with the final section of the catalyst bed is different from that entering the bed.

In a *pulse reactor*, a carrier gas, which may be inert or possibly one of the reactants, flows over the catalyst and small amounts of the other reactant or reactants are injected into the carrier gas at intervals. A pulse reactor is useful for exploratory work but kinetic results apply to a transient rather than to the steady state conditions of the catalyst.

Several alternative modes of operation may be used to avoid the complications of the changing concentrations along the catalyst bed associated with integral flow reactors and each of these has a special name. In a *stirred* flow reactor, effective mixing is achieved within the reactor often by placing the catalyst in a rapidly-rotating basket. If the mixing achieved in this way is efficient, the composition of the mixture in the reactor will be close to that of the exit gases. The same result can be reached by *recirculation* of the gas around a loop containing a fixed bed of catalyst, provided that the rate of recirculation is considerably larger then the rate of flow in and out of the loop. Under these circumstances, a substantial conversion to products can be obtained even though conditions in the bed correspond more closely to those associated with a differential rather than with an integral reactor. Another mode of operation involves a *fluidized bed* in which the flow of gases is sufficient to cause the bed of finely divided particles of catalyst to behave like a fluid. In a fluidized bed, the temperature is uniform throughout, although mixing of gas and solid is usually incomplete. It has special applications in cases where the catalyst has to be regenerated, e.g. by oxidation, after a short period of use. Continuous transfer of catalyst between two vessels (one used as reactor and the other for catalyst regeneration) is possible with a fluidized system. The stirred flow and the recirculation reactors are characterized ideally by very small concentration and temperature gradients within the catalyst region. The term, *gradientless reactor,* may be used to include both types.

All reactors, batch or flow, may be operated in three main ways in regard to temperature. These are *isothermal, adiabatic* and *temperature-programmed.* For the last, in a batch reactor the variation of temperature with time may be programmed, or in a fixed bed reactor the variation of temperature along the length of the bed may be controlled.

When reactors are operated isothermally the batch reactor is characterized by adsorbate concentrations and other aspects of the state of the surface which are constant in space (i.e. uniform within the catalyst mass) but which change with time. In the integral flow reactor with the catalyst at steady state activity, the surface conditions are constant with time but change along the bed. In the gradientless reactor at steady state, the surface conditions are constant in space and, if the catalyst is at a steady state, with time. In the pulse reactor, the catalyst is often not in a condition of steady state, concentrations change as the pulse moves through the bed, and there may be chromatographic separation of reactants and products.

In general, if heterogeneous catalytic reactions are to be conducted isothermally, the reactor design must provide for heat flow to or from the particles of catalyst so as to keep the thermal gradients small. Otherwise, temperatures within the catalyst bed will be non-uniform. The differential reactor and the various forms of the gradientless reactors are advantageous in this regard.

The types of reactors described above can, in principle, be extended to reactions in the liquid phase although the pulse reactor has been little used in such cases.

Reactions in which one reactant is gaseous, the other is in a liquid phase, and the catalyst is dispersed in the liquid phase, constitute a special but not unusual case, for example, the hydrogenation of a liquid alkene catalysed by platinum. A batch reactor is most commonly employed for laboratory scale studies of such reactions. Mass transport from the gaseous to the liquid phase may reduce the rate of such a catalytic reaction unless the contact between the gas and the liquid is excellent (see §1.6).

1.5 *Kinetics of heterogeneous catalytic reactions*

1.5.1 *General terms*

Consider a chemical reaction

$$0 = \Sigma_B \nu_B B,$$

where ν_B is the stoichiometric coefficient (plus for products, minus for reactants) of any product or reactant B. The *extent of reaction* ξ is defined (see §11.1 of the Manual)

$$d\xi = \nu_B^{-1} dn_B,$$

where n_B is the amount of the substance B.

If *rate of reaction* is to have an unambiguous meaning, it should be defined as the rate of increase of the extent of reaction

$$\dot{\xi} = d\xi/dt = \nu_B^{-1} dn_B/dt$$

whereas the quantity dn_B/dt may be called the *rate of formation* (or *consumption*) *of B.*

To facilitate the comparison of the results of different investigators, the rates of heterogeneous catalytic reactions should be suitably expressed and the conditions under which they have been measured should be specified in sufficient detail. If the rate of the uncatalyzed reaction is negligible, the rate of the *catalyzed* reaction may be given as

$$r = \frac{1}{Q} d\xi/dt.$$

If Q, the *quantity of catalyst*, is in mass,

$$r = r_m = \frac{1}{m} d\xi/dt$$

and r_m is the *specific rate of reaction* which may be called the *specific activity of the catalyst* under the specified conditions. If Q is in volume,

$$r = r_v = \frac{1}{V} d\xi/dt.$$

The volume should be that of the catalyst granules excluding the intergranular space. If Q is in area,

$$r = r_a = \frac{1}{A} d\xi/dt$$

where r_a is the *areal*† *rate of reaction.* If the *total* surface area of the catalyst is used, it should be preferably a BET nitrogen area. However, other types of specified areas may be employed, for example, the exposed metal area of a supported metallic catalyst. The exposed metal area is often estimated by selective chemisorption of a suitable sorptive, e.g. hydrogen or carbon monoxide.

The *turnover frequency, N,* (commonly called the *turnover number*) defined, as in enzyme catalysis, as molecules reacting per active site in unit time, can be a useful concept if employed with care. In view of the problems in measuring the number of active sites discussed in §1.2.4, it is important to specify exactly the means used to express Q in terms of active sites. A realistic measure of such sites may be the number of surface metal atoms on a supported catalyst but in other cases estimation on the basis of a BET surface area may be the only readily available method. Of course, turnover numbers (like rates) must be reported at specified conditions of temperature, initial concentration or initial partial pressures, and extent of reaction.

In comparing various catalysts for a given reaction or in comparing various reactions on a given catalyst, it may be inconvenient or impracticable to compare rates at a specified temperature since rates must be measured at temperatures at which they have convenient values. Therefore, it may be expedient to compare the temperatures at which the rates have a specified value.

†The term *areal* meaning per unit area is tentative (see Preface).

In reactors in which the concentrations of reactants and products are uniform in space, the rate is the same on all parts of the catalyst surface at any specified time. In integral flow reactors, however, the rate on each element of the catalyst bed varies along the bed.

1.5.2 *Selectivity*

The term selectivity S is used to describe the relative rates of two or more competing reactions on a catalyst. Such competition includes cases of different reactants undergoing simultaneous reactions or of a single reactant taking part in two or more reactions. For the latter case, S may be defined in two ways. The first of these defines a *fractional selectivity* S_F for each product by the equation

$$S_F = \dot{\xi}_i / \sum_i \dot{\xi}_i.$$

The second defines *relative* selectivities, S_R, for each pair of products by:

$$S_R = \dot{\xi}_i / \dot{\xi}_j.$$

In *shape selectivity*, which may be observed in catalysts with very small pores, the selectivity is largely determined by the bulk or size of one or more reactants. On zeolites, for example, the rate of reaction of alkanes with linear carbon chains may be much greater than that of those with branched chains.

1.5.3 *Rate equations*

Gaseous systems in which all concentrations are uniform in space and in which the reaction is irreversible will be considered first.

The rate $\dot{\xi}$, besides being proportional to the quantity of catalyst, Q, is also in general a function of temperature T and the concentrations c_i or partial pressures p_i of reactants, products and other substances if present:

$$r = \frac{\dot{\xi}}{Q} = f(T, c_i) \quad \text{or} \quad r = f(T, p_i).$$

The statement of this equation is commonly called the *rate equation* or the *rate law*. Frequently, in heterogeneous catalysis, the function f is of the form

$$r = k \prod_i c_i^{a_i}$$

where k is the rate constant which is a function of temperature but not of concentrations and a_i (integral or fractional; positive, negative or zero) is the *order of the reaction* with respect to component i. This form of the rate law is called a *power rate law*. Often, however, a rate expression of different form is used. For example, for a reaction A + B → products, the rate equation might be

$$r = \frac{k K_A K_B c_A c_B}{(1 + K_A c_A + K_B c_B + \sum_n K_n c_n)^2}.$$

This equation can be interpreted in terms of Langmuir adsorption isotherms. It is assumed (see §1.5.4) that both reactants must be adsorbed in order to react and that K_A and K_B are the respective Langmuir adsorption equilibrium constants. The denominator allows for competition for sites between reactants and other substances (diluents, poisons and products) present in the system at concentrations c_n with related adsorption equilibrium constants K_n. A rate law of this type is appropriately called a *Langmuir rate law* although it was made popular by Hinshelwood, Schwab, Hougen, Watson and others. Such rate laws are frequently used for systems in which the adsorptions may not obey the Langmuir adsorption isotherm. Under these circumstances, the rate laws can still provide a useful means of correlating experimental results but the values of the derived constants must be interpreted with caution.

For a single elementary process,

$$k = A \exp(-E/RT),$$

where A is the *frequency factor* and E the *activation energy*. Even though heterogeneous catalytic reactions rarely if ever proceed by a single elementary process, the same relation often applies to the overall rate constant. In such a case, however, A is not a frequency factor but should be called the *pre-exponential factor* and E should be called the *apparent activation energy*.

Sometimes A and E exhibit *compensation*, i.e. they change in the same direction with change in catalyst for a given reaction or with change in reaction for a given catalyst. A special case of compensation called the *θ-rule* occurs when, at least approximately,

$$\ln A = \text{const} + \frac{E}{RT_\theta},$$

where T_θ is the *isokinetic temperature*, the temperature at which all k's would be identical.

These considerations can be extended to reversible processes. They also apply to single phase, liquid systems. For the case, rather common in heterogeneous catalysis, in which one reactant is in a gas phase and the others and the products are in a liquid phase, application of the principles given above is straightforward provided that there is mass transfer equilibrium between gas phase and liquid phase, i.e. the fugacity of the reactant in the gas phase is identical with its fugacity in the liquid phase. In such case, a power rate law for an irreversible reaction of the form

$$\dot{\xi} = k p_g^{a_g} \prod_i c_i^{a_i}$$

may apply where the quantities have the same significance as before except that the gaseous reactant g is omitted from the c_i's and entered as a pressure term with order a_g.

The determination of rate of reaction in a flow system requires knowledge both of the feed rate, v, of a given reactant and of the *fraction converted*, x. The definition of feed rate as the amount of reactant fed per unit time to the inlet of the reactor is consistent with 1.5.1. The rate of reaction is then given by

$$\frac{d\xi}{dt} = v\frac{x}{\nu_B}$$

where ν_B is the stoichiometric coefficient of the reactant of which the fraction x is converted. Alternatively, one may proceed from r_m, r_v, and r_a rather than $d\xi/dt$ by defining the *space velocities*, v_m, v_v, and v_a where the v_i's represent the rate of feed of the given reactant fed per unit mass, volume or surface area of the catalyst. The relation,

$$r_m = v_m\frac{x}{\nu_B}$$

gives the *specific rate of reaction* or, under specified conditions, the *specific activity of the catalyst*. Substitution of v_a or v_v gives the *areal rate of reaction* or the rate divided by volume of the catalyst, respectively. Alternatively, *space times*, τ_m, τ_a, and τ_v, the reciprocals of the space velocities, may be used. "Contact time" and "residence time" are terms which may be misleading for flow systems in heterogeneous catalysis and should be avoided.

1.5.4 *Kinetic aspects of mechanism*

Of general convenience in the treatment of mechanisms are the notions of *rate determining process* or *step* and *most abundant surface intermediate*. The rate determining process is defined, as is usual in kinetics in general, as that single elementary process in the catalytic sequence which is not in equilibrium when the overall reaction is significantly displaced from equilibrium. If the surface of a catalyst has one set of catalytic sites, a particular intermediate is said to be the most abundant surface intermediate if the fractional coverage by that intermediate is much larger than coverages by the other intermediates. Of course, there is no guarantee that either a rate determining process or a most abundant surface intermediate will exist for any particular reaction under a particular set of conditions.

The term *reaction centre* may be used to include both vacant and occupied catalytic sites. The sum of the *surface concentrations of reaction centres* on the surface of a catalyst is a constant L. Thus, if species m at a surface concentration L_m is the most abundant surface intermediate, $L_m + L_v \simeq L$, where L_v is the surface concentration of vacant reaction centres.

Langmuir–Hinshelwood mechanism. This represents a somewhat anomalous use of the term mechanism to specify relative magnitudes of rate constants. In a Langmuir–Hinshelwood mechanism, all adsorption-desorption steps are essentially at equilibrium and a *surface step* is rate determining. Such a surface step may involve the unimolecular reaction of a single adsorbate molecule or the reaction of two or more molecules on adjacent sites with each other. Where the adsorption processes follow Langmuir adsorption isotherms, the overall reaction will follow some kind of a Langmuir rate law (§1.5.3). However, the term Langmuir–Hinshelwood mechanism may cover situations in which Langmuir adsorption isotherms do not apply.

1.5.5 *Non-uniformity of catalytic sites*

A characteristic of a catalytic surface is that its sites may differ in their thermodynamic and kinetic properties. In the kinetic description of catalytic reactions on non-uniform surfaces, a parameter α is frequently used to connect changes in the activation energy of activated adsorption with the enthalpy of the adsorption

$$E_{ads} - E^0_{ads} = \alpha(q - q^0),$$

where E^0_{ads} is the energy of activation and $-q^0$ is the enthalpy of adsorption on the uncovered surface. E_{ads} and q apply to the surface with the same value of θ. In practice the equation may apply only over a restricted range of θ. Sometimes α is defined as in the equation above but in terms of Gibbs energies of activation and adsorption respectively. The name *transfer coefficient* has been used by electrochemists to represent α in another related situation.

1.6 *Transport phenomena in heterogeneous catalysis*

This section will not attempt to cover the more technical aspects of chemical reactor engineering.

A unique feature of heterogeneous catalytic reactions is the ease with which chemical kinetic laws are disguised by various transport phenomena connected with the existence of concentration and/or temperature gradients in the hydrodynamic boundary layer surrounding the catalyst particles (*external gradients*) or in the porous texture of the catalyst particles themselves (*internal gradients*). Additional difficulties arise in batch reactors and in stirred flow reactors if agitation is inadequate to maintain uniform concentrations in the fluid phase. Agitation is particularly critical where one of the reactants is a gas and the catalyst and other reactants and products are in condensed phases for example, in the hydrogenation of a liquid alkene. Here the agitation must be adequate to maintain the fugacity of the dissolved gaseous reactant equal to that in the gaseous phase.

When external gradients correspond to substantial differences in concentration or temperature between the bulk of the fluid and the external surface of the catalyst particle, the rate of reaction at the surface is significantly different from that which would prevail if the concentration or temperature at the surface were equal to that in the bulk of the fluid. The catalytic reaction is then said to be influenced by external mass or heat transfer respectively, and, when this influence is the dominant one, the rate corresponds to a *regime of external mass* or *heat transfer*.

Similarly, when internal gradients correspond to differences in concentration or temperature between the external surface of the catalyst particle and its centre, the rate in the particle is substantially different from that which would prevail if the concentration or temperature were the same throughout the particle. The catalytic reaction is then said to be influenced by internal mass or heat transfer, and, when this influence is the dominant one, the rate corresponds to a *regime of internal mass* or *heat transfer*.

Terms such as *diffusion limited* or *diffusion controlled* are undesirable because a rate may be larger in regimes of heat or mass transfer than in the *kinetic regime* of operation, i.e. when gradients are negligible.

1.7 *Loss of catalytic acitvity*

1.7.1 *Poisoning and inhibition*

Traces of impurities in the fluid to which the catalyst

is exposed can adsorb at the active sites and reduce or eliminate catalytic activity. This is called *poisoning* and the effective impurity is called a *poison*. If adsorption of poison is strong and not readily reversed, the poisoning is called *permanent*. If the adsorption of the poison is weaker and reversible, removal of the poison from the fluid phase results in restoration of the original catalytic activity. Such poisoning is called *temporary*. If adsorption of the poison is still weaker and not greatly preferred to adsorption of reactant, the reduction in rate occasioned by the poison may be called *competitive inhibition* or *inhibition*. Here, of course, the poison may be present in much larger than trace amounts. There are, of course, no sharp boundaries in the sequence permanent poisoning, temporary poisoning, competitive inhibition.

In *selective poisoning or selective inhibition*, a poison retards the rate of one catalysed reaction more than that of another or it may retard only one of the reactions. For example, there are poisons which retard the hydrogenation of olefins much more than the hydrogenation of acetylenes or dienes. Also, traces of sulphur compounds appear selectively to inhibit hydrogenolysis of hydrocarbons during catalytic reforming.

A product of a reaction may cause poisoning or inhibition. The phenomenon is called *self-poisoning* or *autopoisoning*.

1.7.2 *Deactivation—general*

The conversion in a catalytic reaction performed under constant conditions of reaction often decreases with *time of run* or *time on stream*. This phenomenon is called *catalyst deactivation* or *catalyst decay*. If it is possible to determine the kinetic form of the reaction and, thus, to measure the rate constant for the catalytic reaction k, it is sometimes possible to express the rate of deactivation by an empirical equation such as

$$-\mathrm{d}k/\mathrm{d}t = Bk^n,$$

where t is the time on stream, n is some positive constant, and B remains constant during a run but depends upon the temperature and other conditions of the reaction. Alternatively, the decline in k may be assumed to result from elimination of active sites and L may be substituted for k in the preceding equation where L is considered to be the effective concentration of surface centers. It is then common practice to define a *time of deactivation* (or *decay time*) as the time on stream during which k falls to a specified fraction of its original value, often 0.5. Times of deactivation may vary from minutes as in catalytic cracking to years as in hydrodesulphurization.

Catalytic deactivation can sometimes be reversed and the original catalytic activity restored by some special operation called *regeneration*. For example, coked cracking catalyst is regenerated by burning off the coke (see §§1.7.3, 1.9).

If the catalytic reaction is a network of various processes, deactivation can lead to a change in the distribution of products. In such cases, the deactivation not only reduces the overall rate but it changes the selectivity.

1.7.3 *Types of deactivation*

Catalyst deactivation can result from deactivation of catalytic sites by poisoning either by impurities or by products of the catalytic reaction (§1.7.1). Many reactions involving hydrocarbons and particularly those run at higher temperatures lead to the deposition on the catalyst of high molecular weight compounds of carbon and hydrogen which deactivate the catalyst. This phenomenon is called *coking* or *fouling*. Catalysts so deactivated can often be regenerated.

Catalyst deactivation may also result from changes in the structure or in the texture of the catalyst. Changes of this kind are usually irreversible and the catalyst cannot be regenerated. This type of deactivation is often called *catalyst ageing*.

Sintering and *recrystallization*. Catalysts often suffer during use from a gradual increase in the average size of the crystallites or growth of the primary particles. This is usually called *sintering*. The occurrence of sintering leads to a decrease in surface area and, therefore, to a decrease in the number of catalytic sites. In some cases, sintering leads to a change in the catalytic properties of the sites, for example, for catalysts consisting of highly dispersed metals on supports, catalytic properties may change on sintering due to a change in the relative exposure of different crystal planes of the metallic component of the catalyst or for other reasons. Thus sintering leads to a decrease in rate and perhaps also to a change in selectivity. Similar phenomena can occur in oxide catalysts as used in catalytic oxidation. The crystal size increases, or the initial structure of the crystals changes. For example, a binary solid compound may decompose into its components or an amorphous mass may crystallize. These processes may be called *recrystallization*. In some cases the terms sintering and recrystallization may refer to the same process. The removal of surface defects may accompany these processes.

In some cases, as for example in catalytic cracking on silica–alumina, processes similar to those involved in sintering and recrystallization can lead to a change in the texture of the catalyst. Surface areas are diminished and the pore-size distribution is changed.

1.8 *Mechanism of catalytic reactions*

1.8.1 *General*

A chemical reaction proceeds by a set of elementary processes (the Manual, §11.3) which are in series and perhaps also in parallel. These processes start and terminate at species of minimum free energy (reactants, intermediates and products) and each elementary process passes through a state of maximum free energy (the transition state). To specify the mechanism, one must specify the elementary processes. This specifies the intermediates. One must also give the nature (energetics, structure, charge distribution) of the transition state. So much is true for chemistry in general. The special features of mechanism in heterogeneous catalysis are those which involve reactions between sorptives and active sites, reactions among adsorbates, and processes which regenerate active sites to give a type of chain reaction.

In general, only partial approaches to the specification of mechanism as given above have been possible.

Mechanism is sometimes used in different senses. For example, consider the two situations.

$$A + B \rightleftharpoons C \qquad A + B \longrightarrow C$$
$$\text{vs}$$
$$C \longrightarrow D \qquad\quad C \rightleftharpoons D$$

It may be said that the two situations have different mechanisms or that they are two variants of the same mechanism.

1.8.2 *Elementary processes in heterogeneous catalysis*

There are many more types of elementary processes in heterogeneous catalysis than in gas phase reactions. In heterogeneous catalysis the elementary processes are broadly classified as either adsorption–desorption or surface reaction, i.e. elementary processes which involve reaction of adsorbed species. Free surface sites and molecules from the fluid phase may or may not participate in surface reaction steps.

There is no generally accepted classification of elementary processes in heterogeneous catalysis. However, names for a few types of elementary processes are generally accepted and terminology for a partial classification (see M. Boudart, *Kinetics of Chemical Processes.* Chap. 2 (1968)) has received some currency. The particular reactions used below to exemplify this terminology are ones which have been proposed in the literature but some have not been securely established as occurring in nature at any important rate.

Adsorption–desorption. This includes the process of physical adsorption as well as non-dissociative chemisorption.

$$* + NH_3(g) \rightleftharpoons H_3N*$$
$$* + H(g) \rightleftharpoons H*$$

Dissociative adsorption and its reverse, *associative desorption.*

$$2* + CH_4(g) \rightleftharpoons CH_3* + H*$$

The methane might be supposed to react either from the gas phase or from a physisorbed state.

Dissociative surface reaction and its reverse, *associative surface reaction.*

$$2* + C_2H_5* \rightleftharpoons H* + *CH_2CH_2*$$

This involves 'dissociative adsorption' in an adsorbate.

Sorptive insertion. This is analogous to the process of ligand insertion in coordination chemistry.

$$H* + C_2H_4(g) \longrightarrow *C_2H_5$$

This reaction might also be imagined to proceed by adsorption of C_2H_4 followed by ligand migration (an associative surface reaction).

Reactive adsorption and its reverse, *reactive desorption.* This resembles dissociative adsorption but one fragment adds to an adsorbate rather than to a surface site.

$$H_2C{=}CH_2 \;+\; D{-}D(g) \;\rightleftharpoons\; H_2C \overset{CH_2D}{\underset{|}{\diagup}} \qquad D$$

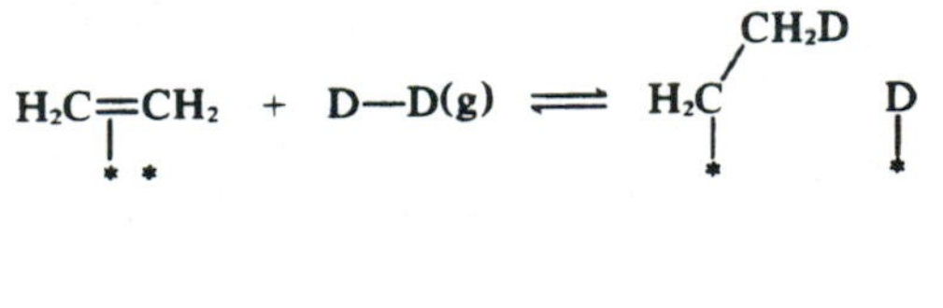

In *abstraction* and *extraction* processes, an adsorptive or adsorbate species extracts an adsorbed atom or a lattice atom respectively.

Abstraction process $*H + H(g) \longrightarrow * + H_2(g)$

Extraction process $O_s^{2-} + CO(g) \longrightarrow 2e + CO_2(g)$

The following elementary process occurring either on one site or, as shown, on two sites is called a *Rideal* or a *Rideal–Eley mechanism*:

$$\begin{matrix} D{-}D(g) & H{-}D(g) \\ H & D \\ *\;\; * & \longrightarrow \quad *\;\;\; * \end{matrix}$$

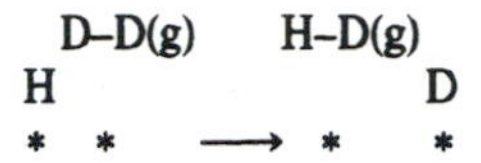

D_2 may also be considered to be in some kind of a weakly adsorbed state. It will be noted that one D atom is never bonded to the surface in any minimum Gibbs energy intermediate. It is recommended that the term Rideal or Rideal–Eley mechanism be reserved for this particular elementary process. However, the term has been used for analogous processes in which there is a reactant molecule and a product molecule of nearly the same energy in the fluid phase or in some weakly adsorbed state and in which one or more atoms are never bonded to the surface. An example is the following elementary process

$$\begin{matrix} H_3C{-}CH{=}CH_2(g) & \longrightarrow & H_2C{=}CH{-}CH_2D(g) \\ D & & H \\ * \qquad\quad * & & * \qquad\qquad * \end{matrix}$$

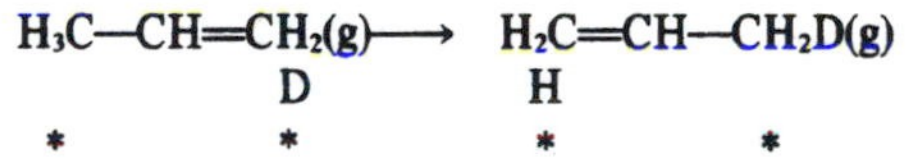

which has been called a *switch* process. The term might well be used generically for similar processes. The term Rideal or Rideal–Eley mechanism has been further extended to include all elementary processes in which a molecule reacts from the fluid phase or from some weakly adsorbed state. Even the sorptive insertion process and the abstraction process illustrated above fall within this extended definition.

1.8.3 *Nomenclature of surface intermediates*

Surface intermediates should be named in ways compatible in so far as possible with chemical nomenclature in general.

Adsorbed species may be treated as surface compounds analogous to molecular compounds. For example, $*H$ may be called surface hydride, $*{=}C{=}O$ may be called a linear surface carbonyl and

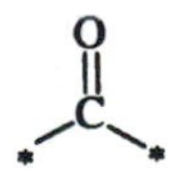

may be called a bridged surface carbonyl. H_2N* may be called a surface amide and H_3C*, a surface methyl or a surface σ-alkyl.

The species $*H$ may also be called an adsorbed hydrogen atom and $*CO$, adsorbed carbon monoxide.

Organic adsorbates pose a particular problem because quite particular structures of some complexity are regularly discussed. A nomenclature is recommended in which the surface is treated as a substituent which replaces one or more hydrogen atoms. The degree of substitution is indicated by *monoadsorbed, diadsorbed,* etc. This terminology does not specify the nature of the chemical bonding to the surface nor does it restrict, *a priori*, the valency of the surface site *. Thus, both of the following species

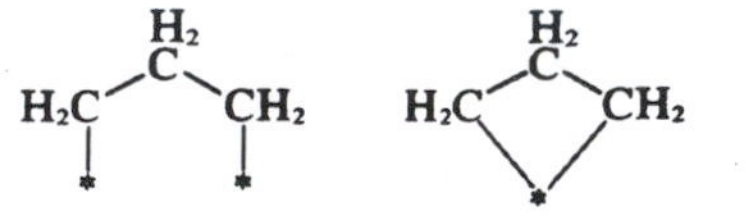

are named 1,3-di-adsorbed propane.   Other examples are:

*CH₃   monoadsorbed methane
*CH₂CH₂CH₃   1-monoadsorbed propane

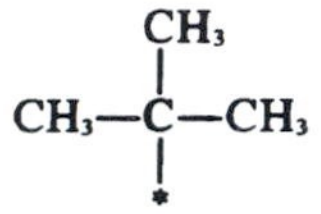

CH₃
|
CH₃—C—CH₃   2-monoadsorbed 2-methylpropane
|
*

*OCH₂CH₃   O-monoadsorbed ethanol
*CH₂CH₂OH   2-monoadsorbed ethanol
CH₃—CH—CH₂—CH—CH₂—CH₃   2,4-diadsorbed
     *   *   hexane

       H   H
   H\ | | /H
      C----C   eclipsed 1,2-diadsorbed ethane
      | |
      *   *

  * = CH—CH₃ or (*)₂CH—CH₃   1,1-diadsorbed
                                     ethane
* = NH or (*)₂NH   diadsorbed ammonia
*COCH₃   1-monoadsorbed acetaldehyde
Species adsorbed as π-complexes are described as π-*adsorbed*:

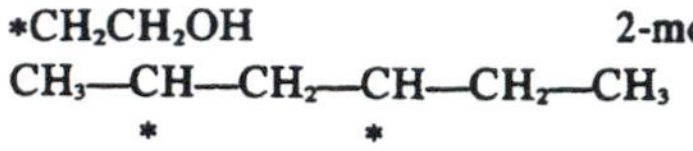

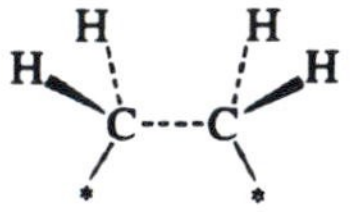

H₂C═CH₂   π-adsorbed ethylene
    ↓
    *

       H
       C
H₂C═C═CH₂ or H₂C⤳⤳CH₂   π-adsorbed allyl
   ↓ ↓   *
   * *

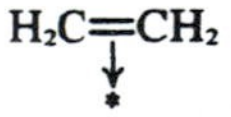

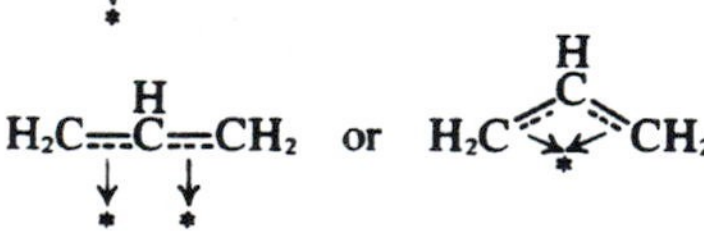

The substitution system of nomenclature should be viewed as showing only how atoms are connected and not as indicating the precise electronic structure. Thus π-adsorbed ethylene is one representation of 1,2-diadsorbed ethane.

Nomenclature based upon the process of formation of a particular adsorbate is to be discouraged. Thus, H* may be 'dissociatively adsorbed hydrogen' but the same species is formed in dissociative adsorption of CH₄, NH₃, H₂O.

1.9 *Nomenclature of catalytic reactions*

In general, a catalytic reaction may be named by adding the adjective "catalytic" to the standard chemical term for the reaction, for example, *catalytic hydrogenation* (or, if clarity demands, *heterogeneous catalytic hydrogenation*), *catalytic hydrodesulphurization, catalytic oxidative dehydrogenation, catalytic stereospecific polymerization*.

In general, special terminology for reactions is to be discouraged. However, certain catalytic processes of technological interest have special names in common use. Where such processes involve the simultaneous occurr-

ence of two or more different chemical reactions, special names for the processes are probably inevitable. Some important examples of such processes of technological interest are:

Catalytic cracking. In this process, a higher boiling cut of petroleum, for example, gas oil, is converted substantially into a lower boiling material of high octane number. Among the processes which appear to be involved are skeletal isomerization of alkanes followed by their cleavage into alkane and olefin, and hydrogen transfer reactions which reduce the amount of olefin formed and which lead to coke and aromatic hydrocarbons.

Catalytic hydrocracking. This is similar to catalytic cracking in its industrial purpose but it is effected under hydrogen pressure and on a catalyst containing an ingredient with a hydrogenating function.

Catalytic reforming. Catalytic reforming is a process for increasing the octane number of naphthas. It involves isomerization of alkanes, dehydrogenation of cyclohexanes to aromatic hydrocarbons, isomerization and dehydrogenation of alkylcyclopentanes, and dehydrocyclization of alkanes.

The following reactions may be mentioned because they are rare except as heterogeneous catalytic reactions and have somewhat specialized meanings in catalysis.

Catalytic methanation. This is a process for removing carbon monoxide from gas streams or for producing methane by the reaction

$$CO + 3H_2 \longrightarrow CH_4 + H_2O$$

Catalytic dehydrocyclization. This is a reaction in which an alkane is converted into an aromatic hydrocarbon and hydrogen, for example,

$$heptane \longrightarrow toluene + 4H_2$$

Catalytic hydrogenolysis. This is ordinarily used for reactions in which $\equiv C\text{--}C\equiv + H_2$ gives $\equiv CH + HC\equiv$, for example,

$$propane + H_2 \longrightarrow ethane + methane$$
$$toluene + H_2 \longrightarrow benzene + methane$$
$$butane + H_2 \longrightarrow 2\ ethane$$

However, it may also be used for cleavage of bonds other than $\equiv C\text{--}C\equiv$, for example,

$$benzyl\ acetate + H_2 \longrightarrow toluene + acetic\ acid$$
$$benzylamine + H_2 \longrightarrow toluene + NH_3$$

Catalytic hydrodesulphurization. This is a process in which, in the presence of hydrogen, sulphur is removed as hydrogen sulphide.

SECTION 2. LIST OF SYMBOLS AND ABBREVIATIONS

2.1 *Catalysis and catalysts*

2.2 *Adsorption*

Area of surface A, A_s, S
Specific surface area a, a_s, s
Surface coverage θ
Area per molecule in
  complete monolayer

of substance *i*	$a_m(i)$
Surface site	*
Ion M^{n+} (or atom M) of adsorbent or catalyst at the surface	M_s^{n+} (or M_s)
Constant in Henry's law	K
Constant in Langmuir's adsorption isotherms	K
Constant in Langmuir's adsorption isotherms for substance i	K_i
Constants in Freundlich isotherms	a, n
Constants in Temkin isotherms	A, B
Constant in BET isotherms	c
Monolayer capacity	n_m
Constants of Roginskii–Zeldovich equation	a, b

2.3 *Composition, structure and texture of catalysts*

2.4 *Catalytic reactors*

2.5 *Kinetics of heterogeneous catalytic reactors*

Stoichiometric coefficient of substance B	ν_B
Extent of reaction	ξ
Rate of catalysed reaction	$\dot{\xi}$
Quantity of catalyst	Q
Specific rate of reaction	r_m
Specific activity of the catalyst	r_m
Rate of reaction per unit volume of catalyst	r_v
Areal rate of reaction	r_a
Turnover frequency (turnover number)	N
Selectivity	S, S_F, S_R
Rate constant	k
Order of the reaction	a_i
Frequency factor	A
Activation energy	E
Isokinetic temperature (Kelvin scale)	T_θ
Fraction converted	x
Feed rate	v
Space velocities	v_m, v_v, v_a
Space times	τ_m, τ_v, τ_a
Sum of surface concentrations of reaction centres	L
Surface concentration of surface intermediate m	L_m
Surface concentration of vacant reaction centres	L_v
Energy of activation for activated adsorption	E_{ads}
Energy of activation for activated adsorption on uncovered surface	E^0_{ads}
(Differential) enthalpy of adsorption	$-q$
(Differential) enthalpy of adsorption on uncovered surface	$-q^0$
Transfer coefficient	α

2.6 *Transport phenomena in heterogeneous catalysis*

2.7 *Loss of catalytic activity*

Constants in equation for rate of deactivation	B, n
Time of run (on stream)	t

2.8 *Mechanism*

2.9 *Nomenclature of catalytic reactions*

SECTION 3. ALPHABETICAL INDEX

Recommendations for Data Compilations and for the Reporting of Measurements of the Thermal Conductivity of Gases[1]

Introductory Remarks

SINCE THE importance of thermal conductivity data for gases is well established, a program for the evaluation of such data was among the early efforts of the Office of Standard Reference Data to develop a National Standard Reference Data System. This program called for evaluators with various interests and backgrounds to match the needs of the various users. Coordination was necessary to provide basic standards of quality and content, and uniformity of presentation. Accordingly, a panel of experts, interested in the program of the Office of Standard Reference Data, was asked what criteria should be established for critical evalutations in this field.

How well data can be evaluated depends in large measure on the method of measurement and of presentation of the original results. Where positive information is lacking on a particular point, a data evaluator must adopt a very cautious attitude to the results in question; therefore, if an author fails to describe his method adequately, his results may not receive the proper attention or weight. To help avoid such an occurrence, the panel has emphasized the experimental features which are most important to the evaluator.

The primary purpose of the panel meeting, coordination of efforts among evaluators of the Office of Standard Reference Data, has already been served. However, it was felt that the results of the meeting might be of interest to other evaluators of data, and indeed to authors of experimental papers in the field, and editors and reviewers concerned with such papers. No claim can be made that the thoughts contained in these considerations are original or particularly subtle, but it is the experience of the panel members involved that at least one of the important features is missing from a large number of potentially important papers. Systematic attention to these considerations in the designing of experiments and the presentation of results would assure the experimenter of the maximum utilization of his results.

These are the reasons for the present communication.

Recommendations for Compilation of Critically Evaluated Data

Presentation of Evaluated Data. The "recommended" or "best" or "definitive" data should be presented in tabular form even if equations have been developed which are thought to fit all or part of the data within their reliability. If graphical representation is desirable, it should be given in addition to the tabular material. If at all feasible, the interval of presentation in the tables should be such as to allow linear interpolation between points.

Units. The International System of Units (SI units) or units approved for use with the SI units should be used. Where desirable to improve communication or to enhance the usefulness to primary recipients, other units can be expressed by indication of conversion factors, inclusion of parallel columns of converted values, or in other suitable supplementary ways. For further information on SI units see "Policy for NBS Usage of SI Units," *NBS Technical News Bulletin*, Vol. 55, No. 1, Jan. 1971 and "ASTM Metric Practice Guide" *NBS Handbook 102* available from the Superintendent of Documents, U. S. Government Printing Office, Washington, D. C. 20402.

Reliability. Explicit quantitative estimates of the reliability of recommended values should be given. These estimates should take account of the precision of experimental measurements and the estimated magnitudes of systematic errors. A good discussion of methods of treating various degrees of imprecision and systematic error is presented by C. Eisenhart in "Expression of the Uncertainties of Final Results," *Science*, Vol. 160, June 14, 1968, p. 1201. Limits of reliability should also be given for any polynomial or graphical representation that is recommended. Consideration or errors necessarily includes those which might be introduced by manipulative procedures such as extrapolation to an axis or by theoretical approximations in addition to those inherent in the experimental measurements.

Discoursive Material. As a general rule, the various decisions which have been made and the reasons for them should be discussed. There are several types. The field covered by the monograph should be explicitly defined. If closely related material, which is usually considered simultaneously with the material at hand, is omitted, it may be advisable to point out the omission and its reason. The theoretical basis for analysis, for conversion of primary measurements of one type to data of another, for extrapolation, approximation, or other manipulative procedures should be dealt with fully. Auxiliary data used should be considered as well as primary data. As far as is feasible, the reasons for selecting specific data should be detailed.

Whether original data need to be tabulated or not is a matter for discretion. If there are relatively few, tablulation is probably desirable. Selected individual results should be tabulated, especially in cases where some of them have been difficult to obtain, or, if the evaluator has had to make unit conversions or other transformations, to put the original results on a comparable basis. In any event, it is usually desirable to compare the recommended values with the individual measurements, at least in cases where there are several sets of measurements. This can often be done very effectively by graphical methods. A preferred graphical method, where the grouping of the results allows it, is the "deviation plot," in which deviations of individual results from the recommended values are plotted against some experimental parameter. All pertinent experimental results should be included in a deviation plot, and the limits of reliability of the recommended values should be shown.

Bibliography. As a general rule the bibliography should be as comprehensive as possible. If certain types of results are being omitted, for example those obtained using a certain method or those made using a certain instrument before it was developed to a given sensitivity, the omissions should be mentioned explicitly in the text along with reasons. Otherwise it is desirable to include reference to pertinent measurements even if they are

---

[1] Considerations of an ad hoc group of specialists assembled by the Office of Standard Reference Data, National Bureau of Standards, consisting of the following members: H. Hanley, NBS Boulder; M. Klein, NBS Washington; P. E. Liley, TPRC Purdue University; S. C. Saxena, TPRC Purdue University (current address, Dept. of Energy Engineering, University of Illinois at Chicago Circle); J. V. Sengers, NBS Washington (current address, Institute for Molecular Physics, University of Maryland); G. Thodos, Northwestern University; and H. J. White, Jr., Office of Standard Reference Data, National Bureau of Standards, U. S. Department of Commerce, Washington, D. C. 20234, to whom inquiries should be directed.

Communicated through S. P. Kezios, Past Senior Technical Editor, JOURNAL OF HEAT TRANSFER. Manuscript received April 1, 1970.

not weighted heavily in determining recommended values. One of the evaluator's tasks is to make it unnecessary for others to search the literature prior to his work. He distills the essence of the literature into his recommended value; but others, approaching with other backgrounds, with other purposes, or in the light of subsequent developments, may wish to rework the data to obtain other numbers or to apply new weighting factors. The evaluator's bibliography and his comments then provide the raw material for their efforts.

Recommendations for the Reporting of Experimental Measurements

It is obvious that experimental papers are not written for the exclusive use of data compilers or evaluators. However, when the experimenter intends to provide definitive data on a given system, he must perforce consider certain factors. If he reports fully on these factors, he eases the evaluator's job and assures maximum utilization of his efforts.

If at all possible, tables of original measurements should be given. When the number of measurements, or the editorial policy of the journal in which the paper is published, precludes publication in tabluar form, the results should be made available either through the American Documentation Institute, directly from the author, or in some other way, and this availability declared in the paper.

The results should be subjected to a detailed error analysis. This analysis should include not only the usual precision measures, but also an analysis of possible systematic errors and of correction factors used.

A full description of experimental details should be given, including specifically the following points:

1 A direct experimental assessment of radiative losses.

2 Experimental proof of the absence of convection.

3 A discussion of parasitic conduction and of the efforts made to estimate its magnitude and correct for it.

4 A discussion of the temperature-gradient measurement including specification of the size of the temperature difference and a discussion of the relation of the measured temperature difference and the gradient in the fluid.

5 A discussion of the method of measuring heat flow and its accuracy.

6 Experimental confirmation that the measured thermal conductivity is independent of the magnitude of the temperature gradient (Fourier's law).

7 The determination of the geometrical constants of the system.

8 The geometry of the temperature field.

9 Accommodation coefficients.

10 If the experimental method is a relative method, the calibration and proof of validity of the method.

11 The purity and composition of the sample.

12 Specification of the state variables, including the temperature, at the position in the cell at which the thermal conductivity is measured.

In addition C_p and the equation of state are desirable although not strictly necessary.

Section 5

Electrochemistry

Table 5.1 contains the recommendations from the Green Book for electrochemistry. Table 5.2 contains additional or extended definitions for relevant symbols or terms taken from reference 20. This reference was used by the authors of the Green Book and much of it is covered in Table 5.1; Table 5.2 contains the remaining areas of reference 20 that we felt were important. Table 5.1 is quite condensed, and for readers who find it confusing or who would like a fuller explanation, we recommend "Manual of Symbols and Terminology for Physiochemical Quantities and Units: Appendix III, Electrochemical Nomenclature" (*20*).

Electrochemical methods are used extensively in analytical chemistry, and an extensive discussion of electroanalytical techniques in Section 8 is devoted to that subject. The separation of papers between this section and Section 8 is somewhat arbitrary, and readers should consider the relevant portions of each section.

"Recommendations for Publishing Manuscripts on Ion-Selective Electrodes" (*21*) is reprinted here. The International Union of Pure and Applied Chemistry, Commission on Analytical Nomenclature produced an earlier paper (*21a*) that discusses substantially the same material in a very similar way; however, one marked difference between the two papers is that the calibration curve is a plot of potential against (negative) log of the ionic activity (or concentration) for the species in question, pa_A. The convention recommended in 1981 for the pa_A, as abscissa, is for the value of a_A to increase from left to right. This difference gives curves of markedly different appearance although of the same technical content. A distinct possibility that the committee will return to the 1975 recommendation exists; however, nothing definitive is known. Readers faced with the problem of which report to use should certainly clearly indicate which convention they have chosen and probably should follow the recommendations of 1981 until the committee makes a definitive statement resolving the dilemma.

For readers interested in pH specifically, reference 22 discusses the definition and measurement of pH, related terminology, and standard reference values. Reference 23 covers the definition and measurement of the absolute electrode potential, gives a table of values for several standard reference electrodes, and recommends values for the standard hydrogen electrode in water and a few other protic solvents. "Recommendations on Reporting Electrode Potentials in Nonaqueous Solvents" (*24*) is reprinted here. Reference 25 is an extensive review article on the development and publication of work with ion-selective electrodes.

Reference 26 covers the publication of transfer coefficients, rate constants, and electrode reaction orders; an appendix to the Green Book (*27*) provides expanded coverage on the nomenclature for transport phenomena in electrolytic systems. As usual, the material in Table 5.1 is condensed. Readers with an interest in the transport properties are advised to consider reference 27 in addition to Table 5.1.

Table 5.1. Electrochemistry. Continued on next page.

Electrochemical concepts, terminology and symbols are more extensively described in [1.i].

Name	*Symbol*	*Definition*	*SI unit*	*Notes*
elementary charge (proton charge)	e		C	
Faraday constant	F	$F = eL$	$\mathrm{C\,mol^{-1}}$	
charge number of an ion	z	$z_B = Q_B/e$	1	1
ionic strength	I_c, I	$I_c = \frac{1}{2}\sum c_i z_i^2$	$\mathrm{mol\,m^{-3}}$	
mean ionic activity	$a_\pm$	$a_\pm = m_\pm \gamma_\pm / m^\circ$	1	2, 3
mean ionic molality	$m_\pm$	$m_\pm^{(\nu_+ + \nu_-)} = m_+^{\nu_+} m_-^{\nu_-}$	$\mathrm{mol\,kg^{-1}}$	2
mean ionic activity coefficient	$\gamma_\pm$	$\gamma_\pm^{(\nu_+ + \nu_-)} = \gamma_+^{\nu_+} \gamma_-^{\nu_-}$	1	2
charge number of electrochemical cell reaction	n, (z)		1	4
electric potential difference (of a galvanic cell)	$\Delta V, E, U$	$\Delta V = V_R - V_L$	V	5
emf, electromotive force	E	$E = \lim_{I \to 0} \Delta V$	V	6
standard emf, standard potential of the electrochemical cell reaction	E°	$E^\circ = -\Delta_r G^\circ / nF$ $= (RT/nF) \ln K^\circ$	V	3, 7
standard electrode potential	E°		V	3, 8
emf of the cell, potential of the electrochemical cell reaction	E	$E = E^\circ - (RT/nF) \times \sum \nu_i \ln a_i$	V	9
pH	pH	$\mathrm{pH} \approx -\lg\left[\frac{c(\mathrm{H^+})}{\mathrm{mol\,dm^{-3}}}\right]$	1	10
inner electric potential	ϕ	$\nabla\phi = -E$	V	11

(1) The definition applies to entities B.

(2) ν_+ and ν_- are the number of cations and anions per formula unit of an electrolyte $A_{\nu_+}B_{\nu_-}$.

Example for $Al_2(SO_4)_3$, $\nu_+ = 2$ and $\nu_- = 3$.

m_+ and m_-, and γ_+ and γ_-, are the separate cation and anion molalities and activity coefficients. If the molality of $A_{\nu_+}B_{\nu_-}$ is m, then $m_+ = \nu_+ m$ and $m_- = \nu_- m$. A similar definition is used on a concentration scale for the mean ionic concentration $c_\pm$.

(3) The symbol $^\bullet$ or $^\circ$ is used to indicate standard. They are equally acceptable.

(4) n is the number of electrons transferred according to the cell reaction (or half cell reactions) as written; n is a positive integer.

(5) V_R and V_L are the potentials of the electrodes shown on the right- and left-hand sides, respectively, in the diagram representing the cell. When ΔV is positive, positive charge flows from left to right through the cell, and from right to left in the external circuit, if the cell is short-circuited.

(6) The definition of emf is discussed on p.53. The symbol E_{MF} is no longer recommended for this quantity.

(7) $\Delta_r G^\circ$ and K° apply to the cell reaction in the direction in which reduction occurs at the right-hand electrode and oxidation at the left-hand electrode, in the diagram representing the cell (see p.53). (Note the mnemonic 'reduction at the right'.)

(8) Standard potential of an electrode reaction, abbreviated as standard electrode potential, is the value of the standard emf of a cell in which molecular hydrogen is oxidized to solvated protons at the left-hand electrode. For example, the standard potential of the Zn^{2+}/Zn electrode, denoted $E^\circ(Zn^{2+}/Zn)$, is the emf of the cell in which the reaction $Zn^{2+}(aq) + H_2 \rightarrow 2H^+(aq) + Zn$ takes place under standard conditions (see p.54). The concept of an *absolute* electrode potential is discussed in reference [26].

(9) $\sum \nu_i \ln a_i$ refers to the cell reaction, with ν_i positive for products and negative for reactants; for the complete cell reaction only mean ionic activities $a_\pm$ are involved.

(10) The precise definition of pH is discussed on p.54. The symbol pH is an exception to the general rules for the symbols of physical quantities (p.5) in that it is a two-letter symbol and it is always printed in roman (upright) type.

(11) E is the electric field strength within the phase concerned.

Table 5.1. Continued. Continued on next page.

Name	*Symbol*	*Definition*	*SI unit*	*Notes*
outer electric potential	ψ	$\psi = Q/4\pi\varepsilon_0 r$	V	12
surface electric potential	χ	$\chi = \phi - \psi$	V	
Galvani potential difference	$\Delta\phi$	$\Delta_\alpha^\beta \phi = \phi^\beta - \phi^\alpha$	V	13
volta potential difference	$\Delta\psi$	$\Delta_\alpha^\beta \psi = \psi^\beta - \psi^\alpha$	V	14
electrochemical potential	$\tilde{\mu}$	$\tilde{\mu}_B{}^\alpha = (\partial G/\partial n_B{}^\alpha)$	$J\,mol^{-1}$	1, 15
electric current	I	$I = dQ/dt$	A	16
(electric) current density	j	$j = I/A$	$A\,m^{-2}$	16
(surface) charge density	σ	$\sigma = Q/A$	$C\,m^{-2}$	
electrode reaction rate constant	k	$k_{ox} = I_a/(nFA \prod_i c_i{}^{n_i})$	(varies)	17, 18
mass transfer coefficient, diffusion rate constant	k_d	$k_{d,B} = \|\nu_B\| I_{l,B}/nFcA$	$m\,s^{-1}$	1, 18
thickness of diffusion layer	δ	$\delta_B = D_B/k_{d,B}$	m	1
transfer coefficient (electrochemical)	α	$\alpha_c = \dfrac{-\|\nu\| RT}{nF} \dfrac{\partial \ln \|I_c\|}{\partial E}$	1	16, 18
overpotential	η	$\eta = E_I - E_{I=0} - IR_u$	V	18
electrokinetic potential (zeta potential)	ζ		V	
conductivity	$\kappa, (\sigma)$	$\kappa = j/E$	$S\,m^{-1}$	11, 19
conductivity cell constant	K_{cell}	$K_{cell} = \kappa R$	m^{-1}	
molar conductivity (of an electrolyte)	Λ	$\Lambda_B = \kappa/c_B$	$S\,m^2\,mol^{-1}$	1
ionic conductivity, molar conductivity of an ion	λ	$\lambda_B = \|z_B\| F u_B$	$S\,m^2\,mol^{-1}$	1, 20
electric mobility	$u, (\mu)$	$u_B = v_B/E$	$m^2\,V^{-1}\,s^{-1}$	1, 11
transport number	t	$t_B = j_B/\Sigma j_i$	1	1
reciprocal radius of ionic atmosphere	κ	$\kappa = (2F^2 I/\varepsilon RT)^{\frac{1}{2}}$	m^{-1}	21

(12) The definition is an example specific to a conducting sphere of excess charge Q and radius r.

(13) $\Delta\phi$ is the electric potential difference between points within the bulk phases α and β; it is measurable only if the phases are of identical composition.

(14) $\Delta\psi$ is the electric potential difference due to the charge on phases α and β. It is measurable or calculable by classical electrostatics from the charge distribution.

(15) The chemical potential is related to the electrochemical potential by the equation $\mu_B{}^\alpha = \tilde{\mu}_B{}^\alpha - z_B F\phi^\alpha$. For an uncharged species, $z_B = 0$, the electrochemical potential is equal to the chemical potential.

(16) I, j and α may carry one of the subscripts: a for anodic, c for cathodic, e or o for exchange, or l for limiting. I_a and I_c are the anodic and cathodic partial currents. The cathode is the electrode where reduction takes place, and the anode is the electrode where oxidation takes place.

(17) For reduction the rate constant k_{red} can be defined analogously in terms of the cathodic current I_c. For first-order reaction the SI unit is $m\,s^{-1}$. n_i is the order of reaction with respect to component i.

(18) For more information on kinetics of electrode reactions and on transport phenomena in electrolyte systems see [27] and [28].

(19) Conductivity was formerly called specific conductance.

(20) It is important to specify the entity to which molar conductivity refers; thus for example $\lambda(Mg^{2+}) = 2\lambda(\frac{1}{2}Mg^{2+})$. It is standard practice to choose the entity to be $1/z_B$ of an ion of charge number z_B, so that for example molar conductivities for potassium, barium and lanthanum ions would be quoted as $\lambda(K^+)$, $\lambda(\frac{1}{2}Ba^{2+})$, or $\lambda(1/3\,La^{3+})$.

(21) κ appears in Debye–Hückel theory. $\kappa^{-1} = L_D$, the Debye length, which appears in Debye–Hückel theory, Gouy–Chapman theory, and the theory of semiconductor space charge.

Table 5.1. Continued. Continued on next page.

Conventions concerning the signs of electric potential differences, electromotive forces, and electrode potentials[1]

(i) *The electric potential difference for a galvanic cell*
The cell should be represented by a diagram, for example:

$$Zn|Zn^{2+} \,\vdots\, Cu^{2+}|Cu.$$

A single vertical bar (|) should be used to represent a phase boundary, a dashed vertical bar (¦) to represent a junction between miscible liquids, and double dashed vertical bars (¦¦) to represent a liquid junction in which the liquid junction potential is assumed to be eliminated. The electric potential difference, denoted ΔV or E, is equal in sign and magnitude to the electric potential of a metallic conducting lead on the right minus that of a similar lead on the left. The emf (electromotive force), also usually denoted E, is the limiting value of the electric potential difference for zero current through the cell, all local charge transfer equilibria and chemical equilibria being established. Note that the symbol E is often used for both the potential difference and the emf, and this can sometimes lead to confusion.

When the reaction of the cell is written as

$$\tfrac{1}{2}Zn + \tfrac{1}{2}Cu^{2+} \rightarrow \tfrac{1}{2}Zn^{2+} + \tfrac{1}{2}Cu, \qquad n=1$$

or

$$Zn + Cu^{2+} \rightarrow Zn^{2+} + Cu, \qquad n=2,$$

this implies a cell diagram drawn, as above, so that this reaction takes place when positive electricity flows through the cell from left to right (and therefore through the outer part of the circuit from right to left). In the above example the right-hand electrode is positive (unless the ratio $[Cu^{2+}]/[Zn^{2+}]$ is extremely small), so that this is the direction of spontaneous flow if a wire is connected across the two electrodes. If however the reaction is written as

$$\tfrac{1}{2}Cu + \tfrac{1}{2}Zn^{2+} \rightarrow \tfrac{1}{2}Cu^{2+} + \tfrac{1}{2}Zn, \qquad n=1$$

or

$$Cu + Zn^{2+} \rightarrow Cu^{2+} + Zn, \qquad n=2,$$

this implies the cell diagram

$$Cu|Cu^{2+} \,\vdots\, Zn^{2+}|Zn$$

and the electric potential difference of the cell so specified will be negative. Thus a cell diagram may be drawn either way round, and correspondingly the electric potential difference appropriate to the diagram may be either positive or negative.

(ii) *Electrode potential* (*potential of an electrode reaction*)
The so-called electrode potential of an electrode is defined as the emf of a cell in which the electrode on the left is a standard hydrogen electrode and the electrode on the right is the electrode in question. For example, for the silver/silver chloride electrode (written $Cl^-(aq)|AgCl|Ag$) the cell in question is

$$Pt|H_2(g, p=p^\circ)|HCl\ (aq, a_\pm = 1) \,\vdots\, HCl\ (aq, a_\pm')|AgCl|Ag$$

A liquid junction will be necessary in this cell whenever $a_\pm'(HCl)$ on the right differs from $a_\pm(HCl)$ on the left. The reaction taking place at the silver/silver chloride electrode is

$$AgCl(s) + e^- \rightarrow Ag(s) + Cl^-(aq)$$

The complete cell reaction is

$$AgCl(s) + \tfrac{1}{2}H_2(g) \rightarrow H^+(aq) + Cl^-(aq) + Ag(s)$$

In the standard state of the hydrogen electrode, $p(H_2) = p^\circ = 10^5$ Pa and $a_\pm(HCl) = 1$, the emf of this cell is the electrode potential of the silver/silver chloride electrode. If, in addition, the mean activity of the HCl in the silver/silver chloride electrode $a_\pm(HCl) = 1$, then the emf is equal to E° for this electrode. The standard electrode potential for $HCl(aq)|AgCl|Ag$ has the value $E^\circ = +0.22217$ V at

---

(1) These are in accordance with the 'Stockholm Convention' of 1953.

Table 5.1. Continued.

298.15 K. For $p^{\circ} = 101325$ Pa the standard potential of this electrode (and of any electrode involving only condensed phases) is higher by 0.17 mV; i.e.

$$E^{\circ}(101\,325\ \mathrm{Pa}) = E^{\circ}(10^5\ \mathrm{Pa}) + 0.17\ \mathrm{mV}$$

A compilation of standard electrode potentials, and their conversion between different standard pressures, can be found in [29]. Notice that in writing the cell whose emf represents an electrode potential, it is important that the hydrogen electrode should always be on the left.

(iii) *Operational definition of* pH [30]

The notional definition of pH given in the table above is in practice replaced by the following operational definition. For a solution X the emf E(X) of the galvanic cell

reference electrode	|	KCl (aq, $m > 3.5\ \mathrm{mol\,kg^{-1}}$)	¦¦	solution X	|	H_2(g)	|	Pt

is measured, and likewise the emf E(S) of the cell that differs only by the replacement of the solution X of unknown pH(X) by the solution S of standard pH(S). The unknown pH is then given by

$$\mathrm{pH(X)} = \mathrm{pH(S)} + (E_S - E_X)F/(RT\ln 10)$$

Thus defined, pH is dimensionless. Values of pH(S) for several standard solutions and temperatures are listed in [30]. The reference pH standard is an aqueous solution of potassium hydrogen phthalate at a molality of exactly $0.05\ \mathrm{mol\,kg^{-1}}$: at 25 °C (298.15 K) this has a pH of 4.005.

In practice a glass electrode is almost always used in place of the $\mathrm{Pt}|\mathrm{H_2}$ electrode. The cell might then take the form

reference electrode	|	KCl (aq, $m > 3.5\ \mathrm{mol\,kg^{-1}}$)	¦¦	solution X	|	glass	|	H^+, Cl^-	|	AgCl	|	Ag

The solution to the right of the glass electrode is usually a buffer solution of KH_2PO_4 and Na_2HPO_4, with $0.1\ \mathrm{mol\,dm^{-3}}$ of NaCl. The reference electrode is usually a calomel electrode, silver/silver chloride electrode, or a thallium amalgam/thallous chloride electrode. The emf of this cell depends on $a(\mathrm{H^+})$ in the solution X in the same way as that of the cell with the $\mathrm{Pt}|\mathrm{H_2}$ electrode, and thus the same procedure is followed.

In the restricted range of dilute aqueous solutions having amount concentrations less than $0.1\ \mathrm{mol\,dm^{-3}}$ and being neither strongly acidic nor strongly alkaline ($2 < \mathrm{pH} < 12$) the above definition is such that

$$\mathrm{pH} = -\lg[\gamma_{\pm} c(\mathrm{H^+})/(\mathrm{mol\,dm^{-3}})] \pm 0.02,$$
$$= -\lg[\gamma_{\pm} m(\mathrm{H^+})/(\mathrm{mol\,kg^{-1}})] \pm 0.02,$$

where $c(\mathrm{H^+})$ denotes the amount concentration of hydrogen ion H^+ and $m(\mathrm{H^+})$ the corresponding molality, and $\gamma_{\pm}$ denotes the mean ionic activity coefficient of a typical uni-univalent electrolyte in the solution on a concentration basis or a molality basis as appropriate. For further information on the definition of pH see [30].

Table 5.2. Phenomena at electrode/electrolyte interfaces at equilibrium. Continued on next page.

5.1 Electric charge divided by area of an interface (electrode) Q
(SI unit $\mathrm{C\,m^{-2}}$)
is given by Lippmann's equation

5.1.1 $$(\partial\gamma/\partial E)_{T,p,\mu_i\ldots} = -Q$$

where γ is the interfacial tension, E is the potential of the electrode with respect to any convenient reference electrode and $\mu_i\ldots$ indicates a set of chemical potentials which is held constant. The choice of these chemical potentials must be specified. T is the thermodynamic temperature and p the external pressure.

If the cell in which E is measured is represented by a conventional diagram (cf. 2.2) with the interface under study forming part of the right-hand electrode, then Q as given by 5.1.1 is the charge on the right-hand side of the interface in the diagram (for a metallic electrode this would be the charge on the metal).

Table 5.2. Continued. Continued on next page.

Q may also be regarded as the quantity of electricity which must be supplied to the interface at thermodynamic equilibrium when its area is increased by unit amount while T, p, E and the set of chemical potentials $\mu_{i \ldots}$ are maintained constant.

At an interface where no charged species can cross the boundary between the two phases (ideal polarized electrode) Q has a unique value. On the other hand, when charge can cross the interface (non-polarizable electrode) Q no longer has a unique value, but depends upon the choice of components whose chemical potentials are held constant in the above definition ($\mu_{i \ldots}$).

The experimental measurement and the theoretical definition of the interfacial tension of a solid electrode is still controversial although the charge Q is measurable.

5.2 Free charge density on the interface (SI unit $\mathrm{C\,m^{-2}}$) σ

is the physical charge density believed to occur on either side of the electrical double layer. It depends upon the model assumed for the interface. In an ideal polarized electrode $Q = \sigma$ if all the components of the interface retain the charges that they have in the adjoining bulk phases. However, if partial charge transfer takes place, σ will differ from Q whether the interface is ideally polarized or not.

In the discussion of models of the double layer, superscripts to σ may be used to denote the location of the charge, e.g. σ^{M} the charge on the metal, σ^{d} the charge on the diffuse layer; subscripts may be used to denote the species contributing the charge, e.g. σ^{s}_{+} the charge on the solution side of the double layer due to cations.

5.3 Potential at the point of zero charge (pzc) (SI unit V) $E_{Q=0}, E_{\sigma=0}$

is the value of the electric potential of the electrode at which one of the charges defined above is zero. The reference electrode against which this is measured should always be clearly stated. It is also important to define the type of charge (see Q and σ in 5.1 and 5.2) which vanishes at the pzc. Except for the simple ideal polarized electrode, each electrode may have more than one pzc. The concentration of the solution should also be specified as well as the nature of the electrode.

5.4 Potential difference with respect to the potential of zero charge (SI unit V) E_{pzc}

5.4.1 $$E_{\mathrm{pzc}} = E - E_{\sigma=0}$$

where the pzc is that for the given electrode in the absence of specific adsorption (other than that of the solvent). Specific adsorption is adsorption which cannot be accounted for by diffuse layer theory.

5.5 Differential capacitance divided by the area of the electrode (SI unit $\mathrm{F\,m^{-2}}$) C

is given by

5.5.1 $$C = (\partial Q/\partial E)_{T,\mu_i \ldots}$$

Various types of differential capacitance corresponding to the charges defined in 5.1 and 5.2 may be defined. Capacities of parts of the double layer may be denoted by superscripts, while components of the capacity contributed by particular species may be indicated by subscripts. As defined here the capacity is an equilibrium property. The differential capacity may also be defined under non-equilibrium conditions. Measured capacities may be dependent on the frequency used. This may be indicated in brackets, e.g. $C(f = 100\,\mathrm{Hz})$ or $C(\omega = 2\pi \times 100\,\mathrm{s^{-1}})$.

5.6 Integral capacitance divided by the area of the electrode (SI unit $\mathrm{F\,m^{-2}}$) K

5.6.1 $$K = Q/(E - E_{Q=0})$$

Integral capacities are related to differential capacities by

5.6.2 $$K = (E - E_{Q=0})^{-1} \int_{E_{Q=0}}^{E} C \,\mathrm{d}E$$

Integral capacities may also be determined between two potentials neither of which is the potential of zero charge.

5.7 Area of the interface (SI unit $\mathrm{m^2}$) A

In all measurements it is desirable that quantities like the charge and the capacity be related to unit, true, surface area of the interface. While this is relatively simple for a liquid–liquid interface, there are great difficulties when one phase is solid. In any report of these quantities it is essential to give a clear statement as to whether they refer to the true or the apparent (geometric) area and, especially if the former is used, precisely how it was measured.

Table 5.2. Continued. Continued on next page.

6. ACTIVITIES IN ELECTROLYTIC SOLUTIONS AND RELATED QUANTITIES

6.1 Mean activity of electrolyte B in solution (number) is given by $a_\pm$

6.1.1 $$a_\pm = \exp[(\mu_B - \mu_B^\ominus)/\nu RT]$$

where μ_B is the chemical potential of the solute B in a solution containing B and other species. The nature of B must be clearly stated: it is taken as a group of ions of two kinds carrying an equal number of positive and negative charges, e.g. $Na^+ + NO_3$ or $Ba^{2+} + 2Cl^-$ or $2Al^{3+} + 3SO_4^{2-}$. ν is the total number of ions making up the group, i.e. 2, 3 and 5 respectively in the above examples. $\mu_B^\ominus$ is the chemical potential of B in its standard state, usually the hypothetical ideal solution of concentration 1 mol kg^{-1} and at the same temperature and pressure as the solution under consideration. The mole must be defined in a way consistent with the group of ions considered above, that is in the third example a mole of aluminium sulphate would consist of 2 mol Al^{3+} + 3 mol SO_4^{2-}

6.2 Stoichiometric mean molal activity coefficient (practical activity coefficient) of electrolyte B (number) is given by $\gamma_\pm$

6.2.1 $$\gamma_\pm = a_\pm/(\nu_+^{\nu_+}\nu_-^{\nu_-})^{\frac{1}{\nu}}(m_B/m^+)$$

where m_B is the molality of B, $m^+ = 1$ mol kg^{-1}, ν_+ is the number of cations and ν_- the number of anions in the chosen group B which is taken as the electrolyte.

6.2.2 $$\nu = \nu_+ + \nu_-$$

The term stoichiometric in the above definitions is used because these definitions are based on the total molality of the solution.

Other types of activity and activity coefficient may be defined in terms of other concentration variables.

6.3 Osmotic coefficient of solvent substance in an electrolytic solution (number) ϕ
is given by

6.3.1 $$\phi = (\mu_A^* - \mu_A)/RTM_A\sum_i m_i$$

where μ_A is the chemical potential of solvent A in the solution containing molality m_i of ionic species i, μ_A^* is the chemical potential of pure A at the same temperature and pressure and M_A is the molar mass of A.

Other types of osmotic coefficient may be defined in terms of other concentration variables.

6.4 Partial molar enthalpy of substance B (SI unit J mol^{-1}) H_B
is given by

6.4.1 $$H_B = -T^2\{\partial(\mu_B/T)/\partial T\}_{p,m_i}$$

6.4.2 $$H_B = H_B^\circ - \nu RT^2(\partial \ln \gamma_\pm/\partial T)_{p,m_i}$$

where H_B° is the partial molar enthalpy of B when the activity of B is unity.

Note that at infinite dilution $\gamma_\pm = 1$ at all temperatures so that the partial molar enthalpy of B in an infinitely dilute solution is

6.4.3 $$H_B^\infty = H_B^\circ$$

6.5 Relative partial molar enthalpy of substance B (SI unit J mol^{-1}) is given by L_B

6.5.1 $$L_B = H_B - H_B^\circ$$

or

6.5.2 $$L_B = -\nu RT^2(\partial \ln \gamma_\pm/\partial T)_{p,m_i}$$

Analogous definitions of relative partial molar enthalpies may be made using mole fraction as the concentration variable, but not with concentration because the composition of a solution of fixed concentration varies with the temperature.

6.6 Partial molar heat capacity at constant pressure (SI unit J K^{-1} mol^{-1}) is given by $C_{p,B}$

6.6.1 $$C_{p,B} = (\partial H_B/\partial T)_{p,m_i}$$

or

Table 5.2. Continued. Continued on next page.

6.6.2 $$C_{p,\mathrm{B}} = C^{\circ}_{p,\mathrm{B}} - \nu R\left(T^2\frac{\partial^2 \ln \gamma_\pm}{\partial T^2} + 2T\frac{\partial \ln \gamma_\pm}{\partial T}\right)_{p,m_i}$$

where $C^{\circ}_{p,\mathrm{M}}$ is the partial molar heat capacity at constant pressure when the activity of B is unity. Note that the partial molar heat capacity of B at infinite dilution is

6.6.3 $$C^{\infty}_{p,\mathrm{B}} = C^{\circ}_{p,\mathrm{B}}$$

6.7 Relative molar heat capacity at constant pressure of substance B (SI unit J K^{-1} mol^{-1}) is given by J_B

6.7.1 $$J_\mathrm{B} = C_{p,\mathrm{B}} - C^{\circ}_{p,\mathrm{B}}$$

or

6.7.2 $$J_\mathrm{B} = -\nu R\left(T^2\frac{\partial^2 \ln \gamma_\pm}{\partial T^2} + 2T\frac{\partial \ln \gamma_\pm}{\partial T}\right)_{p,m_i}$$

7. TRANSPORT PROPERTIES OF ELECTROLYTES

7.1 Electric mobility of species B (SI unit m^2 V^{-1} s^{-1}) is the magnitude of the velocity of migration of B divided by the magnitude of the electric field strength. The frame of reference to which the velocity is referred must be specified. u_B

7.2 Conductivity (formerly called specific conductance) (SI unit S m^{-1} $\{\Omega^{-1}\ m^{-1}\}$) is the reciprocal of the resistivity. $\kappa(\sigma)$

7.3 Cell constant of a conductivity cell (SI unit m^{-1}) K_cell

7.3.1 $$K_\mathrm{cell} = \kappa R$$

where R is the measured resistance of the cell

7.4 Molar conductivity of an electrolyte (SI unit S m^2 mol^{-1} $\{\Omega^{-1}\ m^2\ mol^{-1}\}$) Λ

7.4.1 $$\Lambda = \kappa/c$$

where c is the concentration. The formula unit whose concentration is c must be specified and should be given in brackets, for example

$$\Lambda(\mathrm{KCl}),\ \Lambda(\mathrm{MgCl_2}),\ \Lambda(\tfrac{1}{2}\mathrm{MgCl_2}),\ \Lambda(\tfrac{2}{3}\mathrm{AlCl_3} + \tfrac{1}{3}\mathrm{KCl})$$

The symbol Λ has also been used for the quantity $\kappa/\nu_+ z_+ c$ where ν_+ is the number of cations of charge number z_+ produced in the dissociation of a salt 'molecule' of a given type (cf. 6.1 and 6.2). This quantity has been called the equivalent conductivity. Whatever it is called, its SI unit is the same as that for the molar conductivity. It is recommended that the use of the equivalent conductivity be discontinued.

7.5 Ionic conductivity of ionic species B (SI unit S m^2 mol^{-1} $\{\Omega^{-1}\ m^2\ mol^{-1}\}$) λ_B

7.5.1 $$\lambda_\mathrm{B} = |z_\mathrm{B}|\, F u_\mathrm{B}$$

z_B here is the charge number of the ionic species B. In most current practice it is taken as unity, i.e. ionic conductivity is taken as that of species such as Na^+, $\tfrac{1}{2}Ca^{2+}$, $\tfrac{1}{3}La^{3+}$ etc. To avoid ambiguity the species considered should be clearly stated, e.g. as $\lambda(\tfrac{1}{2}Ca^{2+})$.

7.6 Transport number of the ionic species B (number) in an electrolytic solution t_B

7.6.1 $$t_\mathrm{B} = |z_\mathrm{B}|\, u_\mathrm{B} c_\mathrm{B} \Big/ \sum_i |z_i|\, u_i c_i$$

where c_i is the actual concentration of ionic species i in the solution.

8. KINETICS OF REACTIONS AT ELECTRODES

8.1 Electric current (SI unit A) I_a

A current (of positive electricity) passing from the electrode into the electrolyte is taken as positive.

An electrode at which a net positive current flows is called an *anode*. The chemical reaction which predominates at an anode is an *oxidation*.

An electrode at which a net negative current flows is called a *cathode*. The chemical reaction which predominates at a cathode is a *reduction*.

Table 5.2. Continued. Continued on next page.

This convention is recommended so that, for example, the effective resistance of an electrode interface is positive under normal conditions. The opposite convention is widely used particularly in the electro-analytical literature.

8.2 Partial anodic (cathodic) current of a reaction $I_a(I_c)$
(SI unit A). When a single electrochemical reaction occurs at an electrode (i.e. all other electrochemical reactions may be neglected)

8.2.1 $$I = I_a + I_c$$

I_a being positive and I_c negative. When more than one reaction is significant the reactions may be numbered and the numbers used as subscripts: $I_{1,a}$, $I_{1,c}$, $I_{2,a}$, etc. Then

8.2.2 $$I = \sum_i I_{i,a} + \sum_i I_{i,c}$$

8.3 Exchange current of an electrode reaction (SI unit A) I_0
is the common value of the anodic and cathodic partial currents when that reaction is at equilibrium

8.3.1 $$I_0 = I_a = -I_c$$

For an electrode at equilibrium at which only one reaction is significant $I = 0$. When more than one reaction is significant at a given electrode, subscripts to I_0 may be used to distinguish exchange currents. I is not usually zero when only one of these reactions is at equilibrium.

8.3 Limiting current (SI unit A) I_l
Under usual conditions the steady-state current/potential characteristic has a positive slope. In some circumstances this slope decreases until it becomes approximately zero. This 'plateau' on the characteristic is the limiting current. It arises because the concentration of a reacting species at the electrode approaches zero as a result of the rate of diffusion, convection, chemical reaction in solution or at the interface or of combinations of these effects. A plateau following a region of negative slope in the current/potential characteristic cannot in general be defined as a limiting current.

8.5 Heterogeneous diffusion rate constant (SI unit m s^{-1}) k_d

8.5.1 $$k_d = I_l/nFcA$$

where the limiting current I_l is assumed to be due to the diffusion of a species of concentration c and of diffusion coefficient D. n is the charge number of the cell reaction written so that the stoichiometric coefficient of this species is unity. A is usually taken as the geometric area of the electrode.

8.6 Diffusion layer thickness (SI unit m) δ

8.6.1 $$\delta = D/k_d = nFDcA/I_l$$

where the symbols are subject to the same conditions as described in 8.5. δ has a purely formal significance.

8.7 Mean current density (SI unit A m^{-2}) j

8.7.1 $$j = I/A$$

Other current densities may be similarly defined in terms of the currents defined in 8.2 to 8.4. The nature of the area used in this calculation must be clearly stated (see 5.7). In interpreting the mean current density it is important to know whether the current is uniformly distributed over the electrode interface.

8.8 Mixed potential (SI unit V) E_{mix}
is the value of the potential of a given electrode with respect to a suitable reference electrode when appreciable contributions to the total anodic and/or cathodic partial currents are made by species belonging to two or more different couples, but the total current is zero, i.e.

8.8.1 $$I = \sum_i I_i = 0 \quad (I_i \neq 0)$$

where I_i is the partial current of reaction i

8.8.2 $$I_i = I_{i,a} + I_{i,c}$$

and the summation in 8.8.1 contains appreciable contributions from at least two reactions.

8.9 Overpotential (SI unit V) η
is the deviation of the potential of an electrode from its equilibrium

Table 5.2. Continued. Continued on next page.

value required to cause a given current to flow through the electrode.

Consider a cell with a current I flowing in the external circuit from the cathode C to the anode A. The measured potential difference E(A/C) of A with respect to C can be considered to comprise three components: one associated with processes at the anode, one associated with processes at the cathode and one associated with the conduction of current through the bulk electrolyte, connections to the electrodes, etc. This third component is related to the bulk resistivity: hence, it may be expressed as IR_{cell}, where R_{cell} is the total resistance of the cell excluding resistances due to slow processes at the electrodes.

Measurements are frequently made with a third electrode R, the reference electrode, in the cell. This electrode has a known, invariant, electrode potential and measurements are made under conditions such that a negligible current passes through it. Let the anode A be the electrode under study, i.e. A is the *test* (indicator) electrode and the cathode C is the *auxiliary* (counter) electrode. Then the measured potential of A with respect to R, E contains no component associated with the processes at the auxiliary electrode. It also contains less of the IR component than E(A/C). The remaining contribution from this source is $I\Delta R$ where ΔR is the uncompensated resistance. The value of ΔR can be minimized, but not reduced to zero, by suitable cell design.

The measured quantity E also includes the contribution of the equilibrium potential difference across the reference electrode. This may be eliminated by considering the difference between the measured potential difference when the current is I, $E(I)$ and its value $E(0)$ when the current is zero and the reaction at electrode A is at equilibrium. The difference $E(I) - E(0)$ is characteristic of electrode A only but still includes $I\Delta R$. The quantity characteristic of the slow processes at electrode A is

8.9.1 $$\eta = E(I) - E(0) - I\Delta R$$

and is the overpotential.

Note that, if the reference electrode E is of the same type as the test electrode, but at equilibrium, $E(0) = 0$ and the measured potential difference is equal to $\eta + I\Delta R$.

Sometimes it is possible to express η as a sum of contributions from different stages in the overall electrode reaction, but this is not always valid, for example when the different stages are interdependent.

8.10 Electrode reaction rate constants $k_{\text{ox}}, k_{\text{red}}$

for the oxidizing (anodic) and reducing (cathodic) reactions respectively. (SI units m s^{-1} for a first order reaction: mol$^{1-\Sigma}$ m$^{3\Sigma-1}$ s^{-1} for a reaction of overall order Σ). The rate constants are related to the partial currents by

8.10.1 $$k_{\text{ox}} = I_{\text{a}}/nFA\prod_i c_i^{\nu_i}$$

8.10.2 $$k_{\text{red}} = I_{\text{c}}/nFA\prod_i c_i^{\nu_i}$$

where the product $\prod_i c_i^{\nu_i}$ includes all the species i which take part in the given partial reaction, c_i is the volume concentration of species i and ν_i is the order of the reaction with respect to species i. 8.10.1 and 8.10.2 are not general definitions since not all reaction rates can be expressed in this form. For example the rate of a multistep reaction or a reaction involving adsorbed species may not be expressible in this form.

For an electrode reaction

8.10.3 $$\text{Red} \rightarrow \text{Ox} + ne$$

which is first order in each direction

8.10.4 $$k_{\text{ox}} = I_{\text{a}}/nFAc_{\text{red}}$$

8.10.5 $$k_{\text{red}} = -\ I_{\text{c}}/nFAc_{\text{ox}}$$

and

8.10.6 $$I = AnF(k_{\text{ox}}\, c_{\text{red}} - k_{\text{red}}\, c_{\text{ox}})$$

Here c_{ox} and c_{red} are the volume concentrations of Ox and Red respectively at the site of the reaction (at the electrode surface).

8.11 Cathodic transfer coefficient (number) α_{c}

For a reaction with a single rate-determining step

Table 5.2. Continued.

8.11.1 $$\alpha_c/\nu = -(RT/nF)(\partial \ln |I_c|/\partial E)_{T,p,c_i}\ldots$$

where R is the gas constant and T is the thermodynamic temperature. ν is the stoichiometric number giving the number of identical activated complexes formed and destroyed in the completion of the overall reaction as formulated with the transfer of n electrons. The corresponding anodic transfer coefficient is

8.11.2 $$\alpha_a/\nu = (RT/nF)(\partial \ln I_a/\partial E)_{T,p,c_i}\ldots$$

In 8.11.1 and 8.11.2 I_a and I_c are the partial kinetic currents, i.e. the currents which would flow if mass transport were infinitely fast.

If the mechanism of the reaction is unknown it is preferable to report the experimental quantities α_c/ν or α_a/ν. An unsubscripted α should be used only for a cathodic transfer coefficient.

8.12 Conditional rate constant of an electrode reaction (SI unit, same as k_{ox}, k_{red}) $k^{\ominus}$

is the value of the electrode reaction rate constant at the conditional (formal) potential of the electrode reaction (see 2.7). When α the transfer coefficient is independent of potential.

8.12.1 $$k_c^{\ominus} = k_{ox}/\exp[\alpha_a(E - E_c^{\circ\prime})nF/\nu RT]$$

8.12.2 $$= k_{red}/\exp[-\alpha_c(E - E_c^{\circ\prime})nF/\nu RT]$$

Similar rate constants can be defined using activities in place of concentrations in 8.10.1 and 8.10.2 and the standard electrode potential in place of the conditional potential in 8.12.1 and 8.12.2 This type of rate constant would be called the standard rate constant of the electrode reaction.

8.13 Mean exchange current density (SI unit A m^{-2}) j_0

8.13.1 $$j_0 = I_0/A$$

When the electrode reaction is first order in each direction

8.13.2 $$j_0 = nFk_c^{\ominus}\, c_{ox}^{\alpha_a}\, c_{red}^{\alpha_c}$$

Note that, at equilibrium

8.13.3 $$\alpha_a + \alpha_c = 1$$

8.14 Energy of activation of an electrode reaction at the equilibrium potential (SI unit J mol^{-1}) $U^{\ddagger}$

8.14.1 $$U^{\ddagger} = -R(\partial \ln I_0/\partial T^{-1})_{p,c_i}\ldots$$

8.15 Energy of activation of an electrode reaction at any overpotential η (SI unit J mol^{-1}) $U^{\ddagger}(\eta)$

8.15.1 $$U_a^{\ddagger}(\eta) = -R(\partial \ln I_a/\partial T^{-1})_{p,\eta,c_i}\ldots$$

8.15.2 $$U_c^{\ddagger}(\eta) = -R(\partial \ln |I_c|/\partial T^{-1})_{p,\eta,c_i}\ldots$$

It is also possible to define energies of activation at constant potential E with respect to some reference electrode. However, these quantities include a contribution either from the temperature coefficient of the reference electrode potential, if the reference electrode is at the same temperature as the test electrode; or from the non-isothermal junctions in the cell, if the reference electrode is held at a constant temperature. The energies of activation defined in (8.14.1), (8.15.1) and (8.15.2) are therefore to be preferred since they depend only on the properties of the test electrode.

RECOMMENDATIONS FOR PUBLISHING MANUSCRIPTS ON ION-SELECTIVE ELECTRODES

COMMISSION ON ANALYTICAL NOMENCLATURE*

G. G. GUILBAULT

University of New Orleans, Louisiana, USA

Papers on ion-selective electrodes are concerned with fundamental aspects, developments, appraisal and applications. While it may be difficult to lay down guidelines for publishing manuscripts on fundamental aspects, prospective readers of papers on ion-selective electrode developments, appraisal and applications can more easily reach intelligent decisions on advantages and limitations if attention is given to a set specification. Of prime importance is the objective that all papers should aim at consistency in preferred usages. In this respect special attention is directed to the report of The Analytical Nomenclature Commission of the International Union of Pure and Applied Chemistry on"Recommendation for Terms and Symbols in the Field of Ion-Selective Electrodes." The definitions of most frequently used terms taken from this report are given below, and are followed by summaries of the essential points to be included in papers on new electrode developments and applications respectively. This present report was prepared by a committee consisting of G. G. Guilbault (Chairman), J. D. R. Thomas, R. A. Durst, M. S. Frant, H. Freiser, E. H. Hansen, T. S. Light, G. J. Moody, E. Pungor, G. Rechnitz, N. M. Rice, T. J. Rohm, W. Simon.
It has been circulated in first, second and third draft forms and was discussed in a joint meeting of IDCNS, Commission I.3, V.3, and V.5 in Warsaw. This provisional report was circulated for 12 months, revised in light of all comments, and the final copy was approved at Davos (1979).

ION-SELECTIVE ELECTRODE NOMENCLATURE AND DEFINITIONS

A. DEFINITIONS OF MOST FREQUENTLY USED TERMS

1. Calibration Curve is a plot of the potential (emf) of a given ion-selective electrode cell assembly (ion-selective electrode combined with an identified reference electrode) versus the logarithm of the ionic activity (or concentration) of a given species. For uniformity, it is recommended that the potential be plotted on the ordinate (vertical axis) with the more positive potentials at the top of the graph and that pa_A (-log activity of the species measured, A) or pa_A be plotted on the abscissa (horizontal axis) with highest activity (lowest pa_A) to the left.

2. Limit of Detection. A calibration curve ordinarily has the following shape.

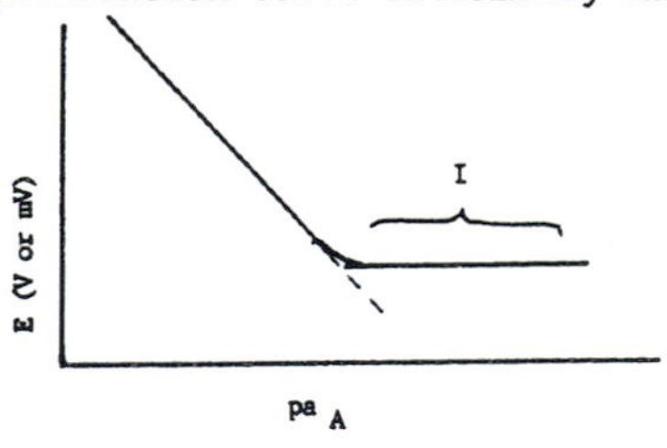

By analogy with definitions adopted in other fields, the limit of detection should be defined as that concentration for which, under the specified conditions, the potential E deviates from the average potential in region I by some stated arbitrary multiple of the standard error of a single measurement of the potential in region I. (i.e. point of extrapolated intersection shown in the illustration).

*Membership of the Commission for 1979-81 is as follows:

Chairman: G. G. GUILBAULT (USA); *Secretary*: G. SVEHLA (UK); *Titular Members*: R. W. FREI (Netherlands); H. FREISER (USA); S. P. PERONE (USA); N. M. RICE (UK); W. SIMON (Switzerland); *Associate Members*: C. A. M. G. CRAMERS (Netherlands); D. DYRSSEN (Sweden); R. E. VAN GRIEKEN (Belgium); H. M. N. H. IRVING (South Africa); G. F. KIRKBRIGHT (UK); D. KLOCKOW (FRG); O. MENIS (USA); H. ZETTLER (FRG); *National Representatives*: A. C. S. COSTA (Brazil); W. E. HARRIS (Canada).

In the present state of the art, and for the sake of practical convenience, a simpler (and more convenient) definition is recommended at this time. The practical limit of detection may be taken as the activity (or concentration) at the point of intersection of the extrapolated lines as illustrated.

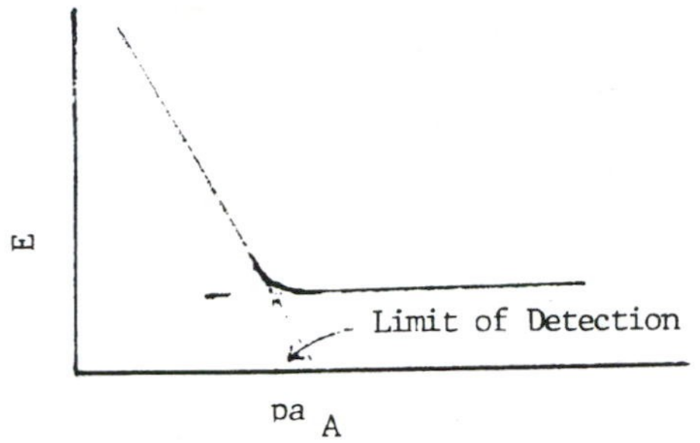

Since many factors affect the detection limit, the experimental conditions used should be reported, i.e., composition of the solution, the history and preconditioning of the electrode, stirring rate, etc.

3. Drift is the slow non-random change with time in the potential (emf) of an ion-selective electrode cell assembly maintained in a solution of constant composition and temperature.

4. Interfering Substance is any species, other than the ion being measured, whose presence in the sample solution affects the measured emf of a cell.

   Interfering Substances fall into two classes. (a) Electrode interferences and (b) Method interferences. Examples of the first class would be those substances which give a response similar to the ion being measured and whose presence generally results in an apparent increase in the activity (or concentration) of the ion to be determined (e.g., Na^+ for the Ca^{++} electrode), those species which interact with the membrane so as to change the chemical composition (e.g. organic solvents for the liquid or poly(vinylchloride (PVC) membrane electrodes). The second class of interfering substances are those which interact with the ion being measured so as to decrease its activity or apparent concentration, but where the electrode continues to report the true activity (e.g., CN^- present in the measurement of Ag^+).

5. The modified Nernst (Nikolsky) Equation for Ion-Selective Electrodes and Definition of $k^{pot}_{A,B}$.

$$E = \text{constant} + \frac{2303\ R\ T}{z_A F} \text{Log}\ [a_A + k^{pot}_{A,B} (a_B)^{z_A/z_B}]$$

   E is the experimentally observed potential of an I.S.E.

   R is the gas constant (and is equal to 8.3144 $J\ K^{-1}\ mol^{-1}$).

   T is the thermodynamic temperature (in K).

   F is the Faraday constant (and is equal to $9.64846 \times 10^4\ C\ mol^{-1}$).

   a_A is the activity of the ion, A (for concentrations measured in mol l^{-1} or molality in mol kg^{-1}).

   a_B is the activity of the interfering ion, B (for concentrations measured in mol l^{-1} or molality in mol kg^{-1}).

   $k^{pot}_{A,B}$ is the potentiometric selectivity coefficient.

   z_A is an integer with sign and magnitude corresponding to the charge of the principal ion, A.

   z_B is an integer with sign and magnitude corresponding to the charge of an interfering ion, B.

   The "constant" term includes the standard potential of the indicator electrode, E^o_{ISE}, the reference electrode potential, E_{Ref}, and the junction potential, E_j (all in millivolts). This equation can only be rigorously derived when $z_A = z_B$.

6. Nernstian Response. An ion-selective electrode is said to have Nerstian response over a given range of activity (or concentration) for which a plot of the potential of such an electrode in conjunction with a reference electrode vs. the logarithm of the ionic activity of a given species (a_A) is linear with a slope of (approximately) $2{,}303RT/z_A F$ ($59.16/z_A$ mV at 25°C).

7. Practical Response Time. The length of time that elapses between the instant at which an ion-selective electrode and a reference electrode are brought into contact with a sample solution (or at which the concentration of the ion of interest in a solution in contact with an ion-selective electrode and a reference electrode is changed) and the first instant when the potential of the cell becomes equal to its steady-state value within 1mV. The experimental conditions used should be stated, i.e., the stirring rate, the composition of solution of which the response time is measured, the composition of the solution to which the electrode was exposed prior to this measurement, the history and preconditioning of the electrode, and the temperature.

The main definition takes in "static response time" while the alternative (within the parentheses) takes in "dynamic response times".

8. Assessment of Selectivity of Electrode

a. Potentiometric Selectivity Coefficient, $k^{pot}_{A,B}$ defines the ability of an ion-selective electrode to distinguish between different ions in the same solution. It is not identical to the similar term used in separation processes. The selectivity coefficient should preferably be evaluated by measuring the response of an ion selective electrode in solutions containing both the primary ion, A, and interfering ion, B (fixed interference method). Alternatively, the separate solution method could be used to calculate the selectivity coefficient. The method is less desirable because it does not approximate as well the conditions under which the electrodes are used.

The activity of the primary ion A and the interferant B at which $k^{pot}_{A,B}$ is determined should always be specified since the value of $k^{pot}_{A,B}$ is not strictly constant but defined by the modified Nernst equation. The smaller the value of $k^{pot}_{A,B}$, the greater the electrode's preference for the primary ion, A.

b. Fixed Interference Method. The emf of a cell comprising an ion-selective electrode and a reference electrode is measured with solutions of constant level of interferant, a_B, and varying activity of the primary ion, a_A. The potential values obtained are plotted versus the logarithm of the activity of the primary ion. The intersection of the extrapolations of the linear portions of this curve (i.e., where the two terms under the log sign in the expanded Nernst equation have an equal contribution) will indicate the values of a_A which are to be used to calculate $k^{pot}_{A,B}$ from:

$$k^{pot}_{A,B} = (a_A / a_B)^{z_A / z_B}$$

9. Separate Solution Method. The emf of a cell comprising an ion-selective electrode and a reference electrode is measured with each of two separate solutions, one containing the ion, A at the activity a_A (but no B), the other containing the ion B, at the same activity $a_B = a_A$ (but no A). If the measured values are E_1 and E_2, respectively, the value of $k^{pot}_{A,B}$ may be calculated from the following equation if the Nernstian equation holds for the indicator ion:

$$\log k^{pot}_{A,B} = \frac{E_2 - E_1}{2303\ RT/z_A F} + \left(1 - \frac{z_A}{z_B}\right) \log a_A$$

This method is not recommended except in those cases where the mixed solution method is not feasible, e.g., non-alignment of emf plots for primary and interferent primary solutions, respectively.

B. CLASSIFICATION OF ION-SELECTIVE ELECTRODES

Electrodes should be classified using the system described in Pure and Applied Chem. 48, 127 (1976). (see also the compendium, pp. 168-175)

C. PAPERS ON NEW OR MODIFIED ELECTRODES

Such papers should include the following essential information:

(a) Constructional details, including the membrane and conditions of its preparation, inner filling solution, internal reference electrode and contacts. Pretreatment should be specified as should the cell assembly - including the reference electrode employed. For commercial electrodes it is adequate to quote the manufacturer's name and model designation.

(b) Calibration range, including slope, detection limit and influence of pH, the latter expressed as a potential/pH diagram for at least two levels of activity of the measured ion, the substance used to effect a pH change and the nature of the acid added.

(c) Stability, including details of storage conditions an drift and information on its susceptibility to attack by chemical agents such as acids, bases, or complexing agents.

(d) Static or dynamic response times with specific information on how these are obtained with respect to solution conditions, stirring rate, pretreatment, etc. Any change in response time with electrode age should be noted.

(e) Interfering substances ought to be classified according to type with selectivity for the primary ion over other counter-ions being computed by the Fixed Interference Method at specified interference levels. The selectivity coefficients should be measured on a newly made electrode and indication given how the coefficients change during electrode lifetime.

(f) Any limitations of the electrode should be clearly stated along with details of operational lifetimes and rejuvenating treatment (which should be evaluated in terms (b) and (e)).

D. PAPERS ON APPLICATIONS FOR DIRECT CONCENTRATION/ACTIVITY MEASUREMENTS

These papers can with advantage refer to papers describing characteristics of the electrode employed and which would fall into category C above. Attention ought also to be given to:

(a) Details of the cell assembly employed and especially of the reference electrode.

(b) Interferences and their possible elimination.

(c) Details of experimental procedure including ionic-strength adjustment and calibrating standards.

(d) A comparison of activity/concentration values obtained with those obtained by traditional or alternative established procedures (if they exist) using appropriate statistical tests.

(e) Recovery tests.

(f) The electrometer or potentiometer used in the emf measurements, as well as its sensitivity, should be specified by manufacturer and model designation.

E. PAPERS ON TITRATION APPLICATIONS

In addition to a mention of papers describing electrode characteristics and attention to items listed in D. special attention must be given to:

(a) Method used for correct end-point location.

(b) Possible interferences.

(c) The possibility of electrode damage during usage.

F. PAPERS ON ON-LINE PROCESS APPLICATIONS AND AUTOMATIC METHODS

Reports on applications to on-line continuous monitoring systems ought to include reference to associated work on electrode characteristics and preliminary experimentation on direct concentration/activity measurement. This, and especially measurements on intermittent samples, will help to characterize possible errors due to the additional parameters of the flowing system. Specific mention should be made of:

(a) Dynamic response-time behavior of the cell.

(b) Full details of cell design including the reference electrode and the supplementary addition of reagents.

(c) Definition of calibration procedure (e.g., on-line, continuous flow or batch, etc.) and frequency of calibration requirements.

RECOMMENDATIONS ON REPORTING ELECTRODE POTENTIALS IN NONAQUEOUS SOLVENTS

COMMISSION ON ELECTROCHEMISTRY*

G. GRITZNER[1] and J. KŮTA[2]†

[1]Institute of Inorganic Chemical Technology,
University of Linz, Austria
[2]Ústav Fyzikálni Chemie a Electrochemie J. Heyrovskeho,
Praha, Czechoslovakia

Another possibility is the use of reference redox systems. The concept of a reference redox system has been developed during studies to find redox systems the electrode potentials of which are only modestly affected by the nature of the solvent. Preferably such variation in the electrode potentials should be smaller than the experimental error.

EXAMPLES OF REFERENCE REDOX SYSTEMS PROPOSED IN LITERATURE

Pleskov (ref.2) was the first to suggest such an idea and proposed the redox systems Rb^+/Rb and $Rb^+/Rb(Hg)$ respectively as such a reference redox system assuming that, because of the large ionic radius of Rb^+, changes in the Gibbs energy of solvation of this ion would be small. It was later found that the changes in the Gibbs energy of solvation of Rb^+ are too large to be ignored (ref.3).

A very thorough study by Strehlow and coworkers on various organometallic complexes has led to the formulation of the following requirements for a suitable reference redox system (ref.3): (i). The ions or molecules forming the reference redox system should preferably be spherical with as large a radius as possible, (ii) the ions should carry a low charge, (iii) the equilibrium at the electrode should be rapid and reversible, (iv) both components of the redox couple should be soluble, (v) no change in the geometry of the ligands should occur, (vi) the redox potential should be in a potential range that is accessible in as many solvents as possible and (vii) both forms should be stable enough to permit potentiometric measurements. Strehlow suggested the systems ferrocene/ferricenium ion (ferrocene is bis-(η-cyclopentadienyl)iron(II)) and cobaltocene/cobalticenium ion (cobaltocene is bis-(η-cyclopentadienyl)cobalt(II)). Besides these two several other reference redox systems have been suggested and used, such as tris(2,2'-bipyridine)iron(I)/tris(2,2'-bipyridine)-iron(0)) (ref.4), bis(biphenyl)chromium(0)/bis(biphenyl)chromium(I) ion (ref.5,6) and redox systems based on polynuclear aromatic hydrocarbons and the respective radical ions (ref.7).

It is not the purpose of this recommendation to discuss the various extrathermodynamic assumptions to obtain single ion properties in nonaqueous solvents, this has been done already in various articles (ref.8-10). The possibility to calculate such data would only be an additional benefit in reporting electrode potential versus such systems. The prime object must be to agree on a procedure to obtain reliable values for electrode potentials in nonaqueous solvents.

RECOMMENDED REFERENCE REDOX SYSTEMS

For pragmatic reasons, namely the ease with which such reference redox systems can be used experimentally, especially in polarography and voltammetry - the two most commonly employed techniques - and the added benefit to use data already reported versus such reference redox systems, it is recommended that electrode potentials in nonaqueous solvents be reported versus a reference redox system.

As mentioned earlier the number of proposed reference redox systems has increased over the years. While each reference redox system may have its merits as an extrathermodynamic assumption, it is necessary to limit ourselves to two such reference redox systems. A greater number would defeat the purpose of this recommendation and would again hinder an easy compilation of electrode potentials in nonaqueous solvents. Although one reference redox

*Membership of the Commission for 1979-83 during which the report was prepared was as follows:

Chairman: R. PARSONS (France) 1979-81; K. E. HEUSLER (FRG) 1981-83; *Vice-Chairman:* A. BARD (USA) 1979-81; S. TRASATTI (Italy) 1981-83; *Secretary:* J. C. JUSTICE (France); *Other Titular Members:* J. KŮTA† (Czechoslovakia); K. NIKI (Japan); *Associate Members:* J. N. AGAR (UK); E. BUDEVSKI (Bulgaria); G. GRITZNER (Austria); H. HOLTAN (Norway); N. IBL† (Switzerland); M. KEDDAM (France); R. MEMMING (FRG); B. MILLER (USA); J. A. PLAMBECK (Canada); Y. SATO (Japan); R. TAMAMUSHI (Japan); *National Representatives:* A. J. ARVÍA (Argentina); B. E. CONWAY (Canada); G. HORÁNYI (Hungary); S. K. RANGARAJAN (India); J. W. TOMLINSON (New Zealand); S. MINČ (Poland); E. MATTSSON (Sweden); A. K. COVINGTON (UK); M. KARSULIN (Yugoslavia) 1979-81; D. DRAŽIC (Yugoslavia) 1981-83.

†Deceased.

system should be the ultimate goal, two are suggested for the present, since not every reference redox system can be used in all solvents. In order to maintain the same experimental conditions the reference redox system is added to the solution studied with exclusion of air and water. Quite often the redox potential of a reference redox system will coincide with the redox potential of the species under study, making the evaluation of data obtained by polarographic and voltammetric techniques difficult. These problems can be circumvented by use of two reference redox systems, the potential differences of which have been determined.

Any selection of two out of several published reference redox systems must be somewhat arbitrary. It is recommended that the two systems ferrocene/ferricenium ion and bis(biphenyl) chromium(0)/bis(biphenyl)chromium(I) ion be used as reference redox systems. Both systems fulfill the requirements for reference redox systems in many solvents. The difference between the respective redox potentials of these two reference systems has been found to be almost constant in a variety of solvents (cf.refs. 11, 15-17; Table 1). The observed deviation from the average value of 1.124 V obtained for 22 solvents was $\pm$12 mV, which is somewhat larger that the experimental error of $\pm$4 mV.

For polarographic and voltammetric measurements where only one form of the redox system has to be added to the solution, ferrocene and bis(biphenyl)chromium tetraphenylborate* should be used.

This recommendation, however, should not discourage further studies on reliable reference electrodes in nonaqueous solvents nor should it discontinue the search for a more nearly "ideal" reference redox system to report standard electrode potentials on a solvent independent scale. It is merely asked that electrode potentials versus one of the two recommended reference redox systems, measured under conditions outlined in this paper, should be included among the experimental data and that the redox behaviour of these two systems should also be investigated for solvents in which such work has not yet been carried out. Besides facilitating the compilation of electrode potentials in nonaqueous solvents the use of a reference redox system in reporting electrode potentials would be of considerable assistance in obtaining further information on the possibility of establishing a scale of standard electrode potentials in nonaqueous solvents and in increasing our understanding of the solvent effects on the behaviour of redoxactive species.

Note: A destinction is made in this recommendation between the expressions "reference electrode" and "reference redox system". The expression reference electrode is used for an electrode with a known, invariant, electrode potential as defined under 8.9 in the Manual of Symbols and Terminology for Physicochemical Quantities and Units, Appendix III - Electrochemical Nomenclature (PAC 37 (1974) 501). The expression reference redox system is reserved for the recommended systems: ferrocene/ferricenium ion and bis(biphenyl)-chromium(0)/bis(biphenyl)chromium(I) ion.

The reference redox systems can be used in arrangements typical for a reference electrode e.g. a platinum wire in contact with an equimolar mixture of ferrocene and ferricenium picrate in the solvent under investigation. Such a setup will be preferred in potentiometric measurements. In polarographic and voltammetric studies however it is sufficient to add one form of the reference redox system (ferrocene or bis(biphenyl)chromium(I) tetraphenylborate) to the solution under study. The other form of the redox couple is formed at the working electrode during the measurement while a current is passed. In case of fast electrode kinetics the redox potential measured will agree to the conditional standard potential for this system. Quite frequently such measurements will be carried out in a three electrode arrangement with reference electrodes other than the recommended reference redox system serving as a reference electrode.

The expression "pilot ion" has been used in the past to specify the technique of adding only one form of the redox couple. The expression pilot ion however does not seem to be fully adequate since in any case it is a redox system which is used as a reference system not one form of the redox couple only.

RECOMMENDATIONS FOR MEASURING ELECTRODE POTENTIALS IN NONAQUEOUS SOLVENTS

It is of prime importance that all measurements by polarographic and voltammetric techniques are made versus reference electrodes the potentials of which do not change during the time of the experiment. If liquid junctions are part of the experimental arrangement it is mandatory that the liquid junction potential remains constant within the experimental error during the time of measurement. The stability of the reference electrode and the liquid junction potential (if included) should be checked carefully. If no variation has been observed in the obtained potential values for the redox system under investigation, a solution containing the same concentration of the supporting electrolyte and the reference redox system should be added to the solution under study with strict exclusion of air and water. After this addition both the studied redox system and the reference redox system should be measured at least twice versus the reference electrode. The potential data of the studied redox system before and after the addition of the reference redox system measured versus the reference electrode should agree within the limits of experimental error. The difference in potentials between the studied redox system and the reference redox system can then be calculated. The data for the two measurements must agree within the experimental error.

---

*Note: Small amounts of bis(biphenyl)chromium tetraphenylborate can be obtained upon request from G. Gritzner, Institut für Chemische Technologie Anorganischer Stoffe, Johannes-Kepler-Universität Linz, A-4040 L i n z / Auhof, Austria

If a stable reference electrode without liquid junction is used in polarographic and voltammetric measurements it is sufficient to measure the redox potential of ferrocene and bis(biphenyl)chromium(I) tetraphenylborate versus such an electrode. For solvents in which these reference redox systems have not been studied, the polarographic and voltammetric behaviour of these systems should also be investigated.

Potentiometric measurements versus the suggested reference redox systems as a reference electrode should be done in arrangements without a liquid junction. Care should be taken that such electrodes yield stable and reproducible data. Since a slow decomposition of the ferricenium ion or the bis(biphenyl)chromium compounds may occur in several solvents, freshly prepared electrodes should be used. As in any studies in nonaqueous media the purity of the solvent under investigation is of utmost importance. Great care should be taken in purifying the solvents; the methods of purification should either be reported or reference given to published procedures. Since traces of water or other impurities may still be present

TABLE 1 Difference between $E_{1/2}$ or $1/2(E_{pa}+E_{pc})$ values,* for ferrocene and bis(biphenyl)chromium tetraphenylborate (ΔE) in tetraethylammonium perchlorate solutions (conc: 0.1 mol l^{-1}) and/or tetrabutylammonium perchlorate solutions (conc: 0.1 mol l^{-1}) in 22 solvents[a]. (T = 298 K)

Solvent	Reference no:	ΔE (Ferrocene-bis(biphenyl)-chromium(I))
1,2-Dichloroethane	(11)	1.13_1 [b]
Dichloromethane	(11)[d]	1.14_8 [b]
		1.15_2 [c]
Nitromethane	(11)	1.11_2
Nitrobenzene	(11)	1.13_0
Acetonitrile	(11)	1.11_8
		1.11_9 [b]
Propylene carbonate	(11)	1.11_4
Butyrolactone	(11)	1.11_2
Acetone	(11)	1.13_0
Methanol	(11)	1.13_4
Ethanol	(11)	1.13_4
N,N-Dimethylformamide	(11)	1.12_7
Methyl-2-pyrrolidinone	(11)	1.12_6
Dimethyl sulfoxide	(11)	1.12_3
Tetramethylene sulfone	(11)	1.11_4
N,N-Dimethylacetamide	(16)	1.13_1 [b]
N,N-Diethylacetamide	(16)	1.13_5
Formamide	(16)	1.12_9 [b]
Trimethyl phosphate	(16)	1.13_0 [b]
Tetramethylurea	(16)	1.13_0 [b]
Hexamethylphosphoric triamide	(16)	1.12_4
N-Methylformamide	(17)	1.13_5
2,2'-Thiodiethanol	(15)	1.12_1 [b]

(a) For further data such as ($E_{3/4}-E_{1/4}$), ($E_{pa}-E_{pc}$), etc. see references 11, 15-17

(b) Tetrabutylammonium perchlorate solution (conc: 0.1 mol l^{-1})

(c) Saturated solution of tetraethylammonium perchlorate

(d) Ferrocene undergoes an irreversible electrode reaction in dichloromethane

*Bis(biphenyl)chromium(I) tetraphenylborate can be studied polarographically on the dropping mercury electrode and by cyclic voltammetry on the stationary platinum electrode. The $E_{1/2}$ and $\frac{1}{2}(E_{pa}+E_{pc})$ values were found to agree within ± 1 mV. The same holds true for ferrocene in those cases where ferrocene can be studied on the dropping mercury electrode.

even upon careful purification of the solvent, the concentration of such impurities should be reported; they may or may not affect the redox behaviour of the investigated species. Their controlled addition to the solution under study will yield such information and should be carried out.

RECOMMENDATION ON REPORTING ELECTRODE POTENTIALS IN NONAQUEOUS SOLVENTS VERSUS A REFERENCE REDOX SYSTEM

It is recommended that electrode potentials of systems under study be reported directly with respect to whichever of the two reference redox systems is chosen, in all solvents and at all temperatures (insofar as possible). Potentiometric measurements should be made in cells without liquid junctions. Salt bridges should be avoided if possible. In any case a detailed description of the apparatus should be given.

Electrode potentials measured by polarographic and voltammetric techniques are generally made in solutions containing a supporting electrolyte. While it is usually assumed (and in many cases true) that tetraalkylammonium perchlorates, most commonly employed as supporting electrolytes, do not significantly affect redox potentials, cases have been reported where altering the cation (ref.12) or the anion (ref.13) of the supporting electrolyte significantly changed the redox potentials of redox active species. During studies of organic compounds leading to radical cations or anions and of the redox behaviour of metal complexes it was noticed that even small amounts of metal cations or anions significantly affected half wave and peak potentials (ref.14). It is therefore mandatory that the nature and the concentration of the supporting electrolyte and of any other compound in solution besides the redox system of interest are reported in the publication.

All data should be reported as "crude" data first, omitting any corrections for ion-pair formation or activity coefficients. While such correction can be made in the further evaluation of the data, the absence of the crude data may prohibit the evaluation as improved models for ion-pair formation or activity coefficients become available.

Abbreviations: It is recommended that the abbreviation Fc be used for the reference redox system ferrocene/ferricenium ion and BCr for bis(biphenyl)chromium(0)/bis(biphenyl)-chromium(I) ion. These abbreviations should be added when symbols for redox potentials are given in the text or in figures, for example $E_{1/2(BCr)}$ or $E^{o}_{(Fc)}$. When specifying the respective reference redox systems, the following abbreviations are suggested: Fc/Fc^{+} and BCr/BCr^{+}. Such abbreviations however should only be used in a publication after specifically defining the names of the compounds to which they refer.

GIBBS ENERGIES OF TRANSFER

Both reference redox systems have also been employed as extrathermodynamic assumptions to calculate Gibbs energies of transfer of single ions. Such data have generally been obtained from polarographic half wave potentials of cations that are reversibly reduced to metal amalgams. It is recommended that the extrathermodynamic assumption employed be indicated by abbreviations such as $\Delta G^{o}_{t(Fc)}$ when reporting Gibbs energies of transfer. Again the meaning of such abbreviations must be defined in each publication.

REFERENCES

1) J.W. Diggle and A.J. Parker, Electrochim. Acta 18, 975 (1973) and references cited therein
2) W.A. Pleskow, Usp. Chim. 16, 254 (1947)
3) Z.M. Koepp, H. Wendt and H. Strehlow, Z. Elektrochemie 64, 483 (1960)
4) N. Tanaka and T. Ogata, Inorg. Nucl. Chem. Letters 10, 511 (1974)
5) A. Rusina, G. Gritzner and A.A. Vlček, Proc. IVth Internat. Congress on Polarography Prague 1966, p. 79
6) G. Gritzner, V. Gutmann and R. Schmid, Electrochim. Acta 13, 919 (1968)
7) D. Bauer and J.P. Beck, Bull. Soc. Chim. France 1252 (1973)
8) O. Popovych, Crit. Rev. Anal. Chem. 1, 73 (1970)
9) D. Bauer and M. Breant, Electroanalytical Chemistry, Vol. 8, p. 282-348 (Marcel Dekker Inc. 1975)
10) A.J. Parker, Electrochim. Acta 21, 671 (1976)
11) G. Gritzner, Inorg. Chim. Acta 24, 5 (1977)
12) G. Gritzner, K. Danksagmüller and V. Gutmann, J. Electroanal. Chem. 72, 177 (1976)
13) T. Fujinaga and I. Sakamoto, J. Electroanal. Chem. 73, 235 (1976)
14) T.M. Krygowski, J. Electroanal. Chem. 35, 436 (1972)
15) G. Gritzner and P. Rechberger, J. Electroanal. Chem. 109, 333 (1980)
16) G. Gritzner and P. Rechberger, J. Electroanal. Chem. 114, 129 (1980)
17) G. Gritzner, J. Electroanal. Chem. 144, 259 (1983)

ELECTRODE REACTION ORDERS, TRANSFER COEFFICIENTS AND RATE CONSTANTS: AMPLIFICATION OF DEFINITIONS AND RECOMMENDATIONS FOR PUBLICATION OF PARAMETERS (Recommendations 1979)

COMMISSION ON ELECTROCHEMISTRY*

ROGER PARSONS

Laboratoire d'Electrochimie Interfaciale, CNRS,
92190 Meudon, France

Electrode reaction rate constants and transfer coefficients have been defined in Appendix III of the Manual of Symbols and Terminology for Physicochemical Quantities and Units (Pure and Applied Chemistry, 37 (1974) 501). In that publication the aim was to provide general definitions in the simplest form. However, it is desirable that further amplification of these definitions should be given in view of the complications and confusions which exist in the literature. A summary of recommendations appears at the end of this document.

1. THE ELECTRODE REACTION

In most steady state electrochemical experiments the primary observation is the relationship between the current and the potential for a single electrode process. The term electrode process is used to describe the totality of changes occurring at or near a single electrode during the passage of current. At a given potential the current is controlled by the kinetics of a number of steps which include the transport of reactants to and from the interface and the interfacial reaction itself. The latter, which is called the electrode reaction, must always include at least one elementary step in which charge is transferred from one phase to the other, but may also involve purely chemical steps within the interfacial region. The charge transfer step, like the electrode reaction is normally written as if the electrons were the carrier of charge across the interface. Consequently this step is frequently referred to as an electron transfer step. Since charge can also be transferred by ions, the more general term charge transfer step is preferred.

The complete kinetic analysis of an electrode reaction would require the determination of the rate constants of each of its steps. Unless the reaction is simple this is a difficult problem almost certainly involving studies as a function of time or frequency. Even when they have used such techniques most studies of electrode reactions up to the present time have provided information on the kinetics of the electrode reaction as a whole and only rarely on its elementary steps. The aim of the present document is to make recommendations about the reporting of kinetic data for the electrode reaction as a whole, which are the quantities normally obtained from steady state studies, although these recommendations may also be used for the results of some time dependent studies.

---

*Titular Members : N. Ibl, Chairman (Switzerland), R. Parsons, Vice Chairman (France), J.C. Justice, Secretary (France), A.J. Bard (USA), K.E. Heusler (FRG), J. Kuta (Czechoslovakia), K. Niki (Japan), S. Trasatti (Italy). Associate Members : R.A. Durst (USA), I. Epelboin (France), M. Froment (France), R. Haase (FRG), H. Holtan (Norway), R. Tamamushi (Japan). National Representatives : J.W. Tomlinson (New-Zealand), S. Minc (Poland), M. Karsulin (Yugoslavia).

*Membership of the Commission (1977-1979) during which the report was prepared is as follows:

Chairman: N. IBL (Switzerland); *Vice-Chairman:* R. PARSONS (France); *Secretary:* J. C. JUSTICE (France); *Titular Members:* A. J. BARD (USA); K. E. HEUSLER (FRG); J. KŮTA (Czechoslovakia); K. NIKI (Japan); S. TRASATTI (Italy); *Associate Members:* R. A. DURST (USA); I. EPELBOIN (France); M. FROMENT (France); R. HAASE (FRG); H. HOLTAN (Norway); R. TAMAMUSHI (Japan); *National Representatives:* J. W. TOMLINSON (New Zealand); S. MINČ (Poland); M. KARSULIN (Yugoslavia).

2. OBSERVABLE OR OPERATIONAL RATE CONSTANTS

The observable electrode reaction rate constant is the constant of proportionality expressing the dependence of the rate of the electrode reaction on the interfacial concentration of the chemical species involved in the reaction. It may be possible to determine this interfacial concentration directly, but more often it must be calculated using a model of the mass transport in the regions adjacent to the interface, as well as chemical reactions at the interface and in the adjoining phases, appropriate to the conditions of the experiment. (Note that the term interfacial concentration is used to denote the local volume concentration at the boundary of the diffusion layer towards the electrode, but not within the electrical double layer) (see 4.1 of Bulletin , Nomenclature for Transport Phenomena in Electrolytic Systems by N. Ibl, Pure and Applied Chemistry, in press). When these regions overlap one, method of applying these models is to vary the mass transport conditions (e.g. rotation speed, potential sweep speed) and then to extrapolate the observed total rate (= current) to a hypothetical state in which all processes in the cell other than the interfacial reaction are not perturbed from their equilibrium state. The great variety of methods for carrying out this procedure will not be discussed here as they are well described in modern text books. It may be noted, however, that many of these methods depend on previous knowledge (or assumption) about the order of the reaction. In some systems, especially when the interfacial reaction is slow, the perturbation from the equilibrium state of processes other than the interfacial reaction may actually be negligible. The dependence of the current on the bulk concentration may then be used directly.

The faradaic current density measures the rate of the interfacial reaction, the proportionality constant being the charge number of the reaction multiplied by the Faraday constant. Thus the study of the current as a function of the interfacial concentration of all reactants at a given potential is equivalent to the determination of the order of the reaction with respect to each reactant. The net rate is the difference between forward and reverse rates, i.e. the net current is the sum of cathodic and anodic partial currents. To simplify the analysis it is preferable to work at potentials sufficiently far from the equilibrium value such that one partial current may be neglected in comparison with the other. Under these conditions many, but by no means all, electrode reactions follow a simple rate law in which the interfacial concentrations of the reactants are raised to a power which is constant and usually integral. This is equivalent to the expressions 8.10.1 and 8.10.2 in Appendix III, i.e. for the anodic partial current

$$I_a = (n/\nu)F A k_{ox} \prod_i c_i^{\nu_{i,a}} \qquad 2.1$$

and for the cathodic partial current

$$I_c = -(n/\nu)F A k_{red} \prod_i c_i^{\nu_{i,c}} \qquad 2.2$$

Here n is the charge number of the electrode reaction as written and ν_i is the order of reaction with respect to reactant i. In Appendix III the two concentration products are written as if they were identical. Here subscripts a and c have been added to the ν_i to indicate that different orders of reaction with respect to each species will be observed in the anodic and cathodic reactions.

The order of reaction ν_i may be positive or negative and need not be integral. It is not necessarily equal to a stoichiometric number in the overall reaction. ν (without subscript) which was omitted in 8.10, is the stoichiometric number giving the number of identical activated complexes formed and destroyed in the completion of the overall reaction as formulated with charge number n (for further explanation see section 3). ν is a positive number. A is the area of the electrode-solution interface over which the reaction occurs uniformly. Experimental arrangements should ensure that the reaction occurs uniformly over the interface. The area A may be a geometrical area or a true area, but if the latter, the way in which this is determined must be clearly stated. Note that, under some conditions, the area available for the reaction may itself be dependent on the potential. For some reactions it is not possible to find a simple rate law of the form of equations 2.1 and 2.2. These are complex reactions, often involving adsorbed species. It is not possible to assign a single overall rate constant to such a reaction. Further analysis into component reactions is necessary.

When the kinetics of the electrode reaction follow equation 2.1 and 2.2 the rate constant can be calculated directly from these equations at the potential of the electrode.

The characteristic feature of an electrode reaction, due to the transfer of charge associated with it, is the dependence of the rate constant on potential. This dependence is expressed by the transfer coefficient in the form

$$\alpha_c/\nu = -(RT/nF)\ (\partial\ \ln k_{red}/\partial E)_{T,p} \qquad 2.3$$

for the cathodic reaction and

$$\alpha_a/\nu = (RT/nF)\ (\partial \ln k_{ox}/\partial E)_{T,p} \qquad 2.4$$

and the quantities α_c/ν and α_a/ν may be considered as the observable transfer coefficients for the cathodic and anodic reactions respectively. If the interfacial reactant concentrations can be held constant experimentally as the potential of the electrode varies then the rate constants in 2.3 and 2.4 may be replaced by the partial currents and equations 8.11.1 and 8.11.2 of Appendix III are obtained. This replacement is valid if the rate law follows 2.1 and 2.2 and may also be valid for some more complex rate laws.

If it is found that the transfer coefficient is independent of potential E, then 2.3 and 2.4 may be integrated simply to obtain

$$k_{red} = k^o_{red}\ \exp[\ -(\alpha_c\ nF/\nu\ RT)\ (E-E^o)] \qquad 2.5$$

$$k_{ox} = k^o_{ox}\ \exp[\ (\alpha_a\ nF/\ \nu RT)\ (E-E^o)] \qquad 2.6$$

where k^o_{red} and k^o_{ox} are the rate constants at some standard potential E^o. The only satisfactory standard potential which yields a standard rate constant without some arbitrary content is the conditional (formal) potential $E^{o'}_c$ on the concentration scale of the electrode reaction whose kinetics are being studied. It follows from the equilibrium condition at this potential that there is a single standard rate constant for the cathodic and anodic reactions which is defined as k^o_c in equations 8.12.1 and 8.12.2 of Appendix III.

Note that there is no justification for describing the coefficient of $(F/RT)\ (E-E^o)$ in the exponent of 2.5 and similar expressions as αn_a where n_a is 'the number of electrons transferred in the rate-determining reaction'. This description is misleading and may be incorrect ; its use should be discontinued.

Equations 8.12.1 and 8.12.2 of Appendix III could be used even if the transfer coefficient is dependent on potential. This procedure is not recommended, but if it is used the result should not be called a conditional (or standard) rate constant. In this case it is preferable to quote the experimental rate constants as functions of potential.

When the interfacial reaction is fast compared to the rate of transport of reactant it may be impossible to make measurements sufficiently far from equilibrium that a single current component can be measured. The accessible quantity is then the slope of the total current with respect to potential at the equilibrium potential which is the effective conductance of the electrode. From 2.5 or 2.6, or more generally from irreversible thermodynamics, it can be shown that

$$(\partial I/\partial E)_{\eta=0} = I_o\ nF/\nu\ RT \qquad 2.7$$

where I_0 is the exchange current. From the concentration dependence of the exchange current it is possible to obtain some information about the orders of reaction and the transfer coefficient. Thus

$$(\partial \ln\ I_o/\partial \ln\ c_B)_{c_i} = \{\alpha_c\ \nu_{B,a} + \alpha_a\ \nu_{B,c}\} \qquad 2.8$$

where $\nu_{B,c}$ and $\nu_{B,a}$ are the orders of the cathodic and anodic reactions respectively with respect to species B (see 2.1 and 2.2) c_B is the concentration of species B and c_i is the concentration of any other reactant. Note that, at equilibrium, the interfacial concentration is equal to the bulk concentration. In the general case it is not possible to disentangle these orders from the transfer coefficients and the stoichiometric number even though at a given potential, in particular the equilibrium potential,

$$\alpha_a + \alpha_c = 1 \qquad 2.9$$

However, in many reactions the situation is sufficiently simple that assumptions may legitimately be made about some of these parameters so that the values of ν, ν_i and α may be found . The conditional rate constant may be obtained from

$$I_o = A(n/\nu)F\, k_c^{\theta}\, \Pi_i\, c_i^{\nu_i} \qquad 2.10$$

where

$$\nu_i = \{\alpha_c\, \nu_{i,a} + \alpha_a\, \nu_{i,c}\} \qquad 2.11$$

3. THE STOICHIOMETRIC NUMBER ν

This quantity is introduced into the equations for the kinetics of complex reactions to allow for the possibility that the rate-determining step may occur more or less than once in the completion of the overall reaction as written. It is clear therefore that ν is arbitrary in the same way that n is arbitrary because it depends on the way in which one chooses to write the equation down. Thus, for example, it is possible to write

$$N_2H_5^+ + 3H^+ + 2e \rightarrow 2NH_4^+ \qquad 3.1$$

with n = 2, or

$$\tfrac{1}{2}N_2H_5^+ + \tfrac{3}{2}H^+ + e \rightarrow NH_4^+ \qquad 3.2$$

with n = 1 and it is well known that this choice leads to no ambiguity in the associated thermodynamics. If it is supposed that the rate determining step in this reaction is the proton transfer

$$NH_3(\text{adsorbed}) + H^+ \rightarrow NH_4^+ \qquad 3.3$$

then this reaction would occur twice (ν = 2) in the completion of 3.1 or once (ν = 1) in the completion of 3.2.

In general the quantity n/ν is not arbitrary and is a true characteristic of the reaction kinetics. This ratio is the charge number which always appears in the kinetic equations. In principle the use of ν could be avoided by choosing n so that ν is always unity but this requires more knowledge about the detailed kinetics of a complex reaction than is often available.

The stoichiometric number ν can be determined once n is chosen if a measurement of the current-potential curve can be made close to, and one far from, equilibrium. The value of I_o obtained from the latter is then used in equation 2.7 to obtain ν. A measurement of the current-potential curve on either side of equilibrium and far from it yields the same information. Both methods assume that the rate determining step is the same in the two regions of measurement. When measurements can be made only close to equilibrium it is not possible to find ν. However, fortunately most such fast reactions are simple.

4. RATE CONSTANTS FOR ELEMENTARY STEPS OR SIMPLE REACTIONS IN THE ABSENCE OF ADSORPTION

Such reactions follow the kinetic laws 2.1 and 2.2 with the concentration products taking rather simple forms. The orders of reaction ν_i are integral and usually 1. Only one or two species appear in these products and no species appears in the product for both anodic and cathodic reaction. The stoichiometric number is unity if the reaction written is the elementary step.

The equations in Section 1 hence take a simple form and for a first order reaction reduce to those given in Appendix III.

There seems no good reason to recommend a special term for the transfer coefficient of an elementary step.

5. RATE CONSTANTS FOR ELEMENTARY STEPS IN THE PRESENCE OF ADSORPTION

When reactants or intermediates are adsorbed, the rate of reaction may no longer be related to the concentration by a simple law like 2.1 or 2.2. The deviation may be due to either entropic or energetic effects or both. Methods for treating these kinetics are at present so diverse and controversial that it is not possible to make general recommendations now.

The situation best understood is that where a reactant is non-specifically adsorbed in the outer Helmholtz plane (inner boundary of the diffuse layer). The effect of such adsorption on electrode kinetics is usually termed the 'Frumkin Effect'. Rate constants, transfer coefficients etc. corrected for this effect are frequently called 'true' rate constants etc. It would be preferable to describe them as 'corrected for the Frumkin Effect', but in any case, if such a correction is carried out, the basis on which it is made should be clearly described.

Physical and chemical adsorption on the electrode surface is usually described by means of an adsorption isotherm and kinetic equations compatible with various isotherms such as the Langmuir, Frumkin, Temkin etc. are known. Rate constants and transfer coefficients calculated on the basis of such an equation should be accompanied by a clear statement of the equation used as well as the evidence for the applicability of this equation.

Note that adsorption effects in electrode kinetics are often potential dependent and they may lead to kinetic parameters which depend on potential and reactant concentration in a complex way.

RECOMMENDATIONS FOR THE PUBLICATION OF PARAMETERS DERIVED FROM ELECTRODE KINETIC DATA

1. Corrections for mass transport and ohmic drop should be as accurate as possible and the method should be clearly described.

2. The conditions by which a steady state is achieved and the time required for this achievement should be described. For measurements in the non-steady state the nature of the time dependent parameters as well as their control or measurement should be described.

3. Potential and concentration dependence of rate should be studied and expressed in terms of a transfer coefficient and reaction orders respectively.

4. If the transfer coefficient is potential dependent, rate constants should be quoted at the potential of measurement.

5. If the transfer coefficient is independent of potential, the conditional rate constant at the equilibrium potential of the reaction being studied should be quoted. The range of potential over which a constant transfer coefficient is observed should be stated.

6. If the equilibrium potential of the reaction under study is not known, the rate constant should be quoted at the potential of measurement rather than at some arbitrary standard potential.

7. Any corrections to kinetic parameters due to adsorption should be clearly described. The term "true" rate constant is not recommended.

Section 6

Colloid and Surface Chemistry

Colloid and surface chemistry deal with systems in which one phase or region is microscopic in at least one dimension. The small size makes measurements more complex and indirect, and, as a result, the definition and determination of quantities relating to the phase or region require care and precise expression. Table 6.1 from the Green Book contains the recommendations for symbols and quantities used in colloid and surface chemistry.

References 25 and 28 discuss symbols and terminology. Reference 25 covers surfaces and colloids. Table 6.1 provides condensed information on surfaces. Reference 28 discusses colloidal systems, surface-active agents, fluid films, sedimentation and other transport processes, and the electrochemistry of colloids. Section 5 covered some of the electrochemical properties. Reference 25 covers heterogeneous catalysis and is printed in Section 4.

This section contains four papers on how to present data in specific areas:

- "Reporting Physisorption Data for Gas/Solid Systems: With Special Reference to the Determination of Surface Area and Porosity" *(29)*
- "Reporting Data on Adsorption from Solution at the Solid/Solution Interface *(30)*
- "Reporting Experimental Pressure-Area Data with Film Balances" *(31)*
- "Reporting Experimental Data Dealing with Critical Micellization Concentrations (C.M.C.'s) of Aqueous Surfactact Systems" *(32)*

Table 6.1 is reprinted with permission from *Quantities, Units and Symbols in Physical Chemistry,* Section 2.14, pages 56–57. Copyright 1988 by the International Union of Pure and Applied Chemistry—All rights reserved.

Table 6.1. Colloid and Surface chemistry. Continued on next page.

The recommendations given here are based on more extensive IUPAC recommendations [1.e–h] and [30a].

Name	*Symbol*	*Definition*	*SI unit*	*Notes*
specific surface area	a, a_s, s	$a = A/m$	$m^2\ kg^{-1}$	
surface amount of B, adsorbed amount of B	$n_B{}^s$, $n_B{}^a$		mol	1
surface excess of B	$n_B{}^\sigma$		mol	2
surface excess concentration of B	Γ_B, $(\Gamma_B{}^\sigma)$	$\Gamma_B = n_B{}^\sigma/A$	$mol\ m^{-2}$	2
total surface excess concentration	Γ, (Γ^σ)	$\Gamma = \sum_i \Gamma_i$	$mol\ m^{-2}$	
area per molecule	a, σ	$a_B = A/N_B{}^\sigma$	m^2	3
area per molecule in a filled monolayer	a_m, σ_m	$a_{m,B} = A/N_{m,B}$	m^2	3
surface coverage	θ	$\theta = N_B{}^\sigma/N_{m,B}$	1	3
contact angle	θ		1, rad	
film thickness	t, h, δ		m	
thickness of (surface or interfacial) layer	τ, δ, t		m	
surface tension, interfacial tension	γ, σ	$\gamma = (\partial G/\partial A_s)_{T,p}$	$N\ m^{-1}$, $J\ m^{-2}$	

(1) The value of $n_B{}^s$ depends on the thickness assigned to the surface layer.
(2) The values of $n_B{}^\sigma$ and Γ_B depend on the convention used to define the position of the Gibbs surface. They are given by the excess amount of B or surface concentration of B over values that would apply if each of the two bulk phases were homogeneous right up to the Gibbs surface. See [1.e], and also additional recommendations on p.57.
(3) $N_B{}^\sigma$ is the number of adsorbed molecules ($N_B{}^\sigma = Ln_B{}^\sigma$), and $N_B{}^m$ is the number of adsorbed molecules in a filled monolayer. The definition to entities B.

Table 6.1. Continued.

Name	*Symbol*	*Definition*	*SI unit*	*Notes*
film tension	Σ_f	$\Sigma_f = 2\gamma_f$	$N\,m^{-1}$	4
reciprocal thickness of the double layer	κ	$\kappa = [2F^2 I_c/\varepsilon RT]^{\frac{1}{2}}$	m^{-1}	
average molar masses				
number-average	M_n	$M_n = \Sigma n_i M_i/\Sigma n_i$	$kg\,mol^{-1}$	
mass-average	M_m	$M_m = \Sigma n_i M_i^2/\Sigma n_i M_i$	$kg\,mol^{-1}$	
Z-average	M_Z	$M_Z = \Sigma n_i M_i^3/\Sigma n_i M_i^2$	$kg\,mol^{-1}$	
sedimentation coefficient	s	$s = v/a$	s	5
van der Waals constant	λ		J	
retarded van der Waals constant	β, B		J	
van der Waals–Hamaker constant	A_H		J	
surface pressure	π^s, π	$\pi^s = \gamma^0 - \gamma$	$N\,m^{-1}$	6

(4) The definition applies only to a symmetrical film, for which the two bulk phases on either side of the film are the same, and γ_f is the surface tension of a film/bulk interface.
(5) In the definition, v is the velocity of sedimentation and a is the acceleration of free fall or centrifugation. The symbol for a limiting sedimentation coefficient is $[s]$, for a reduced sedimentation coefficient s°, and for a reduced limiting sedimentation coefficient $[s^\circ]$; see [1.e] for further details.
(6) In the definition, γ^0 is the surface tension of the clean surface and γ that of the covered surface.

Additional recommendations

The superscript s denotes the properties of a surface or interfacial layer. In the presence of adsorption it may be replaced by the superscript a.

Examples Helmholtz energy of interfacial layer A^s
amount of adsorbed substance n^a, n^s
amount of adsorbed O_2 $n^a(O_2)$, $n^s(O_2)$, or $n(O_2, a)$

The subscript m denotes the properties of a monolayer.

Example area per molecule B in a monolayer $a_m(B)$

The superscript σ is used to denote a surface excess property relative to the Gibbs surface.

Example surface excess amount (or Gibbs surface excess of B) n_B^σ

In general the values of Γ_A and Γ_B depend on the position chosen for the Gibbs dividing surface. However, two quantities, $\Gamma_B^{(A)}$ and $\Gamma_B^{(n)}$ (and correspondingly $n_B^{\sigma(A)}$ and $n_B^{\sigma(n)}$), may be defined in a way that is invariant to this choice (see [1.e]). $\Gamma_B^{(A)}$ is called the *relative* surface excess concentration of B with respect to A, or more simply the relative adsorption of B; it is the value of Γ_B when the surface is chosen to make $\Gamma_A = 0$. $\Gamma_B^{(n)}$ is called the *reduced* surface excess concentration of B, or more simply the reduced adsorption of B; it is the value of Γ_B when the surface is chosen to make the total excess $\Gamma = \sum_i \Gamma_i = 0$.

Properties of phases (α, β, γ) may be denoted by corresponding superscript indices.

Examples surface tension of phase α γ^α
interfacial tension between phases α and β $\gamma^{\alpha\beta}$

Symbols of thermodynamic quantities divided by surface area are usually the corresponding lower case letters; an alternative is to use a circumflex.

Example interfacial entropy per area $s^s(=\hat{s}^s) = S^s/A$

The following abbreviations are used in colloid chemistry:

c.c.c. critical coagulation concentration
c.m.c. critical micellization concentration
i.e.p. isoelectric point
p.z.c. point of zero charge

REPORTING PHYSISORPTION DATA FOR GAS/SOLID SYSTEMS

with Special Reference to the Determination of Surface Area and Porosity

(Recommendations 1984)

COMMISSION ON COLLOID AND SURFACE CHEMISTRY INCLUDING CATALYSIS*

Prepared for publication by the Subcommittee on Reporting Gas Adsorption Data Consisting of

K. S. W. SING (UK, Chairman); D. H. EVERETT (UK); R. A. W. HAUL (FRG); L. MOSCOU (Netherlands); R. A. PIEROTTI (USA); J. ROUQUÉROL (France); T. SIEMIENIEWSKA (Poland)

Reporting physisorption data for gas/solid systems – with special reference to the determination of surface area and porosity

The purpose of this Manual is two-fold: first to draw attention to the problems and ambiguities which have arisen in connection with the reporting of gas adsorption (physisorption) data; second to formulate proposals for the standardisation of procedures and terminology which will lead to a generally accepted code of practice. The proposals are based on, and are in general accordance with, the Manual of Symbols and Terminology for Physicochemical Quantities and Units (1979) and Parts I and II of Appendix II (1972 and 1976).

The first stage in the interpretation of a physisorption isotherm is to identify the isotherm type and hence the nature of the adsorption process(es): monolayer-multilayer adsorption, capillary condensation or micropore filling. The BET method is unlikely to yield a value of the actual surface area if the isotherm is either Type I or Type III; but both Type II and Type IV isotherms are, in general, amenable to the BET analysis, provided that the value of C is neither too low nor too high and that the BET plot is linear in the region of the isotherm containing Point B.

The computation of mesopore size distribution is valid only if the isotherm is of Type IV, but in view of the complexity of most pore systems little is to be gained by the application of an elaborate method of computation. If a Type I isotherm exhibits a nearly constant adsorption at high relative pressure, the micropore volume is given by the amount adsorbed at the plateau. At present there is no reliable procedure available for the computation of the micropore size distribution from a single isotherm.

A check list is recommended to assist authors in the measurement of adsorption isotherms and the presentation of the data in the primary literature.

*Membership of the Commission during the period (1981–85) in which the report was prepared was as follows:

Chairman: 1981–83 J. Lyklema (Netherlands); 1983–85 K. S. W. Sing (UK); *Vice-Chairman*: 1981–85 J. Haber (Poland); *Secretary*: 1981–83 M. Kerker (USA); 1983–85 E. Wolfram (Hungary); *Members*: J. H. Block (FRG; Titular 1983–85, Associate 1981–83); N. V. Churaev (USSR; Associate 1981–85); D. H. Everett (UK; National Representative 1981–85); G. F. Froment (Belgium; National Representative 1981–85); P. C. Gravelle (France; Associate 1981–85); R. S. Hansen (USA; Titular 1981–83); R. A. W. Haul (FRG; National Representative 1981–83); J. W. Hightower (USA; Associate 1983–85); R. J. Hunter (Australia; Associate 1981–85); L. G. Ionescu (Brazil; National Representative 1983–85); A. S. Kertes (Israel; National Representative 1981–85); A. Kitahara (Japan; National Representative 1981–85); J. C. Kuriacose (India; National Representative 1983–85); J. Lyklema (Netherlands; National Representative 1983–85); A. Maroto (Argentina; Associate 1983–85, National Representative 1981–83); S. G. Mason (Canada; National Representative 1981–85); K. Meyer (GDR; National Representative 1981–85); P. Mukerjee (USA; Associate 1981–83); L. G. Nagy (Hungary; National Representative 1981–85); H. van Olphen (Netherlands; Associate 1981–83); J. A. Pajares (Spain; National Representative 1981–83); M. W. Roberts (UK; Titular 1981–83); J. Rouquérol (France; Associate 1983–85); K. S. W. Sing (UK; Associate 1981–83); P. Stenius (Sweden; Titular 1981–85, Associate 1981–83); M. S. Suwandi (Malaysia; National Representative 1983–85); L. Ter-Minassian-Saraga (France; Titular 1983–85, Associate 1981–83); A. Weiss (FRG; National Representative 1983–85); P. B. Wells (UK; Associate 1983–85); E. Wolfram (Hungary; Titular 1981–83).

SECTION 1. INTRODUCTION

Gas adsorption measurements are widely used for determining the surface area and pore size distribution of a variety of different solid materials, such as industrial adsorbents, catalysts, pigments, ceramics and building materials. The measurement of adsorption at the gas/solid interface also forms an essential part of many fundamental and applied investigations of the nature and behaviour of solid surfaces.

Although the role of gas adsorption in the characterisation of solid surfaces is firmly established, there is still a lack of general agreement on the evaluation, presentation and interpretation of adsorption data. Unfortunately, the complexity of most solid surfaces — especially those of industrial importance — makes it difficult to obtain any independent assessment of the physical significance of the quantities derived (e.g. the absolute magnitude of the surface area and pore size).

A number of attempts have been made (see Note a), at a national level, to establish standard procedures for the determination of surface area by the BET-nitrogen adsorption method. In addition, the results have been published (see Note b) of an SCI/IUPAC/NPL project on surface area standards. This project brought to light a number of potential sources of error in the determination of surface area by the gas adsorption method.

The purpose of the present Manual is two-fold: first to draw attention to the problems and ambiguities which have arisen in connection with the reporting of gas adsorption (physisorption) data; and second to formulate proposals for the standardisation of procedures and terminology which will lead to a generally accepted code of practice. The Manual does not aim to provide detailed operational instructions or to give a comprehensive account of the theoretical aspects of physisorption. The determination of the surface area of supported metals is not dealt with here — despite the importance of this topic in the context of heterogeneous catalysis — since this necessarily involves chemisorption processes.

The present proposals are based on, and are in general accordance with, the *Manual of Symbols and Terminology for Physicochemical Quantities and Units* (see Note c) and Parts I and II of Appendix II (see Note d). Although it has been necessary to extend the terminology, the principles are essentially those developed in Part I.

SECTION 2. GENERAL DEFINITIONS AND TERMINOLOGY

The definitions given here are essentially those put forward in Appendix II, Part I, §1.1 and Part II, §1.2. Where a caveat is added, it is intended to draw attention to a conceptual difficulty or to a particular aspect which requires further consideration.

Adsorption (in the present context, positive adsorption at the gas/solid interface) is the enrichment of one or more components in an interfacial layer. *Physisorption* (as distinct from *chemisorption*) is a general phenomenon: it occurs whenever an adsorbable gas (the *adsorptive*) is brought into contact with the surface of a solid (the *adsorbent*). The intermolecular forces involved are of the same kind as those responsible for the imperfection of real gases and the condensation of vapours. In addition to the attractive dispersion forces and the short range repulsive forces, specific molecular interactions (e.g. polarisation, field-dipole, field gradient-quadrupole) usually occur as a result of particular geometric and electronic properties of the adsorbent and adsorptive.

It is convenient to regard the interfacial layer as comprising two regions: the *surface layer* of the adsorbent (often simply called the *adsorbent surface*) and the *adsorption space* in which enrichment of the adsorptive can occur. The material in the adsorbed state is known as the *adsorbate*, as distinct from the adsorptive, i.e. the substance in the fluid phase which is capable of being adsorbed.

When the molecules of the adsorptive penetrate the surface layer and enter the structure of the bulk solid, the term *absorption* is used. It is sometimes difficult, impossible or irrelevant to distinguish between adsorption and absorption: it is then convenient to use the wider term *sorption* which embraces both phenomena and to use the derived terms *sorbent*, *sorbate* and *sorptive*.

Note a. British Standard 4359: Part 1: 1969. Nitrogen adsorption (BET method).

Deutsche Normen DIN 66131, 1973. Bestimmung der spezifischen Oberfläche von Feststoffen durch Gasadsorption nach Brunauer, Emmett und Teller (BET).

Norme Française 11–621, 1975. Détermination de l'aire massique (surface spécifique) des poudres par adsorption de gaz.

American National Standard, ASTM D 3663–78. Standard test method for surface area of catalysts.

Note b. D. H. Everett, G. D. Parfitt, K. S. W. Sing and R. Wilson, *J. appl. Chem. Biotech* 24, 199 (1974).

Note c. *Manual of Symbols and Terminology for Physicochemical Quantities and Units* prepared for publication by D. H. Whiffen, *Pure Applied Chem.*, 51, 1–41 (1979).

Note d. *Part I of Appendix II, Definitions, Terminology and Symbols in Colloid and Surface Chemistry*, prepared by D. H. Everett, *Pure Applied Chem.*, 31, 579–638 (1972).

Part II of Appendix II, Terminology in Heterogeneous Catalysis, prepared for publication by R. L. Burwell, Jr., *Pure Applied Chem.*, 45, 71–90 (1976).

The term *adsorption* may also be used to denote the process in which adsorptive molecules are transferred to, and accumulate in, the interfacial layer. Its counterpart, *desorption*, denotes the converse process, in which the amount adsorbed decreases. Adsorption and desorption are often used adjectivally to indicate the direction from which experimentally determined adsorption values have been approached, e.g. the adsorption curve (or point) and the desorption curve (or point). *Adsorption hysteresis* arises when the adsorption and desorption curves do not coincide.

The relation, at constant temperature, between the amount adsorbed (properly defined in Section 3.2) and the equilibrium pressure of the gas is known as the *adsorption isotherm*.

Many adsorbents of high surface area are porous and with such materials it is often useful to distinguish between the *external* and *internal* surface. The *external surface* is usually regarded as the envelope surrounding the discrete particles or agglomerates, but is difficult to define precisely because solid surfaces are rarely smooth on an atomic scale. A suggested convention is that the external surface be taken to include all the prominences and also the surface of those cracks which are wider than they are deep; the internal surface then comprises the walls of all cracks, pores and cavities which are deeper than they are wide and which are accessible to the adsorptive. In practice, the demarcation is likely to depend on the methods of assessment and the nature of the pore size distribution. Because the accessibility of pores may depend on the size and shape of the gas molecules, the area of, and the volume enclosed by, the internal surface as determined by gas adsorption may depend on the dimensions of the adsorptive molecules (*molecular sieve* effect). The roughness of a solid surface may be characterized by a *roughness factor*, i.e. the ratio of the external surface to the chosen geometric surface.

In the context of physisorption, it is expedient to classify pores according to their sizes:

(i) pores with widths exceeding about 50 nm (0.05 μm) are called *macropores*;

(ii) pores of widths between 2 nm and 50 nm are called *mesopores*;

(iii) pores with widths not exceeding about 2 nm are called *micropores*.

These limits are to some extent arbitrary since the pore filling mechanisms are dependent on the pore shape and are influenced by the properties of the adsorptive and by the adsorbent-adsorbate interactions. The whole of the accessible volume present in micropores may be regarded as adsorption space and the process which then occurs is *micropore filling*, as distinct from surface coverage which takes place on the walls of open macropores or mesopores. Micropore filling may be regarded as a primary physisorption process (see Section 8); on the other hand, physisorption in mesopores takes place in two more or less distinct stages (monolayer-multilayer adsorption and capillary condensation).

In *monolayer adsorption* all the adsorbed molecules are in contact with the surface layer of the adsorbent. In *multilayer adsorption* the adsorption space accommodates more than one layer of molecules so that not all adsorbed molecules are in direct contact with the surface layer of the adsorbent. In *capillary condensation* the residual pore space which remains after multilayer adsorption has occurred is filled with condensate separated from the gas phase by menisci. Capillary condensation is often accompanied by hysteresis. The term capillary condensation should not be used to describe micropore filling because this process does not involve the formation of liquid menisci.

For physisorption, the *monolayer capacity* (n_m) is usually defined as the amount of adsorbate (expressed in appropriate units) needed to cover the surface with a complete monolayer of molecules (Appendix II, Part I, §1.1.7). In some cases this may be a close-packed array but in others the adsorbate may adopt a different structure. Quantities relating to monolayer capacity may be denoted by the subscript m. The *surface coverage* (θ) for both monolayer and multilayer adsorption is defined as the ratio of the amount of adsorbed substance to the monolayer capacity.

The *surface area* (A_s) of the adsorbent may be calculated from the monolayer capacity (n_m^a in moles), provided that the area (a_m) effectively occupied by an adsorbed molecule in the complete monolayer is known.

Thus,

$$A_s = n_m^a . L . a_m$$

where L is the Avogadro constant. The *specific surface area* (a_s) refers to unit mass of adsorbent:

$$a_s = \frac{A_s}{m} .$$

Appendix II, Part I recommends the symbols A, A_s or S and a, a_s or s for area and specific area, respectively, but A_s and a_s are preferred to avoid confusion with Helmholtz energy A or entropy S.

In the case of micropore filling, the interpretation of the adsorption isotherm in terms of surface coverage may lose its physical significance. It may then be convenient to define a *monolayer equivalent area* as the area, or specific area, respectively, which would result if the amount of adsorbate required to fill the micropores were spread in a close-packed monolayer of molecules (see Section 8).

SECTION 3. METHODOLOGY

3.1 Methods for the determination of adsorption isotherms

The many different procedures which have been devised for the determination of the amount of gas adsorbed may be divided into two groups: (a) those which depend on the measurement of the amount of gas removed from the gas phase (i.e. gas volumetric methods) and (b) those which involve the measurement of the uptake of the gas by the adsorbent (e.g. direct determination of increase in mass by gravimetric methods). Many other properties of the adsorption system may be related to the amount adsorbed, but since they require calibration they will not be discussed here. In practice, static or dynamic techniques may be used to determine the amount of gas adsorbed.

In the static volumetric determination a known quantity of pure gas is usually admitted to a confined volume containing the adsorbent, maintained at constant temperature. As adsorption takes place, the pressure in the confined volume falls until equilibrium is established. The amount of gas adsorbed at the equilibrium pressure is given as the difference between the amount of gas admitted and the amount of gas required to fill the space around the adsorbent, i.e. the *dead space*, at the equilibrium pressure. The adsorption isotherm is usually constructed point-by-point by the admission to the adsorbent of successive charges of gas with the aid of a volumetric dosing technique and application of the gas laws. The volume of the dead space must, of course, be known accurately: it is obtained (see Section 3.2) either by pre-calibration of the confined volume and substracting the volume of the adsorbent (calculated from its density), or by the admission of a gas which is adsorbed to a negligible extent (see Section 3.2) Nitrogen adsorption isotherms at the temperature of the boiling point of nitrogen at ambient atmospheric pressure are generally determined by the volumetric method; they provide the basis for the various standard procedures which have been proposed for the determination of surface area (see references in Section 1).

Recent developments in vacuum microbalance techniques have revived the interest in gravimetric methods for the determination of adsorption isotherms. With the aid of an *adsorption balance* the change in weight of the adsorbent may be followed directly during the outgassing and adsorption stages. A gravimetric procedure is especially convenient for measurements with vapours at temperatures not too far removed from ambient. At both high and low temperatures, however, it becomes difficult to control and measure the exact temperature of the adsorbent, which is particularly important in the determination of mesopore size distribution.

In principle, a 'continuous' procedure can be used to construct the isotherm under quasi-equilibrium conditions: the pure adsorptive is admitted (or removed) at a slow and constant rate and a volumetric or gravimetric technique used to follow the variation of the amount adsorbed with increase (or decrease) in pressure. A carrier gas technique, making use of conventional gas chromatographic equipment, may be employed to measure the amount adsorbed provided that the adsorption of the carrier gas is negligible. In all types of measurement involving gas flow it is essential to confirm that the results are not affected by change in flow rate and to check the agreement with representative isotherms determined by a static method.

3.2 Operational definitions of adsorption

To examine the fundamental basis on which experimental definitions depend, consider an adsorption experiment incorporating both volumetric and gravimetric measurements (see Fig. 1). A measured amount, n, of a specified gas (see Note e), is introduced into the system whose total volume, V, can be varied at constant temperature, T. Measurements are made of V, p (the equilibrium pressure) and w (the apparent mass, m, of adsorbent).

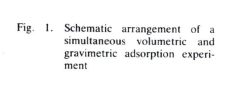

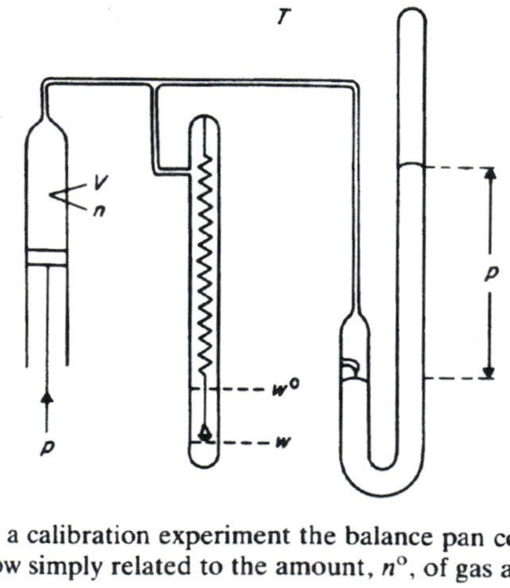

Fig. 1. Schematic arrangement of a simultaneous volumetric and gravimetric adsorption experiment

In a calibration experiment the balance pan contains no adsorbent. The total volume, V^o, of the system is now simply related to the amount, n^o, of gas admitted:

$$V^o = n^o v^g(T,p),$$

Note e. For simplicity, adsorption of a single gas is considered here (cf. Appendix II, Part 1, §1.1.11).

where $v^g(T,p)$ is the molar volume of the gas at T and p, and is known from its equation of state. If the buoyancy effect arising from the balance itself is negligible, the apparent weight will remain constant. Since the gas concentration, $c^o = 1/v^g$, is constant throughout the volume V^o,

$$n^o = \int_{\text{all } V^o} c^o \mathrm{d}V = c^o v^o = \frac{V^o}{v^g}.$$

A mass, m, of adsorbent (weighing w^o in vacuum) is now introduced and the experiment repeated using the same amount of gas. If adsorption is detectable at the given T, p, the volume V will usually be less than V^o and the apparent weight of the adsorbent will increase from w^o to w.

V/v^g is the amount of gas which would be contained in the volume V if the gas concentration were uniform throughout the volume. That the amount actually present is n^o means that local variations in gas concentration must occur: the gas concentration within the bulk of the solid is zero, but is greater than c^o in the interfacial layer. The difference between n^o and V/v^g may be called the *apparent adsorption*

$$n^a\,(\text{app}) = n^o - V/v^g,$$

and is a directly observable quantity. The precision with which it can be measured is controlled only by the experimental precision in T, p and V, and by the reliability with which v^g (and c^o) can be calculated from the equation of state of the gas.

The apparent adsorption may, alternatively, be defined by measuring the amount of gas which has to be added to the system at constant T, p to increase the volume V back to V^o. The apparent adsorption is then equal to the extra amount of gas which can be accommodated in a volume V^o at a given T, p when the solid is introduced. It can, therefore, be expressed in terms of the local deviations of the concentration, c, of adsorptive molecules, from the bulk concentration c^o:

$$n^a(\text{app}) = \int_{V^o} (c - c^o)\mathrm{d}V.$$

If the gas does not penetrate into the bulk solid (i.e. is not *ab*sorbed), the above integral consists of two parts, that over the volume occupied by the solid (V^s) within which $c = o$, and that over the adsorption space plus the gas phase volume, which taken together is denoted by V^g:

$$n^a(\text{app}) = -c^o V^s + \int_{V^g} (c - c^o)\mathrm{d}V.$$

The first term represents the amount of gas excluded by the solid, while the second is the extra amount of gas accommodated because of the accumulation of gas in the neighbourhood of the solid surface. If the adsorption is very weak the first term may exceed the second and the apparent adsorption may be negative ($V > V^o$).

The quantity

$$n^a = n^a(\text{app}) + c^o V^s = \int_{V^g} (c - c^o)\mathrm{d}V$$

is thus equal to the *Gibbs adsorption* (see note f) (surface excess amount of adsorbed substance — see Part I, §1.1.8) when the surface of the solid is taken as the Gibbs dividing surface: it is the difference between the amount of substance actually present in the interfacial layer and that which would be present at the same equilibrium gas pressure in a reference system in which the gas phase composition is constant up to the Gibbs surface, and in which no adsorptive penetrates into the surface layer or the bulk of the solid.

The operational definition of n^a is thus

$$n^a = \int_{V^g} (c - c^o)\mathrm{d}V = n - c^o V^g,$$

where n is the total amount of gas admitted.

The precision with which n^a can be determined depends, not only on the precision of T, p, V and v^g but also on the precision with which V^s (and hence V^g) is known.

The volume of the solid (i.e. the volume enclosed by the Gibbs surface) is often defined experimentally as that volume which is not accessible to a non-sorbable gas (e.g. helium — leading to the ***helium dead-space***). In making this identification it is assumed that the volume available to He atoms is the same as that for molecules

Note f. n_i^σ was used to denote the Gibbs adsorption in Part I, §1.1.8.

of the gas under investigation (which is not true, for example, if the solid acts as a molecular sieve, or if the molecules of the gas are significantly larger than the He atom), and that the solid does not swell under the influence of the adsorbate. Helium adsorption may occur if the solid contains very fine pores (or pore entrances) and the only proof that the adsorption is zero is that the apparent value of V^s is independent of temperature. It is usual to take the high temperature limit of V^s as being the correct value, but if V^s is determined at a temperature widely different from that used in an adsorption experiment, a correction for the thermal expansion of the solid may be required.

Alternatively, V^s may be estimated from the known density of the bulk solid with the implied assumption that this is the same as that of the material of the adsorbent.

The above discussion in terms of the volumetric technique when applied to gravimetric measurements gives for the apparent change in weight

$$\Delta w = w - w_0 = \left[n^a - \frac{V^s}{v^g} \right] M$$

where M is the molar mass of the adsorptive.

Thus $$n^a = \frac{\Delta w}{M} + \frac{V^s}{v^g} .$$

The second term on the right-hand-side is the buoyancy correction which has the same origin as the dead-space correction in a volumetric determination.

An alternative but less useful definition of adsorption is

$$n^s = \int_{V_a} c\,dV$$

where $V_a = \tau A_s$ is the volume of the interfacial layer (see Note g) and c is the local concentration. V_a has to be defined on the basis of some appropriate model of gas adsorption which gives a value of τ the layer thickness. Provided that the equilibrium pressure is sufficiently low and the adsorption not too weak, then

$$n^s = n^a ;$$

the surface excess amount (n^a) and total amount (n^s) of substance in the adsorbed layer become indistinguishable and the general term *amount adsorbed* is applicable to both quantities.

SECTION 4. EXPERIMENTAL PROCEDURES

4.1 Outgassing the adsorbent

Prior to the determination of an adsorption isotherm all of the physisorbed species should be removed from the surface of the adsorbent. This may be achieved by outgassing, i.e. exposure of the surface to a high vacuum — usually at elevated temperature. To obtain reproducible isotherms, it is necessary to control the outgassing conditions (temperature programme, change in pressure over the adsorbent and the residual pressure) to within limits which depend on the nature of the adsorption system. Instead of exposing the adsorbent to a high vacuum, it is sometimes expedient to achieve adequate cleanliness of the surface by flushing the adsorbent with an inert gas (which may be the adsorptive) at elevated temperature. With certain microporous solids reproducible isotherms are only obtained after one or more adsorption-desorption cycles. This problem can be overcome by flushing with the adsorptive and subsequent heating in vacuum.

Where physisorption measurements are to be employed for the determination of surface area and/or porosity, the rigorous surface cleanliness required in chemisorption studies is unnecessary and outgassing to a residual pressure of $\approx$ 10 mPa is usually considered satisfactory. Such conditions are readily achieved with the aid of conventional vacuum equipment — usually a combination of a rotary and diffusion pump in conjunction with a liquid nitrogen trap. The rate of desorption is strongly temperature dependent and to minimize the outgassing time, the temperature should be the maximum consistent with the avoidance of changes in the nature of the adsorbent and with the achievement of reproducible isotherms. Outgassing at too high a temperature or under ultra high vacuum conditions (residual pressure $<$ 1 μPa), as well as flushing with certain gases may lead to changes in the surface composition, e.g. decomposition of hydroxides or carbonates, formation of surface defects or irreversible changes in texture.

For most purposes the outgassing temperature may be conveniently selected to lie within the range over which the thermal gravimetric curve obtained in vacuo exhibits a minimum slope.

Note g. In part I, §1.1.11, the volume of the interfacial layer was denoted by V^s.

To monitor the progress of outgassing, it is useful to follow the change in gas pressure by means of suitable vacuum gauges and, if the experimental technique permits, the change in weight of the adsorbent. Further information on the effect of outgassing may be obtained by the application of temperature programmed desorption in association with evolved gas analysis (e.g. using mass spectrometry).

4.2 Determination of the adsorption isotherm

It is essential to take into account a number of potential sources of experimental error in the determination of an adsorption isotherm. In the application of a volumetric technique involving a dosing procedure it must be kept in mind that any errors in the measured doses of gas are cumulative and that the amount remaining unadsorbed in the dead space becomes increasingly important as the pressure increases. In particular, the accuracy of nitrogen adsorption measurements at temperatures of about 77 K will depend on the control of the following factors:—

(i) Gas burettes and other parts of the apparatus containing appreciable volumes of gas must be thermostatted, preferably to ± 0.1°C. If possible the whole apparatus should be maintained at reasonably constant temperature.

(ii) The pressure must be measured accurately (to ± 10 Pa). If a mercury manometer is used the tubes should be sufficiently wide — preferably ≈ 1 cm in diameter.

(iii) The level of liquid nitrogen in the cryostat bath must be kept constant to within a few millimetres, preferably by means of an automatic device.

(iv) The sample bulb must be immersed to a depth of at least 5 cm below the liquid nitrogen level.

(v) The temperature of the liquid nitrogen must be monitored, e.g. by using a suitably calibrated nitrogen or oxygen vapour pressure manometer or a suitable electrical device.

(vi) The nitrogen used as adsorptive must be of purity not less than 99.9%.

(vii) The conditions chosen for pretreatment of the adsorbent must be carefully controlled and monitored (i.e. the outgassing time and temperature and the residual pressure, or conditions of flushing with adsorptive).

(viii) It is recommended that the *outgassed weight* of the adsorbent should be determined either before or after the adsorption measurements. In routine work it may be convenient to admit dry air or nitrogen to the sample after a final evacuation under the same conditions as those used for the pretreatment.

In the application of a gravimetric technique, (ii) to (viii) must be taken into account and also special attention must be given to the control and measurement of the adsorbent temperature and to the assessment of the buoyancy corrections. Thermal transpiration effects should be allowed for if volumetric or gravimetric measurements are made at low pressure.

SECTION 5. EVALUATION OF ADSORPTION DATA

5.1 Presentation of primary data

The quantity of gas adsorbed may be measured in any convenient units: moles, grams and cubic centimetres at s.t.p. have all been used. For the presentation of the data it is recommended that the amount adsorbed should be expressed in moles per gram of the *outgassed* adsorbent. The mode of outgassing and if possible the composition of the adsorbent should be specified and its surface characterised. To facilitate the comparison of adsorption data it is recommended that adsorption isotherms be displayed in graphical form with the amount adsorbed (preferably n^a in mol g^{-1}) plotted against the equilibrium relative pressure (p/p^o), where p^o is the saturation pressure of the pure adsorptive at the temperature of the measurement, or against p when the temperature is above the critical temperature of the adsorptive. If the adsorption measurements are made under conditions where the gas phase deviates appreciably from ideality (e.g. at high pressure), it is desirable that the isotherms should be presented in terms of gas fugacity rather than pressure. If the surface area of the adsorbent is known the amount adsorbed may be expressed as number of molecules, or moles per unit area (i.e. N^a molecules m^{-2} or n^a mol m^{-2}). Adsorption data obtained with well-defined systems should be given in tabular form, but if this is not possible they should be deposited in an accessible source.

5.2 Classification of adsorption isotherms

The majority of physisorption isotherms may be grouped into the six types shown in Figure 2. In most cases at sufficiently low surface coverage the isotherm reduces to a linear form (i.e. $n^a \propto p$), which is often referred to as the Henry's Law region (see Note h).

The reversible *Type I* isotherm (see Note i) is concave to the p/p^o axis and n^a approaches a limiting value as $p/p^o \rightarrow 1$. Type I isotherms are given by microporous solids having relatively small external surfaces (e.g. activated carbons, molecular sieve zeolites and certain porous oxides), the limiting uptake being governed by the accessible micropore volume rather than by the internal surface area.

Note h. On heterogeneous surfaces this linear region may fall below the lowest experimentally measurable pressure.

Note i. Type I isotherms are sometimes referred to as *Langmuir isotherms*, but this nomenclature is not recommended.

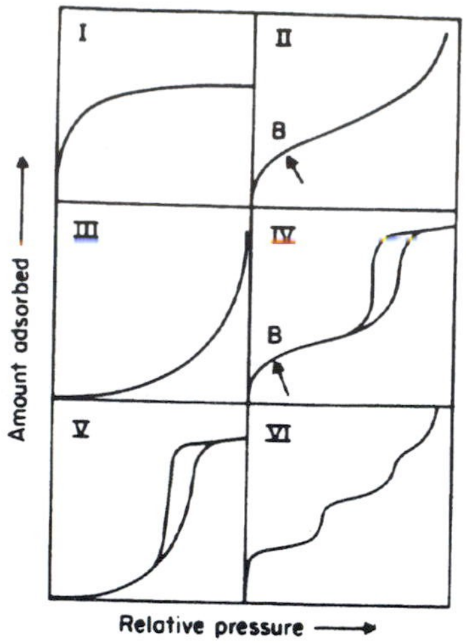

Fig. 2. Types of physisorption isotherms

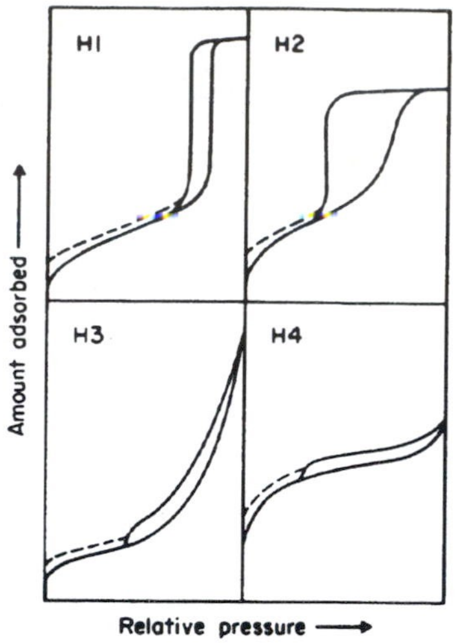

Fig. 3. Types of hysteresis loops

The reversible *Type II* isotherm is the normal form of isotherm obtained with a non-porous or macroporous adsorbent. The Type II isotherm represents unrestricted monolayer-multilayer adsorption. Point B, the beginning of the almost linear middle section of the isotherm, is often taken to indicate the stage at which monolayer coverage is complete and multilayer adsorption about to begin.

The reversible *Type III* isotherm is convex to the p/p° axis over its entire range and therefore does not exhibit a Point B. Isotherms of this type are not common, but there are a number of systems (e.g. nitrogen on polyethylene) which give isotherms with gradual curvature and an indistinct Point B. In such cases, the adsorbate-adsorbate interactions play an important role.

Characteristic features of the *Type IV* isotherm are its hysteresis loop, which is associated with capillary condensation taking place in mesopores, and the limiting uptake over a range of high p/p°. The initial part of the Type IV isotherm is attributed to monolayer-multilayer adsorption since it follows the same path as the corresponding part of a Type II isotherm obtained with the given adsorptive on the same surface area of the adsorbent in a non-porous form. Type IV isotherms are given by many mesoporous industrial adsorbents.

The *Type V* isotherm is uncommon; it is related to the Type III isotherm in that the adsorbent-adsorbate interaction is weak, but is obtained with certain porous adsorbents.

The *Type VI* isotherm, in which the sharpness of the steps depends on the system and the temperature, represents stepwise multilayer adsorption on a uniform non-porous surface. The step-height now represents the monolayer capacity for each adsorbed layer and, in the simplest case, remains nearly constant for two or three adsorbed layers. Amongst the best examples of Type VI isotherms are those obtained with argon or krypton on graphitised carbon blacks at liquid nitrogen temperature.

5.3 Adsorption hysteresis

Hysteresis appearing in the multilayer range of physisorption isotherms is usually associated with capillary condensation in mesopore structures. Such hysteresis loops may exhibit a wide variety of shapes. Two extreme types are shown as *H1* (formerly Type A) and *H4* in Figure 3. In the former the two branches are almost vertical and nearly parallel over an appreciable range of gas uptake, whereas in the latter they remain nearly horizontal and parallel over a wide range of p/p°. In certain respects *Types H2* and *H3* (formerly termed Types E and B, respectively) may be regarded as intermediate between these two extremes. A feature common to many hysteresis loops is that the steep region of the desorption branch leading to the lower closure point occurs (for a given adsorptive at a given temperature) at a relative pressure which is almost independent of the nature of the porous adsorbent but depends mainly on the nature of the adsorptive (e.g. for nitrogen at its boiling point at $p/p^{\circ} \approx 0.42$ and for benzene at 25°C at $p/p^{\circ} \approx 0.28$).

Although the effect of various factors on adsorption hysteresis is not fully understood, the shapes of hysteresis loops have often been identified with specific pore structures. Thus, *Type H1* is often associated with porous materials known, from other evidence, to consist of *agglomerates* (see Note j) or compacts of approximately uniform spheres in fairly regular array, and hence to have narrow distributions of pore size. Many porous adsorbents (e.g. inorganic oxide gels and porous glasses) tend to give Type H2 loops, but in such systems the distribution of pore size and shape is not well-defined. Indeed, the H2 loop is especially difficult to interpret: in the past it was attributed to a difference in mechanism between condensation and

Note j. An *agglomerate* is defined as an assemblage of particles rigidly joined together and an *aggregate* as an assemblage of particles which are loosely coherent.

evaporation processes occurring in pores with narrow necks and wide bodies (often referred to as 'ink bottle' pores), but it is now recognised that this provides an over-simplified picture and the role of network effects must be taken into account.

The Type H3 loop, which does not exhibit any limiting adsorption at high p/p^o, is observed with aggregates (see Note j) of plate-like particles giving rise to slit-shaped pores. Similarly, the *Type H4* loop is often associated with narrow slit-like pores, but in this case the Type I isotherm character is indicative of microporosity.

With many systems, especially those containing micropores, *low pressure hysteresis* (indicated by the dashed lines in Figure 3), may be observed extending to the lowest attainable pressures. Removal of the residual adsorbed material is then possible only if the adsorbent is outgassed at higher temperatures. This phenomenon may be associated with the swelling of a non-rigid porous structure or with the irreversible uptake of molecules in pores (or through pore entrances) of about the same width as that of the adsorbate molecule or in some instances with an irreversible chemical interaction of the adsorbate with the adsorbent.

SECTION 6. DETERMINATION OF SURFACE AREA

6.1 Application of the BET method

The Brunauer-Emmett-Teller (BET) gas adsorption method has become the most widely used standard procedure for the determination of the surface area of finely-divided and porous materials, in spite of the oversimplification of the model on which the theory is based.

It is customary to apply the BET equation in the linear form

$$\frac{p}{n^a \cdot (p^o - p)} = \frac{1}{n^a_m \cdot C} + \frac{(C-1)\, p}{n^a_m \cdot C\, p^o}$$

where n^a is the amount adsorbed at the relative pressure p/p^o and n^a_m is the monolayer capacity.

According to the BET theory C is related exponentially to the enthalpy (heat) of adsorption in the first adsorbed layer. It is now generally recognised, however, that although the value of C may be used to characterise the shape of the isotherm in the BET range it does not provide a quantitative measure of enthalpy of adsorption but merely gives an indication of the magnitude of the adsorbent-adsorbate interaction energy. Thus, in reporting BET data it is recommended that C values are stated, but not converted to enthalpies of adsorption.

A high value of C ($\approx$ 100) is associated with a sharp knee in the isotherm, thus making it possible to obtain by visual inspection the uptake at Point B, which usually agrees with n^a derived from the above equation to within a few per cent. On the other hand, if C is low (< 20) Point B cannot be identified as a single point on the isotherm. Since Point B is not itself amenable to any precise mathematical description, the theoretical significance of the amount adsorbed at Point B is uncertain.

The BET equation requires a linear relation between $p/n^a\,(p^o-p)$ and p/p^o (i.e. the *BET* plot). The range of linearity is restricted to a limited part of the isotherm — usually not outside the p/p^o range of 0.05–0.30. Some adsorption systems give linear (or nearly linear) BET plots over several ranges of p/p^o, but it is only in the region around $\theta = 1$ that the BET plot can be expected to yield the true value of n^a_m. This range is shifted to lower relative pressures in the case of systems having high adsorption energies, especially if the surface is energetically homogeneous, e.g. for the adsorption of nitrogen or argon on graphitised carbon or xenon on clean metal films.

The second stage in the application of the BET method is the calculation of the surface area (often termed the *BET area*) from the monolayer capacity. This requires a knowledge of the average area, a_m (*molecular cross-sectional area*), occupied by the adsorbate molecule in the complete monolayer. Thus

$$A_s\,(\text{BET}) = n^a_m \cdot L \cdot a_m$$

and

$$a_s\,(\text{BET}) = A_s\,(\text{BET})\,/m$$

where A_s (BET) and a_s (BET) are the total and specific surface areas, respectively, of the adsorbent (of mass m) and L is the Avogadro constant.

At the present time, nitrogen is generally considered to be the most suitable adsorptive for surface area determination and it is usually assumed that the BET monolayer is close-packed, giving $a_m\,(N_2) = 0.162$ nm^2 at 77 K. With a wide range of adsorbents it appears that the use of this value leads to BET areas which are within 20 per cent of the true surface areas. The existence of a strictly constant value of $a_m\,(N_2)$ is unlikely, however, and a growing amount of evidence suggests that it may vary by up to about 20 per cent from one surface to another. With other adsorptives, arbitrary adjustments of the a_m value are generally required to bring the BET area into agreement with the nitrogen value. The adjusted values of a_m for a particular adsorptive are dependent on temperature and the adsorbent surface. They may also differ appreciably from the value calculated for the close-packed monolayer on the basis of the density of the liquid or solid adsorptive. In view of this situation and the fact that full nitrogen isotherms may be conveniently measured at temperatures ≈ 77 K, it is recommended that nitrogen should continue to be used for the determination of both surface area and mesopore size distribution (Section 7.3).

The standard BET procedure requires the measurement of at least three and preferably five or more points in the appropriate pressure range on the N_2 adsorption isotherm at the normal boiling point of liquid nitrogen.

For routine measurements of surface areas, e.g. of finely divided or porous industrial products, a simplified procedure may be used involving the determination of only a single point on the adsorption isotherm, lying within the linear range of the BET plot. On many solids the value of C for N_2, is usually sufficiently large (> 100) to warrant the assumption that the BET straight line passes through the origin of the coordinate system. Thus

$$n_m^a = n^a(1 - p/p^o).$$

The validity of the simplifying assumption is usually within the variance of surface area determinations but needs to be checked for the particular system either by calibration against the standard BET procedure or by using surface area reference samples of the same material (see Section 6.2).

It is strongly recommended that in reporting a_s (BET) values, the conditions of outgassing (see Section 4.1), the temperature of the measurements, the range of linearity of the BET plot, the values of p^o, n_m^a, a_m and C should all be stated.

If the standard BET procedure is to be used, it should be established that monolayer-multilayer formation is operative and is not accompanied by micropore filling (Section 8.3), which is usually associated with an increase in the value of C (> 200, say). It should be appreciated that the BET analysis does not take into account the possibility of micropore filling or penetration into cavities of molecular size. These effects can thus falsify the BET surface areas and in case of doubt their absence should be checked by means of an empirical method of isotherm analysis or by using surface area reference samples (see Section 6.2).

For the determination of *small specific surface areas* (< 5 m^2g^{-1}, say) adsorptives with relatively low vapour pressure are used in order to minimise the dead space correction, preferably krypton or xenon at liquid nitrogen temperature. In the case of krypton, use of the extrapolated p^o value for the supercooled liquid tends to give a wider range of linearity in the BET plot and larger monolayer capacities (the difference being < 10%) as compared with those from p^o for the solid. Evaluation of surface areas is further complicated by the difficulty in choosing the value of a_m which is found to vary from solid to solid if compared with the BET nitrogen areas (a_m (Kr) = 0.17–0.23 nm^2; a_m (Xe) = 0.17–0.27 nm^2). Since no generally valid recommendations can be made, it is essential to state the chosen p^o and a_m values.

Ultrahigh vacuum techniques (basic pressure ≈ 10^2 μPa) enable adsorption studies to be made on stringently *clean solid surfaces* whereas degassing under moderate vacuum conditions, as normally applied in surface area determinations, leave the adsorbent covered with a pre-adsorbed layer of impurities and/or the adsorbate. On subsequent adsorption (e.g. of N_2 or noble gases) completion of the physisorbed monolayer is usually reached at $p/p^o \approx 0.1$ whereas on clean surfaces this state occurs at p/p^o values which may be smaller by orders of magnitude. However, as mentioned above, it should be kept in mind that linearity of the BET plot does not in itself provide conclusive evidence for the validity of n_m^a.

Noble gas adsorption is often assumed to be the least complicated form of physisorption. However, on clean solid surfaces the molecular area may depend on the formation of ordered structures of the adsorbate in registry with the adsorbent lattice.

6.2 Empirical procedures for isotherm analysis

In view of the complexity of real solid/gas interfaces and the different mechanisms which may contribute to physisorption, it is hardly surprising to find that none of the current theories of adsorption is capable of providing a mathematical description of an experimental isotherm over its entire range of relative pressure. In practice, two different procedures have been used to overcome this problem. The first approach involves the application of various semi-empirical isotherm equations, the particular mathematical form depending on the range of the isotherm to be fitted and also on the nature of the system. The second procedure makes use of standard adsorption isotherms obtained with selected non-porous reference materials and attempts to explain differences in the isotherm shape in terms of the three different mechanisms of physisorption, i.e. monolayer-multilayer coverage, capillary condensation and micropore filling. In favourable cases, this approach can provide an independent assessment of the *total* surface area (for mesoporous, macroporous or non-porous solids) and an assessment of the *external* area for microporous solids (see Section 8.3). Much discussion has surrounded the choice of the standard isotherm, but it now seems generally accepted that it should be one obtained on a chemically similar adsorbent rather than one having the same value of C as the isotherm to be analysed.

SECTION 7. ASSESSMENT OF MESOPOROSITY

7.1 Properties of porous materials

Most solids of high surface area are to some extent porous. The *texture* of such materials is defined by the detailed geometry of the void and pore space. Porosity, ϵ, is a concept related to texture and refers to the pore space in a material. An *open pore* is a cavity or channel communicating with the surface of a particle, as opposed to a *closed pore*. *Void* is the space or interstice between particles. In the context of adsorption and fluid penetration *powder porosity* is the ratio of the volume of voids plus the volume of open pores to the total volume occupied by the powder. Similarly, *particle porosity* is the ratio of the volume of open pores to the total volume of the particle. It should be noted that these definitions place the emphasis on the accessibility of pore space to the adsorptive.

The *total pore volume*, V_p, is often derived from the amount of vapour adsorbed at a relative pressure close to unity by assuming that the pores are then filled with condensed adsorptive in the normal liquid state.

If the solid contains no macropores, the isotherm remains nearly horizontal over a range of p/p° approaching unity and the total pore volume is well-defined. In the presence of macropores the isotherm rises rapidly near $p/p^{\circ} = 1$ and if the macropores are very wide may exhibit an essentially vertical rise. The limiting adsorption at the top of the steep rise can be identified reliably with the total pore volume only if the temperature on the sample is very carefully controlled and there are no 'cold spots' in the apparatus (which lead to bulk condensation of the gas and a false measure of adsorption in the volumetric method).

The *mean hydraulic radius*, r_h, of a group of mesopores, is defined as

$$r_h = \left(\frac{V}{A_s} \right)_p ,$$

where $(V/A_s)_p$ is the ratio of the volume to the area of walls of the group.

If the pores have a well-defined shape there is a simple relationship between r_h and the *mean pore radius*, r_p. Thus, in the case of non-intersecting cylindrical capillaries

$$r_p = 2r_h \ .$$

For a parallel-sided slit-shaped pore, r_h is half the slit width.

The *pore size distribution* is the distribution of pore volume with respect to pore size. The computation of pore size distribution involves a number of assumptions (pore shape, mechanism of pore filling, validity of Kelvin equation etc.).

7.2 Application of the Kelvin equation

Mesopore size calculations are usually made with the aid of the Kelvin equation in the form

$$\frac{1}{r_1} + \frac{1}{r_2} \doteq - \frac{RT}{\sigma^{lg} v^l} \ln \left(\frac{p}{p^{\circ}} \right)$$

which relates the principal radii, r_1 and r_2, of curvature of the liquid meniscus in the pore to the relative pressure, p/p°, at which condensation occurs; here σ^{lg} is the surface tension of the liquid condensate and v^l is its molar volume. It is generally assumed that this equation can be applied locally to each element of liquid surface.

In using this approach to obtain the *pore radius* or *pore width*, it is necessary to assume: (i) a model for the pore shape and (ii) that the curvature of the meniscus is directly related to the pore width. The pore shape is generally assumed to be either cylindrical or slit-shaped: in the former case, the meniscus is hemispherical and $r_1 = r_2$; in the latter case, the meniscus is hemicylindrical, r_1 = width of slit and $r_2 = \infty$.

Rearrangement of the Kelvin equation and replacement of $\left(\frac{1}{r_1} + \frac{1}{r_2} \right)$ by $\frac{2}{r_k}$ gives

$$r_K = \frac{2\sigma^{lg} v^l}{RT \ln (p^{\circ}/p)}$$

(r_K is often termed the *Kelvin radius*).

If the radius of a cylindrical pore is r_p and a correction is made for the thickness of a layer already adsorbed on the pore walls, i.e. for the *multilayer thickness*, t, then

$$r_p = r_K + t.$$

Correspondingly, for a parallel-sided slit, the slit width, d_p, is given by

$$d_p = r_K + 2t.$$

Values of t are obtained from the data for the adsorption of the same adsorptive on a non-porous sample having a similar surface to that of the sample under investigation.

7.3 Computation of mesopore size distribution

In calculations of the mesopore size distribution from physisorption isotherms it is generally assumed (often tacitly): (a) that the pores are rigid and of a regular shape (e.g. cylindrical capillaries or parallel-sided slits), (b) that micropores are absent, and (c) that the size distribution does not extend *continuously* from the mesopore into the macropore range. Furthermore, to obtain the pore size distribution, which is usually expressed in the graphical form $\Delta V_p/\Delta r_p$ vs. r_p, allowance must be made for the effect of multilayer adsorption in progressively reducing the dimensions of the free pore space available for capillary condensation.

The location and shape of the distribution curve is, of course, dependent on which branch of the hysteresis loop is used to compute the pore size. In spite of the considerable attention given to this problem, in the absence of any detailed knowledge of the pore geometry it is not possible to provide unequivocal general recommendations. In principle, the regions of metastability and instability should be established for the liquid/vapour meniscus in the various parts of a given pore structure, but in practice this would be extremely difficult to undertake in any but the simplest types of pore system.

Recent work has drawn attention to the complexity of capillary condensation in pore networks and has indicated that a pore size distribution curve derived from the desorption branch of the loop is likely to be unreliable if pore blocking effects occur. It is significant that a very steep desorption branch is usually found if the lower closure point of the loop is located at the limiting p/p° (see Section 5.3). In particular, the desorption branch of a Type H2 loop is one that should not be used for the computation of pore size distribution.

It is evident from the above considerations that the use of the physisorption method for the determination of mesopore size distribution is subject to a number of uncertainties arising from the assumptions made and the complexities of most real pore structures. It should be recognised that derived pore size distribution curves may often give a misleading picture of the pore structure. On the other hand, there are certain features of physisorption isotherms (and hence of the derived pore distribution curves) which are highly characteristic of particular types of pore structures and are therefore especially useful in the study of industrial adsorbents and catalysts. Physisorption is one of the few non-destructive methods available for investigating mesoporosity, and it is to be hoped that future work will lead to refinements in the application of the method — especially through the study of model pore systems and the application of modern computer techniques.

SECTION 8. ASSESSMENT OF MICROPOROSITY

8.1 Terminology

It is generally recognised that the mechanism of physisorption is modified in very fine pores (i.e. pores of molecular dimensions) since the close proximity of the pore walls gives rise to an increase in the strength of the adsorbent-adsorbate interactions. As a result of the enhanced adsorption energy, the pores are filled with physisorbed molecules at low p/p°. Adsorbents with such fine pores are usually referred to as *microporous.*

The limiting dimensions of micropores are difficult to specify exactly, but the concept of *micropore filling* is especially useful when it is applied to the primary filling of pore space as distinct from the secondary process of capillary condensation in mesopores.

The terminology of pore size has become somewhat confused because it has been customary to designate the different categories of pores in terms of exact dimensions rather than by reference to the particular forces and mechanisms operating with the given gas-solid system (taking account of the size, shape and electronic nature of the adsorptive molecules and the surface structure of the adsorbent) as well as to the pore size and shape.

The upper limit of 2.0 nm for the micropore width was put forward as part of the IUPAC classification of pore size (see Appendix II, Part I). It now seem likely that there are two different micropore filling mechanisms, which may operate at p/p° below the onset of capillary condensation: the first, occurring at low p/p°, involves the entry of individual adsorbate molecules into very narrow pores; the second, at a somewhat higher p/p°, is a *cooperative* process involving the interaction between adsorbate molecules.

It is recommended therefore that attention should be directed towards the *mechanism* of pore filling rather than to the specification of the necessarily rather arbitrary limits of pore size. Until further progress has been made it is undesirable to modify the original IUPAC classification or to introduce any new terms (e.g. ultrapores or ultramicropores).

8.2 Concept of surface area

In recent years a radical change has been taking place in the interpretation of the Type I isotherm for porous adsorbents. According to the classical Langmuir theory, the limiting adsorption n_p^a (at the plateau) represents completion of the monolayer and may therefore be used for the calculation of the surface area. The alternative view, which is now widely accepted, is that the initial (steep) part of the Type I isotherm represents micropore filling (rather than surface coverage) and that the low slope of the plateau is due to multilayer adsorption on the small external area.

If the latter explanation is correct, it follows that the value of A_s (as derived by either BET or Langmuir analysis (see Note k) cannot be accepted as the true surface area of a microporous adsorbent. On the other hand, if the slope of the isotherm is not too low at higher p/p^o and provided that capillary condensation is absent, it should (in principle) be possible to assess the *external* surface area from the multilayer region.

In view of the above difficulties, it has been suggested that the term *monolayer equivalent area* should be applied to microporous solids. However, the exact meaning of this term may not always be clear and it is recommended that the terms *Langmuir area* or *BET area* be used where appropriate, with a clear indication of the range of linearity of the Langmuir, or BET, plot, the magnitude of C etc. (see Section 6.1).

8.3 Assessment of micropore volume

No current theory is capable of providing a general mathematical description of micropore filling and caution should be exercised in the interpretation of derived quantities (e.g. micropore volume) obtained by the application of a relatively simple equation (e.g. the Dubinin-Radushkevich equation) to adsorption isotherm data over a limited range of p/p^o and at a single temperature. The fact that a particular equation gives a reasonably good fit over a certain range of an isotherm does not in itself provide sufficient evidence for a particular mechanism of adsorption.

The t-method and its extensions provide a simple means of comparing the shape of a given isotherm with that of a standard on a non-porous solid. In the original t-method, the amount adsorbed is plotted against t, the corresponding multilayer thickness calculated from the standard isotherm obtained with a non-porous reference solid. Any deviation in shape of the given isotherm from that of the standard is detected as a departure of the 't-plot' from linearity. For the assessment of microporosity, the thickness of the multilayer is irrelevant and it is preferable to replace t by the 'reduced' adsorption, α_s, defined as $(n^a/n_s^a)_{ref}$ where n_s^a is the amount adsorbed by the reference solid at a fixed relative pressure, $p/p^o = s$. An advantage of this method is that it can be used even when the standard isotherm does not exhibit a well-defined Point B, i.e. when the value of C is low. Once the standard α_s-curve has been obtained for the particular gas-solid system at the given temperature, the α_s-method can be applied in an analogous manner to the t-method. It is essential that the standard isotherms (or t-curves) be obtained with reference solids which are non-porous and of known surface structure. Further, it is strongly recommended that the standard isotherm should be one obtained for the particular adsorption system, and not by choosing a Type II isotherm which happens to have the same C value as the isotherm on a particular microporous solid.

Another procedure which may be used for the assessment of microporosity is the *pre-adsorption method*. In this approach the micropores are filled with large molecules (e.g. nonane), which are not removed by pumping the adsorbent at ambient temperature. In the most straightforward case, this procedure can provide an effective way of isolating the micropores and leaving the external surface available for the adsorption of nitrogen, or another suitable adsorptive.

SECTION 9. GENERAL CONCLUSIONS AND RECOMMENDATIONS

9.1 For evaluation of both the surface area and the pore size distribution from a single adsorption isotherm, nitrogen (at ≈ 77 K) is the recommended adsorptive except with solids of low surface area. If the surface area is relatively low ($a_s < 5$ m^{2}g^{-1}, say), krypton or xenon, also at ≈ 77 K, offer the possibility of higher precision in the actual measurement of the adsorption, but not necessarily higher accuracy in the resultant value of the surface area than could be obtained with nitrogen. When another adsorptive is used it should be calibrated against nitrogen with the aid of carefully selected reference solids.

9.2 For a given system at a given temperature, the adsorption isotherm should be reproducible, but the possibility of ageing of the adsorbent — e.g. through the uptake or loss of water — must always be borne in mind. The reproducibility of the adsorption should be checked whenever possible by measurement of an isotherm on a second sample (of different mass) of the given adsorbent.

9.3 The first stage in the interpretation of a physisorption isotherm is to identify the isotherm type and hence the nature of the adsorption process(es): monolayer-multilayer adsorption, capillary condensation or micropore filling. If the isotherm exhibits low pressure hysteresis (i.e. at $p/p^o < 0.4$, with nitrogen at 77 K) the technique should be checked to establish the degree of accuracy and reproducibility of the measurements.

Note k. In fact, many microporous solids do not give linear BET plots although their Langmuir plots may be linear over an appreciable range of p/p^o.

9.4 The BET method is unlikely to yield a value of the actual surface area if the isotherm is of either Type I or Type III; on the other hand both Type II and Type IV isotherms are, in general, amenable to the BET analysis, provided that the value of C is neither too low nor too high and that the BET plot is linear for the region of the isotherm containing Point B. It is recommended that both the value of C and the range of linearity of the BET plot be recorded. If the value of C is found to be higher than normal for the particular gas-solid system, the presence of microporosity is to be suspected even if the isotherm is of Type II or Type IV; the validity of the BET area then needs checking, e.g. by the α_s-method, in order to ascertain how closely the shape of the isotherm conforms to that of the standard isotherm in the monolayer range.

9.5 The computation of mesopore size distribution is valid only if the isotherm is of Type IV. In view of the uncertainties inherent in the application of the Kelvin equation and the complexity of most pore systems, little is to be gained by recourse to an elaborate method of computation. The decision as to which branch of the hysteresis loop to use in the calculation remains largely arbitrary. If the desorption branch is adopted (as appears to be favoured by most workers), it should be appreciated that neither a Type H2 nor a Type H3 hysteresis loop is likely to yield a reliable estimate of pore size distribution, even for comparative purposes.

9.6 If a Type I isotherm exhibits a nearly constant adsorption at high relative pressure, the micropore volume is given by the amount adsorbed (converted to a *liquid* volume) in the plateau region, since the mesopore volume and the external surface are both relatively small. In the more usual case where the Type I isotherm has a finite slope at high relative pressures, both the external area and the micropore volume can be evaluated provided that a standard isotherm on a suitable non-porous reference solid is available. At present, however, there is no reliable procedure for the computation of micropore size *distribution* from a single isotherm; but if the size of the micropores extends down to molecular dimensions, adsorptive molecules of selected size can be employed as molecular probes.

9.7 The following *check list* is recommended to assist authors in the measurement of adsorption isotherms and the presentation of the data in the primary literature. The reporting of results along generally accepted lines would considerably facilitate the compilation of data in the secondary literature and would thus promote interdisciplinary scientific cooperation (see Note 1).

It is suggested that the following items be checked and the relevant experimental conditions and results reported:–

(i) Characterisation of the sample (e.g. source, chemical composition, purity, physical state, method of sampling).

(ii) Pretreatment and outgassing conditions (e.g. temperature, residual pressure/partial pressures, duration of outgassing, flushing with adsorptive).

(iii) Mass of outgassed sample (m in g).

(iv) Adsorptive (e.g. chemical nature, purity, drying).

(v) Experimental procedure for isotherm determination: method (e.g. volumetric, gravimetric, static, continuous; calibration of dead space or buoyancy). Measurement and accuracy of pressure [p in Pa (or mbar) or p/p^o] and temperature, equilibration times.

(vi) Reproducibility (a) second run, (b) with fresh sample of adsorbent.

(vii) Adsorption isotherm: plot of amount adsorbed (n^a in mol g^{-1} or in mol m^{-2}; or N^a in molecules m^{-2}) versus pressure [p in Pa (or mbar) or p/p^o], statement of measured/calculated p^o value at T.

(viii) Type of isotherm, type of hysteresis, nature of adsorption (monolayer-multilayer adsorption, capillary condensation, micropore filling).

(ix) BET data: adsorptive, temperature (T in K), mathematical procedure used for BET analysis, region of p/p^o and of θ in which the BET plot is linear, single point method, monolayer capacity (n^a_m in mol g^{-1} or N^a_m in molecules m^{-2}), C value, molecular cross-sectional area (a_m in nm^2 per molecule), specific surface area (a_s in m^{2}g^{-1}).

(x) Porosity (ϵ) with reference to powder porosity or particle porosity indicating in the latter case whether open pores, or open plus closed pores, are considered.

(xi) Assessment of mesoporosity (pore width $\approx$ 2–50 nm), method of computation, choice of ad- or desorption branch, p^o value at T and region of p/p^o used, surface tension of liquid adsorptive (σ^{lg} in Nm^{-1} at T), model for pore shape. Correction for multilayer thickness, t-curve: plot of t in nm vs p/p^o (indication whether a standard curve is assumed or an adsorption isotherm determined on a non-porous sample of the adsorbent). Pore size distribution: plot of $\Delta V_p/\Delta r_p$ vs r_p (pore volume V_p per unit mass of adsorbent in cm^3 g^{-1} as calculated with the density ρ^l in g cm^{-3} of the liquid adsorptive, mean pore radius r_p in nm), total pore volume at saturation.

(xii) Assessment of microporosity (pore width < ca 2 nm), method of evaluation, t-plot: amount adsorbed n^a in mol g^{-1} vs multilayer thickness t in nm, α_s-plot: n^a vs $(n^a/n^a_s)_{ref}$, where the suffix refers to a chosen value $s = p/p^o$, Dubinin-Radushkevich plot or pre-adsorption method. Micropore volume per unit mass of adsorbent in cm^{3}g^{-1} as calculated with the density ρ^l of the adsorptive in the normal liquid state, monolayer equivalent area of microporous solid, external surface area.

Note 1 see "Guide for the Presentation in the Primary Literature of Numerical Data Derived from Experiments". Report of the CODATA Task Group on Presentation of Data in the Primary Literature, CODATA Bulletin No. 9 (1973).

Reporting data on adsorption from solution at the solid/solution interface (Recommendations 1986)

COMMISSION ON COLLOID AND SURFACE CHEMISTRY INCLUDING CATALYSIS*

D. H. EVERETT

School of Chemistry, University of Bristol, UK

ABSTRACT This document is a companion to that on Reporting Physisorption Data for Gas/Solid Systems, and is designed to supplement the discussion of adsorption at the solid/solution interface given in the Manual of Symbols and Terminology for Physico-chemical Quantities and Units, Appendix II, part 1: Definitions, Terminology and Symbols in Colloid and Surface Chemistry, Pure Appl. Chem. 31, 579-638 (1972), section 1.1.10.

The definitions of adsorption from solution used here are essentially those adopted in the Manual. A discussion is then given of the operational determination of adsorption from solution, and includes an outline of various available experimental techniques, and recommendations concerning the precautions which need to be taken to ensure reliable results. Recommendations are made regarding the form in which results should be published.

The interpretation of adsorption data in thermodynamic terms is discussed as is the role of adsorption models in elucidating the molecular processes involved in adsorption. Attention is also drawn to the problems associated with the effects of surface heterogeneity on adsorption from solution.

*Membership of the Commission during the period (1979–85) in which the report was prepared was as follows:

Chairman: 1979–83 J. Lyklema (Netherlands); 1983–85 K. S. W. Sing (UK); *Vice-Chairman*: 1979–85 J. Haber (Poland); *Secretary*: 1979–83 M. Kerker (USA); 1983–85 E. Wolfram (Hungary); *Members*: J. H. Block (FRG; Titular 1983–85, Associate 1979–83); N. V. Churaev (USSR; Associate 1979–85); D. H. Everett (UK; National Representative 1979–85); G. F. Froment (Belgium; National Representative 1981–85); P. C. Gravelle (France; Associate 1979–85); G. L. Haller (USA; Associate 1979–81); R. S. Hansen (USA; Titular 1979–83); R. A. W. Haul (FRG; National Representative 1979–83); J. W. Hightower (USA; Associate 1983–85); R. J. Hunter (Australia; Associate 1979–85); L. G. Ionescu (Brazil; National Representative 1983–85); A. S. Kertes (Israel; National Representative 1979–85); A. Kitahara (Japan; National Representative 1979–85); J. C. Kuriacose (India; National Representative 1983–85); J. Lyklema (Netherlands; National Representative 1983–85); A. Maroto (Argentina; Associate 1983–85, National Representative 1981–83); S. G. Mason (Canada; National Representative 1979–85); K. Meyer (GDR; National Representative 1981–85); P. Mukerjee (USA; Associate 1979–83); L. G. Nagy (Hungary; National Representative 1979–85); H. van Olphen (Netherlands; Associate 1979–83); J. A. Pajares (Spain; National Representative 1981–83); M. W. Roberts (UK; Titular 1979–83); J. Rouquérol (France; Associate 1983–85); K. S. W. Sing (UK; Associate 1979–83); P. J. Stenius (Sweden; Titular 1983–85, Associate 1981–83); M. S. Suwandi (Malaysia; National Representative 1983–85); K. Tamaru (Japan; Titular 1979–83); L. Ter-Minassian-Saraga (France; Titular 1983–85, Associate 1979–83); A. Weiss (FRG; National Representative 1983–85); P. B. Wells (UK; Associate 1983–85); E. Wolfram (Hungary; Titular 1979–83).

Considerable assistance in the preparation of this report was given by G. Schay (Hungary), who produced the first draft.

1 INTRODUCTION

The study of adsorption from liquid solutions by solids has expanded rapidly in recent years, and various procedures have been adopted by different workers in presenting and interpreting their work. To ensure that experimental studies contribute most effectively to the understanding of the basic phenomena it is desirable that work from different laboratories be presented in a way which encourages the intercomparison of results and their assimilation into a reliable body of scientific information.

This manual is intended to present a definitive summary of the basis upon which an understanding of the phenomenon of adsorption is founded, to outline various possible experimental techniques for obtaining adsorption isotherms and to consider the ways in which such data can be interpreted. A major objective is to ensure that the data are obtained by reliable techniques and are presented in the literature in a manner that enables them to be interpreted by other workers.

The scope of this report is limited to the reversible adsorption of small non-ionic species by inert solids. It thus excludes adsorption from solutions of strong electrolytes, ion-exchange processes and polymer adsorption. On the other hand, weak electrolytes are not excluded when it can safely be assumed that adsorption of the molecular (uncharged) form predominates. Also excluded are phenomena involving the penetration of the adsorbate into the structure of the adsorbent (e.g. swelling of clay minerals) and adsorption into swollen gels. Note (a)*.

Consideration is in the main restricted to binary solutions or liquid mixtures. Although adsorption from multicomponent systems is of growing importance, e.g. in liquid chromatography and in many practical purification processes, the presentation of a completely general treatment would lead to complications that would obscure some of the underlying features. In very dilute solutions certain approximations to the equations are often justified.

The symbols and terminology used are generally in accordance with the IUPAC Manual of Symbols and Terminology for Physicochemical Quantities and Units, Appendix II, part 1, which deals with Colloid and Surface Chemistry (Note b). The notations for surface excesses in this Manual, although strictly logical tend to be somewhat clumsy and it would seem desirable to adopt simpler abbreviations for the specific excess quantities $n_i^{\sigma(n)}/m$, $n_i^{\sigma(v)}/m$ etc. Alternative symbols that have been used are χ, Ω and n_i^e (the latter particularly by chemical engineers). If authors employ one of these alternatives they should take care to define clearly its relationship to the more explicit quantities $n_i^{\sigma(n)}/m$ etc. The use of Γ for areal excess quantities (e.g. $n_i^{\sigma(n)}/A_s$) is well established. However, since the specific surface areas of particulate solids and porous adsorbents are not always known reliably it is not always possible, nor even desirable, to report experimental data in terms of Γ.

2 DEFINITIONS OF ADSORPTION FROM SOLUTION

2.1 Introduction

The primary experimental observation leading to the concept of adsorption from solution by a solid is that there is a discrepancy between the overall stoichiometric composition and that calculated from the known concentration in the bulk liquid phase and its volume (or mass), assuming that this concentration is uniform throughout the liquid phase. These discrepancies, which for a given component may be positive or negative, are then attributed to non-uniform composition of the liquid phase in the immediate neighbourhood of the phase boundary.

A quantitative mathematical description of the phenomenon of adsorption may be developed either in terms of the concept of a Gibbs dividing surface, or in purely algebraic terms (Note c).

2.2 Use of a Gibbs dividing surface

The concept of a Gibbs dividing surface in its general form is outlined in the Manual, Appendix II. For each interface the adsorption or surface excess of a given component is defined as the difference between the amount of component actually present in the system, and that which would be present (in a reference system) if the bulk concentrations in the adjoining phases were maintained up to a chosen geometrical dividing surface (Gibbs dividing surface, or GDS). In the particular case of a solid/liquid interface in which no component of the liquid phase penetrates into the solid, the situation may be depicted schematically as in figure 1, where the local concentration of a specified component i is plotted as a function of the distance z from a plane solid surface.

---

* Explanations of all Notes are provided in the Section NOTES AND REFERENCES on page 982.

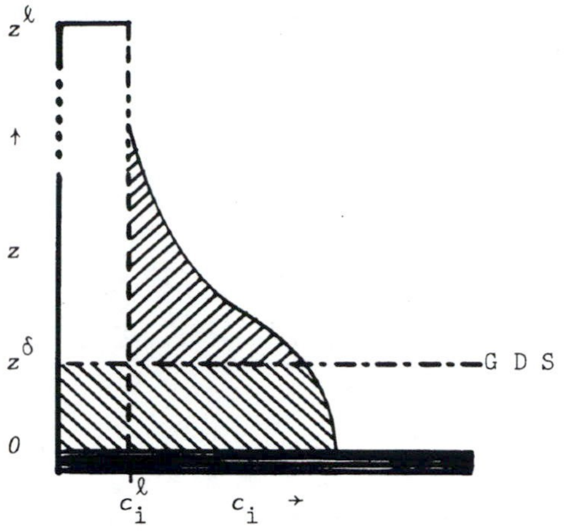

Figure 1: Concentration profile of component i as function of distance z from a plane solid surface (schematic).

If the volume of the liquid phase up to an arbitrarily chosen GDS at z^δ is V^ℓ, then the surface excess, or adsorption, of component i is defined as

$$n_i^\sigma = n_i - V^\ell c_i^\ell , \qquad (1)$$

where n_i is the total amount of i in the system and c_i^ℓ is its bulk concentration in the liquid.[1] (Note d)

The total amount of i is given in terms of the local concentrations by

$$n_i = A_s \int_0^{z^\ell} c_i \, dz ,$$

where A_s is the area of the dividing surface, so that since

$$V^\ell = A_s \int_{z^\delta}^{z^\ell} dz ,$$

$$n_i^\sigma = A_s \left[\int_{z^\delta}^{z^\ell} (c_i - c_i^\ell) dz + \int_0^{z^\delta} c_i dz \right] ; \qquad (2)$$

i.e. n_i^σ per unit area is given by the sum of the areas of the two shaded portions of the diagram. The surface excess concentration (or areal surface excess) is denoted by $\Gamma_i = n_i^\sigma / A_s$.

The value to be ascribed to Γ_i clearly depends on the choice of the location of the GDS:

$$(d\Gamma_i / dz^\delta) = c_i^\ell . \qquad (3)$$

For a binary solution the variation of Γ_1, Γ_2 with z^δ can be represented as in figure 2.

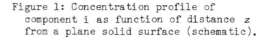

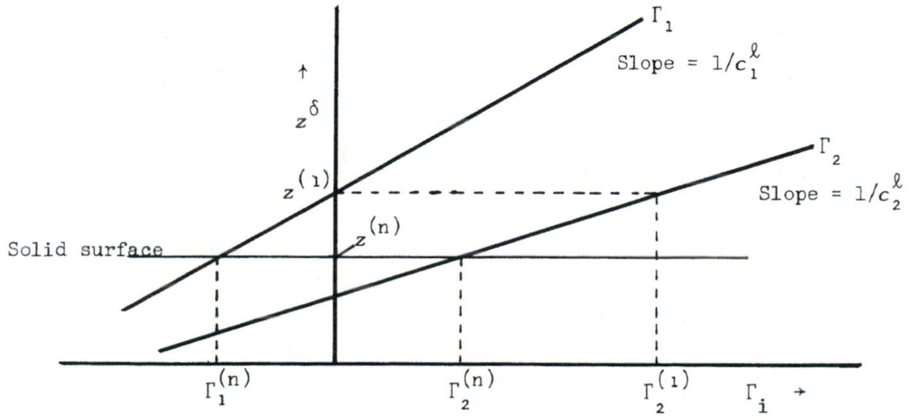

Figure 2: Variation of Γ_1 and Γ_2 as functions of choice of position z^δ of the G.D.S.

To obtain a well-defined measure of adsorption it is necessary to formulate a definition that is independent of the choice of the GDS. Two convenient alternative definitions may be derived as follows.

The surface excesses of 1 and 2 may be written

$$n_1^\sigma = n_1 - V^\ell c_1^\ell, \tag{4a}$$

$$n_2^\sigma = n_2 - V^\ell c_2^\ell, \tag{4b}$$

where V^ℓ is the volume of the liquid phase up to the arbitrarily chosen GDS. Elimination of V^ℓ from these equations leads to the *relative surface excess of 2 with respect to 1:*

$$n_2^{\sigma(1)} = n_2^\sigma - n_1^\sigma(c_2^\ell/c_1^\ell) = n_2 - n_1(c_2^\ell/c_1^\ell) \ . \tag{5}$$

Since the right-hand side of this equation contains only experimental quantities, the left-hand side must be independent of the position of the GDS. On division by A_s, one defines $\Gamma_2^{(1)}$, the *areal relative surface excess of 2 with respect to 1,* by

$$\Gamma_2^{(1)} = n_2^{\sigma(1)}/A_s = \frac{1}{A_s}\left[n_2 - n_1\left(\frac{c_2^\ell}{c_1^\ell}\right)\right] = \Gamma_2 - \Gamma_1\left(\frac{c_2^\ell}{c_1^\ell}\right) = \Gamma_2 - \Gamma_1\left(\frac{x_2^\ell}{x_1^\ell}\right), \tag{6}$$

where x_2^ℓ is the mole fraction of 2 in the bulk liquid.

$\Gamma_2^{(1)}$ as defined by (6) is independent of the choice of GDS. However, if the GDS is chosen such that $\Gamma_1 = 0$, then $\Gamma_2^{(1)}$ is the value of Γ_2 for this particular choice. This is illustrated in figure 2, where $\Gamma_2^{(1)}$ is the value of Γ_2 when z^δ is chosen at $z^{(1)}$. The relation (6) follows immediately from the geometry of this figure.

Alternatively, the surface excess of 2 and the total surface excess may be written

$$n_2^\sigma = n_2 - V^\ell c_2^\ell \ , \tag{7a}$$

$$n^\sigma = n^0 - V^\ell c^\ell \ , \tag{7b}$$

where n^0 is the total amount of substance present and c^ℓ the total concentration. Eliminating V^ℓ leads to the *reduced surface excess:*

$$n_2^{\sigma(n)} = n_2^\sigma - n^\sigma(c_2^\ell/c^\ell) = n_2^\sigma - n^\sigma x_2^\ell = n_2 - n^0 x_2^\ell \ . \tag{8}$$

Again the right-hand side is an experimental quantity so that the left-hand side is independent of the GDS. On division by A_s, the *areal reduced surface excess*, denoted by $\Gamma_2^{(n)}$, is defined as

$$\Gamma_2^{(n)} = n_2^{\sigma(n)}/A_s = \Gamma_2 - \Gamma x_2^\ell = \frac{1}{A_s}(n_2 - n^0 x_2^\ell). \tag{9}$$

$\Gamma_2^{(n)}$ as defined in equation (9) is independent of the GDS. However, if the GDS is defined such that $\Gamma = \Gamma_1 + \Gamma_2 = 0$, then $\Gamma_2^{(n)}$ is the value of Γ_2 for this particular choice. It follows therefore that

$$\Gamma_1^{(n)} = -\Gamma_2^{(n)}. \tag{10}$$

This is also illustrated in figure 2 where $\Gamma_2^{(n)}$ is the value of Γ_2 when z^δ is chosen at $z^{(n)}$, which also co-incides with the surface of the solid.

The relationship between $\Gamma_2^{(1)}$ and $\Gamma_2^{(n)}$ follows if in equation (6) we insert $\Gamma_1^{(n)} = -\Gamma_2^{(n)}$, in place of Γ_1 and Γ_2 respectively, whence

$$\Gamma_2^{(1)} = \Gamma_2^{(n)}/x_1^\ell \ . \tag{11}$$

The importance of this equation arises because in many experimental methods it is $\Gamma_2^{(n)}$ that is measured, while $\Gamma_2^{(1)}$ is the quantity which occurs in the fundamental thermodynamic equations.

For dilute solutions $x_1^\ell \to 1$ and $\Gamma_2^{(1)} \to \Gamma_2^{(n)}$.

If the composition of the bulk solution is expressed in mass fractions w_1^ℓ, w_2^ℓ then the surface excess masses of the components are given by

$$m_2^{\sigma(1)} = m_2^{\sigma} - m_1^{\sigma}\left(\frac{w_2^{\ell}}{w_1^{\ell}}\right) = m_2 - m_1\left(\frac{w_2^{\ell}}{w_1^{\ell}}\right) , \qquad (12a)$$

$$m_2^{\sigma(m)} = m_2^{\sigma} - m^{\sigma} w_2^{\ell} = m_2 - m^{o} w_2^{\ell} , \qquad (12b)$$

with

$$m_2^{\sigma(1)} = m_2^{\sigma(m)}/w_1^{\ell} . \qquad (12c)$$

Equations (9) and (12b) lead immediately to a prescription for measuring the reduced surface excess amount or reduced surface excess mass. In each case the first term on the right of the last equality sign is the total amount, or total mass, of component 2. This is known from the composition (x_2^o or w_2^o) of the solution before equilibration with the solid sample:

$$n_2 = n^o x_2^o; \quad m_2 = m^o w_2^o .$$

Hence

$$n_2^{\sigma(n)} = n^o(x_2^o - x_2^{\ell}) = n^o \Delta x_2^{\ell} , \qquad (13a)$$

$$m_2^{\sigma(m)} = m^o(w_2^o - w_2^{\ell}) = m^o \Delta w_2^{\ell} . \qquad (13b)$$

Alternatively, since $n^o = n_1 + n_2$, and $m^o = m_1 + m_2$,

$$n_2^{\sigma(n)} = n_2 x_1^{\ell} - n_1 x_2^{\ell} , \qquad (14a)$$

$$m_2^{\sigma(m)} = m_2 w_1^{\ell} - m_1 w_2^{\ell} . \qquad (14b)$$

These two quantities are related by

$$m_2^{\sigma(m)} = n_2^{\sigma(n)}(w_1^{\ell} M_2 + w_2^{\ell} M_1) = n_2^{\sigma(n)} \frac{M_1 M_2}{x_1^{\ell} M_1 + x_2^{\ell} M_2} , \qquad (14c)$$

where M_1 and M_2 are the molar masses of the two components.

In most earlier work and in much recent work, adsorption from solution is reported in terms of liquid volumes and volume concentrations. If V^o is the initial volume of liquid (Note e)

$$n_2^{\sigma(v)} = V^o \Delta c_2^{\ell} . \qquad (15)$$

Provided that the partial molar volumes, v_1 and v_2 of the components are constant in the concentration interval considered, then the following relation holds:

$$n_2^{\sigma(v)} = \left[1 + (v_1 - v_2)c_2^{\ell}\right] n_2^{\sigma(n)} . \qquad (16)$$

v_1 and v_2 are in general concentration dependent (except for perfect or ideal dilute solutions). However, if (v_1 - v_2) is small enough (say less than $10^{-1} dm^3 mol^{-1}$) and the equilibrium solution sufficiently dilute (say $c_2^{\ell} < 10^{-1} mol\ dm^3$) the second term in the brackets on the right-hand side of (16) will be practically negligible and the two measures of the surface excess will be approximately equal. Since, however, we are often interested in the course of the adsorption isotherm at higher concentrations, or in the case of completely miscible systems over the whole concentration range, the use of $n_2^{\sigma(v)}$ is to be discouraged especially since the collection of data necessary for the representation of the data according to equations (13a or b) is hardly more cumbersome from the experimental point of view.

2.3 Special case of pure liquids (Note f)

In the case of a pure liquid, despite its low compressibility, the variation of density near a solid surface can be detected and measured. The total volume V of a system consisting of solid and pure liquid is different from (usually less than) that calculated assuming a constant liquid density. If the densities of bulk solid (ρ^{sol}) and liquid (ρ^{ℓ}) are known then an excess volume (usually negative) can be defined as:

$$V^{\sigma} = V - V^{sol} - V^{o} = V - m^{sol}/\rho^{sol} - m^{\ell}/\rho^{\ell} ; \qquad (17)$$

here m^{sol} is the mass of solid and V^{sol} its volume calculated from the bulk density, and m^{ℓ} is the mass of liquid. Alternatively the excess mass is

$$m^{\sigma} = m^{\ell} - (V - V^{sol})\rho^{\ell} , \qquad (18)$$

and the areal excess mass

$$\Gamma = \{m^{\ell} - (V - V^{sol})\rho^{\ell}\}/A_s . \qquad (19)$$

If the compressibility of the solid is much less than that of the liquid, these excesses can be attributed to changes in the density of the liquid in close proximity to the solid surface. Such changes must also occur in the case of mixtures, but they do not affect the definitions in equations (13).

3 OPERATIONAL DETERMINATION OF ADSORPTION FROM SOLUTION

3.1 Methodology

The basic experimental method of determining adsorption from solution takes, in effect, the solid surface as the GDS, and measures $\Gamma_2^{(n)}$.

The simplest method follows from equation (13a). A sample of liquid containing an amount n^0 at a mole fraction x_2^0 is equilibrated with a mass m of solid and the final mole fraction x_2^ℓ is measured. The reduced surface excess of component 2 is then given by equation (13a):

$$n_2^{\sigma(n)} = n^0 \Delta x_2^\ell \ . \qquad (13a)$$

The specific reduced surface excess and the areal reduced surface excess are then

$$n_2^{\sigma(n)}/m = n^0 \Delta x_2^\ell / m \ , \qquad (20)$$

and

$$\Gamma_2^{(n)} = n^0 \Delta x_2^\ell / (m a_s), \qquad (21)$$

where a_s is the specific surface area of the solid. Similar equations, following from equation (13b) apply if mass and mass fractions are measured. The alternative method using equation (15) may be used if the solutions are sufficiently dilute so that $\Gamma_2^{(v)} \simeq \Gamma_2^{(n)}$. It is not recommended for use at higher concentrations.

Experiments are repeated with different concentrations of liquid to build up an adsorption isotherm (specific reduced surface excess isotherm, sometimes called the 'composite isotherm') in which $n_2^{\sigma(n)}/m$ or $\Gamma_2^{(n)}$ is given as a function of x_2^ℓ.

3.2 Practical considerations and precautions

(i) In selecting systems for the study of adsorption from solution it is important to check that the adsorbent does not swell, dissolve or otherwise deteriorate in contact with the solution, and that the adsorption is reversible with respect to changes of temperature and/or composition of the liquid phase. If these criteria are not met, then caution is required in the interpretation of the results.
(ii) In certain circumstances it may be necessary to eliminate or allow for adsorption on, or reaction of the components of the liquid with, the walls of the experimental apparatus. Particular care is needed if a filter is used to separate the solid from the liquid.
(iii) If the surface of the solid is contaminated with a soluble constituent, then previous rinsing or leaching of the adsorbent with a suitable solvent may remove the contamination, but in unfavourable cases this may alter the structure of the surface and/or its specific surface area.
(iv) It is preferable that the adsorbent should be outgassed before use, bearing in mind the precautions outlined in 'Reporting Physisorption Data for Gas/Solid Systems'(Note g).
(v) In some techniques each measurement is made with a fresh sample of adsorbent. Consequently the homogeneity of the adsorbent must be checked carefully. The homogeneity of commercial adsorbents may sometimes be poor, and a comparatively large number of replicate measurements on randomised samples may be needed to obtain sufficiently reliable results. Such replications should all be carried out at the same liquid/solid ratio. In many cases proper attention must be paid to the method of sampling the adsorbent to obtain a representative sample.
(vi) The purity of the components of the solution and the avoidance of contamination during preparation and handling are essential. When working with organic media, contamination with water can have a drastic effect on the measurements.
(vii) After equilibration, the adsorbent together with the adsorbate bound to it must be separated from the bulk equilibrium liquid by sedimentation, centrifugation or filtration. The separation must be carried out at the same temperature as that at which equilibrium was established, and for experiments far from ambient temperatures special techniques (e.g. a thermostatted centrifuge) are needed.
(viii) The supernatant liquid has to be subjected to chemical analysis to obtain Δx_2^ℓ. Any analytical method which is sensitive and accurate enough can be used, but in practice optical methods are most frequently employed e.g. refractometry, colorimetry or spectrophotometry. The latter can also be used for uncoloured substances that can be transformed into coloured ones by the addition of suitable reagents. By using radio-labelled adsorptives, measurement of changes of radioactivity can also be employed.

In the case of volatile liquids, appropriate measures have to be taken at all stages in the experiment to minimise losses by evaporation which may change the concentration of the liquid. This is a particular problem if the concentration of the equilibrium liquid is determined using a conventional refractometer (i.e. Abbé or Pulfrich). It should also be noted, in work aiming at high precision, that the refractive index of liquid mixtures can be affected significantly by dissolved air. To avoid the necessity of the conversion of volume fractions to mole fractions, it is advisable to determine the calibration curve for any analytical method with solutions prepared on a weight/weight basis.

(ix) The liquid/solid ratio has to be chosen appropriately. The smaller the ratio the greater will be the resulting change Δx_2^ℓ and thus the accuracy.

(x) It is important to check that **equilibrium** has been achieved: this may take as little as 1-2h. with non-porous adsorbents, but may extend to 1-2 days with porous adsorbents since diffusion within pores cannot be influenced by agitation. A reduction in grain size of the adsorbent may increase the speed of equilibration, but may also alter the adsorption capacity either by increasing the accessibility of the pores, or increasing the external surface area.
(xi) If the excess isotherm is determined with the objective of finding the best conditions for the practical use of the adsorbent, exhaustive treatment with the solvent to be used subsequently is recommended. Great care and circumspection is needed if the aim of the investigation is the intercomparison of the adsorption behaviour of an adsorbent with different solutions in various solvents.

3.3 Immersion method (Note h)

The traditional method of determining adsorption from solution is to add a known mass of solid to a measured amount of solution of known composition in a convenient container which is then sealed and equilibrated, usually with agitation, in a thermostat. Violent agitation which may lead to abrasion of the solid particles is to be avoided. A sample of supernatant liquid is withdrawn and analysed to obtain the change in mole fraction or of concentration.

This method, although widely used in the past, and still popular for less precise work, is tedious and suffers from a number of important disadvantages for accurate work.

Among the factors that have to be borne in mind when using this method are the following.

(i) Preliminary work is needed to establish the time needed to establish equilibrium.
(ii) Outgassing of the components of the solution and of the solid, and their mixing out of contact with the atmosphere requires an elaborate technique.
(iii) Problems also arise in the separation of adsorbent and supernatant liquid if experiments are carried out at temperatures other than ambient, or if the system is sensitive to atmospheric contamination.
(iv) Since each experiment is usually made with a fresh sample of adsorbent, random sampling errors may become important.
(v) In choosing an appropriate liquid/solid ratio, this must be large enough to ensure the retrieval of a sample of bulk liquid sufficient for analysis- preferably for duplicate or replicate determinations.

3.4 Circulation method (Note i)

Many of the problems associated with the classical immersion method may be eliminated if the procedure is carried out in the absence of air and equilibrium is achieved by circulating the liquid over a sample of solid, the concentration of the liquid being monitored continuously by passage through a flow refractometer or other convenient concentration measuring device. A differential method in which solution of the initial concentration is circulated through the reference cell of the refractometer is particularly convenient.

Among the advantages of such a technique are the following.

(i) The adsorption cell may be accurately thermostatted and the temperature dependence of the adsorption determined by varying the cell temperature without the need to refill the apparatus.
(ii) The same sample of adsorbent can be used throughout, and the constancy of its properties checked from to time. If the components of the liquid are volatile the adsorbent can be outgassed under controlled conditions before and between measurements, while if one or both of the components is of low volatility the adsorbent can be contained in a demountable (e.g.stain less steel) cell and can be washed with a suitable solvent, replaced and outgassed.
(iii) The solutions can be made up from thoroughly purified and outgassed components on a vacuum line and transferred in vacuo to the outgassed measuring line.
(iv) The approach to equilibrium can be monitored continuously.
(v) Calibrations can be associated with each experiment by injecting samples of the non-preferentially adsorbed component into the reference circuit.

A number of relatively minor limitations of this method remain. Thus an accurate value of n^{o} is needed , so that the quantitative accuracy of the transfer of solution from the preparation cell into the apparatus must be checked. Care has to be taken to ensure that there are no stagnant regions in the circulation system where the solution concentration may not be at the equilibrium value. Consideration also has to be given to the design of suitable pumps that will neither adsorb the solution components, contaminate the system, nor be corroded when used with aggressive solutions. Difficulties may also arise if the solid pack is not readily permeable to the circulating liquid. Some materials may tend to gel in contact with the solution and in such cases the method may not be practicable. The method also becomes less accurate in very dilute solution.

3.5 Chromatographic method (Note j)

In this technique the solid adsorbent is used as the column packing and measurements are made of the concentrations of ingoing and outgoing solution. The difference between them Δc_2 may again be monitored conveniently using a differential refractometer. The volume of liquid V, passing through the column must also be measured, and the experiment continued until the inlet and outlet concentrations are equal. The reduced surface excess at this concentration is given by

$$n^{\sigma(n)} = \int_0^{V_f} \Delta c_2 \, dV \quad , \tag{22}$$

where V_f is a value greater than that needed to bring Δc_2 back to zero. This technique is particularly useful when working with very dilute solutions, although it may have the disadvantage that large volumes of solution are used up in each run.

3.6 Slurry method (Note k)

This variant on the immersion method overcomes the difficulty of rigorous separation of the supernatant. The sample after equilibration is centrifuged and a weighed sample of the slurry is taken and analysed. In effect, use is made of equation (6):

$$n_2^{\sigma(1)}/m = \frac{1}{m}\left[n_2 - n_1\left(\frac{c_2}{c_1}\right)\right] = \frac{n_1}{m}\left[\frac{n_2}{n_1} - \frac{c_2}{c_1}\right] , \tag{23}$$

where m is the mass of solid in the slurry sample, n_1 and n_2 the amounts of components 1 and 2 in the slurry and c_1 and c_2 the concentrations in the bulk solution. If the liquid/solid ratio in the equilibration step is large, then c_1 and c_2 will be the same as those in the original solution. By having a small liquid/solid ratio in the slurry n_1/m is small and consequently the term in brackets is correspondingly large and can be determined accurately.

3.7 Null method (Note l)

An alternative procedure is to equilibrate an amount n^0 of initial solution of mole fraction x_2^0 with a mass m of solid, and then add to the system an amount Δn^a of a solution of mole fraction x_2^a such that the final concentration of solution returns to x_2^0. The total amount of component 2 in the system is $(n^0 x_2^0 + \Delta n^a x_2^a)$. If the mole fraction in the liquid phase were constant up to the solid surface the amount of 2 present would be $(n^0 + \Delta n^a)x_2^0$. The reduced surface excess amount of 2 is therefore

$$\begin{aligned} n_2^{\sigma(n)} &= (n^0 x_2^0 + \Delta n^a x_2^a) - (n^0 + \Delta n^a)x_2^0 \\ &= \Delta n^a (x_2^a - x_2^0) , \end{aligned} \tag{24}$$

or

$$\Gamma_2^{(n)} = \frac{\Delta n^a}{m a_s}(x_2^a - x_2^0) . \tag{25}$$

In the particular case in which pure component 2 is added

$$\Gamma_2^{(n)} = \frac{\Delta n_2^a}{m a_s}(1 - x_2^0) . \tag{26}$$

It follows, from (11), that for a binary system the relative adsorption of component 2 is given by

$$\Gamma_2^{(1)} = \frac{\Delta n_2^a}{m a_s} \tag{27}$$

i.e. Δn_2^a is a direct measure of the relative adsorption. (Note m)

Experimentally this procedure is easily realised using a circulation technique. It is in fact not necessary to inject exactly the correct amount of pure component 2 to bring the solution concentration back to x_2^0, since the injection of several aliquots enables the required quantity to be obtained either by interpolation or extrapolation.

The important feature of this method is that it does not require a knowledge of n^0, nor is it necessary to calibrate the detection system. It retains all the advantages of the circulation method, but in addition it has the major advantage that it is unnecessary to know the amount of solution with which the solid is equilibrated.

3.8 Radioactive method for low surface areas

Certain specialised techniques have been developed to meet specific problems. For example, the radioactive method may be used for the adsorption, or co-adsorption, of radio-chemically labelled substances from dilute solution at the surface of a thin extended solid sample transparent to the radiation emitted by the labelling nuclides. One side of the solid is equilibrated with the solution, and the other side faces an appropriate detector. If co-adsorption is to be studied then it is necessary to use specifically labelled co-adsorptives.

3.9 Other methods

Adsorption from solution may also be studied by a variety of other techniques such as various forms of spectroscopy (i.r., u.v., n.m.r., e.s.r.), neutron scattering, and ellipsometry. These provide important information on the molecular state of the interfacial region, but do not usually lead to a strictly defined measure of the amount adsorbed.

3.10 Guidance on choice of methods

The choice of method of measuring adsorption from solution by solids with high specific surface areas depends on the type of system being studied and the objectives of the work. The immersion method is often chosen because of the simplicity of the apparatus needed and the use of standard laboratory techniques. It may be the preferred method for a preliminary study in which relatively few experimental points are needed to establish the general pattern of behaviour. However, even in such cases care must be taken when using volatile liquids and especially when the preferential adsorption is small: it appears that some early work is in serious error because of failure to take adequate precautions.

The slurry method has a number of advantages in that the total amount of the components in the slurry, and not their concentrations, are measured. The chromatographic method is particularly useful when working with dilute solutions, where the circulation method becomes less accurate. For work of highest accuracy, especially if temperature coefficientsare to be measured, one or other form of circulation apparatus is recommended. This method, however, involves the use of more sophisticated equipment including vacuum pumps, suitable liquid-circulating pumps, and some form of flow-through detector. However, equipment of this kind is readily constructed from the standard components which are now available for high performance liquid chromatography. The nul method isa development of the circulation method and is to be particularly recommended for future work. The main types of system to which circulation methods are inapplicable are those in which the solid tends to form a gel in contact with the solution.

4 EVALUATION OF ADSORPTION DATA

4.1 Presentation of primary data

Adsorption data are most commonly presented in the form of *specific reduced surface excess isotherms* in which $n_2^{\sigma(n)}/m$, conveniently expressed in mmol g^{-1}(i.e.mol kg^{-1}), or $n_2^{\sigma(m)}/m$ in mg g^{-1}, is plotted against the mole fraction or mass fraction of the equilibrium bulk solution. Whenever possible, tabulated data should also be provided, or deposited in a readily accessible library or data store. It is particularly important that the preferentially adsorbed component should be clearly indicated. The tabulated data should be those derived directly from experiment and not those interpolated from a smoothed graph. Information should also be provided on the relevant details of the particular technique employed.

In all cases the following should be reported:

(i) Characterisation of the adsorbent: chemical identity or commercial name and provenance, grain size, specific surface area (if necessary before and after pretreatment) and method of determination(Note g),mode of pretreatment, and, in the case of porous adsorbents the pore volume and pore size distribution.
(ii) Characterisation of the solution components: chemical identity, provenance, degree of purity as supplied (e.g. analytical reagent, etc.), further purification steps, characteristic physical properties (e.g. refractive index,boiling and/or melting point,n.m.r.spectrum) chromatographic test of purity, check on absence of traces of water when this is relevent.
(iii) Description of the experimental method: details of precautions to eliminate sources of error indicated in Section 3.2 above.
(iv) details of the method of sampling, temperature control, analytical method, number of replicate runs and their reproducibility.

4.2 Classification of adsorption isotherms

(i) completely miscible systems
Most specific reduced surface excess isotherms measured over the whole concentration range for completely miscible liquids fall broadly into one of two classes, the so-called inverted U-shape and the S-shape isotherms (figure 3a and 3b).

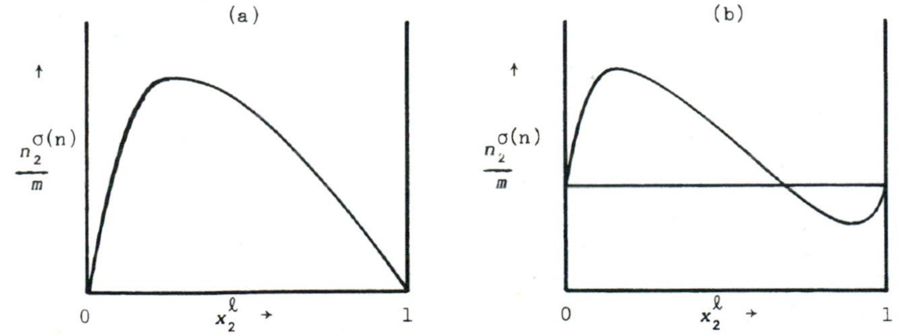

Figure 3: The two main classes of specific reduced surface excess isotherms

Variations occur in the detailed features (e.g. the length of the linear segment around the inflexion point in 3b, or the sharpness of the maximum in 3a.) A more detailed classification is possible (Note n), but the various sub-groups merge into one another, leaving as the main distinction whether or not the preferentially adsorbed component is the same over the whole concentration range, or whether there is a reversal of sign of the adsorption. In the latter case the point of intersection of the isotherm with the abscissa is called an *azeotropic point*, at which the relative composition of the surface layer is identical with that of the bulk liquid.

(ii) dilute solutions

In dilute solutions, especially when the preferentially adsorbed component is of limited solubility, the surface excess isotherms may exhibit the extreme forms shown in figures 4a and 4b.

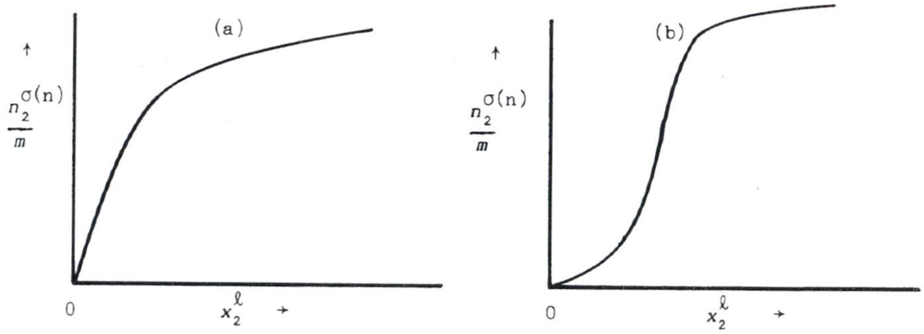

Figure 4: The two extreme types of adsorption isotherm from dilute solution

Transition from the type shown in figure 4b to that in figure 4a is often observed as the temperature is raised. Once again more detailed classifications have been proposed (Note o). At higher concentrations, if the solubility is high enough, these isotherms tend towards the shape shown in figure 3a.

(iii) special cases

In a number of special cases more complex behaviour may be observed, e.g. a point of inflexion may appear on the higher concentration limb of figure 3a, or the curve may show two maxima.

(iv) composite and individual isotherms

Specific reduced surface excess isotherms are often referred to as composite isotherms to distinguish them from so-called 'individual isotherms' which purport to give the adsorption of each component separately. As pointed out below (Section 6) the latter class of isotherms can only be calculated on the basis of some model of the interfacial region, and have no place in the primary presentation of experimental data.

5 INTERPRETATION OF ADSORPTION DATA: THERMODYNAMIC METHODS

5.1 Interfacial tension of the fluid/solid interface (Note p)

A thermodynamic analysis of adsorption from solution leads to the following equation (Gibbs equation) relating the so-called interfacial tension, σ, of the fluid/solid interface, defined by $(\partial G/\partial A_s)_{T,p,n_1,n_2}$ (where G is the Gibbs energy of the whole system), to the adsorption:

$$(\partial\sigma/\partial\mu_2)_{T,p} = -\Gamma_2^{(1)} = -\Gamma_2^{(n)}/x_1^\ell \,, \qquad (28)$$

where μ_2 is the equilibrium chemical potential of component 2, which is the same in both liquid and interfacial regions.

Since

$$\mu_2 = \mu_2^* + RT\ln x_2^\ell\gamma_2^\ell \,, \qquad (29)$$

where the asterisk refers to pure component 2, and γ_2^ℓ is its activity coefficient at the mole fraction x_2^ℓ,

$$d\mu_2 = RT\, d\ln x_2^\ell\gamma_2^\ell \,,$$

and

$$d\sigma = -\frac{RT\Gamma_2^{(n)}}{x_1^\ell}\, d\ln x_2^\ell\gamma_2^\ell \,.$$

On integration

$$\sigma - \sigma_2^* = -\frac{RT}{a_s}\int_{x_2^\ell=1}^{x_2^\ell} \frac{n_2^{\sigma(n)}/m}{x_1^\ell x_2^\ell\gamma_2^\ell}\, d(x_2^\ell\gamma_2^\ell) \,. \qquad (30)$$

Integration across the whole mole fraction range gives $\sigma_1^* - \sigma_2^*$. If a_s is not known, then the only quantity that can be calculated in this way is $(\sigma - \sigma_2^*)a_s$. The integration is most conveniently carried out graphically from smoothed curves of $(n_2^{\sigma(n)}/m)/(x_1^\ell x_2^\ell \gamma_2^\ell)$ against $x_2^\ell \gamma_2^\ell$.

The following important considerations must be borne in mind:
(i) Adsorption measurements must be made accurately over the whole concentration range since it is necessary to extrapolate the curves both to pure component 2 and to infinite dilution of that component. This is particularly important at low concentrations when the bulk solution shows substantial deviations from ideal behaviour since the abscissa of the graph is $x_2^\ell \gamma_2^\ell$. For example, in the case of ethanol(1) + heptane(2) $\gamma_2^\ell \approx 12$ at a mole fraction of 0.02, so that an adsorption measurement at this concentration will appear on the graph at $x_2^\ell \gamma_2^\ell = 0.24$.
(ii) The calculation is critically dependent on accurate knowledge of the activity coeff-ients of the bulk solution, and it is the frequent absence of such information that makes a reliable thermodynamic analysis difficult or impossible. Constancy of the a_i's is also implied.

5.2 Enthalpy and entropy of immersion (or wetting) (Note q)

The following equations enable the enthalpies and entropies of immersion to be calculated from the interfacial tensions derived according to the methods outlined in Section 5.1 as a function of temperature:

$$\Delta_w \hat{h} - \Delta_w \hat{h}_2^* = \frac{\partial}{\partial(1/T)} \left[\frac{\sigma - \sigma_2^*}{T} \right], \qquad (31)$$

$$\Delta_w \hat{s} - \Delta_w \hat{s}_2^* = \frac{\partial}{\partial T}(\sigma - \sigma_2^*), \qquad (32)$$

or

$$\Delta_w \hat{s} - \Delta_w \hat{s}_2^* = \frac{1}{T}\{(\sigma - \sigma_2^*) - (\Delta_w \hat{h} - \Delta_w \hat{h}_2^*)\}. \qquad (33)$$

The notations $\Delta_w \hat{h}$ and $\Delta_w \hat{s}$ refer to the enthalpy and entropy changes associated with the immersion of unit area of solid in the liquid.

Enthalpies of immersion may be determined independently by calorimetry and comparison of the values of the above differences obtained by the two methods provides a check on the reliability of the experimental methods and the methods of analysis of the adsorption data.

5.3 Interpretation

The thermodynamic quantities obtained by the above methods are independent of any physical model of the nature of the adsorption process. They can thus form the basis upon which the predictions of various theories can be compared with experiment.

It is important to observe that the thermodynamic quantities obtained are all relative to those of the reference liquid, in this case component 2. It is, in principle, possible to relate them to the properties of the clean solid surface if information on the vapour adsorption of component 2 is known accurately, but this is unfortunately not generally available.

If component 2 is only partially soluble in component 1, then the latter is conveniently taken as the reference component. By interchanging the suffixes in equation (30), and remembering that $n_1^{\sigma(n)} = - n_2^{\sigma(n)}$, values of $\sigma - \sigma_1^*$ can be calculated.

6 INTERPRETATION OF ADSORPTION DATA: USE OF ADSORPTION MODELS

6.1 General

A complete theory of adsorption at the solid/liquid interface will involve a detailed discussion of the shape of the concentration profile and of the orientations of molecules in the vicinity of the surface. Although progress is being made, none of the theories so far developed is expressed in a form that can be compared directly with experimental measurements.

At the present time, therefore, it is necessary to employ greatly simplified models which although not always physically realistic, nevertheless are useful in the correlation of experimental data.

In using such theories, it is important to bear their limitations in mind. In special cases they may give useful information about the molecular state of the interface, but they must not be used uncritically.

6.2 Surface phase model

The most commonly employed model of adsorption from solution (figure 5) approximates the concentration profile (the dashed curve) by a step function. In effect the liquid volume, containing a total amount of substance, n, is split into two parts within each of which the composition is constant: V^s in which the mole fraction is x_2^s defines the so-called 'surface phase' (Note r) and contains an amount of substance n^s, while V^ℓ is the bulk homogeneous liquid of mole fraction x_2^ℓ containing $n^\ell = n - n^s$.

The reduced surface excess in this model may be expressed in any of the following ways:

$$n_2^{\sigma(n)} = (x_2^s - x_2^\ell)n^s = n_2^s - n^s x_2^\ell = n_2^s x_1^\ell - n_1^s x_2^\ell \quad (34)$$

If the adsorption of 2 is large enough, then at sufficiently low equilibrium concentrations $n^s x_2^\ell$ may become negligibly small so that $n_2^{\sigma(n)}$ can be equated to n_2^s, the amount of 2 bound by adsorption to the interface. Nothing can be said, however, concerning n_1^s, the amount of solvent present in the interfacial layer without making assumptions about the structure of the latter.

Two main models of the adsorbed phase may be considered, namely the layer model when the concept of surface area can be given a clear meaning, and the pore filling model appropriately applied to porous, especially microporous, materials where the notion of surface area becomes blurred. (Note g).

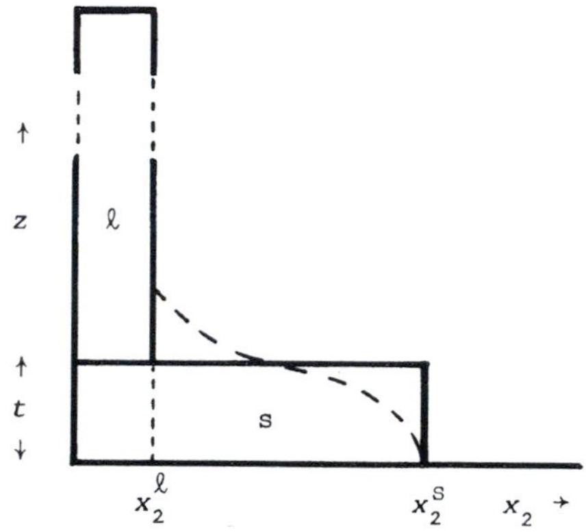

Figure 5: 'Surface phase' model in which the continuous concentration profile (dashed curve) is approximated by a step function separating the liquid phase (ℓ) from the surface phase (s) consisting of t molecular layers

6.3 The layer model

It is assumed that the adsorbed phase consists of t layers of molecules on a plane smooth homogeneous surface. The condition that the surface is always completely covered is that

$$n_1^s a_1 + n_2^s a_2 = A_s , \quad (35a)$$

or that

$$x_1^s a_1 + x_2^s a_2 = A_s/n^s , \quad (35b)$$

where a_1 and a_2 are respectively the partial molar areas of components 1 and 2; they are approximately equal to a_1^o/t and a_2^o/t where a_1^o and a_2^o are the molar cross-sectional areas of the molecules. It has to be emphasised that in the case of molecules of markedly asymmetrical shape these effective cross sectional areas will depend of the orientation of the molecules with respect to the surface. As this orientaion may vary along the isotherm, the values of a^o do not necessarily remain constant, nor can it always be assumed that t is constant.

Subject to these restrictions, the mole fraction of (2) in the surface phase is given by

$$x_2^s = \frac{t x_2^\ell + a_1^o \Gamma_2^{(n)}}{t - (a_2^o - a_1^o)\Gamma_2^{(n)}} . \quad (36)$$

A widely used assumption is that $t = 1$, i.e. that the surface phase consists of a monolayer. There are strong arguments, however, partly intuitive but more precisely based on thermodynamic arguments, supporting the view that in general there must be a gradual transition in composition from that of the first layer adjoining the solid surface to that of the bulk liquid. Consequently the values attributed to x_1^s and x_2^s should more generally be regarded as mean values in the thickness t.

In many cases, however, mainly of marked preferential adsorption of one component, the monolayer model seems to be a satisfactory approximation. In these circumstances, and if the molecules are of about the same size, equation (36) reduces to

$$x_2^s = x_2^\ell + a\Gamma_2^{(n)} , \quad (37)$$

where a is the common value of a_1^o and a_2^o.

The functions $x_1^s(x_2^\ell)$ and $x_2^s(x_2^\ell)$ are often called the 'individual isotherms' for components 1 and 2 respectively, referred to in Section 4.2(iv). Before accepting a monolayer model it is essential to check its consistency by confirming that (a) values of x_i^s calculated assuming a monolayer do not exceed unity, and (b) x_i^s always increases with x_i^ℓ i.e. $(\partial x_i^s/\partial x_i^\ell) > 0$. If the data fail to satisfy either of these criteria, then a minimum thickness of the surface layer may be estimated by repeating the calculation of x_i^s with increasing values of t until both criteria are satisfied.

The separation of the components brought about by adsorption can be characterised by the *separation factor*, S, defined as

$$S = \frac{x_2^s x_1^\ell}{x_2^\ell x_1^s} \quad . \tag{38}$$

$S > 1$ indicates preferential adsorption of component 2.

Rearrangement leads to

$$x_2^s = \frac{x_2^\ell / x_1^\ell}{1/S + x_2^\ell / x_1^\ell} \quad , \tag{39}$$

which is formally an expression of the Langmuir type with the variable x_2^ℓ / x_1^ℓ going from $0 \to \infty$ over the whole concentration range. This represents a true Langmuir isotherm only if S is constant.

If it is assumed that the molecules are of the same size and that both the bulk and surface phases behave ideally, then S is equal to the *adsorption equilibrium constant* K_a and equations (37) and (38) with (35b) and $a_1 = a_2 = a$, lead to

$$\frac{x_1^\ell x_2^\ell}{n_2^{\sigma(n)}/m} = \frac{m}{n^s}\left(x_2^\ell + \frac{1}{(K_a - 1)} \right) \quad . \tag{40}$$

This equation provides, in principle, a means of finding K_a and n^s/m from a graph of the left hand side against x_2^ℓ. When n^s/m is known, the specific surface area of the solid may be calculated:

$$a_s = (n_s/m)\, a \quad . \tag{41}$$

In many cases, when the graphical representation of (40) is satisfactorily linear, the values of a_s derived in this way are in good agreement with those obtained by the BET method from nitrogen gas adsorption. In these instances adsorption from solution provides an alternative method of measuring specific surface areas. Even when equation (40) is not followed, other methods of using the data to estimate surface areas may often be applied (Note **t**).

Deviations of K from constancy may be formally associated with non-ideality of one or both of the two phases:

$$K_a = \frac{x_2^s \gamma_2^s x_1^\ell \gamma_1^\ell}{x_1^s \gamma_1^s x_2^\ell \gamma_2^\ell} = S \frac{\gamma_2^s \gamma_1^\ell}{\gamma_1^s \gamma_2^\ell} \quad , \tag{42}$$

where γ_i^s is the activity coefficient of i in the surface phase.

It is often useful to calculate the surface activity coefficients from the thermodynamically derived equation, applicable to the surface phase model (Note **u**):

$$\ln \gamma_i = \ln \frac{x_i^\ell \gamma_i^\ell}{x_i^s} - (\sigma - \sigma_i^*) a_i /(RT) \quad , \; i = 1,2, \tag{43}$$

where $(\sigma - \sigma_i^*)$ is obtained using equation (30) and x_i^s from equation (36) or (37). The activity coefficients so derived may be compared with those for the bulk liquid. It must be stressed, however, that the concept of surface activity coefficients only has any meaning in terms of the surface phase model, and that the values calculated for these coefficients depend on what assumptions are made concerning t and a_i. These assumptions must always be stated explicitly.

6.4 Pore filling model

In the case of adsorbents with narrow pores, especially micropores, where the clear meaning of the concept of surface area, and the picture of mono- or multilayer coverage becomes blurred or even meaningless (Note **g**), it is more appropriate to consider the material in the pore volume V_p as the adsorbed phase. In this case one must analyse the data in terms of <u>specific</u> rather than <u>areal</u> quantities. However, in interpreting adsorption data for such systems one must bear in mind that molecular sieving effects may complicate the phenomenon.

In the absence of such complicating factors the condition for complete filling of the pores (which replaces the condition (35) for complete filling of the surface) is

$$n_1^s v_1^s + n_2^s v_2^s = V_p \quad , \tag{44}$$

where v_1^s and v_2^s are the partial molar volumes of the components in the pore space. In this picture, equilibrium bulk liquid is not supposed to be present within the pores. The phenomenological definition of the reduced surface excess is still given by equation (8). Again since the composition of the liquid contained in the pores may not be uniform throughout the whole volume of the pore, the mole fractions characterising the composition of the adsorbed phase are to be understood as mean values, in much the same way as explained in connection with the multilayer model.

For molecules of the same size, the amount of material which can be accommodated in the pore spacc is

$$n^s = V_p/v \qquad (45)$$

where v is the common molar volume. Equation (40) still applies and it should therefore be possible to derived values of n^s and hence of V_p from adsorption isotherms in the same way as surface areas can be obtained for non-porous materials.

6.5 Other models

The simple surface phase model presented here for mixtures of molecules of the same size, may be developed in various ways. These include:
(i) extension to the case in which the ratio of the areas occupied by molecules of types 2 and 1 is r , when to maintain complete filling of the surface, equation (42) takes the form

$$K_a = \left(\frac{x_2^s \gamma_2^s}{x_2^\ell \gamma_2^\ell}\right)\left(\frac{x_1^\ell \gamma_1^\ell}{x_1^s \gamma_1^s}\right)^r \; ; \qquad (46)$$

no simple linearisation of the form (46) is possible, even when all the activity coefficients are unity;
(ii)theories of the behaviour of surface activity coefficients;
(iii) multilayer theories in which the single step function discussed above (figure 5) is replaced by a series of steps;
(iv) statistical mechanical theories and computer calculations of the concentration profiles;
(v) theories incorporating the effects of surface heterogeneity (see Section 7);
(vi) theories of adsorption from solution by zeolites, where molecular sieving may play an important role.

7 SURFACE HETEROGENEITY

As outlined in Section 6, adsorption from solution is often interpreted in terms of a layer model, assuming the surface to be an ideally smooth homogeneous plane, characterised by constant values of the energies of adsorption of the two components at all points on the surface. This implies that K_a is constant over the surface. However, few solid surfaces are perfectly uniform and planar, and it is important to understand how surface heterogeneity and roughness affects adsorption behaviour. A major problem is that of distinguishing between deviations from ideal behaviour arising from these factors, and those associated with non-ideality of the surface phase caused by interactions between adsorbed molecules, by molecular size differences and by orientation effects. It may not, even in principle, be possible to make such a separation since the influence of intermolecular interactions depends on whether the heterogeneity is randomly distributed or associated with different patches of the surface. Attempts to derive information on surface heterogeniety from measurements of adsorption from solution require the introduction of assumptions concerning both the nature of the adsorbed phase and the spacial distribution of the heterogeneity, e.g. it may be assumed either that the adsorbed phase is ideal, or that it deviates from ideality in the same way as the bulk solution, while the heterogeneity may be described in terms of various distribution functions. If the bulk phase is ideal it may be justified to assume that the adsorbed phase is also, so that heterogeneity effects dominate the behaviour. At the moment there is no independent check on the validity of such assumptions, and it is necessary to resort to fitting of experimental data to test alternative theoretical models. However, only rarely are such data of high enough precision to lead to a unique solution. It has been suggested that studies of the temperature dependence of adsorption, or calorimetric measurements of enthalpies of adsorption may help to resolve this problem, but this possibility has yet to be tested.

It is therefore essential, in presenting an analysis of data in terms of a model of a heterogeneous surface, to speeify clearly what assumptions are involved. The resulting conclusions should also be examined critically to check that they do not conflict with other evidence. For example, one should be suspicious if strong heterogeneity is indicated for a surface that on the basis of other evidence (e.g. vapour adsorption or electron microscopy) is thought to be essentially homogeneous (e.g. graphitised carbon black). Similarly the validity of the analysis may be in doubt if the same surface appears to exhibit widely different degrees of heterogeneity based on adsorption measurements using different liquid mixtures. This will be particularly so if in the bulk these mixtures deviate from ideality to different extents, and if there is no expectation on chemical grounds for specific differences in the interactions of the molecules involved with the surface. On the other hand, if surfaces have been made deliberately heterogeneous (e.g. clays which have been ion exchanged to different extents with cationic surfactants) then clearly this fact must be reflected in the interpretation of the results.

In general, the situation with respect to the influence of surface imperfections on adsorption from solution has yet to be resolved by further work, both theoretical and experimental. Future experimennts should include both adsorption and calorimetric studies, and must seek the highest attainable precision, since, as with the problem of vapour adsorption on heterogeneous surfaces, the calculated energy distribution functions are very sensitive to experimental errors in the measured isotherm.

8 GENERAL CONCLUSIONS AND RECOMMENDATIONS

8.1 In presenting the results of measurements on adsorption from solution, the raw data should be given in terms of the specific reduced surface excess $n_i^{\sigma(n)}/m$ or $m_i^{\sigma(m)}/m$ as a function of the equilibrium liquid mole fraction (x_i^{ℓ}). The data may be presented graphically or in tabular form, but if published only as graphs, the numerical data should be available either from the authors or from a readily accessible source.

8.2 If the surface area of the solid is known (and stated in the paper) then the data may be expressed as areal reduced surface excesses, $\Gamma_i^{(n)}$.

8.3 The experimental method employed should be adequately described: a list of details which should be included in given in Section 4.1.

8.4 In presenting an analysis of the results the methods used and the assumptions involved should be stated explicitly.

8.5 If a thermodynamic analysis is presented, full details of the sources of information on the activity coefficients of the bulk solution should be given, and if the values adopted are different from those already published, they should be given either in a table or by an interpolation formula.

8.6 If the analysis is made in terms of the layer model the interpretation of large deviations from simple behaviour should be made with caution in view of the generally unrealistic physical assumptions involved.

NOTES AND REFERENCES

(a) Adsorption at the *surface* of a solid may also lead to changes in volume but these are usually small and are not discussed here.

(b) Manual of Symbols and Terminology for Physicochemical Quantities and Units, Appendix II, part 1, Definitions, Terminology and Symbols in Colloid and Surface Chemistry, prepared for publication by D.H.Everett. Pure Appl. Chem.,31, 579-638 (1972).

(c) The algebraic method makes no specific reference to a dividing surface, but involves a general thermodynamic discussion of the degrees of freedom of the whole system. However, it leads to the same operational equations for the adsorption (equations 6 and 9) as those based on the use of a dividing surface. See, for example, R.S.Hansen, J.Phys.Chem.66,410(1962); F.C.Goodrich, Trans.Faraday Soc.64,3403(1968); F.C.Goodrich in Surface and Colloid Science (E.Matijevic and F.Eirich, Eds.) Wiley-Interscience, New York(1969), vol.1, p.1; G.Schay, ibid,vol.2,p.155.

(d) In the framework of the so-called algebraic method V^{ℓ} is thought of as the volume of an arbitrarily chosen amount of bulk liquid.

(e) This definition is ambiguous in practice. V^0 is usually the volume of solution measured before immersion: it should, strictly, be the volume of the equilibrium solution. The definition in terms of initial volume is exact only when the volume is unchanged on adsorption. See A.V.Vernov and A.A.Lopatkin, Zhur.Fiz.Khim.55,428(1981)(Russ.J.Phys.Chem.55, 240(1981)).

(f) S.G.Ash and G.H.Findenegg, Spec.Disc.Faraday Soc. 1,105(1970).

(g) Reporting Physisorption Data for Gas/Solid Systems, prepared for publication by K.S.W. Sing, Pure Appl.Chem.,57, 603 (1985).

(h) J.J.Kipling, Adsorption from Solutions of Nonelectrolytes, Academic Press, London and New York, 1965, chap. 2; G.D.Parfitt and P.C.Thompson, Trans.Faraday Soc.,67,3372(1971).

(i) E.Kurbanbekov, O.G.Larionov, K.V.Chmutov and M.D.Yudelevich, Zhur.Fiz.Khim.43,1630(1969) (Russ.J.Phys.Chem.43,916(1969)); S.G.Ash, R.Bown and D.H.Everett, J.Chem.Thermodynamics 5, 239(1973).

(j) G.Schay, L.G.Nagy and G.Racz, Acta Chimica Acad.Sci.Hung.71,23(1972); S.C.Sharma and T.Fort, J.Coll.Interface Sci.43,36(1973); H.L.Wang, J.L.Duda and C.J.Radke, J.Coll.Interface Sci.66,153(1978).

(k) C.Nunn, R.S.Schechter and W.H.Wade, J.Coll.Interface Sci.80,598(1981).

(l) C.Nunn and D.H.Everett, J.Chem.Soc.Faraday Trans.I,79,2953(1983).

(m) This equation is essentially equivalent to that discussed by C.Wagner (Nach.Akad.Wiss. Göttingen II,Math.Phys.Kl.1973,37;cf.G.N.Lewis and M.Randall, Thermodynamics,2nd.edition, revised by K.S.Pitzer and L.Brewer,McGraw-Hill,New York(1961),p.479):

$$\Gamma_2^{(1)} = (\partial n_2/\partial A_s)_{int,V,n_1}$$

where int means that the intensive variables are kept constant.

(n) e.g. G.Schay and L.G.Nagy, J.Chim.Phys.1961,140.

(o) e.g. C.H.Giles, T.H.MacEwan, S.N.Nakhwa and D.Smith, J.Chem.Soc.1960,3973.

(p) The use of the term ***surface tension*** when applied to interfaces involving a solid phase has been the subject of much discussion since only in very special cases is it possible to devise a means of measuring the surface tension of a solid by mechanical means. Gibbs avoided this problem in the case of a fluid/solid interface by calling σ "the superficial tension of the fluid in contact with the solid" thus implying that the solid is inert and unaffected by the presence of the liquid. By adopting the term "interfacial tension of the fluid/solid interface" for the quantity defined by $(\partial G/\partial A_s)_{T,p,n_1,n_2}$ the role of interactions between the solid and fluid is acknowledged. σ as defined here plays the same part as the surface tension of a liquid in determining thermodynamic equilibrium.

As noted in Section 1, it is often not possible to assign a reliable value to A_s. In such cases it is recommended that equ.(30) be used to calculate $(\sigma - \sigma_2^*)a_s$. Then σa_s may be called the *specific free energy of immersion.*

(q) If A_s is not known then equs.(31-33) should be used to calculate $(\Delta_w\hat{h} - \Delta_w\hat{h}_2^*)a_s$ and $(\Delta_w\hat{s} - \Delta_w\hat{s}_2^*)a_s$. $\Delta_w\hat{h}.a_s$ and $\Delta_w\hat{s}.a_s$ are then, respectively, the *specific enthalpy of immersion* and *specific entropy of immersion.* The direct calorimetric technique of measuring enthalpies of immersion requires careful analysis to ensure that appropriate correction terms are allowed for (e.g.for stirring and bulb-breaking)(see e.g. D.H.Everett, A.G.Langdon and P.Maher, J.Chem.Thermodynamics,16, 981,(1984).

(r) See e.g. J.Davis and D.H.Everett in Specialist Periodical Reports, Colloid Science, Royal Society of Chemistry, London, vol.4,1983,p.85.

(s) The term "surface phase" is not strictly justified since unlike bulk phases (which are autonomous) the properties of surface phases depend on the interactions with adjacent phases (i.e. they are not autonomous). Consequently they are not phases in the sense of the phase rule.

(t) See e.g. G.Schay in Proc.Int.Symp.on Surface Area Determination,1969 (D.H.Everett and R.H.Ottewill,Eds.)Butterworth, London,1970,p.272.

(u) See ref.(r) p.86.

LIST OF SYMBOLS

A_s	area of surface or interface
a_s	specific surface area = A_s/m
a_i	partial molar area of component i
a_i^o	molar cross sectional area of component i
c_i^ℓ	concentration of component i in bulk liquid
c_i	local concentration of component i
$\Delta_w H$	enthalpy of immersion of solid in a given solution
$\Delta_w\hat{h}$	areal enthalpy of immersion of solid in a given solution = $\Delta_w H/A_s$
$\Delta_w\hat{h}_i^*$	areal enthalpy of immersion of solid in pure component i = $\Delta_w H_i^*/A_s$
K_a	adsorption equilibrium constant
m	mass
m_i	mass of component i
m^{sol}	mass of solid
m^o	total mass of liquid
m^σ	excess mass
$m_i^{\sigma(1)}$	relative mass adsorption of component i with respect to component 1
$m_i^{\sigma(m)}$	reduced mass adsorption of component i
M	molar mass of component i
n	amount of substance
n_i	amount of component i
n^o	total amount of liquid
n_i^σ	surface excess amount of component i
$n_i^{\sigma(1)}$	relative adsorption of component i with respect to component 1 = relative surface excess of component i with respect to component 1
$n_i^{\sigma(n)}$	reduced adsorption of component i = reduced surface excess of component i

$n_i^{\sigma(n)}/m$	specific reduced adsorption of component i
$n_i^{\sigma(n)}/A_s$	areal reduced adsorption of component i = $\Gamma_i^{(n)}$
n^s	total amount of substance in 'surface phase'
n_i^s	amount of component i in 'surface phase'
S	separation factor
$\Delta_w S$	entropy of immersion of solid in a given solution
$\Delta_w \hat{s}$	areal entropy of immersion of solid in a given solution = $\Delta_w S/A_s$
$\Delta_w s_i^*$	areal entropy of immersion of solid in pure component i
t	thickness of adsorbed layer (in molecular layers)
V	volume
V^ℓ	total volume of liquid up to the Gibbs Dividing Surface (Section 2.2); volume ascribed to the bulk liquid in the surface phase model(Section 6.2)
V^o	initial volume of liquid
V^σ	excess surface volume
V^f	volume of solution passed in chromatographic process
V_p	pore volume
V^{sol}	volume of solid
v	molar volume
v_i^s	partial molar volume of component i in adsorption space
w	mass fraction
w_i^ℓ	mass fraction of component i in bulk liquid
w_i^o	initial mass fraction of component i in liquid
x	mole fraction
x_i^ℓ	mole fraction of component i in bulk liquid
x_i^o	initial mole fraction of component i in liquid
x_i^s	mole fraction of component i in 'surface phase'
z	distance normal to a surface
z^δ	z co-ordinate of Gibbs dividing surface
z^ℓ	z co-ordinate of boundary of liquid phase
γ_i^ℓ	activity coefficient of component i in bulk liquid
γ_i^s	activity coefficient of component i in 'surface phase'
Γ	areal adsorption(areal surface excess) = n^σ/A_s
$\Gamma_i^{(1)}$	areal relative adsorption of component i with respect to component 1 = areal relative surface excess of component i with respect to component 1 = $n_i^{\sigma(1)}/A_s$
$\Gamma_i^{(n)}$	areal reduced adsorption of component i = areal reduced surface excess of component i = $n_i^{\sigma(n)}/A_s$
$\Gamma_i^{(v)}$	areal surface excess of component i on volume basis = $n_i^{\sigma(v)}/A_s$
μ_i	chemical potential of component i
μ_i^*	chemical potential of pure component i
ρ^ℓ	density of liquid
ρ^{sol}	density of solid
σ	surface or interfacial tension of the fluid/solid interface
σ_i^*	surface or interfacial tension of the pure component i/solid interface.

REPORTING EXPERIMENTAL PRESSURE-AREA DATA WITH FILM BALANCES

(Recommendations 1984)

COMMISSION ON COLLOID AND SURFACE CHEMISTRY INCLUDING CATALYSIS*

LISBETH TER-MINASSIAN-SARAGA

Laboratoire de Physico-Chimie des Surfaces
et des Membranes, CNRS, Paris 06, France

Abstract - The present guide provides the principles for reporting pressure-area data obtained with film balances for spread films. It contains six sections, a check-list covering the basic features discussed in the guide and a list of symbols. In setting up the present guide, the main concern of Commission I.6 was to specify criteria for the reliability and reproducibility of reported results. Except for one modification, the nomenclature used is that recommended in (Ref.1).
Although the guide deals only with the presentation of pressure-area data of spread films, it applies or may be extended to other properties of spread films.
Section I deals with the films and film balances in general. Two film properties : the total area and surface pressure are defined and several experimental conditions which may affect the results are mentioned.
The concept of the monolayer state for spread films and criteria for reliability of the data are introduced and discussed in Section II. Recommendations for the description of experimental conditions, of the apparatus and of the procedures involved are dealt with in Sections III and IV. In Section V, the presentation of the (numerical) values of the data and their units is considered.
Several applications of the film balance are briefly discussed in Section VI.
The check-list covers the basic features discussed in the present guide.

*Membership of the Commission during the period (1981–1985) in which the report was prepared was as follows:

Chairman: 1981–83 J. Lyklema (Netherlands); 1983–85 K. S. W. Sing (UK); *Vice-Chairman*: 1981–85 J. Haber (Poland); *Secretary*: 1981–83 M. Kerker (USA); 1983–85 E. Wolfram (Hungary); *Members*: J. H. Block (FRG; Titular 1983–85, Associate 1981–83); N. V. Churaev (USSR; Associate 1981–85); D. H. Everett (UK; National Representative 1981–85); G. F. Froment (Belgium; National Representative 1981–85); P. C. Gravelle (France; Associate 1981–85); R. S. Hansen (USA; Titular 1981–83); R. A. W. Haul (FRG; National Representative 1981–83); J. W. Hightower (USA; Associate 1983–85); R. J. Hunter (Australia; Associate 1981–85); L. G. Ionescu (Brazil; National Representative 1983–85); A. S. Kertes (Israel; National Representative 1981–85); A. Kitahara (Japan; National Representative 1981–85); J. C. Kuriacose (India; National Representative 1983–85); J. Lyklema (Netherlands; National Representative 1983–85); A. Maroto (Argentina; Associate 1983–85, National Representative 1981–83); S. G. Mason (Canada; National Representative 1981–85); K. Meyer (GDR; National Representative 1981–85); P. Mukerjee (USA; Associate 1981–83); L. G. Nagy (Hungary; National Representative 1981–85); H. van Olphen (Netherlands; Associate 1981–83); J. A. Pajares (Spain; National Representative 1981–83); M. W. Roberts (UK; Titular 1981–83); J. Rouquérol (France; Associate 1983–85); K. S. W. Sing (UK; Associate 1981–83); P. Stenius (Sweden; Titular 1983–85, Associate 1981–83); M. S. Suwandi (Malaysia; National Representative 1983–85); L. Ter-Minassian-Saraga (France; Titular 1983–85, Associate 1981–83); A. Weiss (FRG; National Representative 1983–85); P. B. Wells (UK; Associate 1983–85); E. Wolfram (Hungary; Titular 1981–83).

Reprinted with permission from *Pure and Applied Chemistry,* Volume 57, Number 4, 1985, pages 621–632.

INTRODUCTION

The aim of the present guide is to provide principles for reporting pressure-area data obtained with film balances. It is divided as follows.

Section I contains : a) a general description of films at fluid-fluid interfaces ;
b) a general description of the film balance and the definitions of two measurable film properties : the total area A and the surface pressure π^s ;
c) a list of experimental conditions which may affect the results.

In Section II we discuss a_i the average area per molecular unit in spread layers (see Note a) and the film thickness. We also deal with the origins of various errors in the values of a_i and of the surface pressure π^s. Ways of assessing reliable values of a_i and π^s are provided in this section. The macromolecular films are dealt with in a distinct subsection.

Sections III and IV contain recommendations for the description of the experimental conditions, of the apparatus and of the procedures involved in the study.

Section V deals with the presentation of the (numerical) values of a_i and π^s and with the corresponding units. It includes a description of the various π^s vs a_i diagrams and of the relevant nomenclature.

In Section VI several applications of the film balance are briefly discussed (see Note b).

A check-list covering the basic features discussed in the present guide and a list of symbols concludes it.

SECTION I. GENERAL CONSIDERATIONS

Films at fluid-fluid interfaces

In the context of the present report, films are defined as thin layers of any interfacially active substance accumulated at interfaces. The mode of accumulation of a substance depends on the solubility of the substance in the adjacent fluid phases : films may form either spontaneously by *adsorption* from one or both of the adjacent bulk solutions or by *deposition* or *spreading* directly onto the interface.

Film balances have been designed for the study of area and pressure of spread films. For them, the total number n_i^s of deposited molecules i and the film area A are primary data (for A, see Fig. 1 and Section below). They are generally not suitable for measuring the pressure of adsorbed films.

The *average area per molecule* in the film is $a_i = (A/N_i^s)$. The *surface concentration* or the film density is $\Gamma_i = (n_i^s/A)$.

Film balance

The film balance (Fig. 1) is a device which provides simultaneously two types of data :

- the *total film area* A, which can be varied ; this area is enclosed by non-wettable materials (B, trough, f, thread, ribbon in Fig. 1) which prevent escape of the film ;
- the *net force* F exerted on the mobile non-wettable float which separates the film from a clean surface (Fig. 1). The float f is attached to the trough by flexible floating non-wettable threads or ribbons which permit freedom of motion of the float while preventing leakage.

The float should not deform the meniscus so that contact angle hysteresis and tilting of the float are avoided. This can be achieved with very thin floats.

Note a. In the present context a_i is used instead of $a_{i,s}$ [see (Ref.1)].

Note b. The principles enunciated here would also apply to presentation of data obtained from studies of films by techniques other than those treated in the document e.g. surface viscosity and potential measurements, ellipsometry, radiotracer studies.

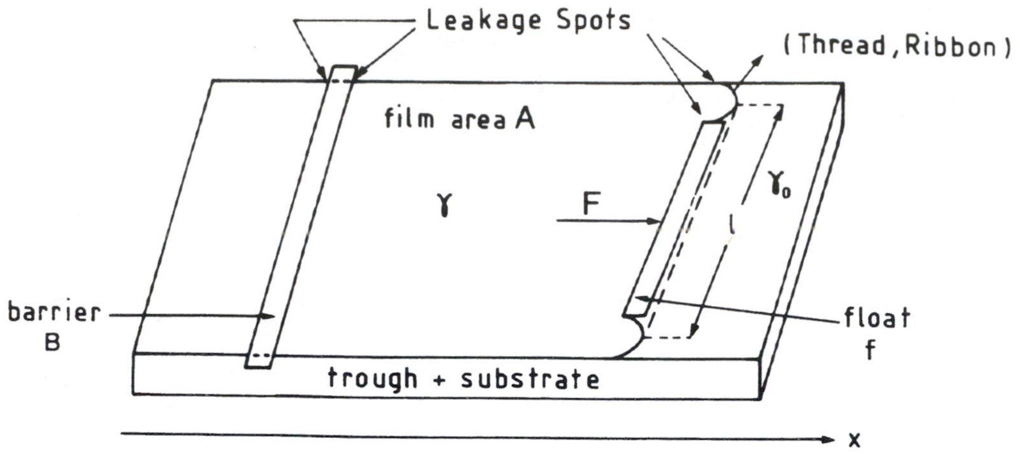

Fig. 1. Schematic diagram of a film balance.

Let γ be the interfacial tension in the film compartment and γ_0 that of the clean interface. For a rectangular film, when the mobile barrier of length l has been displaced parallel to itself, isothermally and reversibly, by an infinitesimal distance δx, the net work of displacement is $F\delta x$. During this process the area increases by an amount $\delta A = l\delta x$. The net work of displacement is $(\gamma_0 - \gamma)l\delta x$. The condition of mechanical equilibrium for the mobile barrier leads to the following equation :

$$\Delta\gamma = \gamma_0 - \gamma = \frac{F}{l} . \quad (1)$$

When the interface is clean, γ_0 is the interfacial tension of the two non-miscible fluids. In that case the difference $\Delta\gamma$ is called the film surface pressure :

$$\Delta\gamma = \gamma_0 - \gamma = \pi^s . \quad (2)$$

If the barrier separates two film-covered surfaces, the value of $\Delta\gamma$ corresponds to the difference between two surface pressures.

Sources of errors

The factors and conditions listed below may lead to inaccuracies in the values of n_i^s, A, F and γ_0 :

- presence of surface active contaminants in the trough, the studied components, the bulk fluids and the spreading solvent ; film contamination by the spreading solvent ;
- incomplete spreading or non-uniform distribution of the spread molecules over the interface ;
- incomplete mechanical relaxation of the films either *under* dynamic conditions (expansion or compression), or *subsequent* to a change in the film area (static metastable state) ;
- inaccurate measurement of the temperature at the interface.

Contamination and incomplete spreading or non-uniform film distribution *on a molecular scale* can lead to ill-defined values of a_i and π^s, useless for further interpretation. Values of π^s obtained for films that are not in mechanical and thermal equilibrium with their surroundings are useless for interpretation in terms of equilibrium thermodynamics.

SECTION II. PROPERTIES OF STABLE, WELL-SPREAD, UNIFORM AND NON-CONTAMINATED FILMS

This section deals with well-spread films and elaborates criteria to ascertain that the spread films are uniform and non-contaminated.

Average areas per molecule

Small molecules. The primary experimental quantities needed to calculate a_i are A and n_i^s. In some cases radiolabelled substances have been used to estimate n_i^s.

The values of A, which include the contributions originating in the loops and the displacement of the float, are known with an accuracy which depends in part on the ratio of substrate area to perimeter and on the wettability of the trough, barrier and float materials : the relative contribution of the meniscus area to A increases when A decreases and should be taken into account.

The accuracy of n_i^s is mainly determined by the accuracies of the volume and concentration of solution spread.

Reproducibility may be poorer or drifting values may be obtained when losses occur through leakage from the film compartment or through dissolution into the adjacent bulk phases. Leakage can take place between the trough and the barrier or the float (see Fig. 1). For films at the air-water interface leakage can be detected by the movement of talcum powder sprinkled on the surface at the suspected leakage spot. To avoid the dissolution of films into water, concentrated solutions of inorganic electrolytes are sometimes used as the substrate. Then the effect of the electrolytes on film properties should be considered. At oil-water interfaces it is difficult to distinguish loss of film material by leakage from that by dissolution. When dissolution occurs, reproducibility of the rate $(d\pi^s/dt)_A$ or $(dA/dt)_{\pi^s}$ may be considered a reliable test for absence of leakage.

If leakage and rate of dissolution are very small, the relative loss $(\Delta n_i^s/n_i^s)$ may be negligible. This criterion is often satisfied by the spread "insoluble" films defined in (Ref. 1), p.584.

When no loss of film molecules occurs, the following test may help to establish whether or not the film is well-spread and in stable equilibrium.

A well-spread film, at equilibrium on a given area A, has the same surface pressure when formed by any of the three following techniques :

- the film is spread from a solution of known concentration on a large initial area which is subsequently compressed to A ;
- the film is spread from a solution of known concentration on the given A ;
- the film is spread on the area A from a solid of known mass.

Such a film does not display hysteresis upon successive compression and expansion cycles. However, metastable monolayers may exist which do not satisfy the first three requirements for equilibrium,yet do not display hysteresis on limited compression and expansion.

The second technique above is useful when one studies the penetration of a radio-labelled, substrate-soluble constituent, into spread insoluble films.

The third technique is best. It can be used only with rapidly spreading virtually insoluble substances, when available, because it provides pure films not contaminated by the spreading solvents. Volatile non-polar solvents are the least contaminating for films spread on aqueous substrates. They may not always be appropriate solvents, so that polar solvents have to be used. A test for film contamination is available (see section below).

In a well-spread pure film of small spherical molecules, a_i cannot be smaller than the molecular cross section in the crystalline state.

In general, spread films consist of asymmetric or symmetric elongated molecules. In their crystals, several distinct molecular cross sections may be defined depending on the crystalline face considered. Usually, a spread film formed by elongated asymmetric molecules with their long axes normal to the interface is considered monomolecular as long as a_i is larger than the *smallest* molecular cross section in the crystalline state which serves as a reference value. If substrate molecules penetrate the film and separate its molecules, the minimum values $a_m(i)$, which is the area per molecule in a complete monolayer of substance i (Ref. 1), is larger than this reference value. If data exist for these reference values they should be quoted for comparison.

Macromolecules. Spread films of macromolecules are handled in the same way as spread films of small molecules. In several respects, spread films of macromolecules behave differently from those of low molecular mass substances. Therefore the principles and criteria discussed in section *Small molecules* apply only to a limited extent to polymeric materials. Experience with macromolecular films is still not very extensive so that no definite recommendations can be given. The following properties of macromolecules and the ensuing effects on the behaviour of spread films deserve attention.

- Synthetic polymers are usually polydisperse, i.e. the films contain molecules of different masses.

- The conformation at the interface of molecules of unequal size may be different.

- Some natural polymers are homodisperse but may retain part of their three-dimensional conformation upon spreading.

- Molecular relaxation processes in macromolecular films are often very slow so that non-equilibrium states are easily frozen in.

- Hysteresis, or in general the observation that film properties depend on spreading history, is more common with macromolecular than with low molecular mass films. Because of these considerations in reporting spread film data of macromolecules, it is important to give full details including the molecular mass distribution, the nature of the spreading solvent and the effect of film preparation and history on the results. For spread macromolecular films a quantity a_i can be mathematically computed. It is usually expressed per monomeric unit. Its physical meaning is simple only in a few favourable cases.

Surface pressure in well-spread mechanically relaxed monolayers

The force F and effective length l of the float (as shown in Fig. 1) lead to the film surface pressure π^S according to (2). Strictly speaking, the measured $\Delta\gamma$ value in (1) or (2) equals the difference between the interfacial tension on each side and contiguous to the float. When the area A is varied by a stepwise movement of the barrier, the surface pressure relaxation may be monitored for each step as a function of time and the thermodynamic stable equilibrium value of π^S may be obtained. It is the same at the float and at the barrier and corresponds to static or mechanical equilibrium conditions of the film and the substrate.

In contrast, under dynamic conditions A is varied continuously. A surface pressure difference $\Delta\pi^S$ may occur, between barrier and float, coupled to substrate movement. Then a state of non-equilibrium may occur such that the pressure measured by the float does not represent the film surface pressure everywhere. In automatic film balances, the continuous film compression or expansion process may induce such non-equilibrium conditions of the film. Then, for a given value of a_i different values of π^S may be measured on compressing and on expanding respectively the film area (hysteresis). The difference $\Delta\pi^S$ decreases when the speed of the barrier decreases. Phase transition may also be involved in such effects.

Compression and expansion isotherms

The function relating the surface pressure π^S to the molecular area a_i for films formed by any of the three spreading techniques mentioned under *Average areas per molecule* above, at a given temperature are named compression or expansion isotherms when A is decreased or increased respectively.

For well-spread, uncontaminated films, continually in stable equilibrium, the isotherms are independent of the techniques of film formation and of both the concentrations and the spread volumes of spreading solutions. For such films compression and expansion are reversible processes.

Figs. 2a & 2b represent typical compression and expansion isotherms for various films. The isotherms may display plateaus and/or kinks (transitions).

For one-component spread films the plateaus correspond to first order transitions either between two coexisting films or between a film and a coexisting condensed bulk phase. The sections between the transitions shown in the Figs. 2a and 2b, correspond to the vapour V, liquid L and solid S state of the film in analogy with the bulk phases. This nomenclature is not in use for the isotherms of macromolecular films.

Films and transitions. The S monolayers are characterized by a seemingly linear variation of π^S with a_i in contrast to the monolayers L or V. The transition V $\rightleftarrows$ L is characterized by a_i^V, a_i^L, π^{LV} which are the partial areas in the coexisting films and the equilibrium transition surface pressure, respectively. For the V $\rightleftarrows$ S transition the corresponding parameters are a_i^V, a_i^S, π^{SV}. Extrapolation of the linear part of the isotherm $\pi^S(a_i)$ for an S film to $\pi^S = 0$ provides the molecular area a_{io}^S of the component i in a corresponding solid, close-packed monolayer. Evidently $a_{io}^S > a_{m(i)}$. It is stressed that the analogous extrapolation for L films is not accurate.

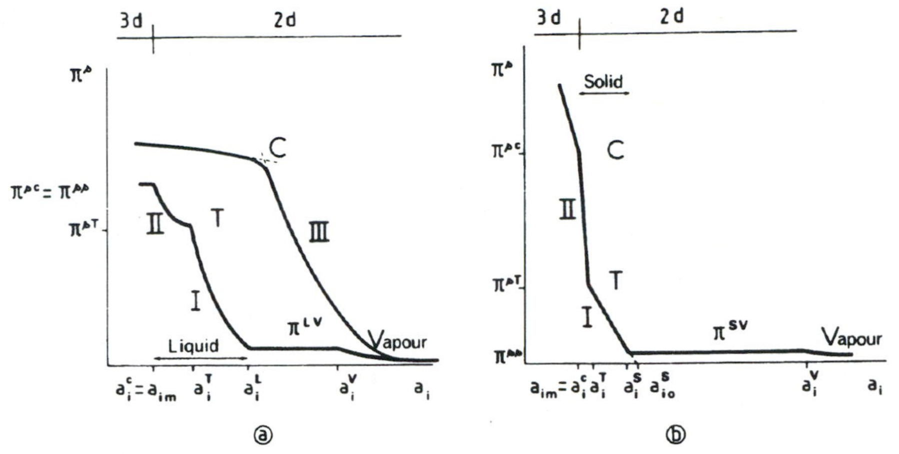

Fig. 2. Pressure-area isotherms for spread films of small molecules. π^{LV}, π^{SV} 1st order transitions ; I, II, III film states ; T = 2nd order transitions ; C collapse. See also list of symbols.

The condensed S or L films may display polymorphism. The various forms are characterized by distinct compressibilities (Fig. 2a I & II, Fig. 2b I & II). Highly or moderately compressible L films are respectively named *expanded* or *condensed*.

The presence or absence (see Fig. 2a), and the location of plateaus or kinks in the isotherms obtained with films depend on both the chemical nature and the structural characteristics of the film material as well as on various experimental conditions (temperature, substrate composition, pH, ionic strength, etc..).

The two plots shown in Fig. 2a and the one in Fig. 2b are typical of films of small molecules under various experimental conditions.

When the values of π^{LV} and π^{SV} are smaller than 1 mN m^{-1} microbalances sensitive to within several μN m^{-1} are used. They provide the isotherms of the monolayers in the vapour state V and the values π^{LV} and π^{SV}.

In general, even these microbalances do not reveal any condensation phenomena in the spread films of macromolecules at low surface pressure.

Uniformity of molecular distribution in films. Often, the uniformity of the molecular distribution in films is taken for granted. However, if uniformity is not ensured, the interpretation of the isotherms and film polymorphism in terms of purely molecular properties becomes obscure, particularly for solid films of small molecules and for macromolecular films.

Both for films comprised of small molecules or macromolecules, complementary studies have sometimes revealed a non-uniform distribution of the molecules at both macroscopic and microscopic levels. Surface potential studies have shown the presence of macroscopic islands in the range of molecular areas corresponding to the plateaus designated by π^{LV} and π^{SV} in Fig. 2a & Fig. 2b, respectively. Electron microscopy of transferred films has revealed a "microporosity" of solid films in the range of surface pressures π^{SV} to π^{ST} (Fig. 2b).

Although the isotherms of macromolecular films may not display kinks or plateaus, surface potential studies have revealed in some cases macroscopic non-uniform distribution of molecules at low surface pressures.

Coexistence of film and bulk phases. The limit of the spread film state is named the collapse point (a_i^c, π^{SC} in Figs. 2a & 2b). This is the transition between the film and the bulk phase. Below a_i^c the film is unstable. At the point of collapse, folds, fibers, liquid and gel droplets, or solid particles may coexist with the film at the interface. For collapsed films, π^{SC} may (Fig. 2b) or may not (Fig. 2a, curve II) vary with a_i when $a_i < a_i^c$.

On the other hand, when macroscopic droplets or solid particles of the film forming substance are deposited upon the interface, a coexistent spread film is obtained at a final equilibrium spreading pressure π^{SS} independent of the apparent molecular area. If $\pi^{SC} = \pi^{SS}$ spreading and collapse are reversible processes. Then the film is in a state which satisfies the criterion of equilibrium stability and the reported results have thermodynamic significance (see also Section *Small molecules* above).

Generally, L and V films satisfy the criterion of stability, viz. the surface pressure of their collapsed films is independent of the molecular area when $a_i < a_i^c$ (Fig. 2a) and $\pi^s < \pi^{ss}$ for $a_i > a_i^c$.

For some films spread from a solution and then compressed $\pi^{sc} > \pi^s > \pi^{ss}$. They are metastable, collapse irreversibly and will not respread. Solid films of low molecular weight substances and soluble or insoluble (in the substrate) spread macromolecules may form such metastable films. For solid films of small, elongated, asymmetric molecules on first compression $a_i^c = a_m$ (i) defined in Section *Small molecules* above. A value of a_m (i) cannot be defined in all the cases of macromolecular films (see Section *Macromolecules* above). This value of a_m (i) depends on the molecular conformation in the spread film.

The general conclusion of this section is that, once the film contamination and the accuracy of the results have been assessed, information on the state of equilibrium of the spread films may be obtained from studies of reproducibility of total area A at given pressures of films spread using different techniques, reproducibility of surface pressure π^s of a film on expansion and compression, equality of the spreading pressure and collapse pressure or $\pi^{ss} = \pi^{sc}$.

Equilibrium films need not be uniform at a molecular level. Coexisting gaseous and liquid films are clear examples of film non-uniformity. In such cases, the film balance combined with surface potential studies can provide qualitative information on the macroscopic heterogeneity. Similar heterogeneities may be present close to the kinks shown in the Figs. 2a & 2b. Then direct observation techniques, such as electron microscopy, may provide additional information of film uniformity. Interpretation of average molecular areas in structural terms requires generally that the film be both in equilibrium and uniform.

SECTION III. MATERIALS AND TEMPERATURE

The description of the film material should include chemical composition, distribution of molecular mass, molecular conformation, degree of hydration, melting or softening point. Preparation and purification methods must be described. The purity and purification methods of the spreading solvents and of the liquid substrate as well as their composition should be given. The pH might be reported if the liquid substrate is water. Traces of surface active contaminants in the spreading solvents may considerably modify γ (but not γ_0) and π^s. For acidic surfactants, traces of Ca^{2+} act as severe contaminants ; for basic surfactants the same applies to CO_2.

A spreading film compresses any surface active impurity that may inadvertently be present in the substrate. This may cause significant errors in the results. If prior to spreading rapid compression of the "clean" surface does not lead to any measurable surface pressure this fact could be taken as a criterion of substrate purity. The mobility of talcum powder, upon blowing air, throughout the surface is another criterion.

The temperature of the film depends on the temperature(s) of the adjacent phases and of any heat or material transfer between them. In the case of an aqueous substrate and an air phase of low humidity, water evaporation may produce a local (lower) temperature, different from that of the bulk phases. On the other hand, if the air is supersaturated with vapour, condensation would lead to an interfacial temperature higher than that of the bulk phases. In the case of two incompletely miscible liquids the evolution towards mutual saturation may again produce a different local interfacial temperature. Therefore, precautions must be taken to ensure thermal equilibrium in the system.

Contamination of the film by the spreading solvent may occur. A dependence of the results π^s (a_i) on the initial spreading area and on the spreading solution volume and concentration is then observed.

For polymeric substances the nature of the spreading solvent determines the molecular conformation inside the spreading solution and may affect the spread film isotherm. Complete knowledge of the composition of the mixed spreading solvent is important. When macromolecules or small molecules soluble in the substrate are spread, dissolution of film material into the adjacent phases must be prevented during the spreading period.

SECTION IV. APPARATUS AND PROCEDURES

The description of the apparatus should include :

a) the trough and film frame, if any (material, pretreatment, construction and size). Due to the meniscus, film areas smaller than about 10 cm^2 involve inaccuracies which may be estimated from surface tension and meniscus shape. A comment on minimizing vibration, a brief description of the enclosure for minimizing contamination through the gas phase, temperature variation and the effects of draught are required.

b) Barrier and float (material, pretreatment and dimensions) including the shape and material of the float, shape and material of the thread and ribbon loops which prevent film leakage. A comment might be included that describes the connection of the thread and ribbon loops to the float and to the trough walls.

c) The equipement and devices for spreading the film and measuring the horizontal force F (Eq. 1).

d) The film barostat device used for kinetic experiments at constant pressure.

The description of experimental procedures should include :

a) The cleaning of the apparatus and of the substrate surface, technique of depositing the spreading solution, time allowed for solvent evaporation or dissolution in the substrate, film compressing technique and relative rate of film compression.

b) The method used to test for leakage and for stability or equilibrium of the films.

c) The operations between film spreading and the starting time of the measurements.

d) The method of checking film contamination by the spreading solvent.

Extrapolation methods, if any, to obtain the film spreading pressure at time t_0 of monolayer deposition should be described and justified if possible.

SECTION V. RESULTS AND ACCURACY

Insoluble, one-component films

The accuracy of measuring the spread volumes, concentration of the spreading solution, float length, force and surface pressure should be reported.

The plots π^s vs. a_i should be shown with error bars when a stepwise compressing technique is used. With an automatic compressing technique, a full compression - expansion cycle should be reported. It is recommended that tables of a_i and π^s values be included in the paper unless numerical analysis is used to fit the experimental points in which case the sensitivity of the fit to the values assigned to the parameters should be assessed. The various values of π^s and a_i characterizing the film transitions are important parameters and should be reported.

Units

Areas A are expressed in sq.meters (m^2) and forces in newtons (N). Surface tension and surface pressure are expressed in N m^{-1} or J m^{-2} ; molar areas in m^2 mol^{-1} and molecular areas in nm^2 $molecule^{-1}$. The recommended units of molar or molecular surface concentrations are mol m^{-2} and molecule nm^{-2} respectively. For polydisperse polymeric substances the units are : m^2 (mol of monomer)$^{-1}$ or nm^2 $monomer^{-1}$. For substances of ill-defined composition, areas may be expressed as specific areas in m^2 mg^{-1} and the surface concentration in mg m^{-2}.

The former unit of surface pressure dyn cm^{-1} is equivalent to mN m^{-1}. The former unit of molecular area $Å^2$ (molecule)$^{-1}$ is 100 times smaller than one nm^2 (molecule)$^{-1}$.

SECTION VI. APPLICATIONS

Mixed films

Spread mixed films are obtained by spreading a solution of two or more film-forming substances in a volatile solvent. When none of the substances dissolves sighificantly into the substrate the mixed film may be compressed and dealt with experimentally in the same way as a one-component film. If i is one component of the mixed film and x_i^s its mole fraction in the film, the average area per molecule a of the spread mixture is equal to $a = \Sigma_i x_i^s \bar{a}_i$, where $\bar{a}_i = (\partial A/\partial n_i^s)_{\pi^s}$ is the partial area per molecule of constituent i in the mixed film at a given surface pressure and temperature. Even more than for one-component films, it should be emphasized that the uniformity of the distribution of constituents on a molecular scale cannot be assessed unambiguously by trough experiments only.

Built-up films

Film balances with pressure control devices can be used to transfer insoluble spread films onto solid vertical plates (Langmuir-Blodgett films).

Relevant additional information to be provided includes :

- description of the set-up for building-up the films ;
- nature and size of the solid plate ;
- pretreatment of the solid plate (cleaning and polishing) ;
- speed of vertical movement of the plate ;
- the number of layers deposited ;
- molecular area ratios in the film built-up on the solid and in that spread on the liquid surface ;
- orientation of the polar part of the molecules in the first and in the last deposited layers.

Kinetics of processes in films

Chemical reactions such as hydrolysis, isomerization, oxidation and photochemical processes can occur in either soluble or insoluble films. In addition, dissolution occurs in soluble films.

Studies of these processes are often carried out at a controlled film pressure. Some controlling device automatically drives the compressing barrier while A is recorded as a function of time t. Since π^s is constant the rate of the process per unit film area is expressed as the relative film area change $|k| = |\partial \ln A/\partial t|_{\pi^s}$ and varies with the surface pressure, temperature and substrate properties. The accuracy of surface pressure measurement and of its control should be reported.

BIBLIOGRAPHY

1. D.H. Everett, IUPAC, Manual of Symbols and Terminology for Physico-Chemical Quantities and Units. Appendix II : Definitions, Terminology and Symbols in Colloid and Surface Chemistry, Pure Applied Chem. 31, 583 (1972).

APPENDIX. CHECK-LIST

The following check-list is recommended to assist authors in the presentation of pressure-area data obtained with film balances. It is suggested that the following items be checked and the relevant experimental conditions and results be described and/or reported :

a) the trough, barrier, float, flexible ribbon or thread loops ;
b) surface pressure measuring device ;
c) surface temperature measuring device ; also other temperature conditions ;
d) origin, purity (or purification) of the substances forming the film, the spreading solvent, the liquid substrate ;
e) the composition of the liquid substrate and its pH (for aqueous substrates) ;
f) cleaning procedures for : the apparatus, the water surface (before spreading) ;
g) contamination of the clean surface (by the spreading solvent too) ;
h) film leakage (at barrier and loop ends) ;
i) film contamination by the spreading solvent ;
j) film solubility ;
k) film hysteresis (expansion - compression) ;
l) equality of film collapse and spreading pressures ;

m) presentation of the data : accuracy, reproducibility, units ;

n) assessment of a_i values for macromolecules.

LIST OF SYMBOLS

S = solid-like film ;
L = liquid-like film ;
V = vapour-like film ;
SV = coexisting solid-like and vapour-like films (plateau in pressure-area plot) ;
LV = coexisting liquid-like and vapour-like films (plateau in pressure-area plot) ;
T = transition in S or in L films (kink in pressure-area plot) ;
C = collapse point ;
n_i^S = amount of component i in the interfacial layer (Ref. 1) ;
A = area of films in the present context (see also Ref. 1) ;
a_i = average area per molecule of substance i in pure films of i ;
a = average area per molecule in mixed films ;
$\overline{a}_i$ = partial area per molecule of substance i in mixed films ;
a_i^S, a_i^L, a_i^V = values of a_i in coexisting SV and LV films ;
a_i^T = value of a_i at the kink point T ;
a_{io}^S = average area per molecule of substance i in S films at $\pi^S = 0$ (extrapolated value) ;
a_i^C = value of a_i at the collapse point ;
a_m (i) = area per molecule in complete monolayer of substance i (Ref. 1) ;
x_i^S = mol fraction of i in the film ;
l = length of the float (Fig. 1) ;
F = force on the float (Fig. 1) ;
γ_o = interfacial tension of clean interface ;
γ = interfacial tension of the film-occupied interface ;
Γ_i = surface (excess) concentration (Ref. 1) ;
π^S = film or surface pressure (Ref. 1) ;
π^{SC}, π^{SS} = collapse and equilibrium spreading pressures ;
π^{ST}, π^{SV}, π^{LV} = equilibrium film pressure for the transitions T, SV and LV ;
t = time (in the present context ; see also Ref. 1) ;
$k = (1/A)(dA/dt)_{\pi^S}$ = relative rate of a process studied at constant pressure.
N_i^S = number of molecules i in the interfacial layer

REPORTING EXPERIMENTAL DATA DEALING WITH CRITICAL MICELLIZATION CONCENTRATIONS (c.m.c's) OF AQUEOUS SURFACTANT SYSTEMS

K. J. MYSELS

General Atomic Co., San Diego,
USA

P. MUJERJEE

University of Wisconsin, USA

The IUPAC Manual of Symbols and Terminology makes the following statements about micelles and c.m.c.'s [Appendix II, Part I, 1.6 (see Note b)].

> Surfactants in solution are often association colloids, that is, they tend to form *micelles*, meaning aggregates of colloidal dimensions existing in equilibrium with the molecules or ions from which they are formed.
>
> There is a relatively small range of concentrations separating the limit below which virtually no micelles are detected and the limit above which virtually all additional surfactant forms micelles. Many properties of surfactant solutions, if plotted against the concentration, appear to change at a different rate above and below this range. By extrapolating the loci of such a property above and below this range until they intersect, a value may be obtained known as the *critical micellization concentration* (critical micelle concentration), symbol c_m, abbreviation c.m.c. As values obtained using different properties are not quite identical, the method by which the c.m.c. is determined should be clearly stated.

C.m.c. data are determined by a variety of methods and are used in a wide range of areas and specialties. Intercomparison, compilation, and evaluation of results are, therefore, of great importance and require high standards of reporting in the primary literature. This paper suggests criteria for the presentation of experimentally determined c.m.c. values and offers some caveats about the meaning and significance of the data. Attention paid to the considerations here presented should not only help in obtaining significant and reliable c.m.c. data, but also improve their comparability and facilitate their evaluation.

A listing of some 70 methods of determining c.m.c.'s encountered in the literature up to 1966 has appeared recently (see Note c). Although additional methods such as those based on NMR have been developed since that time, there seems to be no need for a further review, nor does it seem necessary to discuss here the reasons for systematic differences between some of the methods, which are described in the same reference.

It may be worth noting, however, that a return to the early definition of the c.m.c. as the concentration at which there is the first perceptible appearance of micelles (as shown by the beginning of a deviation from behavior attributable to nonmicellar species) is not compatible with the above definition and is strongly discouraged. The value of the c.m.c., by this early definition, is very dependent on the sensitivity of the method used. In addition this early definition is inappropriate for extrapolating properties of micelles to "infinite dilution of micelles," which the IUPAC definition allows.

Note a. At the time of the approval of this document (1977), the membership of the Commission was: *Chairman:* Dr. K. J. Mysels (USA); *Vice-Chairman:* Prof. R. Haul (GFR); *Secretary:* Prof. J. Lyklema (Netherlands); *Titular Members:* Prof. R. L. Burwell (USA), Prof. R. Hansen (USA), Dr. V. B. Kazansky (USSR), Prof. C. Kemball (UK), Prof. M. W. Roberts (UK); *Associate Members:* Prof. R. M. Barrer (UK), Prof. G. Ertl (GFR), Prof. J. Haber (Poland), Prof. P. Mukerjee (USA), Dr. E. Terminassian-Saraga (France), Dr. I. I. Tretiakov (USSR), Dr. H. van Olphen (USA), Prof. E. Wolfram (Hungary); *National Representatives:* Prof. D. H. Everett (UK), Dr. K. Morikawa (Japan), Prof. W. Schirmer (DDR).
Correspondence Address: Prof. J. Lyklema, Lab. for Physical and Colloid Chemistry, State Agricultural University, De Dreijen 6, 6703 BC Wageningen, The Netherlands.

Note b. Published in *Pure Appl. Chem.*, Vol. 31, No. 4 (1972), pp. 577-638.

Note c. Mukerjee, P., and K. J. Mysels, *Critical Micelle Concentrations of Aqueous Surfactant Systems*, NSRDS-NBS-36; Superintendent of Documents, U.S. Government Printing Office, Washington, D.C. USA, 1971. Now available from National Technical Information Center, 5285 Port Royal Rd., Springfield, VA 22161, $9.50.

1. GENERAL

The following principles are basic to the presentation of data:

(a) The work should be described in sufficient detail as regards equipment, conditions, and procedure to allow the results to be reproduced and the quality of the work to be appraised.

(b) The results should be presented in a form that will permit them to be reworked by others to a reasonable extent.

The discussion below does not deal with all possible systems and methods but presents only points of general importance.

2. EXISTENCE OF A c.m.c.

The existence of a c.m.c. is reflected in a change in the concentration dependence of some property of the solution over a narrow range. The reverse is, however, not necessarily true, and real or apparent changes in such dependence need not correspond to a c.m.c. Hence it is important that there be evidence that indeed below the c.m.c. region the solution is mainly monomeric whereas above the c.m.c. the number of properly defined micelles increases rapidly with concentration. Such evidence may be provided directly by the measurement itself, as in light scattering where the micellar weight is determined simultaneously, or indirectly from the structure of the surfactant and previous experience as for a long chain amphipathic compound in aqueous media. For other cases such evidence should be explicitly presented.

The definition of the c.m.c. implies in particular that for a single surfactant or aggregating species there can be only one c.m.c. A transition of the structure of micelles above the c.m.c., even though distinct, should not be designated as a second c.m.c.

An essential part of the definition of the c.m.c. is that the transition occurs over a relatively small range of concentration. It is only then that the c.m.c. can be defined with some accuracy and only then that the c.m.c. value can be used to decide whether a solution of a given concentration will be mainly micellar or mainly monomeric. If the transition range is broad, then a specific value within it may still be determined by any of a number of procedures, but the concordance between the values obtained by different methods and by different investigators becomes poor and the difference between the properties of the solution below and above the value obtained less clear. In other words, the "critical" part of the c.m.c. gets lost, the clear indication of solution composition above and below the c.m.c. is not obtained, and what is more serious, a misleading description of the system may be given.

When only one surfactant is present, "small range" corresponds, in line with the laws of chemical equilibrium, to relatively "large" micelles. Though the terms remain somewhat arbitrary, in this context "small range" would correspond to less than about 20% of the c.m.c. and "large" to more than 20 monomers in the micelle. Thus, in systems showing the "small range," when a micelle with, for example, an average degree of aggregation of 50 become detectable, smaller micelles containing less than 20 or 30 monomers are not present to a significant extent although all the aggregates are the result of stepwise self-association. In many cases, although this is not essential, larger micelles containing more than, say, 70 monomers will also be absent. There are other systems in which self-association of solutes also occurs by the stepwise association of monomers to oligomers and multimers, but in such a manner that an average degree of aggregation of 20 or more will correspond to a wide distribution of aggregate sizes and to the presence of considerable amounts of smaller aggregates. Such systems may contain micelles but do not show a "small concentration range" having the meaning and significance normally attached to c.m.c. values defined above. Though it may be useful to extend the term micelle to such smaller aggregates, it does seem meaningful to extend the c.m.c. concept to systems in which the range is much wider than stated. Thus, systems not showing the "narrow range" should not be assigned c.m.c.'s.

The "relatively small range of concentrations" for the transition corresponds to a relatively sharp change of the slope on a plot of a measured property vs. concentration. It happens, however, that this sharpness becomes obscured by the curvature of the lines on both sides of the c.m.c. It is then often useful to plot $\Delta X/\Delta c$ vs. the average concentration c. Here Δ indicates the difference between two neighboring experimental points and X a property of the solution which is roughly additive with respect to monomers and to micelles and increases (or decreases) with their concentrations. Examples of X are the conductivity of the solution, or its density or its refractive index. Quantities Y, averaged over one or more species such as molar conductivity or NMR line shift, can be transformed into the additive type by multiplying by c. Thus $\Delta X/\Delta c = \Delta(Yc)/\Delta c$. The sharpness of the transition is indicated by the steepness of that part of the curve that connects the $\Delta X/\Delta c$ plots of the mainly monomeric and the mainly micellar states.

In the case of mixtures of surfactants which form mixed micelles, the situation is much more complicated and the "small range" can also become large because the composition of the micelles changes with total concentration. This situation requires a particularly careful description of the criteria used and an appreciation of the fact that although only large micelles may form, the "micellization concentration" may become an ill-defined range instead of a "critical" one. This is especially so when the components of the mixture have by themselves widely differing c.m.c.'s. Here plots of the $\Delta X/\Delta c$ type can be very useful in indicating the onset of micelle formation.

Listed below are some phenomena unrelated to a c.m.c. which may cause marked changes in the concentration dependence of some property:

A basic surfactant reacts with all the carbon dioxide in the distilled water solvent, or a soap with all the ions responsible for any hardness of water.

The color of an indicator changes when an acid surfactant reaches a concentration giving a suitable pH.

The surfactant, or its association product with an impurity, reach a solubility limit.

The concentration of dimers in a monomer-dimer equilibrium changes as a function of total concentration.

A property varies roughly linearly with concentration and the logarithm of its value is plotted against concentration in the region in which the value of this property is unity. Figure 1 shows a related example for an absorbing system obeying Beer's law. Experimental uncertainties and lack of data points in the transition region can further increase the apparent sharpness of the change of slope.

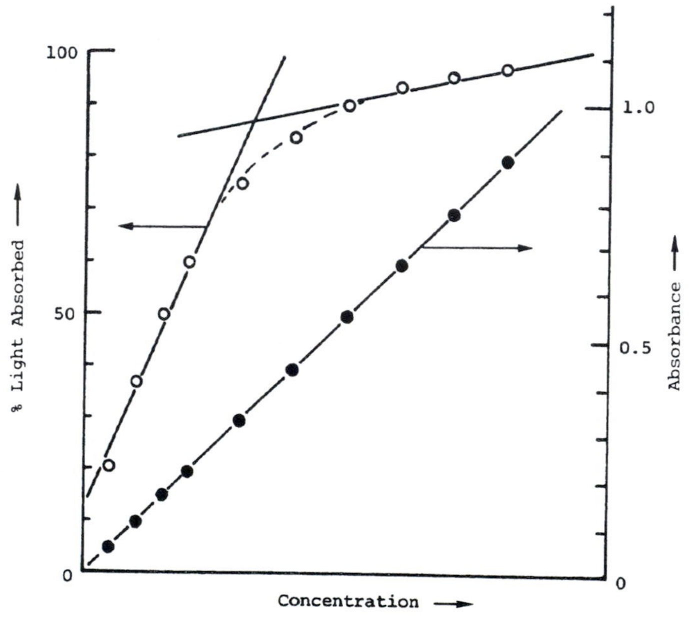

Fig. 1. Same optical data for a compound obeying Beer's law, plotted as % light absorbed and as absorbance. The former could suggest a fictitious c.m.c.

It should also be noted that all directly measurable equilibrium properties of a solution must be continuous functions of the concentration. Their derivatives with respect to concentration may show large changes over a narrow concentration range which may be mistaken for a discontinuity. Any report of discontinuity should therefore be regarded with utmost suspicion.

3. MATERIALS

An important characteristic of the c.m.c. of most surfactants is that it may be sensitive to some impurities and not to others, and that the sensitivity can depend greatly on the surfactant and on the method of determination. Thus the presence of hydrophilic impurities such as sugar generally has little effect whereas that of oleophilic ones such as dodecanol tends to lower the c.m.c., the determining factor being the partition of the impurities between micelle and solvent. The surface tension method seems to be very sensitive to traces of surface active impurities whereas conductivity methods are much less so. Salts of weak acids are sensitive to carbon dioxide in the system and to solution pH, and may form products of hydrolysis that are surface active, whereas salts of strong acid or non-ionics are much less affected by these factors. Hence the possible presence of impurities likely to affect the results in the particular system and the particular method used must receive special attention both in the procedure employed and in the description given.

The c.m.c. should not normally be used as a criterion of identity or purity because of the difficulties of determining it accurately, its low sensitivity to some impurities, and the relatively narrow range in which the c.m.c.'s of many surfactants lie. On the other hand, any significant difference between an observed c.m.c. and that reliably reported (see Ref. 1) is a good reason to doubt the purity or identity of a compound.

Publication of c.m.c. values of systems which contain substantial and unspecified fractions of impurities is undesirable. In particular, the value of data for weak acids, for which the influence of carbonic acid from air, water and reagents is not specified or kept to a minimum, is highly questionable. Needless to say, the above is not intended to discourage the study of well characterized mixed systems or of the effect of known additives upon the c.m.c.

4. APPARATUS AND METHODS

Relative but precise measurements of a property of the solution often suffice for the determination of a c.m.c. This should be kept in mind when describing the instrumental method. In particular, the degree of linearity of instrumental response should be mentioned. Of course, high accuracy is desirable in any published data and should be documented if achieved.

If commercial apparatus is used, it is not sufficient to describe it by a commercial trade name. Rather, its pertinent nature and characteristics must be given either directly or by reference to the scientific literature. It should be remembered that the manufacturer has the right to change specifications without changing names and has no obligation to supply, and continue to supply indefinitely, information about his product.

Concentration changes have to be measured precisely to obtain the concentration dependence whereas the absolute concentration has to be known accurately for the c.m.c. itself. Since c.m.c. values are seldom accurate to better than 0.1%, it is the avoidance of gross errors such as overlooking the presence of water of crystallization that is most important in this respect. Thus final sample conditioning, the molecular weight used, and the method of preparing the solution should be reported, and attention should be paid to a clear statement of units used.

The temperature (or its limits) should be stated explicitly. In some cases, for example with nonionic surfactants, particularly when close to their cloud points, special precautions to assure uniformity and constancy of temperature during measurements are required and should be reported.

5. REPORTING OF RESULTS

There is no substitute for a numerical value given by the author as his best summary of the results. Graphical presentation of c.m.c. values as a function of some variable is not a substitute unless the precision of the c.m.c. is very low. A graphical summary always places a considerable burden on the reader interested in a numerical value. A graph of the variation with concentration of the property leading to the c.m.c. is not a good substitute either, but such a graph can be a great help in judging the value of the numerical result quoted, in addition to being often of intrinsic interest. It is recognized, however, that when more than few c.m.c.'s are reported, it is not always practical to document them all in this way.

The report should include an indication of the reproducibility estimated by the author, specifying the number and the extent of independence of the experiments; e.g., whether starting with different raw materials in the preparation of the sample at one extreme, or making separate dilutions from the same stock solution at the other.

Since the c.m.c. is obtained by extrapolation of two, generally linear, trends from above and from below the c.m.c., the range used in the extrapolation should be reported either in words or in a figure. In precise determinations, any points within the transition range have to be disregarded in the determination of these trends, and therefore, this transition range should also be indicated.

Section 7

Photochemistry

Although the Green Book doesn't have a section on photochemistry, Table 7.1 contains the section on radiation, which is relevant to this section and also to Section 10. "Recommended Standards for Reporting Photochemical Data" (*33*) (reprinted here) gives recommendations for publishing data from a variety of types of photochemical measurements. Photochemistry and radiation chemistry are also covered briefly in reference 15.

Table 7.1. Electromagnetic radiation. Continued on next page.

The quantities and symbols given here have been selected on the basis of recommendations by IUPAP [3], ISO [4.g], and IUPAC [17, 18, 19].

Name	*Symbol*	*Definition*	*SI unit*	*Notes*
wavelength	λ		m	
speed of light				
in vacuum	c_0		$m\ s^{-1}$	
in a medium	c	$c=c_0/n$	$m\ s^{-1}$	
wavenumber in vacuum	$\tilde{\nu}$	$\tilde{\nu}=\nu/c_0=1/n\lambda$	m^{-1}	1
wavenumber (in a medium)	σ	$\sigma=1/\lambda$	m^{-1}	
frequency	ν	$\nu=c/\lambda$	Hz	
circular frequency, pulsatance	ω	$\omega=2\pi\nu$	s^{-1}, rad s^{-1}	
refractive index	n	$n=c_0/c$	1	
Planck constant	h		J s	
Planck constant/2π	$\hbar$	$\hbar=h/2\pi$	J s	
radiant energy	Q, W		J	2
radiant energy density	ρ, w	$\rho=Q/V$	$J\ m^{-3}$	2
spectral radiant energy density				2
in terms of frequency	ρ_ν, w_ν	$\rho_\nu=d\rho/d\nu$	$J\ m^{-3}\ Hz^{-1}$	
in terms of wavenumber	$\rho_{\tilde{\nu}}$, $w_{\tilde{\nu}}$	$\rho_{\tilde{\nu}}=d\rho/d\tilde{\nu}$	$J\ m^{-2}$	
in terms of wavelenglth	ρ_λ, w_λ	$\rho_\lambda=d\rho/d\lambda$	$J\ m^{-4}$	

(1) The unit cm^{-1} is generally used for wavenumber in vacuum.

(2) The symbols for the quantities *radiant energy* through *irradiance* are also used for the corresponding quantities concerning visible radiation, i.e. luminous quantities and photon quantities. Subscripts e for energetic, v for visible, and p for photon may be added whenever confusion between these quantities might otherwise occur. The units used for luminous quantities are derived from the base unit candela (cd), see chapter 3.

Example radiant intensity	I_e, SI unit: W sr^{-1}	
luminous intensity	I_v, SI unit: cd	
photon intensity	I_p, SI units: $s^{-1}\ sr^{-1}$	

Table 7.1. Continued. Continued on next page.

Name	*Symbol*	*Definition*	*SI unit*	*Notes*
Einstein transition probabilities				3
spontaneous emission	A_{nm}	$dN_n/dt = -A_{nm}N_n$	s^{-1}	
stimulated emission	B_{nm}	$dN_n/dt = -\rho_{\tilde{\nu}}(\tilde{\nu}_{nm}) \times B_{nm}N_n$	$s\ kg^{-1}$	
stimulated absorption	B_{mn}	$dN_n/dt = \rho_{\tilde{\nu}}(\tilde{\nu}_{nm})B_{mn}N_m$	$s\ kg^{-1}$	
radiant power, radiant energy per time	Φ, P	$\Phi = dQ/dt$	W	2
radiant intensity	I	$I = d\Phi/d\Omega$	$W\ sr^{-1}$	2
radiant exitance, (emitted radiant flux)	M	$M = d\Phi/dA_{source}$	$W\ m^{-2}$	2
irradiance, (radiant flux received)	E, (I)	$E = d\Phi/dA$	$W\ m^{-2}$	2, 4
emittance	ε	$\varepsilon = M/M_{bb}$	1	5
Stefan–Boltzmann constant	σ	$M_{bb} = \sigma T^4$	$W\ m^{-2}\ K^{-4}$	5
first radiation constant	c_1	$c_1 = 2\pi h c_0^2$	$W\ m^2$	
second radiation constant	c_2	$c_2 = hc_0/k$	K m	
transmittance, transmission factor	τ, T	$\tau = \Phi_{tr}/\Phi_0$	1	6
absorptance, absorption factor	α	$\alpha = \Phi_{abs}/\Phi_0$	1	6
reflectance, reflection factor	ρ	$\rho = \Phi_{refl}/\Phi_0$	1	6
(decadic) absorbance	A	$A = -\lg(1-\alpha_i)$	1	7
napierian absorbance	B	$B = -\ln(1-\alpha_i)$	1	7
absorption coefficient				
(linear) decadic	a, K	$a = A/l$	m^{-1}	8
(linear) napierian	α	$\alpha = B/l$	m^{-1}	8
molar (decadic)	ε	$\varepsilon = a/c = A/cl$	$m^2\ mol^{-1}$	8, 9
molar napierian	κ	$\kappa = \alpha/c = B/cl$	$m^2\ mol^{-1}$	8, 10

(3) Note that $E_n > E_m$, $E_n - E_m = hc\tilde{\nu}_{nm}$, and $B_{nm} = B_{mn}$, in the definitions. The coefficients B are defined here using energy density $\rho_{\tilde{\nu}}$ in terms of wavenumber; they may alternatively be defined using energy density in terms of frequency ρ_ν, in which case B has SI units $m\ kg^{-1}$.

(4) The irradiance, with SI unit $W\ m^{-2}$, is often also called the intensity and denoted with the symbol I. This is particularly true in discussions involving collimated beams of light, as in applications of the Lambert–Beer law for spectrometric analysis.

(5) The emittance of a sample is the ratio of the flux emitted by the sample to the flux emitted by a black body at the same temperature; M_{bb} is the latter quantity.

(6) If scattering and luminescence can be neglected, $\tau + \alpha + \rho = 1$. In optical spectroscopy internal properties (denoted by a subscript i) are defined to exclude surface effects and effects of the cuvette, so that if scattering and luminescence can be neglected $\tau_i + \alpha_i = 1$. This leads to the customary form of the Lambert–Beer law $\Phi_{tr}/\Phi_0 (= I_{tr}/I_0) = \tau_i = 1 - \alpha_i = \exp(-\kappa cl)$.

(7) The definitions given here relate the absorbance A or B to the internal absorptance α_i; see note (6). However the subscript i on the absorptance α is often omitted.

(8) l is the absorbing path length, and c is the amount (of substance) concentration.

(9) The molar decadic absorption coefficient ε is frequently called the 'extinction coefficient' in published literature. Unfortunately numerical values of the 'extinction coefficient' are often quoted without specifying units; the absence of units usually means that the units used are $mol^{-1}\ dm^3\ cm^{-1}$. See also [20]. The word 'extinction' should properly be reserved for the sum of the effects of absorption, scattering and luminescence.

(10) κ may be called the molar absorbance cross-section, and κ/N_A the cross-section per molecule, denoted σ; this terminology is more common in physics than chemistry.

Table 7.1. Continued.

Name	*Symbol*	*Definition*	*SI unit*	*Notes*
absorption index	k	$k = \alpha/4\pi\tilde{\nu}$	1	
complex refractive index	$\hat{n}$	$\hat{n} = n + ik$	1	
molar refraction	R, R_m	$R = \frac{(n^2-1)}{(n^2+2)} V_m$	$m^3\ mol^{-1}$	
angle of optical rotation	α		1, rad	11

(11) The optical rotatory power of a solute in solution may be specified by a statement of the type

$$\alpha(589.3\ \text{nm},\ 20\,°\text{C},\ \text{sucrose},\ 10\ \text{g dm}^{-3}\ \text{in}\ H_2O,\ 10\ \text{cm}) = +66.470°$$

The same information may also be conveyed by quoting the specific optical rotatory power of the material, usually denoted $[\alpha]_\lambda^\theta$, where

$$[\alpha]_\lambda^\theta = \alpha/\gamma l$$

Here α is the angle of optical rotation, γ the mass concentration, and l the path length. Thus γl is the mass concentration per unit area in the beam path. (Amount concentration c might be more appropriate than mass concentration γ, but in practice it is usually γ that is used.) The specified wavelength λ (frequently the sodium D line) and Celsius temperature θ are written as a subscript and superscript respectively to the specific rotatory power $[\alpha]$. For pure substances $[\alpha]_\lambda^\theta$ is defined similarly by

$$[\alpha]_\lambda^\theta = \alpha/\rho l$$

where ρ is the mass density.

Specific optical rotatory powers are customarily called *specific rotations*. For pure liquids and solutions they are usually quoted as numerical values in units deg $dm^{-1}\ g^{-1}\ cm^3$, where deg is used to symbolize 1° of plane angle. (Sometimes this unit is wrongly described as deg dm^{-1}.) For solids the unit deg $mm^{-1}\ g^{-1}\ cm^3$ is usually used (sometimes wrongly described as: deg mm^{-1}). Neither notation follows the rules of quantity calculus (see chapter 1) because the symbol $[\alpha]$ is usually used to denote the numerical value in specified units rather than the value of the physical quantity itself.

RECOMMENDED STANDARDS FOR REPORTING PHOTOCHEMICAL DATA

(Recommendations 1983)

COMMISSION ON PHOTOCHEMISTRY*
A. A. LAMOLA[1] and M. S. WRIGHTON[2]

[1]Bell Laboratories, Murray Hill, New Jersey 07974, USA
[2]Massachusetts Institute of Technology, Cambridge, Mass. 02139, USA

The assessment or reinterpretation of experimental results requires that the reports contain sufficient information about experimental procedures and conditions. This is, of course, especially important in order to allow repetition of experiments by other investigators. While some experimental practices are so standard or common as to be assumed, many are not. It is therefore, important that the crucial experimental parameters be included in published reports.

The purpose of this document is to recommend a comprehensive set of experimental parameters and data manipulation methods that should be reported in published accounts of photochemical investigations and attendant spectroscopic studies.

It should be understood that quantitative data concerning all the parameters recommended for inclusion in a report may not be required. For example, references to instrumental settings or operating conditions may be sufficient. In other cases limiting values may suffice, e.g., "the pulse width was less than 1μs."

I. Photochemical Reactions

A. Conditions

1. Reactants and addends (sensitizers, quenchers, buffers, etc.)
    a. Purities and methods of purification
    b. Concentrations or partial pressures
    c. Relevant spectral properties
2. Solvent, including method of purification
3. Temperature
4. Gaseous atmosphere and methods of manipulation (degassing method, composition of flushing gases)
5. Description of reaction vessel
    a. Material, e.g., type of glass
    b. Geometry
    c. Relevant optical properties (transmission cutoff wavelength, etc.)
    d. Additional equipment (stirrer, etc.)

*Membership of the Commission during which the report was prepared (1979–83) was as follows:
Chairman: 1979–81 K. SCHAFFNER (FRG); 1981–83 F. C. DE SCHRYVER (Belgium); *Secretary*: 1979–81 F. C. DE SCHRYVER (Belgium); 1981–83 A. A. LAMOLA (USA); *Members*: S. E. BRASLAVSKY (FRG, Associate 1981–83); D. F. EATON (USA, Associate 1981–83); Z. R. GRABOWSKI (Poland, Titular 1979–81); C. HÉLÈNE (France, Associate 1979–81); K. N. HOUK (USA, Titular 1979–83); H. IWAMURA (Japan, Associate 1979–83); M. G. KUZMIN (USSR, Associate 1981–83); J. MICHL (USA, Associate 1979–81, Titular 1981–83); M. OTTOLENGHI (Israel, Associate 1979–81); C. SANDORFY (Canada, Associate 1979–83); K. TOKUMARU (Japan, Titular 1979–83); D. W. TURNER (UK, Associate 1979–83); D. G. WHITTEN (USA, Associate 1979–81); U. P. WILD (Switzerland, Associate 1979–81); F. WILKINSON (UK, Associate 1981–83); M. S. WRIGHTON (USA, Titular 1979–83); *National Representatives*: A. M. OSMAN (Arab Rep. of Egypt); O. P. STRAUSZ (Canada); K. LEMPERT (Hungary); J. W. VERHOEVEN (Netherlands).

6. Exciting radiation
   a. Spectrum
      i. Excitation source and condition of operation
      ii. Dispersing or filtering elements (bandpass)
      iii. Linewidth (for laser lines)
   b. Intensity (radiant flux; fluence rate) at sample ($Jm^{-2}s^{-1}$; photons $m^{-2}s^{-1}$) and area irradiated
   c. Irradiation dose (fluence) (Jm^{-2})
      i. If continuous irradiation, state radiant flux (intensity) and time.
      ii. If intermittant irradiation, state radiant flux, cycle duration and number of cycles.
      iii. If pulsed irradiation, state dose (fluence) per pulse, pulse width, repetition rate, number of pulses.

B. Quantum Yields

1. Definition of yield reported:

$$\Phi = \frac{\text{number of moles transformed (reacted, produced, etc.) in a process}}{\text{number of moles of photons absorbed}}$$

In the classical definition, the substance absorbing the exciting radiation is identical to that undergoing the photochemical (or photophysical) process of interest. In a complicated reaction mixture more than one substance, and perhaps not the reactant of interest, may absorb the excitation radiation. The reference photon absorption (the denominator in the equation above) must then be clearly defined. For example, in a sensitized reaction a quantum yield based upon the number of substrate molecules reacted per photon absorbed by the sensitizer may be defined. When a clear definition cannot be made, it is preferred that the term "apparent quantum yield" be used. It may not be possible in some cases, for example, in biological systems, to determine the absorbed dose. An experimentally determined efficiency based upon the exposed dose may be defined. Such a quantity should not be called a quantum yield. It is, of course, commonly possible to determine the efficiencies of individual primary processes and to define a quantum yield based upon the number of occurrences of the process of interest per molecule in the excited state which may undergo that process. THE AUTHOR SHOULD CLEARLY INDICATE WHAT YIELD IS BEING REPORTED.

2. "Units"
   Should be reported as a fraction and not as a percent (e.g., 0.5 rather than 50%)
3. Conditions
   a. See section I.A.
   b. Other necessary data include:
      i. Wavelength and bandwidth, or spectrum of exciting radiation
      ii. Intensity (radiant flux) of exciting radiation
      iii. Actinometric or dosimetric methods
      iv. Computational methods if complex (corrections for absorption of radiation by substances other than reagent of interest)
      v. Error estimate and number of determinations

C. Chemical Yields

1. Definition
   The yield should be clearly defined: Isolated yield refers to material that is actually isolated. Detected yield refers to quantity of material that can be detected by some analytical method, e.g., gas chromatography
2. Conditions
   See section I.A.

3. Variation of yield

   a. Dependence upon intensity (radiant flux) of exciting radiation

   b. Dependence upon wavelength of exciting radiation

   c. Dependence upon concentrations of relevant substances

   d. Dependence upon conversion (extent of reaction)

D. Kinetics

1. Conditions
   See section I.A.

2. Data display
   Data points and error limits should be displayed in plots

3. Analysis

   a. Model or mechanistic scheme upon which the analysis of kinetics is based should be clearly indicated

   b. Fitting methods used should be reported

   c. Statistics and error analysis should be included

II. Absorption Spectroscopy

A. Display

1. Energy or wavelength on abscissa (x-axis)
   A display linear in energy is preferred with the wavenumber ($\tilde{\nu}$) (cm^{-1}) the preferred scale. The preferred scale direction is "red to right," i.e., decreasing energy or increasing wavelength goes left to right. The measure of wavelength is nm (nanometer).

2. Measure of absorptivity on ordinate (y-axis)
   Options are:

   a. ϵ, molar absorption coefficient (molar absorptivity) (m^2mol^{-1} is the preferred unit; $lmol^{-1}cm^{-1}$ has been commonly used. Note that $m^2mol^{-1} = 10\ lmol^{-1}cm^{-1}$. SPECIFY UNITS).

   b. Absorbance (A) (internal absorbance or decadic internal absorption) ($\log_{10}T^{-1}$, where T = internal transmittance or fractional transmission). This refers specifically to the attenuation of radiation by absorption within the specimen.

   c. Attenuance ($\tilde{A}$) ($\log_{10}\tau^{-1}$ where τ is the transmittance)
   This refers specifically to the attenuation of transmitted radiation by all mechanisms including scatter.

3. If spectrum was measured point by point, the data points and error limits should be displayed.

B. Conditions

1. See section I.A.

2. Instrumentation should be described.

3. Spectral resolution should be defined.

4. Photometric errors and errors due to stray, scattered and emitted radiation should be assessed.

C. Transient absorption and flash kinetics

1. See sections I.A., I.D., and II.A.3.

2. Instrumentation should be described

   a. Pulse width (and pulse rate if applicable)

   b. Time resolution

   c. Wavelengths and bandwidths of exciting and monitoring radiation beams

   d. Intensities of exciting and monitoring radiation beams (energy per pulse for laser or flash sources)

   e. Geometry, cross-sections, etc. of exciting and monitoring radiation beams

   f. Relative polarizations of exciting and monitoring radiation beams (laser sources)

3. Deconvolution method

D. Action spectra

1. See sections I.A. and II.A.
2. Intensity (radiant flux) dependence
3. Concentration dependence (extrapolation to "zero" concentration for comparison with absorption spectra)

III. Luminescence Spectroscopy

A. Display

1. Energy or wavelength on abscissa (a linear energy scale is preferred, see section II.A.1.). (If a spectrum on a linear wavelength scale is to be transformed to one on a linear energy scale remember that $d\nu = \lambda^{-2}d\lambda$.)
2. Relative measure of photon flux on ordinate
   a. If point by point, show points
   b. When comparing spectra, normalize integrated intensities (radiant fluxes) to reflect relative luminescence yields. (It is best to use a linear energy scale for the abscissa.)
   c. State what corrections were applied to the experimental spectrum (i.e., corrections for instrumental spectral response, dispersion of solvent refractive index, polarization, stray and scattered excitation radiation, bleaching, etc.).
3. Define time gate
   a. "Total" emission vs "time-resolved" emission
   b. Time gate for time-resolved fluorescence

B. Conditions

1. See section I.A.
2. Description of instrument
3. Wavelength and bandwidth of exciting radiation
4. Intensity (radiant flux) of exciting radiation
5. Appropriate description of exciting radiation intensity-time profile
6. Polarization of exciting radiation
7. Method of calibration of excitation and emission wavelength determination
8. Description of excitation spectrum

C. Kinetics (luminescence decay)

1. See sections I.D. and II.C.
2. Emission wavelength and bandwidth
3. Methods used for deconvolution and analysis

D. Luminescence yield

1. See sections I.A., I.B., and III.B.
2. Reference substance defined if relative measure (include yield assumed for reference substance)
3. Dosimetry defined if absolute measure
4. Dependence upon wavelength of exciting radiation
5. Concentration dependence
6. Dependence upon intensity (radiant flux) of exciting radiation
7. State what corrections were made for reabsorption effects, etc.

E. Excitation spectra

1. Define geometry
2. Absorbance of sample
3. Emission wavelength and bandwidth
4. State what corrections were made for photon flux variations of spectrum of exciting radiation.
5. State what corrections were made for extraneous absorbance, etc.

F. Chemiluminescence

1. Conditions, see sections I.A.1 to I.A.5

2. Spectra

   a. Instrumentation

   b. State what corrections were made for nonconstant reaction rate during spectral scan.

3. Intensity (radiation flux) and Quantum Yield

   a. Reaction rate dependence

   b. Dosimetry (calibration)

IV. Abstract

Because the abstract of a published report is essentially what is reprinted by abstracting services, etc., it is important for effective data retrieval that quantitative data for important findings be placed in the abstract. Such data might include quantum yields, rate constants, equilibrium constants, and spectral features such as wavelengths of maxima.

Section 8

Analytical Chemistry

This section differs from the others devoted to specific technical areas. Each of the other sections refers to a specific class of methodologies, a group of techniques closely related through their underlying chemistry. This section is essentially defined by the purpose of providing quantitative analysis and might encompass any or all techniques that can be used successfully. As a result, the material in this section is more diverse than that in the other sections and is broken into smaller subsections, each similar to the other specific sections of the book. Some subsections are closely related to other sections, such as electrochemistry, thermodynamics, or spectroscopies, although the influence of the steady focus on analysis is usually apparent.

No table in the Green Book is devoted specifically to analytical chemistry, although all of the tables reprinted here are relevant to analytical techniques to which they are closely related scientifically. Also papers on symbols, nomenclature, and terminology for specific groups of analytical techniques are reprinted in this section.

The practitioners of analytical chemistry are aware of the needs and difficulties of comparing values measured by different techniques, and of the resulting need for care, precision, and completeness in expressing results. The use of statistical methods for obtaining precision and accuracy in the statement of the uncertainty of experimental results is highly developed, and statistical methodology has its own extensive literature that we will not cover here. Reference 34 provides a brief set of recommendations for the presentation of the results of chemical analysis. Reference 35 discusses recommendations on the usage of "selective", "selectivity", "specific", and "specificity" in the analytical literature.

Papers of an analytical and philosophical nature on the characteristics of the methods of analytical chemistry cover such topics as the definition of upper and lower limits of sensitivity, calibration, bias, and interference. These papers carry strong implications as to how information on various steps of the process should be presented but give no recommendations for specific types of measurements. The series of papers by Wilson (*36*) and the paper by Chalmers (*37*) are such papers.

Analytical chemists have also, on occasion, incorporated statements about requirements for proper publication into review papers on the design and development of analytical techniques. This is a laudable step, although this incorporation makes it difficult to locate these papers for the purposes covered here unless the title contains the information. The paper by Rossotti (*39*) and the paper by Mark (*40*) are such papers.

The rest of this section will be divided into subsections, each covering a few closely related categories of chemical analysis.

Classical Analytical Chemistry

This category might be classified as "wet chemistry", techniques that involve separation from solution in one way or another. Three papers are reprinted here:

- "Recommendations for Publication of Papers on Precipitation Methods of Gravimetric Analysis" (*38*),
- "Design and Publication of Work on Stability Constants" (*39*),
- "Development and Publication of New Methods in Kinetic Analysis" (*40*)

Thermal Analysis

"Nomenclature for Thermal Analysis IV" (*41*) modifies and consolidates the previous papers in the series (*41a, 41b*) and supersedes still earlier efforts. "Suggestions for Reporting

Dynamic Thermogravimetric Data" (*42*) is reprinted here. Reference 43 (reprinted here) discusses differential thermal analysis.

Chromatography

"Recommendations on Nomenclature for Chromatography" (*44*) (reprinted here) resulted from an extended effort to obtain a unified nomenclature for all branches of chromatography, and superseded a number of earlier recommendations for gas, ion-exchange, and liquid–liquid chromatography. It is reprinted because it is not covered in the Green Book and because it gives clear quantitative statements of the various forms of data met in this field. Reference 45 contains recommendations for publication of any new method involving ion-exchange or ion-exchange chromatography.

Electroanalytical Chemistry

"Recommended Terms, Symbols, and Definitions for Electroanalytical Chemistry" (*46*) is reprinted because electoanalytical chemistry is not covered in the Green Book and because reference 46 contains clear definitions of the quantitative terms involved in the field. "Conditional Diffusion Coefficients of Ions and Molecules in Solution: An Appraisal of the Conditions and Methods of Measurement" (*47*) is reprinted here.

Surface Analysis Techniques

"General Aspects of Trace Analytical Methods—4. Recommendations for Nomenclature, Standard Procedures and Reporting of Experimental Data for Surface Analysis Techniques" (*48*) covers reporting of data taken by using physical techniques for the analysis of solid surfaces. Various methods involving bombardment of the surface with high-energy particles in a vacuum are considered.

Spectroscopies

Spectroscopies cover a family of techniques involving excitation of a sample and observation of resulting radiation. Many spectroscopic techniques are used for analytical purposes. "Nomenclature and Spectral Presentation in Electron Spectroscopy Resulting from Excitation by Photons" (*49*) discusses electron spectroscopy for chemical analysis (ESCA) (as it is commonly called), and contains a statement on the nomenclature involving such techniques. "Recommendations for Publication of Papers on Methods of Molecular Absorption Spectrophotometry in Solution Between 200 and 800 nm" (*50*) is also reprinted here.

References 51 a–h and 52, on nomenclature, symbols, and units for spectrochemical analysis, make full use of the Green Book and coordinated and related tables prepared by the International Union of Pure and Applied Physics (IUPAP), and are strictly limited to terminology. Terms are given in detail for each step of the spectroanalytical process in question. The audience for the papers is a wide range of users, including industrial and clinical technicians and chemists.

RECOMMENDATIONS FOR PUBLICATION OF PAPERS ON PRECIPITATION METHODS OF GRAVIMETRIC ANALYSIS

COMMISSION ON ANALYTICAL NOMENCLATURE*
G. F. KIRKBRIGHT
University of Manchester Institute of Science & Technology, Manchester, UK

Many papers continue to be published concerned with precipitation methods of gravimetric analysis. These may be devoted to the elucidation of the fundamental processes involved in precipitation or to the development of new precipitants for inorganic or organic species or to the application of complexing agents to extend the possibilities of gravimetry for the separation and determination of metal ions. The role of gravimetric analysis in the provision of referee methods for establishment of the composition of standards used for calibration in instrumental methods of analysis remains important.

The fundamental operation in gravimetric analysis is the quantitative precipitation of the component to be determined in a form that is free from contaminants and easy to separate from the mother liquor. The precipitate must either itself be a stoichiometric compound that is suitable for weighing, i.e. involatile, non-hygroscopic, non-efflorescent, and inert to reaction with air, or one that is easily convertible into such a compound by drying or ignition. Although these requirements seem simple, it is difficult to keep the error by dissolution losses, contamination of the precipitate, etc., below 0.1–0.2% which, in most cases are the upper limits required in practice.

This report is based on a paper by Erdey, Polos and Chalmers (1) and it proposes a standard set of requirements for the development and publication of a gravimetric method and makes recommendations concerning the date which should be incorporated into such a paper.

Information pertaining to the following should be given:

1. Sample and Preparation of Sample for Analysis

(a) Description of chemical composition of sample matrix.
(b) Appropriate sample weight (based on homogeneity of sample and concentration of component to be determined).
(c) Necessary drying conditions for solid samples.
(d) Dissolution procedure for solid samples and evidence of complete dissolution.

2. Precipitant and other reagents

(a) Purity and stability of solid reagents used to form the insoluble species that is to be weighed.
(b) Any procedures required for the purification of solid reagent(s).
(c) The method of preparation, composition, formulae and properties of new reagents in such detail as to allow their production and characterisation by other workers. Source of commercially available reagents.
(d) Required purity and stability of other reagents used (buffers, solvents, wash-liquid(s), standard solutions, etc).
(e) Stability of precipitant in solution under laboratory conditions (i.e. effect of daylight, temperature, oxygen, CO_2).

3. Conditions for Sample Solution

(a) Chemical composition of the solution before precipitation.
(b) Values of permissible range of absolute amount and concentration of the component to be determined.
(c) Values of permissible range of pH, temperature and volume of sample solution before precipitation.

4. Method of Precipitation

(a) pH, volume, temperature and concentration of precipitant reagent solution. Permissible limits for these variables, e.g. pH 5± 0.2.
(b) Order of addition of reagents.

*Membership of the Commission for 1979-81 is as follows:

Chairman: G. G. GUILBAULT (USA); *Secretary:* G. SVEHLA (UK); *Titular Members:* R. W. FREI (Netherlands); H. FREISER (USA); S. P. PERONE (USA); N. M. RICE (UK); W. SIMON (Switzerland); *Associate Members:* C. A. M. G. CRAMERS (Netherlands); D. DYRSSEN (Sweden); R. E. VAN GRIEKEN (Belgium); H. M. N. H. IRVING (South Africa); G. F. KIRKBRIGHT (UK); D. KLOCKOW (FRG); O. MENIS (USA); H. ZETTLER (FRG); *National Representatives:* A. C. S. COSTA (Brazil); W. E. HARRIS (Canada).

Reprinted with permission from *Pure and Applied Chemistry,* Volume 56, Number 7, 1984, pages 939–944. Copyright 1984 by the International Union of Pure and Applied Chemistry—All rights reserved.

(c) Rate of addition of precipitant (or details of method of conducting precipitation from homogeneous solution).
(d) Final pH, time and temperature of digestion.
(e) Temperature for filtration.
(f) Type of filter.
(g) Composition of wash-liquid, volume and number of washes.
(h) Temperature(s) and duration of drying or ignition.
(i) Any other special technique or precautions required during precipitation, filtration, washing, drying or ignition.
(j) Recommended method for cleaning filtration apparatus.

5. Properties of Precipitate

(a) Statement of expected nature of precipitate (e.g. gelatinous, crystalline, etc.) and estimate of particle size or particle size distribution if applicable.
(b) Values of solubility of precipitate in (i) moth liquor, and (ii) wash liquid, to enable estimate of completeness of precipitation and loss on washing, respectively.
(c) Thermogravimetric data for precipitate if available.

6. Method of Calculation of Analytical Result

The stiochiometric factor (gravimetric conversion factor) should be given for the weighing form. This is expressed as the ratio of the relative atomic (or molecular) mass of the species determined to the relative molecular mass of the weighing form.

7. Selectivity

(a) Any systematic deviations in the analytical result arising from the presence of other components (which thus constitute 'interference') should be measured for the possible interfering species likely to be present in sample matrices at the appropriate concentration levels. Concentrations which cover the full range of likely ratios of interfering species to analyte species should be investigated at high and low analyte concentrations. The existence of systematic deviation (interference) should be defined in relation to the precision of the complete analytical procedure (i.e. when the error is greater than e.g. 2 or 3 times the standard deviation of the analytical result obtained in the absence of the other component).
(b) Limiting permissible concentrations of those elements found to interfere should be reported.
(c) When two or more species, which do not interfere individually, are present together with the component to be determined, checks should be made that there is no slight interference from each which is additive or subtractive.
(d) Recommendations should be given for minimising the effects of interfering species, e.g. by use of masking agents or preliminary separation of either the species to be determined or interfering species.

8. Precision

(a) The reproducibility of the complete analytical procedure under optimum conditions and in separation procedures or other sample pretreatment (e.g. wet or dry ashing) should be reported. The relative standard deviation for the complete method of analysis, expressed as a decimal fraction, should be given at both high and low concentrations of the component determined. The number of measured values from which the relative standard deviation is derived must be stated in each case.

9. Accuracy

(a) The results of tests for losses, recovery and contamination when the method is applied to the analysis of standard or reference samples or simulated sample solutions should be reported.
(b) Results should be compared with those obtained by other methods in analysis of standard or reference samples. Estimate of systematic errors.

10. Applications

The applicability of the recommended procedures for various specified matrices should be described together with results for samples actually examined.

11. Assessment

There should be a realistic assessment of the new method in terms of precision, accuracy, speed, cost, selectivity, simplicity, etc., compared to existing methods.

New methods for the gravimetric determination of simple inorganic anionic and cationic species should be proposed only if they show promise of superiority over the best existing methods. The particular advantages and disadvantages of the method proposed should be mentioned. It is stressed that the above requirements represent the minimum acceptable amount of information which should be included in any report of a new gravimetric method of analysis.

REFERENCES

1. L.Erdey, L.Polos and R.A.Chalmers, Talanta, (1970), 17, 1143.

DESIGN AND PUBLICATION OF WORK ON STABILITY CONSTANTS

HAZEL S. ROSSOTTI
Inorganic Chemistry Laboratory, Oxford University, Oxford

Many discussions of stability constants start by an exhortation to consider either "the formation of N complexes BA_n (where $n \leq N$ and charges are omitted for clarity) by the addition of n ligands A to a central group B" or, even more depressingly, "the formation of complexes of general composition $B_qA_pH_jM_mX_x$ $(H_2O)_w$ in an aqueous solution containing the bulk electrolyte MX." This latter formula is usually simplified by assuming that, since the concentrations of M, X and water hardly vary, the (average) values of m, x and w are fixed for a particular set of values of q, p and j. The contributions of those species which make up the medium are then thankfully ignored, and the complexes are written as $B_qA_pH_j$, where q and p may have values of zero, or of any positive whole number. The value of j is also integral, and is positive for species which act as Brønsted acids, but negative for those which contain one or more hydroxyl groups.

The discussions usually continue by defining an overall stability constant, which is merely the equilibrium constant for the formation of each particular complex from the appropriate components B, A and H (or OH). Obviously, the concentration of each complex in a given solution depends on the value of this constant, expressed as a concentration quotient, and on the equilibrium concentrations of B, A and H, and if it were possible to measure the concentrations of the complex, and of the components in equilibrium with it, the value of the stability constant could be obtained.

The very large literature on the determination of stability constants (see, *e.g.* ref. 1) has grown up because it is seldom, if ever, possible to obtain direct measurements of the equilibrium concentrations of all species which are present in a solution in which complexes are formed. The equilibrium concentrations of the components B, A and H depend not only on the fractions of these species which are consumed in the course of complex formation, but also on the *total* concentrations of these components. An increase in the total concentration of a ligand will, by the law of mass action, always increase the concentration of free ligand, albeit sometimes by a very small amount. The equilibrium concentration of one of the components can therefore be related to the stability constants of the complexes and to the analytical composition of the system. Conversely, measurement of the variation of the free concentration of but one species with the overall composition of the solution can, in principle, lead to values of the stability constants.

Any change in the extent of complex formation in a solution is reflected by changes in several aspects of its behaviour. Variation of suitable properties with the composition of the solution may therefore be exploited to give information about the complexes which are formed. It is sometimes possible to study a property which depends only on the concentration of one species, and so provides a method of measuring the equilibrium concentration of that species. For example, a potentiometric cell can sometimes be designed so that its e.m.f. can give a measure of the concentration of uncomplexed metal ion, and the total concentration of a metal ion in an organic solvent which has been equilibrated with an aqueous solution is often proportional to the concentration of uncharged metal complex in the aqueous phase.

Interpretation is much more difficult when several species contribute to the property which is being measured. Thus the distribution of a metal ion between a cation-exchange resin and a solution depends on the concentration of the free metal ion and of any cationic complex, together with the various partition coefficients of the cations between the two phases.

Spectrophotometric measurements may also be difficult to interpret, since the absorbance of the solution depends on the concentrations, and molar absorptivities, of all species which absorb radiation of the wavelength used. However, if addition of the ligand to a metal ion results in a marked change in absorbance, because of the formation of a single complex, the concentration of this species may be determined spectrophotometrically. Methods in which the concentration of a single species can be followed are deservedly more popular than those in which several species contribute to the quantity being measured. The latter are, indeed, best restricted to those situations in which the former cannot be used, but some studies seem to have been motivated by a desire to show that a particular technique can be used to measure stability constants, rather than by a resolve to obtain "good" experimental results in the most efficient possible manner.

Although the general formula $B_qA_pH_j$ embraces conventional polybasic acids ($q = 0$, $p = 1$, $j \geq 0$) and molecular complexes of organic species ($j = 0$), the term "stability constant" normally refers to the formation of a complex in which the central group, B, is a metal ion. The ligand, A, may be a simple anion or neutral molecule (*e.g.*, Cl^-, NH_3, C_2H_2) or a multidentate organic anion or molecule of considerable complexity (*e.g.*, polyamines and aminopolycarboxylate ions). With the exception of the method-hunters, the majority of workers have measured stability constants by means of some form of potentiometry, and it is with this method that the present paper is mainly, but not exclusively, concerned. An attempt will be made to answer such questions as: "on what assumptions is the experimental method based?"; "how can we tell if they are justified?"; "what should we do if it emerges that they are not?"; "which method is appropriate for which system?"; "how can experimental results best be processed?"; "what do the stability constants mean?"; how may the results best be reported?"; and, of course, "why should one want to measure stability constants anyway?"

SURVEY OF EXPERIMENTAL METHODS

Methods for measuring stability constants may be classified according to whether or not any of the species $B_qA_pH_j$ exists in more than one phase. Within each group of techniques, further classification may be based on the way in which the measured property varies with the concentrations of the species being studied. The techniques most commonly used[2] are shown in Table 1.

Table 1. Some methods for measuring stability constants

	All species $B_qA_pH_j$ in homogeneous solution	At least one species $B_qA_pH_j$ distributed between two phases
Experimental output (fairly) directly related to concentrations of individual species.	Classical analysis **Potentiometry** Polarography Amperometry	**Solvent extraction** Solubility Vapour pressure
Output depends in general on concentrations of more than one species; but in favourable cases method can give single concentrations.	**Spectrophotometry** Kinetics	**Ion-exchange**
Output always depends on the concentrations of two or more species	Colligative properties Conductivity	

Many of the methods listed are of very restricted application. Classical chemical analysis, for example, can only be used on very inert systems: most complexes are so labile that addition of an analytical reagent displaces the equilibrium before, say, a precipitate can be separated or an absorbance measured. Determination of concentration by measuring distribution between solution and vapour is restricted to systems where only one species is volatile, but it has been used to study complexes of metal ions with small, uncharged ligands such as ammonia and acetylene. Ebullioscopy and cryoscopy are, of course, restricted to a single temperature for any particular system. Changes in the colligative properties of a solution, and in the electrical conductivity, are usually swamped by the presence of a bulk electrolyte. If the solubility of a sparingly soluble solid, *e.g.*, BA_c, is measured, the constancy of the solubility product $[B][A]^c$ precludes independent variation of the concentrations of free metal ion and free ligand. The use of amperometry and polarography is restricted to those systems in which the metal ion can be reversibly reduced at a dropping mercury electrode; but these systems can better be studied potentiometrically by using amalgam electrodes.

Measurements which depend on the concentrations of several species are often difficult to interpret. Colligative properties, which depend only on the total number of solute species, can only be interpreted unambiguously in simple systems. On the other hand, techniques such as spectrophotometry and ion-exchange may lead to ambiguous (or, at least, imprecise) results just because each species contributes to the property to a different extent, and so increases the number of parameters needed to describe the system. Thus the expression for the absorbance of a solution involves the molar absorptivities of all absorbing species in addition to stability constants and concentration variables. The interpretation of kinetic measurements is often particularly complicated. The rate of exchange between say, Fe(II) and Fe(III) in a solution of their complexes, can be described only by introducing a rate constant for each pair of species between which exchange is possible. Techniques of this class are most useful for systems in which it can be firmly established that only one species contributes to the property measured.

The rest of this paper will deal only with the four techniques printed in boldface in Table 1. The general principles of careful equilibrium work will be illustrated with references to potentiometry, since it is the method most commonly used: and potentially the best. An analyst might however, complain that since potentiometry can only be used for complexes which are appreciably soluble in the medium, the technique involves the use of just those solvents which he tries to avoid. Solvent extraction will also be considered in some detail, since it can often be used when potentiometry is inapplicable. Spectrophotometry and ion-exchange will be discussed in the light of how to use apparently unpromising techniques to the best advantage.

Systems of simple, mononuclear complexes

The simplest systems of complexes are those in which only one series of complexes is formed and only one of the integers q, p and j varies. Examples are:

(*i*) Oligomers B_q ($p = 0, j = 0$) formed by self-association of such species as S_2 in the vapour phase, or phenol in organic solvents. (*ii*) Acids H_jA ($q = 0, p = 1$) in solutions which contain no complexing metal ions. (iii) Complexes BA_n ($q = 1, j = 0$; p is variable, conventionally written as $p = n$ when $q = 1$). These species are formed by combination of non-basic ligands, such as halide ions, with simple metal ions such as Hg^{2+} and Fe^{3+}; similar complexes are formed by combination between these ligands and composite, but infinitely stable, central groups such as

$$Hg_2^{2+}(q = 2, j = 0),\ VO^{2+}(q = 1, j = -2)^* \text{ or } UO_2^{2+}(q = 1, j = -4)^*$$

The great majority of those species which act as ligands for metal ions can also act as Brønsted bases and combine with at least one proton; and so it is usually necessary to deal with systems in which at least the two series of species BA_n and H_jA coexist. In many studies it is assumed that no further complexes are formed (*i.e.*, that q can have values of only 0 and 1, such that, when $q = 0, p = 1$; and when $q = 1, j = 0$).

Considering, for the moment, only the series BA_n and H_jA, we may express the total (analytical) concentrations B, A and H as

$$B = \sum [BA_n] \quad (1)$$

$$A = \sum [H_jA] + \sum n[BA_n] \quad (2)$$

$$H = \sum j[H_jA] + [H^+] - [OH^-] \quad (3)$$

(For alkaline solutions, H may be negative.) The summations run from zero to the maximal values $n = N$ and $j = J$, inclusive.

The equilibrium concentrations $[BA_n]$ and $[H_jA]$ may be expressed in terms of the equilibrium concentrations, b, a and h of B. A and H, together with $(N + J)$† overall stability constants

$$\beta_n = [BA_n]/ba^n \quad (4)$$

$$^H\beta_j = [H_jA]/h^ja \quad (5)$$

The quantities β_n and $^H\beta_j$ are concentration quotients, and may be treated as parameters only if the activity coefficients of the species involved remain constant over the whole range of concentrations used. Introduction of the stability constants into the mass-balance expressions (1–3) gives

$$B = \sum \beta_n ba^n \quad (6)$$

$$A = \sum {}^H\beta_j h^j a + \sum n\beta_n ba^n \quad (7)$$

$$H = \sum j\,{}^H\beta_j h^j a + h - K_w h^{-1} \quad (8)$$

where $K_w = h[OH^-]$ is the stoichiometric ionic product of water.

Values of $^H\beta_j$ for the formation of the acids H_jA are usually determined in solutions which contain no B. Measurement of h leads to the average number, $\bar{j}$, of protons bound to each A. For $B = 0$, we may write

* VO^{2+} cannot be distinguished from $V(OH)_2^{2+}$, nor UO_2^{2+} from $U(OH)_4^{2+}$ by *equilibrium* studies in *aqueous* solution.

† For $n = 0$ or $j = 0$ the constants β_n and $^H\beta_j$ become unity.

$$\bar{j} = \frac{\sum j[H_jA]}{\sum [H_jA]} = \frac{H - (h - K_w h^{-1})}{A} \tag{9}$$

Thus $\bar{j}$ can easily be obtained from measurements of h, H and A (provided that $h \gg K_w h^{-1}$, or that K_w has been determined under the exact conditions used). By combining equations (7)–(9) we can also express $\bar{j}$ as

$$\bar{j} = \frac{\sum j^{H}\beta_j h^j}{\sum {}^{H}\beta_j h^j} \tag{10}$$

which is a polynomial in the single variable h. Since the coefficients of equation (10) are the required values of ${}^{H}\beta_j$, these may be obtained from the experimental function $\bar{j}(h)$, and hence from measurements of h, H and A. The most suitable way of solving equations derived from equation (10) depends on how many acids are formed, and to what extent they coexist.

The average number, $\bar{n}$, of ligands A bound to B can be analogously defined as

$$\bar{n} = \frac{\sum n[BA_n]}{\sum [BA_n]} \tag{11}$$

Again, this average degree of complex formation may be related to experimental quantities. Thus, from equations (2), (3) and (9)

$$\bar{n} = \frac{A - [H - (h - K_w h^{-1})]/\bar{j}}{B} \tag{12}$$

If values of ${}^{H}\beta_j$ (and K_w) have been determined *under exactly the same conditions*, the value of $\bar{j}$ can be calculated for any value of h. Combination of the values of h and $\bar{j}$ with the analytical concentrations B, A and H then gives the value of $\bar{n}$.

The variation of $\bar{n}$ with a, like the function $\bar{j}(h)$, can be expressed as a simple polynomial, in which the coefficients are the required stability constants. Thus, from equations (4) and (11) we obtain

$$\bar{n} = \frac{\sum n\beta_n a^n}{\sum \beta_n a^n} \tag{13}$$

where the sole variable is the concentration, a, of free ligand. The value of a may also be obtained by measuring h in a solution of known B, A and H. Since, from equation (2)

$$\sum [H_jA] = A - \sum n[BA_n] \tag{14}$$

no further information is needed to enable us to calculate a from the expression

$$a = \frac{A - \bar{n}B}{\sum {}^{H}\beta_j h^j} \tag{15}$$

which follows from equations (6), (7), (11) and (14). If the ligand A does not combine with protons, ${}^{H}\beta_j = 0$ for all values of j above zero. The denominator of equation (15) then becomes unity, and equation (12) takes the simple form

$$\bar{n} = \frac{A - a}{B} \tag{16}$$

The experimental function $\bar{n}\,(a)$ can then be obtained from measurements of a, A and B.

Equations (15) and (16) become trivial unless the value of $(A - \bar{n}B)$ differs appreciably from A: and this condition may be unfulfilled if the metal ion is present only in low concentration. It may, however, be possible to measure an alternative concentration variable, such as the fraction, α_c, of B which is present in the form of a particular species BA_c. (The concentrations most commonly measured are those of the uncomplexed metal ion, and of the electrically neutral complex.) From equation (2)

$$\alpha_c = \frac{[BA_c]}{B} = \frac{[BA_c]}{\sum [BA_n]} = \frac{\beta_c a^c}{\sum \beta_n a^n} \tag{17}$$

the quantity α_c, like $\bar{n}$, is a function of a only, and the stability constants are calculable from the measurements of α_c and a. If $A \gg \bar{n}B$, approximate values of

$$a \approx \frac{A}{\sum {}^{H}\beta_j h^j} \tag{18}$$

may be combined with measurements of α_c and h to give preliminary values of β_n which may then be refined. The values of a from equation (18) are combined with the rough values of β_n to give values of $\bar{n}\,(A, h)$, which are fed into equation (15) to give better values of a. Refined values of β_n are obtained from the experimental values of α_c together with the improved values of a, and so on, until convergent values of the stability constants are obtained.

The procedure outlined above requires that: (*a*) precise values of B, A and H are known for all the solutions used; (*b*) values of the concentration quotients ${}^{H}\beta_j$ and β_n are constant; (*c*) the *concentration* of H^+ (or, as appropriate, of A, or BA_c) can be measured; (*d*) no complexes other than H_jA or BA_n can be detected; (*e*) the stability constants calculated from the experimental functions $\bar{j}(h)$, $\bar{n}(a)$ or $\alpha_c(a)$ are compatible with the primary measurements.

These points will now be considered in more detail.

Total analytical concentrations

Solutions are normally prepared by dilution of standard stock solutions, which have been made up from highly purified components. Stock, and diluted, solutions should be stored at the constant temperature at which measurements will be made. The stock solutions (including those prepared by weighing-out of reagent grade materials) should always be analysed, preferably by two widely differing methods. The analysis of stock solutions is often considered to be one of the most tedious aspects of the determination of stability constants. But since the values of B, A and H form the basis of all subsequent calculations, the utmost care should be taken with the initial analyses. Failure to obtain "good" stability constants is often traceable, after much elapsed time, to inadequate analysis of stock solutions.

Dilute solutions should be prepared by using Grade A volumetric glassware. (Some workers may feel that the effort involved in recalibration of Grade A glassware is amply repaid by their resulting sense of virtue: but it is questionable whether it leads to any significant improvement in the values of the stability constants.) In order to allow for any temperature change which may occur during dilution, slightly less than the required volume of diluent should be added in the first instance, and the dilution completed by topping up to the mark only when the bulk of the diluted solution has returned to the constant temperature of the surroundings.

In addition to conventional methods of analysis, there are a number of useful tricks of the equilibrium chemist's trade. One is Gran's method for determining end-points of a variety of potentiometric titrations.[3–5] It can be used to determine concentrations of weak acids, strong acids, and even of strong acids in the presence of a readily hydrolysable metal ion such as Fe^{3+}. Solutions which contain both hydrogen ions and metal ions M^{z+} may also be analysed by running an aliquot (v l.) through a cation-exchange resin in the hydrogen form. Since $v(H + zB)$ moles of hydrogen ion emerge from the resin, separate determination of the initial concentration B of metal gives the required value of H.

Occasionally, the analysis of a solution of an organic base, such as a carboxylate ion, can be carried out potentiometrically[5,6] in the same operation as the measurement of $\bar{j}$. A volume V of a solution of say NaA, of initial concentration A_i is titrated with a volume v of strong acid of concentration H_i, and the free hydrogen ion concentration is measured potentiometrically. From equations (8) and (9), the value of $\bar{j}$ is given by

$$\bar{j} = \frac{H - h}{A} = \frac{H_i v - h(V + v)}{A_i V} \quad (19)$$

Since, for a monobasic acid, $\bar{j}$ tends to a value of $\bar{j}_{max} = 1$ at high acidities, the required value of A_i is obtained as

$$A_i = \lim_{h \to \infty} [H_i v - h(V + v)]V^{-1} \quad (20)$$

Constancy of stability constants

The stoichiometric stability constant β_n may be related to the standard free energy change $\Delta G°$ for the formation of BA_n from its components by the expression

$$-RT \ln \beta_n = \Delta G° + RT \ln \frac{\gamma_n}{\gamma_0 \gamma_A^n} \quad (21)$$

The activity coefficients γ_n and γ_A of the species BA_n and A refer to the same (usually molar) concentration scale as do the values of β_n. An exactly analogous expression can be written for ${}^{H}\beta_j$. So the concentration quotients β_n and ${}^{H}\beta_j$ only behave as parameters if both the temperature and the activity coefficient term $\gamma_n \gamma_0^{-1} \gamma_A^{-n}$ are held constant.

Temperature control is little problem. For work at "room" temperature the whole laboratory can often be thermostatically controlled at, say, 25° ± 1°. The temperature of the thermostat tank in which the measurements are done can then usually be kept steady to $\pm 0{\cdot}1^\circ$.

Control of activity coefficients is less easy. Only at exceedingly low ionic strengths are activity coefficients independent of the ion-size parameter, even for ions of the same charge-type. It is usually preferable to use solutions in which there is a high, almost constant, concentration of a bulk electrolyte which plays the smallest possible part in the equilibria which are being studied and has little, if any, additional effect on the measurements. The perchlorate ion is a popular bulk anion on account of its very low tendency to form complexes. The background cation is usually sodium, which complexes only weakly with most ligands and forms a very soluble perchlorate, but lithium perchlorate, although less soluble, has advantages in potentiometric studies of protonic equilibria, on account of the similarity between the H^+ and Li^+ ions.

The concentration of bulk electrolyte can be held "constant" in a number of slightly different ways. Let us suppose that the background electrolyte is $3M$ sodium perchlorate and that we wish to study complex formation between a metal ion M^{z+} and an anionic ligand A^{y-}. In order to keep the number of types of species as low as possible, it is best to add the metal ion in the form of $M(ClO_4)_z$ and the ligand as Na_yA. There is then the choice of designing the solutions so that, for example, $[Na^+] = 3M$, or $[ClO_4^-] = 3M$ or that ($[Na^+] \nless 3M$ and $[ClO_4^-] \nless 3M$). Table 2 gives details of the concentrations of metal ion solution $[M^{z+}] = B_i$ and ligand solution $[A^{y-}] = A_i$ in the three cases mentioned, and shows how the concentrations change when the solutions are mixed.

It is convenient to define the standard state as the hypothetical molar state in the appropriate ionic medium. Thus as B_i and A_i decrease to zero, the activity coefficients of metal ion, ligand and complexes tend to unity. In moderately concentrated solutions of metal ion and ligand, the activity coefficients may differ appreciably from unity, to an extent which depends on the exact composition of the solution. For studies of weak complexes, fairly high concentrations of ligand will be needed. Care should therefore be taken to see that solutions are always prepared so that the concentrations are consistent with the (explicit) conditions which have been chosen for the ionic medium.

At the beginning of an investigation, belief in the constancy of the activity coefficient quotient must be an act of faith. But the faith can sometimes be put to the test. If no more than two complexes are formed in any concentration range, the formation curve, $\bar{j}$ (log h) or $\bar{n}$ (log a) is symmetrical about its mid-point; and if a single complex is formed, the formation curve is of unique shape. It is most unlikely that a formation curve would be fortuitously symmetrical, let alone of the required shape. Such theoretical behaviour is strongly indicative of the constancy of the activity coefficient quotients, and hence also of the stability constants.

When three or more complexes coexist, interpretation of experimental results may be ambiguous. It is possible that measurements made up to a high free ligand concentration are equally compatible with, say, the formation of complexes BA–BA_4, with constant activity coefficient quotients, or with the existence of only three complexes, with variable activity coefficient quotients. Particular care must therefore be taken when interpreting measurements made in media which differ markedly from that referring to standard state.

Further complications arise if the system is heterogeneous; means of supposedly controlling activity coefficients in aqueous solutions cannot be applied to ion-exchange resins, or to solutions in organic solvents (see p. 821). However, we shall optimistically assume that, at least in aqueous solutions, the law of mass action is valid in terms of concentrations. The next problem is that of determining the concentration variables experimentally.

Potentiometric determination of concentrations

Potentiometric studies of chemical equilibria are usually carried out by means of the cell

−		+
reference half-cell ‖	solution in which equilibria take place |	probe for metal, ligand, or hydrogen ions

(I)

although cells without liquid junction have occasionally been used. Metal, and amalgam, electrodes can act as probes for a variety of metal ions. In skilled hands, the latter electrodes are usually the more satisfactory: they are more readily reversible provided that all traces of oxygen have been excluded. The concentrations of a few simple anionic ligands, such as halides and sulphate, can be measured with "electrodes of the second kind" such as $Ag/AgCl/Cl^-$. The range of probes for metal ions and simple anions has been greatly extended by the introduction of a variety of specific-ion electrodes.[7] Although some of these function satisfactorily, even in the presence of a high concentration of background electrolyte, considerable development is needed before they can be widely used for very precise work.

Further extension of the range of probes for metal and ligand ions can be achieved by introducing additional species. The ratio of $[Fe^{3+}]$ to $[Fe^{2+}]$ can be determined by means of a redox electrode, and if the Fe(II) is known to be uncomplexed, the value of $[Fe^{3+}]$ can be obtained. This redox system has been used to determine the concentration of free fluoride ions in a solution containing Al(III) fluoride complexes: the stability constants of the Fe(III)–F^- complexes were known and those of the Fe(II)–F^- complexes were negligible. The use of auxiliary species, added only to facilitate measurement, greatly extends the scope of potentiometry, but each reagent added carries with it additional parameters, or assumptions, or both, and thereby decreases the precision of the final stability constants.

Electrodes which respond to hydrogen ions are used much more frequently than those which respond to metal ions or ligands. The hydrogen gas electrode is, however, unpopular except for extremely rigorous work. It comes to equilibrium inconveniently slowly, and cannot be used in the presence of substances which oxidize hydrogen or poison the catalytic surface of the electrode. On the other hand, the quinhydrone electrode is a quick and convenient probe for hydrogen ions, but quinhydrone is no exception to the general caveat about auxiliary species. At values of $pH > 7$ or of $pH < 1$, it acts as an acid or base respectively, and so displaces the very equilibria it was introduced to probe. Quinhydrone may be an added nuisance in studies of metal complexes, since it combines with some metal ions, such as Cu(II). An overwhelming majority of studies of protonic equilibria in systems A, H and A, B, H have been carried out with a glass electrode, thereby avoiding the introduction of any auxiliary species. The e.m.f. of cell (I), which contains a glass electrode in the right-hand half-cell, is given by

$$E = E^{\circ\prime} + RTF^{-1} \ln h + E_j \qquad (22)$$

The term $E^{\circ\prime}$ (which includes the half-cell potential of the reference electrode, the standard potential and asymmetry potential of the glass electrode, and the supposedly constant term $RTF^{-1} \ln \gamma_H$) should be constant over a period of at least several hours. The term E_j represents the overall diffusion potential generated at any liquid–liquid junction within the cell: its value should be kept as low as possible by avoiding gross concentration gradients. The same ionic medium should therefore be used for the two half-cell solutions and the salt-bridge, as in the cell:

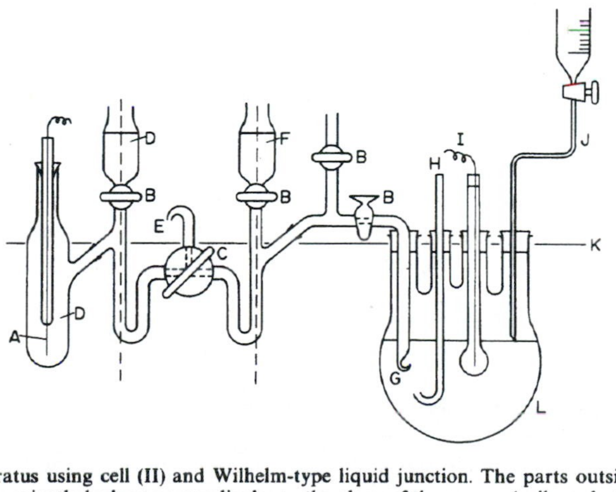

Fig. 1. Apparatus using cell (II) and Wilhelm-type liquid junction. The parts outside the dotted lines may conveniently be bent perpendicular to the plane of the paper. *A*, silver electrode; *B*, two-way stop-cocks; *C*, three-way stop-cock; *D*, reference solution; *E*, to waste; *F*, bridge solution; *G*, J-shaped liquid junction; *H*, inlet for nitrogen stirring; *I*, glass electrode; *J*, burette; *K*, level of thermostat liquid; *L*, vessel containing test solution. (Reproduced with permission, from ref. 5.)

		reference solution			test solution		
−							+
Ag(s)	|	Ag^+ 0·01*M* ClO_4^- 3·00*M* Na^+ 2·99*M*	| $NaClO_4$ 3·00*M* |		H^+ *hM* Na^+ (3 − *h*)*M* ClO_4^- 3·00*M*	|	GE

(II)

A convenient cell incorporates a "Wilhelm" salt-bridge, see Fig. 1. The ubiquitous "calomel electrode" which incorporates both the reference half-cell and the salt-bridge, is much less satisfactory. There will be large non-reproducible concentration gradients across the junction (*a*) in the cell

$$\overset{-}{\mathrm{Hg(l)}} \;\Big|\; \mathrm{Hg_2Cl_2(s)} \;\Big|\; \text{satd. KCl} \;\overset{(a)}{\Big|}\; \text{test solution} \;\Big|\; \overset{+}{\mathrm{GE}}$$

(III)

Moreover, equilibria involving metal ions may be disturbed by interaction between these ions and chloride ions which have leaked across the junction.

Determination of h by measurement of the e.m.f. of cell (II) requires values of $E^{\circ\prime}$ and E_j for substitution into equation (22); and since $E^{\circ\prime}$ includes the variable asymmetry potential of the glass electrode, its value must be determined afresh for each set of measurements. At the start of an investigation the value of $(E^{\circ\prime} + E_j)$ must be determined for a set of solutions in which h is known, as in the titration of strong acid with strong base. (Only one of these need be of known concentration, since Gran's procedure[3–5] may be used to give a precise location of the end-point). When only a small fraction of the background cation is replaced by H^+, the value of $(E^{\circ\prime} + E_j)$ is often a constant, indicating that E_j is negligible. For a larger degree of replacement, however, $(E^{\circ\prime} + E_j)$ varies with h. Very fortunately, the variation is linear, so that values of $E^{\circ\prime}$ and $x + E_j h^{-1}$ may be obtained as the intercept and slope respectively. The value of x depends only on the ionic medium (*i.e.*, on the solvent, and on the nature and concentration of the bulk electrolyte): it is unaffected by the presence of metal ion and ligand provided that the background salt is indeed present in large excess. Once the value of x for a particular medium has been obtained the constant value of

$$E^{\circ\prime} = E - RTF^{-1}\ln h - xh \tag{23}$$

may readily be determined as a prelude to each set of measurements. Titration of strong acid with strong base, or of ionic medium with strong acid is convenient. Although $E^{\circ\prime}$ may vary by a few tenths of a millivolt from day to day, larger variations may indicate deterioration in the health of the glass electrode and should be viewed with suspicion.

The value of h in the test solution can be obtained from the measured e.m.f. and the known values of $E^{\circ\prime}$ and x by means of equation (23); but in solutions which are so acidic that E_j is not negligible, an approximate value of h must first be obtained by setting $x = 0$, and then refined by successive approximation.

A glass electrode can be analogously used as a probe for hydroxyl ions, provided that it has a negligible "sodium error", *i.e.*, that the value of $(E + RTF^{-1}\ln[OH^-])$ is a linear function of $[OH^-]$. The intercept of such a plot gives $E^{\circ\prime}_{OH}$ while the slope gives $y = E_j[OH^-]$. For a given medium, the difference $(E^{\circ\prime}_{OH} - E^{\circ\prime}_{H})$ gives $RTF^{-1}\ln K_w$, and the value of y is numerically smaller than that of x, and of opposite sign.

Half-cells which respond to metal ions or to simple ligands may be calibrated as concentration probes in the same way as glass electrodes, but the values of E°, as defined by *e.g.*, $(E - RTF^{-1}\ln[M^{z+}]^{1/z})$ or $(E + RTF^{-1}\ln[Cl^-])$ are acceptable parameters: they neither change with time, nor need to be "corrected" by a term attributable to junction effects.

Although measurements with glass electrodes normally involve the use of pH-meters, the assembly should always be calibrated as a concentration probe, as described above, rather than "standardized" with a buffer of conventionally agreed pH. Since the pH of the test solution is operationally defined as

$$\mathrm{pH} - \mathrm{pH}_s = \frac{E - E_s}{RTF^{-1}\ln 10}$$

where E_s is the e.m.f. of cell (III) when the test solution consists of a standard buffer of pH_s, the difference $(\mathrm{pH}-\mathrm{pH}_s)$ can only give the log of the ratio, h_s/h, of the hydrogen ion concentrations in the two solutions if (*a*) the values of E_j for the two cells are either negligible or identical, and (*b*) the values of γ_H in the two solutions are identical or calculable. If the test solution contains a high concentration of background electrolyte, and if the standard buffer is prepared as recommended, it is most unlikely that either condition referred to in (*b*) is fulfilled. Standardization of a pH-meter with a buffer of fixed pH becomes even more difficult if the glass electrode is to be used in a mixed aqueous–organic solvent; but the electrode can be calibrated as a concentration probe in any medium which contains H^+ ions.

The titration procedure may be made more convenient if various tricks-of-the-trade are employed. Stirring, for example, can be carried out with a stream of nitrogen, which has been freed from oxygen and presaturated with solvent by being bubbled through a sample of the ionic medium. It is often elegant, and sometimes essential, to keep either A or B constant during a titration, so that functions $\bar{j}(h)_A$ and $\bar{n}(a)_B$ may be obtained. This is readily accomplished by having more than one titrant. For example, a solution of initial concentrations B_i, A_i, H_i may be titrated with equal volumes of a solution of sodium hydroxide and of metal ion solution of concentration $2B_i$. In this way, the value of H can be varied over a wide range, while that of B is kept at a constant value of B_i.

Ingenious procedures can be devised so that as much information as possible can be obtained from one set of measurements, based on a single determination of $E^{\circ\prime}$ with no subsequent transfer of the glass electrode between one solution and another. For example, one composite titration in a study of the system B, A, H might consist of the following stages.

(*a*) Titration of an aliquot of standard strong acid with a solution of sodium bicarbonate. The concentration of the base is obtained by Gran's method, and the value of $E^{\circ\prime}$ from equation (22).

(*b*) Addition of an aliquot of a solution of B and H, and titration with the bicarbonate solution. This gives the concentration of acid in the metal ion solution, together with the pH at which the metal ion undergoes appreciable hydrolysis or precipitation. (The bicarbonate ion is often preferable to the hydroxyl ion as a titrant for solutions containing metal ions, since it is less likely to give local hydrolytic precipitation.)

(*c*) Addition of an aliquot of either (i) ligand in acid, followed by titration with base, or (ii) ligand in neutral solution, followed by titration with acid. Alternatively, the metal ion solution from (*b*) may be titrated with a solution of ligand, often in the form of an HA–A buffer.

Potentiometry can yield values of $\bar{n}$ for many systems in which $0{\cdot}05M > B > 0{\cdot}0001M$ and, for some metal ions, values of α_0 within the concentration range $0{\cdot}05M > B > 0{\cdot}001M$. Before we discuss the processing of the measurements, we shall look briefly at some other experimental methods.

Liquid-liquid extraction

Determination of stability constants by this technique normally involves distribution of an uncharged metal complex BA_c between an aqueous solution (containing bulk electrolyte) and an organic solvent which is almost immiscible with it. The uncharged complexing reagent H_gA may also be distributed between the two phases, but charged species BA_n and H_jA are usually confined to the aqueous phase. Since the metal is often present only in tracer quantities, the total metal concentration in each phase is usually measured radiometrically. The total concentrations, B_0 and A_0, of B and A in the organic phase are related to the partition coefficients P_c and P_g of BA_c and H_gA by the expressions

$$B_0 = [BA_c]_0 = P_c[BA_c] \qquad (25)$$

$$A_0 = [H_gA]_0 + c[BA_c]_0 = P_g[H_gA] + cP_c[BA_c] \qquad (26)$$

where $[\,]_0$ and $[\,]$ refer to concentrations in the organic and the aqueous phase respectively. The distribution ratio, q, of metal is then

$$q = \frac{B_0}{B} = \frac{P_c[BA_c]}{B} = P_c\alpha_c \qquad (27)$$

The value of a can be obtained from a knowledge of the total quantity of A in the system, the volumes of the two phases, the values of P_g, β_j, and the measured hydrogen ion concentration of the aqueous phase at equilibrium. Thus the function $P_c\alpha_c(a)$ can be obtained from measurements of q and h, after equilibrium has been established.

Determination of q involves measuring B_0 and B in uncontaminated samples of each phase. Direct radiometric liquid-counting of such samples of a β-emitter does not, however, lead to a value of q, because the two phases have slightly different self-absorption characteristics. The value of B_0 is best obtained from the difference between the initial and final counts of the aqueous phase, assuming 100% mass balance. Values of B_0 obtained in this way should then be plotted against those obtained by direct counting of the organic phase. A linear plot, of slope slightly different from unity, should be obtained; and any points which lie markedly off the line should be discarded. Alternatively, the radioactive solutes may be precipitated from aliquots of the various solutions, and redissolved in a fixed volume of a particular solvent. Since all counts then refer to the same medium, the mass balance may readily be checked, provided that the equilibrium volumes of two phases are known. This method is especially suited to the determination of very high, or very low, values of q by use of unequal volumes of the two phases.

The value of h may be determined by using the equilibrated aqueous phase as test solution in cell (II). The value of $E^{\circ\prime}$ for the cell must previously have been established by using a solution of standard acid in the appropriate ionic medium, and checked with a second standard acid solution. Since the electrode must be washed and dried before being immersed in the test solution it is essential to check, by means of one of the standard solutions, that the value of $E°$ has not changed during this transfer.

Although q and h can be obtained with less precision than is afforded by potentiometric titrations, the results are acceptable if sufficient care is taken. The assumption of constant partition coefficients (tantamount to supposing activity coefficients to be constant in both phases) should not overtax credulity, provided that concentrations of solutes in the organic phase are low and that these solutes are uncharged species (rather than ion-pairs). The assumption is vindicated by the fact that a good description of distribution data can usually be obtained by introducing only one parameter in addition to the stability constants.

Ion exchange

The distribution ratio of a metal ion between a cation-exchange resin and an aqueous solution is given by

$$q = \sum P_n \alpha_n \tag{28}$$

where the summation includes terms for all cationic species. Interpretation is much more difficult than for distribution between two liquids, except in the special case where the free metal ion is the only cation formed; but even then, there are considerable difficulties inherent in the technique. Since it is impossible to wash the resin free from aqueous solution without displacing the partition equilibria, the quantity of metal ion adsorbed on the resin can only be obtained from the difference between the initial and final aqueous solutions. Activity coefficients in the resin are highly sensitive to the nature of the cations. Values of q can only be compared if they refer to constant load (of B) on the resin and so for each required point q, a set of values $q(B)_a$ must be measured, and interpolated to give q at one value of B. The method therefore demands a much greater expenditure of time and effort than does solvent extraction; and the results obtained are probably less precise and more difficult to interpret. The technique is therefore not recommended.

Spectrophotometry

This might seem to be an unpromising technique, because the apparent molar absorptivity ε of B in a solution containing complexes BA_n gives $\sum \varepsilon_n \alpha_n$ where ε_n is the individual molar absorptivity of BA_n. If, as often happens, the values of ε_n are appreciable for all species BA_n at the wavelength used, computation of the $(2N + 1)$ parameters ε_n and β_n is difficult, and the results obtained are of low precision. Nonetheless, the technique has marked advantages over, say, ion-exchange. The value of ε can be measured easily and reliably, and the values of ε_n are true parameters and are unchanged by minor variations in the composition of the solution. So in the special situation where only one absorbing species (BA_c) exists, spectrophotometry provides a convenient and precise method for measuring α_c, and forms a useful complement to potentiometry. Moreover, if the complex has a high absorptivity only trace concentrations of B are needed. So spectrophotometry can be used to study some systems in which low solubilities preclude the use of potentiometry.

A value of a is, of course, also needed and is usually obtained by combining the experimental value of h with the known value of A. (The approximation $A \sim a$ is valid only for non-basic ligands in the presence of tracer concentrations of metal ions, a situation seldom encountered in spectrophotometric work.) Unless the metal is present in tracer concentrations, measurements of B, A, a or B, A, h are needed to give values of $\bar{n}$, a, and these measurements can yield values of β_n without recourse to spectrophotometry.

Much early, and regrettably, some recent spectrophotometric work has made use of the "method of continuous variations" in which absorbance is measured as a function of AB^{-1} in solutions of constant $(A + B)$. If a single complex is formed, its composition and sometimes its stability constant can be obtained from the position of the extremum on the graph. But the interpretation is ambiguous unless a single 1:1 complex is formed, and the method is not to be recommended.

A forthcoming review paper deals with spectrophotometric methods *in extenso*.[9]

TREATMENT OF RESULTS

Checking the general formula

The measured quantities $\bar{n}$, α_c, q and ε are all unique functions of a if, and only if, complex formation between B and A is restricted to the simple, mononuclear species BA_n. If, however, polynuclear species $B_qA_p (q > 1)$ are present the value of

$$\frac{A - a}{B} = \frac{\sum p\beta_{qp}b^q a^p}{\sum q\beta_{qp}b^q a^p} \tag{29}$$

will depend on both b and a; and an experimental plot of $(A - a)B^{-1}$ against log a will produce separate curves for different values of B. If, on the other hand, the complexes are mononuclear, but include hydrolysed or protonated complexes BA_nH_j ($j \neq 0$) the value of $(A - a)B^{-1}$ will be independent of b, but will be a function of both a and h. (Dependence on both a and h will be observed even in the absence of mixed species if the two series of complexes $B(OH)_{-j}$ and BA_n are formed.)

Before attempting to calculate stability constants it is therefore essential to obtain measurements for solutions with at least two, widely spaced, values of B in order to check whether or not polynuclear species are formed. It is advisable to plot all experimental values of $(A - a)B^{-1}$ (or α_c, q or ε, as appropriate) as a function of log a so that the results may be easily surveyed. The possibility of combination of H^+ or OH^- with any species containing B must be similarly checked by plotting $(A - a)B^{-1}$ against log a for systems of widely differing h or A. [Since, for basic ligands, a varies with both A and h, coincidence of functions $\bar{n}$ (log $a)_A$ demonstrates the absence of protonated or hydroxo-complexes as effectively as does coincidence of the functions $\bar{n}$ (log $a)_h$.]

Computation of stability constants

"If at least N sets of values of $\bar{n}$, a or $\alpha_{c.a}$ are available, equations (13) or (17) may be solved for the required stability constants β_n'' is a typical conclusion to a brief account of experimental aspects of stability constant work. It assumes, of course, that only species BA_n are present: but questions arise as to the best way of handling the experimental data when, as is usually the case, more than N sets of measurements are available.

Large numbers of methods, good and bad, have been described[1,10] for obtaining stability constants, not only from measurements of $\bar{n}$, a and α_c, a, but also from functions q, a and ε, a which contain one or more unknown parameters in addition to the stability constants. One criterion of a "good" method is that it makes full use of the data. There is little point in carefully determining a whole titration curve and then calculating values of β_1 and β_2 from only two points on it. Conversely, the constants obtained can never be more precise than the primary data, however elegant the method used for calculating them. Satisfactory stability constants can be obtained only be combining careful, well-designed experiments with rigorous computational technique.

Whatever the complexity of the system, the problem is one of fitting experimental values of $\bar{n}(a)_{B,h}$ or $\alpha_c(a)_{B,h}$ with parameters which give as good a description as possible of the measurements. If the only complexes formed are BA_n, the data can be represented in two dimensions, since $\bar{n}$ and α_c are functions only of a. If mixed, or polynuclear, complexes are formed, the data must represent a surface in three, or more, dimensions.

Data of high precision are probably best processed by electronic computation, regardless of the type of complexes which are formed; and the advantages of such methods increase with the complexity of the system. Available programs,[10] include those for treating errors (systematic, correlated and random). Systems containing complexes BA_n, even in the presence of BHA, can be treated by linear least-squares techniques. If polynuclear complexes, or two or more mixed complexes, are present, more elaborate search techniques, such as "pit-mapping" must be used. LETAGROP VRID, the most sophisticated of these programs, is of very wide applicability, but it would be foolish to use so sledge-hammer a procedure unless the system were fairly complicated.

If the system can be described by only one, or two, parameters, electronic computers offer little advantage over graph paper, unless the measurements are exceedingly precise. Graphical methods are of two main types.

(*i*) *Linear, non-logarithmic plots*. Up to two parameters may be obtained from a single plot. Further parameters can be squeezed (with decreasing precision) from the measurements by successive extrapolations.

(*ii*) *Curve-fitting methods*, usually based on a semi-logarithmic plot involving one normalized variable. If a single complex BA is formed, the curves $\bar{n}$ (log a) and α_c (log a) are of unique shape. The value of β_1 can be obtained by matching the experimental points with the theoretical curves $\bar{n}$ (log $\mathbf{a}$) and α_c (log $\mathbf{a}$) where $\mathbf{a} = \beta_1 a$ is the normalized value of a.

For systems of two or more complexes, the curves are not of unique shape: but when $N = 2$, the parameter $\beta_1^2\beta_2^{-1}$ can be obtained from the shape of the experimental curve, and that of β_2 from its position on the log $\mathbf{a}$ axis. The "projection strip" method[1,10] is particularly convenient.

Curve-fitting procedures are of little use when more than three species (*e.g.*, B, BA and BA_2) coexist, but if the formation curve shows a plateau at an integral value of $\bar{n}$, measurements of $\bar{n}$ above this value may be treated separately from those below it. In many systems H_jA (*e.g.*, H^+–PO_4^{3-}) the formation of each species occurs in widely separated steps, each of which may be treated as for a monobasic acid. Similarly, since mercury(II) halides HgX_2 exist as the sole complex over a very wide range of free halide ion concentration, sets of data in the regions $\bar{n} < 2$ and $\bar{n} > 2$ may be treated independently of each other.

Graphical methods have been much criticized by the purists on account of their subjectivity, both in the values of the parameters and in their limits of error. But the judgment of an experienced worker may be but little inferior to the machinations of a computer; and when waiting-time is taken into account, graphical methods may well be quicker (as well as cheaper). It is often easier to tell from a graph, rather than from tabulated print-out, when all is not well. Dud points, and systematic errors, obtrude like sore thumbs. A glance at the formation curve may suggest that it starts, improbably, at $\bar{n} = 0{\cdot}0005$ rather than at $\bar{n} = 0$; or that it is not symmetrical about $\bar{n} = 1$, as it should be, if the only complexes present are BA and BA_2. A good fit between experimental points and a theoretical curve is an effective way of demonstrating that the calculated parameters give a satisfactory description of the system; curve-fitting procedures provide an automatic comparison of the two functions.

Data should, ideally, be processed by two different procedures, and whatever the method used to calculate stability constants, the values obtained should *always* be checked* by substitution into equations (13) and (17) to ensure that they represent the experimental data satisfactorily (see Figs. 2–4). Nevertheless, however good the experimental and theoretical techniques, and the compatibility between them, stability constants can never be unequivocally "right", but only "compatible with the data" within certain limits. This distinction may seem ludicrously pedantic for precise work on, say, a monobasic acid

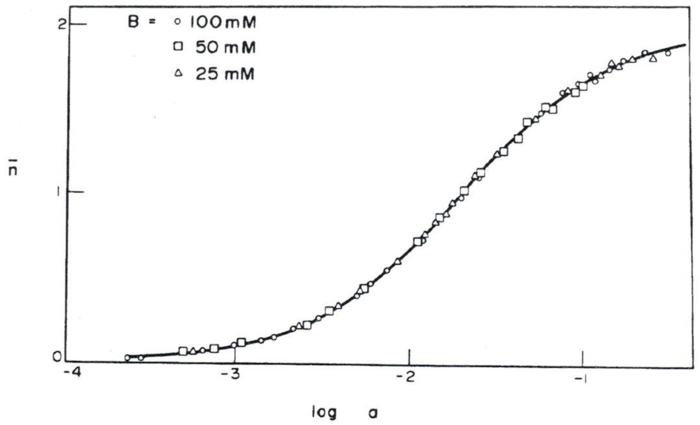

Fig. 2. Formation curve $\bar{n}$ (log a) for the mononuclear copper(II) methoxyacetate system at 25°C in $3M$ $NaClO_4$.[12] The points represent experimental values, and the curve is calculated for $\log \beta_1 = 2{\cdot}01$ and $\log \beta_2/\beta_1 = 1{\cdot}33$. (Reproduced with permission, from ref. 5.)

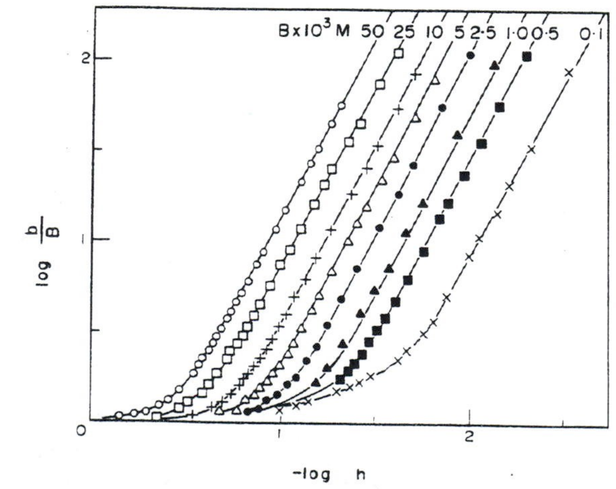

Fig. 3. Values of log b/B as a function of log h for hydrolysed bismuth(III) perchlorate solutions in $3M$ $NaClO_4$ at 25°C. The curves are calculated by using only two parameters β_{qpj} for the species $B_qA_pH_j$, *viz.* $\log \beta_{1.0.-1} = 1{\cdot}58$ and $\log \beta_{6.0.-12} = 0{\cdot}33$. (Reproduced with permission from ref. 13.)

* Not only by the author, but also by the referee.

over the whole range $0 \leq \bar{j} \leq 1$, but the distinction increases in importance with the complexity of the system. It is an essential corrective to complacency about interpretation of measurements on systems containing mixed, or polynuclear, complexes, or even about stability constants of simple mononuclear complexes obtained by techniques such as spectrophotometry, which involve several additional parameters.

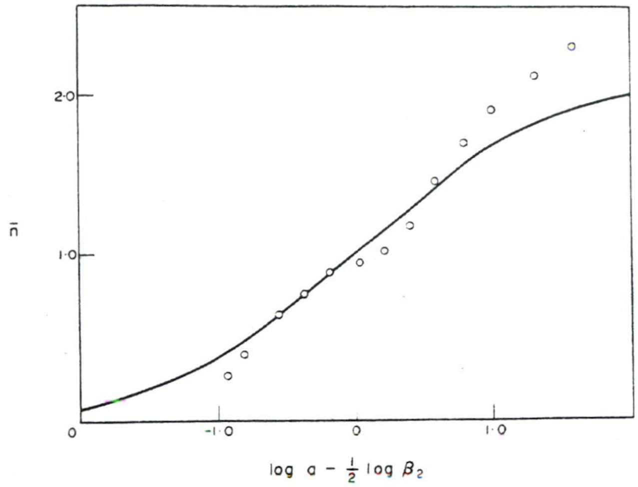

Fig. 4. Illustration of a type of formation curve encountered all too frequently, even in the 1970's. Points: experimental values submitted for publication. Curve: calculated from authors' constants "obtained by least-squares method".

Table 2. Molarities of some differently prepared ionic media

	Solution 1. Metal ion			Solution 2. Ligand			Mixture of v_1 ml of solution 1 and v_2 ml of solution 2			
	B	$[Na^+]$	$[ClO_4^-]$	A	$[Na^+]$	$[ClO_4^-]$	B	A	$[Na^+]$	$[ClO_4^-]$
$[Na^+] = 3$	B_i	3	$3 + zB_i$	A_i	3	$3 - yA_i$	$\frac{v_1 B_i}{v_1 + v_2}$	$\frac{v_2 A_i}{v_1 + v_2}$	3	$3 + \frac{zB_i v_1 - yA_i v_2}{v_1 + v_2}$
$[ClO_4^-] = 3$	B_i	$3 - zB_i$	3	A_i	$3 + yA_i$	3	$\frac{v_1 B_i}{v_1 + v_2}$	$\frac{v_2 A_i}{v_2 + v_2}$	$\frac{yA_i v_2 - zB_i v_1}{v_1 + v_2}$	3
$[Na^+] \nless 3$ $[ClO_4^-] \nless 3$	B_i	3	$3 + zB_i$	A_i	$3 + yA_i$	3	$\frac{v_1 B_i}{v_1 + v_2}$	$\frac{v_2 A_i}{v_1 + v_2}$	$3 + \frac{yA_i v_2}{v_1 + v_2}$	$3 + \frac{zB_i v_1}{v_1 + v_2}$

Display of results

Values of stability constants, represented as numbers in a table, may not seem very interesting; graphs[14] have greater impact.

We have seen that a combined plot of experimental points $\bar{n}$, $\log a$ and the calculated formation curve is useful for convincing the researcher and others that the values obtained for the stability constants are respectable. But a value of $\bar{n} = 1$ may, of course, mean that $\alpha_1 = 1$; or that $\alpha_0 = 0{\cdot}5$, $\alpha_1 = 0$ and $\alpha_2 = 0{\cdot}5$. A better feel for the system may be obtained if the stability constants are used to construct distribution plots for the various species present.

In simple, mononuclear systems, the distribution of species can be represented by two-dimensional diagrams (see Fig. 5). Plots of α_n ($\log a$) for values of $0 \leq n \leq N$ probably provide the best insight into the system, but distribution diagrams in the form of a set of plots of $\sum_0^n \alpha_n$ against $\log a$ have also been used.

If mixed or polynuclear complexes are formed, values of α_n depend on two or more variables. The distributions of species in such systems can be represented as surfaces in three or more dimensions. Two-dimensional graphs can be constructed by cutting sections through the surface at constant values of one, or more, of the variables. A system which contains species BA_nH_j may, for example, be represented as a series of plots of α_n against $\log a$ at a number of values of pH.

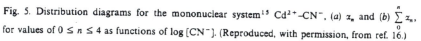

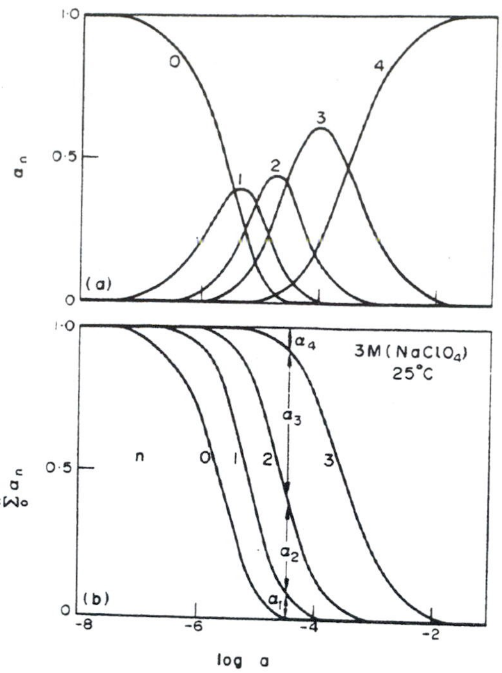

Fig. 5. Distribution diagrams for the mononuclear system[15] Cd^{2+}–CN^-, (a) α_n and (b) $\sum_0^n \alpha_n$, for values of $0 \leq n \leq 4$ as functions of log $[CN^-]$. (Reproduced, with permission, from ref. 16.)

Publication

Plots providing comparisons of theoretical and experimental functions and those showing the distribution of the various complexes, should always be drawn, if only for the edification of the research group. Presentation of the work to a wider audience (whether of examiners or journal-readers) requires, in addition, a clear statement of the following points.

1. The temperature of the measurements.
2. The ionic medium (including a statement as to which species, if any, is held at constant concentration).
3. The credentials of any stock solutions. This should include the source and purity of starting materials (especially of any synthesized) and a report on the analysis of the solutions.
4. The apparatus.
5. The design and execution of the experiments.
6. Methods of converting measurements, e.g. of E, A, B, H or of q, E, A, B, H into secondary functions such as $\bar{n}$, a or α_c, a. The values of any necessary parameters, such as β_j and P_c should be quoted, together with a report of how they were obtained. (These values must, of course, refer to conditions identical to those used for the measurements.)
7. The methods used for obtaining values of stability constants together with (defined) limits of error, from the secondary functions.
8. The numerical values of stability constants, and of the limits of error, together with the concentration scale to which they apply.
9. A critical comparison of these values with any which have been previously reported for the same system.

The object of the work

Determination of good stability constants clearly involves a great deal of painstaking work. Why should people take all this trouble? (Some, of course, don't, to the obvious detriment of their results.)

Equilibrium constants are measured for two main types of reason, to which the over-simplified labels "applied" and "pure" may be attached. Although the two categories overlap, the "applied" situations are basically those in which values of equilibrium constants are needed to calculate the distribution of species in a particular solution. Thus values of ${}^{H}\beta_j$ are needed in order to calculate the concentration of A in equilibrium with protonated species at a particular pH. Values of β_n may be needed in order to design satisfactory separative, or other analytical, procedures. Observed values of ΔH or ε for a solution containing several complexes can yield values of ΔH_n and ε_n for the individual species if the concentration of each species can be calculated.

For "applied" purposes of this type, the most useful quantities are probably the concentrations $[B_qA_pH_j]$ as a function of log h at fixed values of B and A. Calculation involves solving equations (6), (7) and (8) for b and a and, except for very simple systems, is best performed by an electronic computer.

"Pure" problems are those designed to increase insight into the nature of chemical behaviour, rather than into the properties of one particular set of solutions. The values obtained have provided evidence for the existence of a strange collection of polynuclear hydroxo-complexes of metal ions, and have led to many attempts to correlate the stability of complexes with the properties of the metal ion, the ligand and, occasionally, the solvent.

The usefulness of work on stability constants is not restricted to studies of chemical behaviour *in vitro*. Interactions between metal ions and a wide variety of ligands, ranging in complexity from cyanide to haem, have long been of interest to biochemists. More recently, stability constants have extended their domain to become the concern of soil scientists and oceanographers.

Since the stability constants give a measure of ΔG for the formation of a complex, the variation of β_n with temperature, leads in principle to values of ΔH_n and ΔS_n for the formation of the separate complexes, but over the narrow range of temperatures which can be used, changes in β_n may be little larger than experimental error. Moreover, the method is based on the assumption that ΔH is independent of temperature. Values of ΔH_n are better obtained by combining calorimetric values of ΔH with the appropriate values of α_n, which are calculated from the stability constants. These values of ΔH_n may then be combined with the stability constants to give values of ΔS_n. This type of work is only successful if all the measurements are of the highest precision.

Acknowledgement—The author is very grateful to St. Anne's College, Oxford for financial support.

REFERENCES

1. F. J. C. Rossotti and H. Rossotti, *The Determination of Stability Constants*. McGraw-Hill, New York, 1961.
2. L. G. Sillén and A. E. Martell, eds. *Stability Constants of Metal-ion Complexes*. Chemical Society, London, 1964.
3. G. Gran, *Analyst*, 1952, **77**, 61.
4. F. J. C. Rossotti and H. Rossotti, *J. Chem. Educ.*, 1965, **42**, 375.
5. H. S. Rossotti, *Chemical Applications of Potentiometry*. Van Nostrand, London, 1969.
6. F. J. C. Rossotti, unpublished work.
7. G. J. Moody and J. C. R. Thomas, *Talanta*, 1972, **19**, 623.
8. L. Sommer, T. Sepel and V. M. Ivanov, *Talanta*, 1968, **15**, 949, and refs. therein.
9. W. A. E. McBryde, *ibid.*, 1974, **21**, 0000.
10. F. J. C. Rossotti, H. S. Rossotti and R. J. Whewell, *J. Inorg. Nucl. Chem.*, 1971, **33**, 2051.
11. F. J. C. Rossotti, H. Rossotti and L. G. Sillén, *Acta Chem. Scand.*, 1956, **10**, 203.
12. J. D. E. Carson and F. J. C. Rossotti, unpublished work.
13. Å. Olin, *Acta Chem. Scand.*, 1957, **11**, 1445.
14. L. G. Sillén in I. M. Kolthoff, P. J. Elving and E. B. Sandell, eds., *Treatise on Analytical Chemistry*, Part I, Vol. Interscience, New York, 1959.
15. I. Leden, *Svensk Kem. Tidskr.*, 1944, **56**, 31.
16. F. J. C. Rossotti in J. Lewis and R. G. Wilkins eds., *Modern Coordination Chemistry*. Interscience, New York, 1960.

Zusammenfassung—Es wird eine kritische Übersicht über die Methoden zur Bestimmung von Stabilitätskonstanten gegeben. Es werden Empfehlungen zur Wahl der Arbeitsweise, zur genauen Planung der Experimente, zur Berechnung der Konstanten und zur Wiedergabe der Ergebnisse in der Veröffentlichung ausgesprochen.

Résumé—On passe en revue de manière critique les méthodes de détermination des constantes de stabilité. On fait des recommandations sur le choix de la technique, l'étude détaillée des expériences, le calcul des constantes et la présentation des résultats pour publication.

DEVELOPMENT AND PUBLICATION OF NEW METHODS IN KINETIC ANALYSIS

HARRY B. MARK, JR.

Department of Chemistry, University of Cincinnati, Cincinnati, Ohio 45221, U.S.A.

Although the concept of using analytical methods based on kinetics or reaction rates goes back over fifty or sixty years to the early literature in biochemistry, radiochemistry and gas-phase diffusion,[1] and there has been extensive use of enzymatic and other catalytic reactions[2] in analysis, especially in clinical applications,[3] it was not until 1951 in the work of Lee and Kolthoff[4] that the broad inherent possibilities and advantages of kinetic-based analytical methods with respect to conventional equilibrium methods in many chemical situations were pointed out clearly to chemists. This work stimulated considerable fundamental development and special application research by Siggia,[5-7] Reilley,[8-10] Guilbault[3,11] and others,[1,9,12-17] particularly in the area of *in situ* analysis of complex systems. There have been several books,[1-3,18,19] monograph chapters[9,13-15,20] and reviews[11,16,21-23] published during the last ten years which summarize the research efforts in this field.

It is somewhat surprising in view of the potentialities of kinetic-based techniques and the considerable amount of fundamental method development published, that there has not, except for continued massive use in clinical assays, been any significant application of kinetic-based techniques to routine and/or practical analytical problems. However, there are two main reasons for this lack of application of kinetic techniques. First of all, almost all commercial instrumentation for chemical measurement is designed for steady-state or equilibrium measurement and does not perform satisfactorily when used for quantitative time-dependent measurement. Thus, analysts often found it necessary to modify or completely build the instrumentation for analytical rate-based determinations. Secondly, it is felt that many of the fundamental method-development papers reported during the past ten years did not fully explain applicability and limitations of the individual methods and, thus, it was not clear to other analytical chemists how they could modify or adapt these methods to their particular system or analysis problem.

A very striking example of this occurred at a recent symposium on "reaction-rate methods of analysis." After one talk on the principles of a particular method for the simultaneous *in situ* analysis of closely related mixtures, someone in the audience made the following comment: "I tried this method on mixtures of amines. The rate constants, ratio of rate constants and ratio of reactant concentrations were ideal according to the discussion in this talk. However, the analytical results were terrible. Why?"

The first problem, instrumentation, has essentially been overcome. Now that inexpensive analogue and digital integrated circuitry and minicomputer systems are readily available, considerable research on the basic concepts of stable accurate instruments for kinetic measurement has been reported[15,20,24-32] and automated quantitative instruments are beginning to become available (or easily built). However, the general guidelines for carrying out research to develop kinetic-based analytical procedures and for reporting the data and results have never been given in one place. Thus, this paper will attempt to present recommendations concerning important points that should be covered and investigated during a research project in this area.

MATHEMATICAL BASIS OF KINETIC METHODS OF ANALYSIS

The past few years have seen the development of a very large number of methods for the calculation of the initial concentration(s) of the species of interest from reaction rate data. These methods involved, in general, the simple mathematical manipulation and rearrangement of the usual differential or integral forms of the classical reaction-rate equations to put them in a convenient form for the calculation of initial concentration(s) of the unknown reactant(s). With few exceptions most of these mathematical treatments or methods have been reported within the past 5–10 years. Such methods can be classified into two main categories: methods for a single species and methods for the simultaneous *in situ* analysis of mixtures. Within each of these two categories, the methods can be subdivided according to the kinetic order of the reactions employed: pseudo zero-order or initial-rate methods, first-order and pseudo first-order methods, and second-order techniques. The principles, mathematical treatments, applicability and limitations of these methods have been discussed in the book by Mark, Rechnitz and Greinke[1] and the monograph chapters by Mark, Papa and Reilley,[9] Blaedel and Hicks,[13] Pardue,[15] and Crouch.[20]

Although these various methods are generally classified by most investigators as zero-order (initial reaction-rate), first-order, or second-order reaction-rate methods,[1,9,20,23] the actual mechanisms for the chemical reactions employed in virtually all of these methods (except those involving radiochemical decay reactions[1] and catalytic reactions[20]) are bi-molecular reactions of the type:

$$A + R \underset{k_b}{\overset{k_f}{\rightleftharpoons}} P \tag{1}$$

[where A is the species of analytical interest, R is the added reagent, and P is the product (or products)] and the general differential rate expressions have the form:

$$-\frac{d[A]_t}{dt} = -\frac{d[R]_t}{dt} = \frac{d[P]_t}{dt} = k_f[A]_t[R]_t - k_b[P]_t \tag{2}$$

(where $[A]_t$, $[R]_t$, and $[P]_t$ represent the concentrations of these species at any time t). Thus, the nomenclature zero-, first- and second-order actually refers to the experimental conditions under which the rate measurements are made and/or the relative concentrations of the reactants A and R.

If the rate data are taken only during the initial 1–2% completion of the total reaction, then the concentrations of A and R during this initial reaction period are virtually unchanged and equal to the initial concentrations $[A]_0$ and $[R]_0$ respectively and the reverse reaction can be ignored, as only a negligible amount of product is formed. Thus the differential rate expression (2) simplifies, under these initial rate conditions, to a pseudo zero-order form:

$$\left(\frac{d[P]_t}{dt}\right) \sim k_f[A]_0[R]_0 \sim \text{Constant} \tag{3}$$

[Note that only the rate of initial change of concentration of product(s) is followed experimentally, as the change in concentration of A or R is small and, therefore d[A]/dt or d[R]/dt cannot be measured accurately.] If the reaction (1) is run under such conditions that the initial concentration of one of the reactants (either A or R) is very large compared to the other, then either $[A]_t$ or $[R]_t$ will remain virtually unchanged as the reaction proceeds to equilibrium and can be considered equal to its initial concentration. Also, the reverse reaction can usually be neglected as the large excess of one of the reactants drives the reaction to virtual completion. Under these conditions the reaction is a pseudo first-order reaction and the differential rate expression takes the form

$$-\frac{d[A]_t}{dt}\left(\text{or } -\frac{d[R]_t}{dt}\right) \sim k_f[R]_0[A]_t \text{ (or } k_f[A]_0[R]_t) \tag{4}$$

$$\sim k_f'[A]_t \text{ (or } k_f' [R]_t) \tag{5}$$

where $k_f' = k_f[R]_0$ (or $k_f[A]_0$). If the rate of reaction (1) is measured during a significant degree of completion under conditions where $[A]_0$ is of the same order of magnitude as $[R]_0$, then the method is called a second-order method and the exact differential rate expression (2) must be employed in the data analysis. Note also that only when the reaction mechanism is virtually irreversible can the reverse reaction be ignored in equation (2). Also, for the special case $[R]_0 = [A]_0$, a modified form of the calculation of initial concentrations must be employed.[33]

It is obvious from the discussion above that any complete research and/or report on a kinetic-based analytical procedure must carefully take into account the degree of approximation made in the use of equation (2) with respect to the time period of data measurement and relative initial concentrations of reactants and, in some cases, the reversibility of the reaction(s). For example, care must be taken when using a pseudo first-order method when the initial concentrations of the unknown varies over several orders of magnitude. The error introduced in assuming the validity of the pseudo first-order approximation of equation (2) is a function of $[A]_0$. Although the reaction mechanism for enzyme and other catalysed reactions and, hence, the rate equations, seem somewhat more complex,[1,2,20] similar assumptions, simplifications and therefore, restrictions in validity are made on the rate-measurement techniques employed in the analytical use of these systems.[20,34]

Within each of the zero-order, first-order and second-order classifications of reaction-rate methods there are a large variety of different methods of display and/or mathematical manipulation of the reaction-rate data or equations in order to calculate the initial concentration(s) of the species being determined. The selection of the technique of calculation can have very significant effects on the accuracy of the analysis.[1,9,20,34,35] For example, Crouch[20,34] has pointed out that the kinetic role of the determinand species in methods employing first-order and enzyme or other catalysed reactions has a strong effect on the choice of measurement of the reaction-rate data. It was shown that a fixed-time approach[1,13,15] is superior for pseudo first-order reaction conditions and for the determination of the substrate in any type of pseudo zero-order catalysed reactions. However, in the cases where the enzyme or other catalyst is to be determined (under pseudo zero-order conditions), the variable-time procedure[1,13,15] yields better results. The choice of rate data reduction method in the simultaneous *in situ* analysis of mixtures is even more critical with respect to accuracy of the analyses. There have been numerous results in the literature[1,4,8,9,12,35,36] which have shown that both the relative and absolute values of the rate constants and the initial concentrations of the species to be determined dictate the choice of data treatment for the calculation of the initial concentrations. Furthermore, within the mathematical framework of each of these calculation procedures, there are generally optimum and/or limited times at which rate data should be taken in order to minimize the effects of random and absolute error in measurement.[1,9,12,35] The choice of procedure and optimization of the measurement is very complex and no simple rules can be given here. However, very detailed discussions of this problem are available.[1,9,12,35] Thus, any research on the development of new analytical methods should include a very comprehensive analysis of the choice of data measurement and reduction procedures.

THE REACTION MECHANISM

In most equilibrium-based analytical methods, the success or failure of a determination is not affected by the reaction mechanism provided that the reaction is either quantitative or the measured parameter at equilibrium is linearly proportional to initial concentration of the species of interest. This is not the case in reaction-rate methods. Any report of a new kinetic method should include, if possible, a *complete* study of the mechanism of the reaction(s) involved in the procedure (some reactions, such as catalytic reactions, are so complicated that complete elucidation of the mechanism would be impossible). It should also include a detailed study of the effects of typical sample matrix components, which can act as catalysts, induce side-reactions, alter the activity of the reactants, *etc.* The rates and rate constants for chemical reactions are very sensitive to low concentrations of "spectator" species, *etc*, and hence, samples containing the same true initial composition of the species of interest but coming from different sources can very often give quite different apparent concentrations. Unless the experimenter is completely aware of the total reaction mechanism and all possible factors that can affect either the activation energy and the reaction path, erroneous analytical results could be obtained under certain circumstances which may not be detected. A detailed investigation of the simultaneous *in situ* analysis of binary amine mixtures illustrates this point.[37]. (It should be mentioned here that most systems will be less error-prone than the example given.) It was found that rate constants for the reaction of many individual organic amines with methyl iodide in acetone solvent had ideal differentiation and range of absolute values when the reactions were conducted under pseudo first-order conditions with respect to the methyl iodide. The data reduction method of Roberts and Regan[38] and the simple and accurate data acquisition system employing a continuously recording conductivity instrument[39,40] seemed to indicate that almost ideal analytical results would be attained. However, it was found, just as had been reported by the analyst who commented at the symposium mentioned in the introduction to this paper, that most binary mixtures gave very poor results when analysed by this procedure.

A careful examination of the reaction mechanism of the system revealed that there were numerous unexpected sources of error. First, it was found that the acetone used as a solvent underwent a Schiff-base type reaction with some primary amines at a rate comparable to the methylation reaction of the method. This resulted in the variation of the composite rate constant, K^*,[37,38] during the course of the reaction of mixtures containing such primary amines, as shown in curve 1 of Fig. 1. As the calculation of initial concentration from K^* by the method of Roberts and Regan[38] requires K^* to remain constant over the total reaction time, serious errors are introduced by this side-reaction. As added water would suppress the Schiff-base reaction, acetone–water mixtures were then used as the solvent systems. While this improved the results obtained for several amine mixtures, those containing high molecular weight amines still gave poor results. For example, K^* is not constant over the whole reaction for n-butylamine–tributylamine mixtures, as shown in curve 2 of Fig. 1. In this case the addition of water results in the formation of small suspended micelles of amine in the solution, which alters the reaction rates. Thus, it was necessary to go to a less

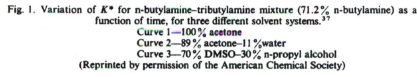

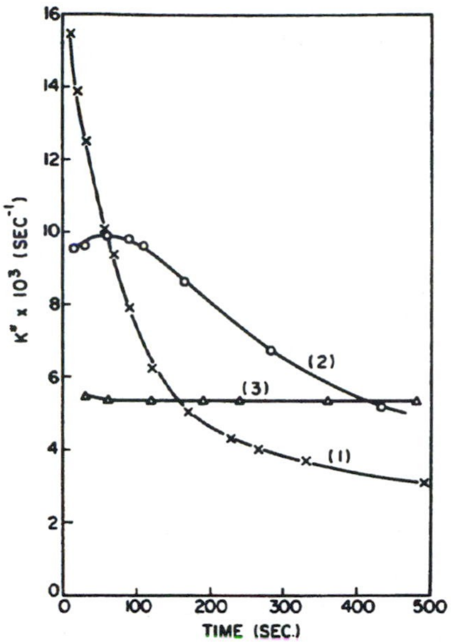

Fig. 1. Variation of K^* for n-butylamine–tributylamine mixture (71.2% n-butylamine) as a function of time, for three different solvent systems.[37]
Curve 1—100% acetone
Curve 2—89% acetone–11%water
Curve 3—70% DMSO–30% n-propyl alcohol
(Reprinted by permission of the American Chemical Society)

basic solvent system, DMSO–n-propyl alcohol, which had no significant tendency to react with primary amines[41] and in which most amines are soluble.[37] Curve 3 of Fig. 1 shows that K^* for the n-butylamine–tributylamine mixture is then constant and, thus, accurate initial concentrations are obtained.[37]

However, certain amine mixtures, generally those containing methylamime, still gave poor analyses. It was found that such mixtures were slightly hygroscopic and samples often contained traces of water. This trace water acted as a catalyst in the methylation reaction. Thus, the rate constants obtained from the reactions of standard dry amine solutions and used in the calculation of initial concentrations of unknown mixtures containing traces of water were not valid for the unknown mixture reactions. The addition of water to the solvent system (10% or more) was found to swamp out these variations of rate constants in unknown amine mixtures. Recall, however, that the addition of water can result in solubility complications for some amines. This investigation also revealed that a synergistic effect, probably caused by changes in the activity of the amines by the build-up of product, resulted in variation of K^* during the reaction when the amine mixture concentration was high. Other sources of error arose from the reaction of the reagent methyl iodide with the DMSO solvent system on storage, and loss of measurement sensitivity when conducting impurities were present in the unknown.[37]

The point here is that no simple general reaction procedure for this method can be given that is universally applicable to all types of amine mixtures. This is probably true in general for any procedure for a kinetic-based method. Thus, in reporting a new method, it is necessary to have investigated the chemistry of the reactions employed so that other analysts trying to use the method for their samples can anticipate possible sources of errors resulting from the chemical nature and/or composition of their samples.

INSTRUMENTATION FACTORS

Of course, the accuracy and precision of the instrumental and/or experimental measurement is equally important to both kinetic-based and equilibrium analytical methods and should be discussed in any publication concerning these methods. However, there are a few special instrumental and/or experimental factors that are of critical importance in the kinetic-based methods and quantitative information concerning them should be included in any report about a new method. Those factors particularly important to reaction rate methods are listed below.

Instrumental stability

Although high-frequency noise in an instrument used in measuring reaction rates can be a problem, it can generally be eliminated by simple electronic filtering, as the frequency of the noise is generally very high compared to the rate of change of the signal, which is proportional to the rate of reactions employed in analytical procedures.[23,32,42] Weichselbaum *et al.*,[32,42] Pardue *et al.*,[15,24,25] and Crouch[20] have discussed in detail the circuitry, design criteria and magnitude of the effects of low-frequency drift on analytical results, in spectrophotometric rate-measurement. The principles discussed are applicable to any type of instrumental measurement.

Linearity of transducer response

Most transducers which convert chemical concentration into an electrical signal have a non-linear response. For example, potential and optical transmission are not proportional to concentration. In general, this non-linearity is easily and simply corrected in equilibrium analytical measurements. However, Weichselbaum *et al.*,[32,40] and Crouch[20] have shown that it is considerably more difficult to linearize instrumentally the response–concentration function in reaction rate methods and often this transformation can introduce significant errors into the analytical results.[32,42] For example, Weichselbaum *et al.* have shown that simple non-linear feedback elements employed in log-response operational amplifier circuits are not sufficiently accurate in the transformation of transmittance into absorbance, to be used for many analytical purposes. They found it necessary to go to a very elaborate servo-mechanical system to circumvent this problem. Crouch[20,34] in an extensive analysis of the various data measurement and reduction procedures in kinetic methods has shown that the variable-time approach can be used advantageously in the case of non-linear response, as the measured reaction rate in this procedure is linearly proportional to *initial* concentration of the species of interest in spite of the fact that the actual transducer response is not proportional to concentration. This is because the time required to reach a fixed concentration level is the parameter measured and, thus, linearity of the overall response–concentration curve is not necessary. Thus, it is not necessary to resort to instrumental linearization of signal response when the variable-time method is used. The point here is that in some cases the instrument and not the chemical reaction can dictate the method used.

Rise-time of the instrument

As mentioned above, high-frequency noise can often be eliminated by simple electronic filtering. However, caution is necessary and the investigator should be very familiar with the actual effective rise-time of the instrument used, over all operating ranges. Obviously, a fixed filter time constant cannot be applicable over a large range of initial concentrations as the rate of the reaction at any time is a function of initial concentration.[32] Thus, a report of a new rate method should include a quantitative evaluation of the rise-time of the instrument under all conditions of damping employed and a comparison with the maximum reaction rates measured under each setting of the filter time constant.

Data acquisition mode

Most reaction rate methods reported[1,2,15,20] utilize only a small number of actual data points (from one to about four) in the calculation of the initial concentration(s) of the species of interest. Clearly this approach throws away a considerably amount of data that could be used advantageously. For example, Margerum and co-workers have discussed the parameters affecting the accuracy of the simultaneous analysis of mixtures by pseudo first-order reaction methods.[30,31,43,44] Their results point out that a continuous data-utilization approach to the problem is the superior data-analysis method in several examples of the use of reaction rate methods for analyses. In the early development of reaction rate methods for analysis of closely related mixtures, the procedures used chemical reactions which were not suitable for continuous automatic measurement of the entire rate-response curves.[1,9] Also, all calculations at that time in the development of the reaction rate methods attempted to limit the number of data points taken and to predetermine the optimum times, *etc*, for taking this minimal amount of data.[1,9] However, in recent years the advances in the field of electronic circuitry and computer technology has had a tremendous influence on the design of instrumentation for kinetic analysis. These instrumental advances have also had a strong influence on the principles of and approaches to differential rate methods. Margerum and co-workers (and also Weichselbaum and co-workers[32,42]) have shown conclusively that the use of built-in computation systems allows continuous analysis of the entire reaction-rate data curve throughout the course of the reaction. The data are thus processed with both ensemble-averaging and smoothing routines in real time. Experimental results and detailed error analysis study have shown that this approach to data acquisition, reduction, and display leads to a much greater accuracy and precision of the analytical results.[30,31,43,44] In fact, good results can be obtained for fast reaction differential rate analyses where the usual finite or minimal data-point methods fail completely.[43,44] Thus, any new report of a reaction rate method should include a comparison of results calculated from both a finite and a continuous number of data points. This will warn others that certain methods will fail under limited data point computation.

Temperature control

The rate of a chemical reaction is considerably more sensitive to temperature variation than the position of the equilibrium (provided the formation constant is very large and the reaction can be considered to be "quantitative"). Thus, temperature control is critical in reaction rate analytical methods and quantitative data concerning this control must be given in any publication describing a new method. Also, the effect of temperature fluctuation on the accuracy and precision of the analytical results should be discussed.

Two factors in temperature control must be considered. The accuracy of the temperature control in the thermostated jacket of the reaction cell is the initial consideration, of course.[45,46] However, as chemical reactions are either exothermic or endothermic, Pardue and Rodriguez,[24] Weichselbaum *et al.*,[47] and Feil *et al.*[48] have shown that rapid temperature exchange and equilibration of the reaction solution with the thermostated cell jacket is also an extremely important consideration in obtaining analytical-quality rate data.

Automation of operations

It is very obvious that automated control of solution mixing, measurement sequences, *etc*, will minimize time-measurement errors compared to those arising from manual control and solution handling and, hence, will increase the accuracy of a given reaction rate procedure.[23,25,29,32] Thus, a report of a new rate method should cover the variation of the analytical results under varying increase in automatic control, going from hand operation to complete computer control of operation and optimization.[29] Thus, again there will be warning about the experimental limits and operational controls under which the method is applicable. It would also be valuable to include a quantitative discussion of the relative merits and limitations of *continuous flow*[13,15] and *discontinuous-sampling* automated systems.[20,50–54]

CONCLUSIONS

Useful general guidance on the information that should be included in a published paper on an analytical method can be found in articles by Wilson,[55,56] Chalmers,[57,58] and others.[59,60,61] The specific requirements unique for the development and description of kinetic based analytical methods are discussed in detail above and are summarized here for convenience and emphasis. The points listed are those that *must* be examined and reported in publication of a new method.

1. *Mathematical basis of the method.* The complete reaction order, relative concentrations of reactants, identification of species followed during reaction, mathematical basis of data reduction and assumptions or simplifications should all be discussed. The selection of optimum conditions, such as the times of data collection, must be justified. Evidence must be given for the choice of data-reduction procedure.
2. *The reaction mechanism.* A complete description of the mechanism and reaction-order of all reactions involved in the kinetic method must be given wherever possible, as it is important to demonstrate the effect of any potential impurities on the reaction mechanism or path. The stability of reagent–solvent systems as well as synergistic effects related to concentration are also important.
3. *Trace impurities as catalysts or inhibitors.* It is necessary to examine the effects of trace amounts of all conceivable impurities that might be expected to be present in the types of samples that the method is designed for.
4. *Instrumental factors.* Any publication must include data concerning the long-term stability of the instrumentation employed, the linearity of the transducer response over the concentration ranges to be expected and its effect on the data-reduction procedure, the rise-time of the instrument with respect to desired data-taking rates, the number of data points taken and utilized in the data-reduction procedure, the temperature control of the reaction solution under dynamic conditions, and the effect of automation on accuracy of analyses.

The following points must also be covered in the publication of any new kinetic method. These points were not specifically discussed above in this paper as they are essential to the report on any type of new analytical method.

5. *Accuracy and precision.* As the accuracy and precision of kinetic-based methods is, in general, less than that of equilibrium techniques, it is essential that a statistical analysis of repetitive analyses be given. It would also be necessary to give statistical evidence in support of the choices of optimum measurement parameters. Simple theoretical considerations alone are generally not sufficient as they often neglect unsuspected effects.[1,9,10]

6. *Applications.* Procedures suitable for various sample matrices should at least be discussed. Also, matrices that are at least potentially unsuited for the method should be pointed out.
7. *General.* The value of the method must be realistically compared, in terms of speed, accuracy, selectivity, simplicity, cost, *etc*, with other methods for the same species.

Acknowledgement—This work was supported in part by the National Science Foundation, grant No. GP-27216.

REFERENCES

1. H. B. Mark, Jr., G. A. Rechnitz and R. A. Greinke, *Kinetics in Analytical Chemistry*, Wiley-Interscience, New York, 1968.
2. K. B. Yatsimerskii, *Kinetic Methods of Analysis*, Pergamon, Oxford, 1966.
3. G. G. Guilbault, *Enzymatic Methods of Analysis*, Pergamon, Oxford, 1970.
4. T. S. Lee and I. M. Kolthoff, *Ann. N.Y. Acad. Sci.*, 1951, **67**, 1484.
5. S. Siggia and J. G. Hanna, *Anal. Chem.*, 1961, **33**, 896.
6. J. G. Hanna and S. Siggia, *J. Polym Sci.*, 1962, **56**, 297.
7. *Idem*, *Anal. Chem.*, 1962, **34**, 547.
8. R. G. Garmon and C. N. Reilley, *ibid.*, 1962, **34**, 600.
9. H. B. Mark, Jr., L. J. Papa and C. N. Reilley, *Advances in Analytical Chemistry and Instrumentation*, Vol. 2, C. N. Reilley, Ed., pp. 255–385. Wiley-Interscience, New York, 1963.
10. L. J. Papa, H. B. Mark, Jr., and C. N. Reilley, *Anal. Chem.*, 1962, **34**, 1443.
11. G. G. Guilbault, *Crit. Rev. Anal. Chem.*, 1970, **1**, 377.
12. H. B. Mark, Jr., R. A. Greinke and L. J. Papa, *Proc. Soc. Anal. Chem. Conf., Nottingham, England*, 1965, p. 490.
13. W. J. Blaedel and G. P. Hicks, *Advances in Analytical Chemistry and Instrumentation*, Vol. 3, C. N. Reilley, Ed., p. 105. Wiley-Interscience, New York, 1964.
14. J. Janata, *Computers in Chemistry and Instrumentation*, Vol. 3, J. S. Mattson, H. B. Mark, Jr. and H. C. MacDonald, Jr., Eds., p. 209. Dekker, New York, 1972.
15. H. L. Pardue, *Advances in Analytical Chemistry and Instrumentation*, Vol. 7, C. N. Reilley and F. W. McLafferty, Eds., p. 141. Wiley-Interscience, New York, 1968.
16. G. A. Rechnitz, *Anal. Chem.* 1964, **36**, 453R; 1966, **38**, 513R; 1968, **40**, 455R.
17. E. O. Schmalz and G. Geiseler, *Z. Anal. Chem.*, 1962, **188**, 241, 253; 1962, **190**, 222, 233.
18. H. V. Bergmeyer, Ed., *Methods of Enzymatic Analysis*, 2nd Ed., Verlag Chemie, Mannheim, 1965.
19. R. Ruyssen and E. L. Vandenriesche, Eds., *Enzymes in Clinical Chemistry*, Elsevier, Amsterdam, 1965.
20. S. R. Crouch, *Computers in Chemistry and Instrumentation*, Vol. 3, J. S. Mattson, H. B. Mark, Jr., and H. C. MacDonald, Jr., Eds., p. 107. Dekker, New York, 1972.
21. G. G. Guilbault, *Anal. Chem.*, 1966, **38**, 527R; 1968, **40**, 459R; 1970, **42**, 334R.
22. R. A. Greinke and H. B. Mark, Jr., *ibid.*, 1972, **44**, 295R.
23. H. B. Mark, Jr., *Talanta*, 1972, **19**, 717.
24. H. L. Pardue and P. A. Rodriguez, *Anal. Chem.*, 1967, **39**, 901.
25. H. L. Pardue and S. N. Deming, *ibid.*, 1969, **41**, 986.
26. G. E. James and H. L. Pardue, *ibid.*, 1969, **41**, 1618.
27. G. D. Hicks, A. A. Eggert and E. C. Toren, Jr., *ibid.*, 1970, **42**, 729.
28. A. A. Eggert, G. P. Hicks and J. E. Davis, *ibid.*, 1971, **43**, 736.
29. S. N. Deming and H. L. Pardue, *ibid.*, 1971, **43**, 192.
30. B. G. Willis, W. H. Woodruff, J. R. Frysinger, D. W. Margerum and H. L. Pardue, *ibid.*, 1970, 42, 1350.
31. B. G. Willis, J. A. Bittikofer, H. L. Pardue and D. W. Margerum, *ibid.*, 1970, **42**, 1340.
32. T. E. Weichselbaum, W. H. Plumpe, Jr., R. E. Adams, J. C. Hagerty and H. B. Mark, Jr., *ibid.*, 1969, **41**, 725.
33. C. N. Reilley and L. J. Papa, *ibid.*, 1962, **34**, 801.
34. J. D. Ingle, Jr. and S. R. Crouch, *ibid.*, 1971, **43**, 697.
35. R. A. Greinke and H. B. Mark, Jr., *ibid.*, 1967, **39**, 1577.
36. S. Siggia and H. B. Mark, Jr., *ibid.*, 1963, **35**, 405.
37. R. A. Greinke and H. B. Mark, Jr., *ibid.*, 1966, **38**, 1001.
38. J. D. Robert and C. Regan, *ibid.*, 1952, **24**, 360.
39. C. N. Reilley, *J. Chem. Ed.*, 1962, **39**, A853.
40. L. J. Papa, J. H. Patterson, H. B. Mark, Jr. and C. N. Reilley, *Anal. Chem.*, 1963, **35**, 1889.
41. W. O. Ranky and D. C. Nelson, *Organic Sulfur Compounds*, Vol. 1, N. Kharasch, Ed., Chap. 17. Pergamon, Oxford, 1961.
42. T. E. Weichselbaum, W. H. Plumpe, Jr. and H. B. Mark, Jr., *Anal. Chem.*, 1968, **40**, 108A.
43. D. W. Margerum, J. B. Pausch, G. A. Nysseu, and G. F. Smith, *ibid.*, 1949, **41**, 233.
44. J. B. Pausch and D. W. Margerum, *ibid.*, 1969, **41**, 226.
45. M. Van Swaay, *J. Chem. Educ.*, 1969, **46**, A515.
46. *Idem*, *ibid.*, 1969, **46**, A565.
47. T. E. Weichselbaum, R. E. Smith, and H. B. Mark, Jr., *Anal. Chem.*, 1969, **41**, 1913.
48. P. D. Feil, D. G. Kubler, and D. J. Wells, Jr., *ibid.*, 1969, **41**, 1908.
49. M. K. Schwartz and O. Bodansky, *Methods of Biochemical Analysis*, Vol. XI, D. Glick, Ed., p. 211. Wiley-Interscience, New York, 1963.
50. H. V. Malmstadt and G. P. Hicks, *Anal. Chem.*, 1960, **32**, 445.
51. H. V. Malmstadt and H. L. Pardue, *ibid.*, 1962, **34**, 299.
52. A. C. Javier, S. R. Crouch and H. V. Malmstadt, *ibid.*, 1969, **41**, 239.
53. S. N. Deming and H. L. Pardue, *ibid.*, 1971, **43**, 192.
54. K. A. Mueller and M. F. Burke, *ibid.*, 1971, **43**, 641.
55. A. L. Wilson, *Talanta*, 1970, **17**, 21.
56. *Idem*, *ibid.*, 1970, **17**, 31.
57. R. A. Chalmers, *Crit. Revs. Anal. Chem.*, 1970, **1**, 217.
58. *Idem*, *Proc. Soc. Anal. Chem.*, 1969, **6**, 85.
59. L. Erdey, L. Pólos and R. A. Chalmers, *Talanta*, 1970, **17**, 1143.
60. G. J. Moody and J. D. R. Thomas, *ibid.*, 1972, **19**, 623.
61. C. F. Kirkbright, *ibid.*, 1963, **13**, 1.

Recommendations for Reporting Thermal Analysis Data

Sir: Because thermal analysis involves dynamic techniques, it is essential that all pertinent experimental detail accompany the actual experimental records to allow their critical assessment. This was emphasized by Newkirk and Simons (1) who offered some suggestions for the information required with curves obtained by thermogravimetry (TG). Publication of data obtained by other dynamic thermal methods, particularly differential thermal analysis (DTA), requires equal but occasionally different detail, and this letter is intended to present comprehensive recommendations regarding both DTA and TG.

In 1965 the First International Conference on Thermal Analysis (ICTA) established a Committee on Standardization charged with the task of studying how and where standardization might further the value of these methods. One area of concern was with the uniform reporting of data, in view of the profound lack of essential experimental information occurring in much of the thermal analysis literature. The following recommendations are now put forward by the Committee on Standardization, in the hope that authors, editors, and referees will be guided to give their readers full but concise detail. The actual format for communicating these details, of course, will depend upon a combination of the author's preference, the purpose for which the experiments are reported, and the policy of the particular publishing medium.

To accompany each DTA or TG record, the following information should be reported:

1. Identification of all substances (sample, reference, diluent) by a definitive name, an empirical formula, or equivalent compositional data.
2. A statement of the source of all substances, details of their histories, pretreatments, and chemical purities, so far as these are known.
3. Measurement of the average rate of linear temperature change over the temperature range involving the phenomena of interest.
4. Identification of the sample atmosphere by pressure, composition, and purity; whether the atmosphere is static, self-generated, or dynamic through or over the sample. Where applicable the ambient atmospheric pressure and humidity should be specified. If the pressure is other than atmospheric, full details of the method of control should be given.
5. A statement of the dimensions, geometry, and materials of the sample holder; the method of loading the sample where applicable.
6. Identification of the abscissa scale in terms of time or of temperature at a specified location. Time or temperature should be plotted to increase from left to right.
7. A statement of the methods used to identify intermediates or final products.
8. Faithful reproduction of all original records.
9. Wherever possible, each thermal effect should be identified and supplementary supporting evidence stated.

In the reporting of TG data, the following additional details are also necessary:

10. Identification of the thermobalance, including the location of the temperature-measuring thermocouple.
11. A statement of the sample weight and weight scale for the ordinate. Weight loss should be plotted as a downward trend and deviaitons from this practice should be clearly marked. Additional scales (e.g., fractional decompostion, molecular composition) may be used for the ordinate where desired.
12. If derivative thermogravimetry is employed, the method of obtaining the derivative should be indicated and the units of the ordinate specified.

When reporting DTA traces, these specific details should also be presented:

10. Sample weight and dilution of the sample.
11. Identification of the apparatus, including the geometry and materials of the thermocouples and the locations of the differential and temperature-measuring thermocouples.
12. The ordinate scale should indicate deflection per degree Centrigrade at a specified temperature. Preferred plotting will indicate upward deflection as a positive temperature differential, and downward deflection as a negative temperature differential, with respect to the reference. Deviations from this practice should be clearly marked.

Members of the Committee on Standardization of ICTA are: Professor C. Mazieres (France), Professor T. Sudo (Japan), Mr. R. S. Forsyth (Sweden), Mr. H. G. Wiedemann (Switzerland), Dr. I. S. Rassonskaya (U.S.S.R.), Mr. C. J. Keattch (United Kingdom), and Dr. P. D. Garn (United States). Other delegates to the Committee inclucde Professor L. G. Berg (U.S.S.R.), Dr. R. C. Mackenzie (United Kingdom), Dr. J. P. Redfern (United Kingdom), and Dr. S. Gordon (United States). The chairman is Dr. H. G. McAdie from Canada.

H. G. McAdie Ontario Research Foundation 43 Queen's Park Crescent East Toronto, 5, Canada

Received for review January 18, 1967. Accepted February 10, 1967.

(1) A. E. Newkirk and E. L. Simons, *Talanta,* **10,** 1199 (1963).

Suggestions for reporting dynamic thermogravimetric data

SIR:

THE increasing use of dynamic thermogravimetry has emphasised the need for workers in this field to use methods of reporting that, while concise, still contain all of the data that are essential to a critical evaluation and possible future reinterpretation of the results by other investigators. We have read many articles that could have included important additional information at the cost of no more than two or three lines of type; indeed, the space devoted to thermograms in published articles is often largely wasted, because the supporting data are incomplete.

A thermogram produced by a reliable recording thermobalance often shows certain features that are deliberately or inadvertently ignored in the reported discussion of it, either because their significance is not readily apparent, or because they are not considered germane to the principal line of investigation. This is particularly true of complex thermograms, which cannot easily be interpreted in complete detail, and for thermograms showing small weight changes of uncertain significance.

It is important that the published thermogram be as faithful a reproduction as possible of the original recording, so that any additional information is available to other workers. A well-prepared photocopy of the original record often makes the best illustration because it can be critically evaluated, even at the reduced scale of the usual journal reproduction, provided the experimental conditions are given. When it is necessary to trace the original thermogram, care should be taken to avoid either losing important detail or introducing inflections and curvatures not in the original.

The following eight pieces of supporting information should be given for each thermogram:

1. Identification of the substance examined by a definitive name, or an empirical formula, or equivalent compositional data; and the source of the substance.
2. The sample weight, and a weight scale for the ordinate.
3. Furnace, heating rate.
4. Atmosphere.
5. Size, shape, and material of the container.
6. Methods used to identify intermediates and final products.
7. Identification of the thermobalance, including the location of the thermocouple used for temperature measurements.
8. Identification of the abscissa scale in terms of time, furnace temperature, or sample temperature.

The best way to report this information is a matter of taste, and depends upon the purpose for which the thermogram is being shown. Thermograms of precipitates of interest in gravimetric analysis are used mainly to show whether plateaus of constant weight exist for the precipitate, for any intermediates, and for the final product of pyrolysis. The weight scale is usually given either directly or as a bar (parallel to the ordinate) whose length represents a known weight. Thermograms of hydrates and similar co-ordination compounds are often shown with the number of co-ordinated molecules as ordinate. For comparative purposes, such as studies of varieties of commercial plastics, it may be desirable to use as ordinate the percentage weight change, or the weight fraction of sample remaining. If there is a reasonable doubt about the interpretation, the weight scale should be shown as the ordinate, and the suggested interpretation should be indicated by fiducial marks and a formula near the point of interest. An open grid of lines at the major scale divisions is also helpful, particularly near plateaus, in revealing gentle changes in slope.

ARTHUR E. NEWKIRK
EDWARD L. SIMONS

General Electric Research Laboratory
P.O. Box 1088
Schenectady, New York, U.S.A.
21 August 1963

RECOMMENDATIONS ON NOMENCLATURE FOR CHROMATOGRAPHY

RULES APPROVED 1973

ANALYTICAL CHEMISTRY DIVISION
COMMISSION ON ANALYTICAL NOMENCLATURE

Nearly ten years ago the Division of Analytical Chemistry approved a set of recommendations for the nomenclature of Gas Chromatography[1]. Since then the Commission on Nomenclature has been endeavouring to produce a unified nomenclature applicable to all forms of separation processes, and proposals have now been made for Liquid–Liquid Distribution[2] and for Ion Exchange[3]. In the present proposals prepared for the Commission by Dr D. Ambrose, Professor E. Bayer and Professor O. Samuelson, the work has been extended to all forms of chromatography. For the sake of uniformity, compromises have inevitably had to be made, as a result of which, for example, there are some changes from the recommendations in Ref. 1. Account was taken, in the drafting, of other relevant proposals[4–7].

It is recommended that quantities should be expressed in the units (or their multiples or submultiples) of the International System of Units, or in the units approved for use with the International System[8]; in particular, that physical dimensions, e.g. of columns, should be so expressed. It is to be noted that the symbol T relates to thermodynamic temperatures and should not be used to represent temperatures expressed on the Celsius scale[9].

1 CHROMATOGRAPHY

A method, used primarily for separation of the components of a sample, in which the components are distributed between two phases, one of which is stationary while the other moves. The stationary phase may be a solid, or a liquid supported on a solid, or a gel. The stationary phase may be packed in a *column*, spread as a *layer*, or distributed as a *film*, etc.; in these definitions *chromatographic bed* is used as a general term to denote any of the different forms in which the stationary phase may be used. The mobile phase may be gaseous or liquid.

2 PRINCIPAL METHODS

2.1 Frontal chromatography
A procedure for chromatographic separation in which the sample (liquid or gas) is fed continuously into the chromatographic bed.

2.2 Elution chromatography
A procedure for chromatographic separation in which an *eluent* (see Item 8.6) is passed through the chromatographic bed after the application of the sample.

2.3 Displacement chromatography
An elution procedure in which the *eluent* contains a compound more effectively retained than the components of the sample under examination.

3 CLASSIFICATION ACCORDING TO PHASES USED

In this classification, the first word specifies the mobile phase and the second the stationary phase. A liquid stationary phase is supported on a solid.

3.1 Gas chromatography (GC)

3.1.1 Gas–liquid chromatography (GLC)

3.1.2 Gas–solid chromatography (GSC)

3.2 Liquid chromatography (LC)

3.2.1 Liquid–liquid chromatography (LLC)

3.2.2 Liquid–solid chromatography (LSC)

3.2.3 Liquid–gel chromatography

In gas chromatography the distinction between gas–liquid and gas–solid may be obscure because liquids are used to modify solid stationary phases, and because the solid supports for liquid stationary phases affect the chromatographic process. For classification by the phases used, the term relating to the predominant effect should be chosen. Liquid–gel chromatography includes gel-permeation and ion-exchange chromatography.

4 CLASSIFICATION ACCORDING TO MECHANISMS

4.1 Adsorption chromatography
Separation based mainly on differences between the adsorption affinities of the components for the surface of an active solid.

4.2 Partition chromatography
Separation based mainly on differences between the solubilities of the components in the stationary phase (gas chromatography), or on differences between the solubilities of the components in the mobile and stationary phases (liquid chromatography).

4.3 Ion-exchange chromatography
Separation based mainly on differences in the ion-exchange affinities of the components.

4.4 Permeation chromatography
Separation based mainly upon exclusion effects, such as differences in molecular size and/or shape (e.g. molecular-sieve chromatography) or in charge (e.g. ion-exclusion chromatography). The term *gel-permeation chromatography* is widely used for the process when the stationary phase is a swollen gel. The term *gel-filtration* is not recommended.

4.5 Other mechanisms
In addition to Items 4.1 to 4.4, there exist many techniques based upon other mechanisms. Examples are ligand-exchange, formation of charge-transfer complexes, and bio-specific sorption, e.g. formation of enzyme–substrate and antigen–antibody complexes. Classification according to mechanism should be avoided unless the predominant mechanism is known. In many instances more than one mechanism is involved.

5 CLASSIFICATION ACCORDING TO TECHNIQUES USED

All types of chromatography can be classified according to Section 3 by the phases used or according to Section 4 by mechanism, but the terms in this section specify techniques and may provide a more useful characterization of the process.

5.1 Column chromatography (CC)

5.2 Open-tube chromatography (see Item 8.4)

5.3 Paper chromatography (PC)

5.4 Thin-layer chromatography (TLC)
Chromatography carried out in a layer of adsorbent spread on a support, e.g. a glass plate.

5.5 Filament chromatography

6 TERMS FOR SPECIAL TECHNIQUES

6.1 Temperature-programmed chromatography
A procedure in which the temperature of the column is changed systematically during a part or the whole of the separation.

6.2 Flow-programmed chromatography
A procedure in which the rate of flow of the mobile phase is changed systematically during a part or the whole of the separation.

6.3 Salting-out chromatography
A procedure in which a non-sorbable electrolyte is added to the eluent to modify the distribution equilibria of the components to be separated.

6.4 Selective elution
An elution procedure in which a specific eluent is used, e.g. a complexing agent that forms stable non-sorbable complexes with one or a group of the compounds to be separated, but affects the other components only to a negligible extent.

6.5 Stepwise elution
An elution procedure in which two or more eluents of different composition are used in succession to elute the components in a single chromatographic run.

6.6 Gradient elution
An elution procedure in which the eluent composition is changed continuously.

6.7 Two-dimensional chromatography
A procedure applied in paper chromatography and thin-layer chromatography in which the components are caused first to migrate in one direction, and subsequently in a direction at right angles to the first one. The two elutions are usually carried out with different eluents.

6.8 Reversed-phase chromatography
A term of historical interest in liquid–liquid chromatography referring to an elution procedure in which the stationary phase is non-polar, e.g. paper treated with hydrocarbons or silicones.

7 TERMS RELATING TO THE METHOD IN GENERAL

7.1 Chromatogram
A graphical or other presentation of detector response, effluent concentration, or other quantity used as a measure of effluent concentration, versus effluent volume or time. The term is also applied to the layer or paper after separation has occurred.

7.2 Elution curve
A chromatogram, or part of a chromatogram, recorded when elution techniques are used.

7.3 Chromatograph (verb)
To separate by chromatography.

7.4 Chromatograph (noun)
The assembly of apparatus for carrying out chromatographic separation.

7.5 Elute
To chromatograph by elution chromatography. This term is preferred to the term *develop*, which has been used in paper chromatography and thin-layer chromatography. The process of elution may continue until the components have left the chromatographic bed.

7.6 Extract
To recover a compound from a chromatographic zone by treatment with a solvent.

7.7 Zone
A region in a chromatographic column or layer where one or more components of the sample are located.

7.8 Spot
A zone in paper and thin-layer chromatography of approximately circular appearance.

7.9 Starting point or line
The point or line on a chromatographic layer where the substance to be chromatographed is applied.

7.10 Baseline
The portion of a chromatogram recorded when only eluent or carrier gas emerges from the column.

7.11 Peak

The portion of a differential chromatogram (See Item 8.20) recording the detector response or eluate concentration (See Item 8.18) while a component emerges from the column (*Figure 1*). If separation is incomplete, two or more components may appear as one *unresolved peak.*

7.12 Elution band

Synonymous with peak.

7.13 Tailing

Asymmetry of a peak such that, relative to the baseline, the front is steeper than the rear. In paper chromatography and thin-layer chromatography, the distortion of a zone showing a diffuse region behind the zone in the direction of flow.

7.14 Fronting

Asymmetry of a peak such that, relative to the baseline, the front is less steep than the rear. In paper chromatography and thin-layer chromatography, the distortion of a zone showing a diffuse region in front of the zone in the direction of flow.

7.15 Step (on an integral chromatogram)

The portion of an integral chromatogram (See Item 8.21) recording the amount of a component, or the corresponding change in the signal from the detector as the component emerges from the column (*Figure 1*).

7.16 Step height (on an integral chromatogram)

The distance (KL, *Figure 1*), perpendicular to the time or volume axis, through which the baseline moves as the result of a step on an integral chromatogram (See Item 8.21).

7.17 Internal standard

A compound added to a sample in known concentration, for example, for

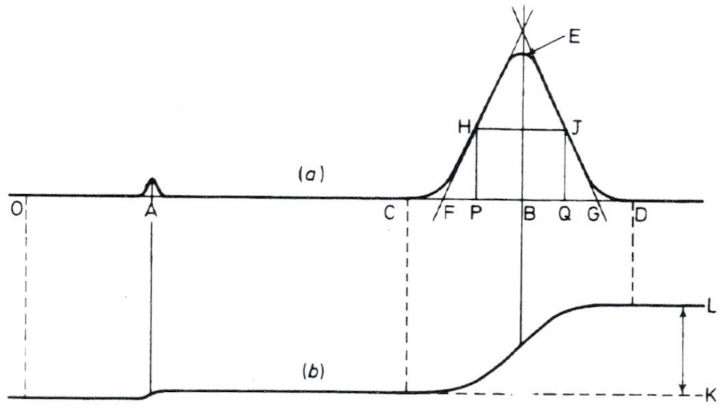

Figure 1. (a) Differential, and (b) integral chromatograms

the purpose of eliminating the need to measure the size of sample in quantitative analysis.

7.18 Marker

A reference substance chromatographed with the sample to assist in identifying the components.

8 TERMS RELATING TO THE SEPARATION PROCESS AND THE APPARATUS

8.1 Column

The tube that contains the stationary phase, and through which the mobile phase passes.

8.2 Packing

The active solid, stationary liquid plus solid support, or swollen gel put in the column. The term *packing* refers to the conditions existing before the chromatographic run is started (i.e. to the material introduced into the column) whereas the *stationary phase* (see Item 8.8) refers to the conditions during the run.

8.3 Packed column
A column filled with packing.

8.4 Open tubular column
A column, usually of capillary dimensions, in which the column wall, a liquid or an active solid supported on the column wall acts as the stationary phase.

8.5 Mobile phase
The phase that is moving in the chromatographic bed. It includes the fraction of the sample present in this phase.

8.6 Eluent
The liquid or gas entering the chromatographic bed and used to effect a separation by elution.

8.7 Carrier gas
The term normally used for the *eluent* in gas chromatography.

8.8 Stationary phase
The non-mobile phase in the chromatographic bed, on which the separation depends. For example, in gas–solid chromatography and liquid–solid chromatography the active solid is the stationary phase, and in gas–liquid and liquid–liquid chromatography the liquid, but not the solid support, is the stationary phase.

8.9 Active solid
A solid with sorptive properties.

8.10 Modified active solid
An active solid, the adsorptive properties of which have been changed by treatment with a gas, liquid or another solid.

8.11 Solid support
A solid that holds the stationary liquid phase.

8.12 Support plate
The plate that supports the thin layer in thin-layer chromatography.

8.13 Gradient layer or gradient packing
A layer or column packing with continuous change of property affecting the separation, e.g. pH gradient.

8.14 Sample injector
A device by which a sample is introduced into the eluent (carrier gas) or the column.

8.15 Bypass injector
A sample injector by means of which the eluent (carrier gas) may be temporarily diverted through a sample chamber so that the sample is carried to the column.

8.16 Chamber saturation
Uniform distribution of the eluent vapour throughout the chamber prior to chromatography.

8.17 Layer equilibration
Saturation of the stationary phase with the mobile phase via the vapour phase.

8.18 Eluate
The effluent from a chromatographic bed emerging when elution is carried out.

8.19 Detection
The process by which the presence of chromatographically separated substances is recognized.

8.20 Differential detector
A detector whose response is dependent on the instantaneous difference in composition between the column effluent and the eluent (carrier gas).

8.21 Integral detector
A detector whose response is dependent on the total amount of a component that has passed through it.

8.22 Solvent front
The front line of the eluent.

8.23 Solvent migration-distance
The distance travelled by the solvent front.

8.24 Separation temperature (often column temperature in column chromatography)
The temperature of the chromatographic bed.

8.25 Injection temperature
The temperature at the injection point.

8.26 Initial and final temperatures
The range of separation temperatures in temperature-programmed chromatography.

9 TERMS RELATING TO QUANTITATIVE EVALUATION AND THE THEORY OF CHROMATOGRAPHY

9.1 Column volume, X
The volume (empty) of the part of a column that contains the packing. It is recommended that the column dimensions be given as the inner diameter and the height or length of the column occupied by the stationary phase under the applied chromatographic conditions. If swelling changes occur, the conditions under which the height is determined should be specified.

9.2 Bed volume
Synonymous with *column volume* for a packed column.

9.3 Interstitial volume, V_I
The volume occupied by the mobile phase in the packed section of a column. In gas chromatography the gas occupying the interstitial volume expands to a volume V_I/j at the outlet pressure, where measurements are normally made (see Item 9.11).

9.4 Interstitial fraction, ε_I
The interstitial volume per unit volume of a packed column:

$$\varepsilon_I = V_I/X$$

9.5 Volume of the stationary phase, V_S
The volume of the stationary liquid phase or of the active solid or of the gel in the column. The volume of any solid support is not included.

9.6 Stationary-phase fraction, ε_S
The volume of the stationary phase per unit volume of a packed column:

$$\varepsilon_S = V_S/X$$

9.7 Phase ratio
The ratio of the volume of the mobile phase to that of the stationary phase in a column.

9.8 Hold-up volume, V_M
The volume of eluent required to elute a component the concentration of which in the stationary phase is negligible compared to that in the mobile phase. The *hold-up volume* corresponds to the distance OA, *Figure 1*, and includes any volumes contributed by the sample injector and the detector.

9.9 Gas hold-up volume, V_M
Synonymous with *hold-up volume* in gas chromatography. The volume of carrier gas (eluent) is specified at the same temperature and pressure as the total retention volume (see Item 9.23).

9.10 Dead volume, V_d
The volume between the effective injection point and the effective detection point, less the column volume *X*.

9.11 Pressure-gradient correction-factor, *j*
A factor, applying to a homogeneously filled column of uniform diameter, that corrects for the compressibility of the mobile phase; the values of the measured quantities obtained after multiplication by the factor *j* are independent of the pressure drop in the column. In practice, these quantities include contributions arising from the column inhomogeneities making up the dead volume but since these are small in comparison with retention volumes the consequent errors are normally ignored. In gas chromatography, if p_i, p_o are respectively the pressures of the carrier gas at the inlet and outlet of the column:

$$j = \frac{[3\ (p_i/p_o)^2 - 1]}{[2\ (p_i/p_o)^3 - 1]}$$

9.12 Peak base
The interpolation, in a differential chromatogram, of the baseline between the extremities of the peak (the line CD, *Figure 1*).

9.13 Peak area
The area (CHEJD, *Figure 1*) enclosed between the peak and the peak base.

9.14 Peak maximum
The point on the peak at which the distance to the peak base, measured in a direction parallel to the axis representing detector response, is a maximum (E, *Figure 1*).

9.15 Peak height
The distance between the peak maximum and the peak base, measured in a direction parallel to the axis representing detector response (the distance BE, *Figure 1*).

9.16 Peak width
The segment of the peak base intercepted by tangents to the inflection points on either side of the peak (the distance FG, *Figure 1*) projected onto the axis representing time or volume if the baseline is not parallel to this axis.

9.17 Peak width at half height
The length of the line parallel to the peak base that bisects the peak height and terminates at the intersections with the two limbs of the peak (the distance PQ, *Figure 1*) projected onto the axis representing time or volume if the baseline is not parallel to this axis.

9.18 Volumetric flowrate, F_c
The volumetric flowrate of the mobile phase ($cm^3\ min^{-1}$). In gas chromatography, the flowrate is normally specified at the column temperature and outlet pressure, although the measurement may be made at ambient temperature and must be corrected accordingly (and possibly also for water vapour present in the flowmeter).

9.19 Nominal linear flow, *F*
The volumetric flowrate of the mobile phase divided by the area of the cross section of the column ($cm\ min^{-1}$), i.e. the linear flowrate in a part of the column not containing packing.

9.20 Interstitial velocity, *u* (u_o at the outlet pressure in gas chromatography)
The linear velocity of the mobile phase inside a packed column calculated as the average over the entire cross section. This quantity can, under idealized conditions, be calculated from the equation,

$$u = F/\varepsilon_1$$

9.21 Mean interstitial velocity of the carrier gas, $\bar{u}$
The interstitial velocity of the carrier gas multiplied by the pressure-gradient correction-factor:

$$\bar{u} = Fj/\varepsilon_1$$

9.22 Retention volumes

Retention measurements (and measurements of hold-up volume and peak width) may be made in terms of times, e.g. t_R, t'_R analogous to V_R, V'_R (see Items 9.23 and 9.25), or chart distances as well as volumes. If flow and recorder speeds are constant, the volumes are directly proportional to the times and chart distances. The definitions given here are drawn up in terms of volume, and it is recommended that theoretical discussion should be couched in the same terms whenever possible. However, the proportionality between volumes, times and chart distances is implied in the references to *Figures 1* and *2*.

9.23 Total retention volume, V_R

The volume of eluent (carrier gas) entering the column between the injection of the sample and the emergence of the peak maximum of the specified component (OB, *Figure 1*). It includes the hold-up volume. In gas chromatography the volume of carrier gas is specified at the outlet pressure and the temperature of the column.

Note:

The word *total* in this definition allows *retention volume* to be used as a general term when specification of a particular quantity is not required.

9.24 Peak elution volume, $\bar{V}$

The volume of eluent entering the column between the start of the elution and the emergence of the peak maximum. The term applies only to liquid chromatography. It does not include the effluent obtained when the sample is introduced into the column nor the volume of the detector, if used.

Sometimes the column is washed with a liquid, before the elution is started, but after application of the sample, to displace components that are not retained. The effluent obtained during the washing process is not included in the peak elution volume unless the solutes are moved during the washing (see Item 6.5).

9.25 Adjusted retention volume, V'_R

The total retention volume less the hold-up volume (corresponding to the distance AB, *Figure 1*), i.e.

$$V'_R = V_R - V_M = \bar{V} - V_I$$

9.26 Net retention volume, V_N

The adjusted retention volume multiplied by the pressure gradient correction factor:

$$V_N = jV'_R$$

9.27 Specific retention volume, V_g

The net retention volume per gram of stationary liquid, active solid or solvent-free gel. In liquid chromatography, except when conducted at very high pressures, the compression of the mobile phase is negligible, and the adjusted and net retention volumes are identical; the specific retention volume is then the adjusted retention volume per gram of stationary liquid, active solid or solvent-free gel. It is recommended that, when appropriate, authors specify the drying conditions.

9.28 Relative retention, $r_{A/B}$

The adjusted retention volume of a substance related to that of a reference compound obtained under identical conditions. If subscripts A and B refer to the substance and the reference compound respectively, then

$$r_{A/B} = \frac{V_{g,A}}{V_{g,B}} = \frac{V_{N,A}}{V_{N,B}} = \frac{V'_{R,A}}{V'_{R,B}}$$

Note that $r_{A/B}$ is not equal to $V_{R,A}/V_{R,B}$ nor $\bar{V}_A/\bar{V}_B$.

9.29 Retention temperature

The column temperature (see Item 8.24) when the peak maximum for a component has been reached in temperature-programmed chromatography.

9.30 R_f value

The ratio of distance travelled by the centre of a zone to the distance simultaneously travelled by the mobile phase. In paper and thin-layer chromatography, R_f may be determined from the distance moved by the eluent front.

9.31 R_B value

The ratio of the distance travelled by a zone to the distance simultaneously travelled by a reference substance B.

9.32 Distribution constant, K_D

The ratio of the concentration of a component in a single definite form in the stationary phase to its concentration in the same form in the mobile phase at equilibrium. Both concentrations are calculated per unit volume of the phase.

This term is recommended in preference to *partition coefficient* which has been used with the same meaning.

In chromatography a component may be present in more than one form; these forms are generally not specified (and may not be known), and it will therefore usually be more appropriate for specification of conditions in the column to use one of the following terms, which are defined by the analytical concentration (or amount) of the component, the analytical concentration (or amount) referring to its total concentration (or amount) without regard to its possible existence in associated or dissociated forms.

9.33 Concentration distribution ratio, D_c

The ratio of the analytical concentration of a component in the stationary phase to its analytical concentration in the mobile phase:

$$D_c = \frac{\text{amount of component/cm}^3\text{ of stationary phase}}{\text{amount of component/cm}^3\text{ of mobile phase}}$$

9.34 Distribution coefficients, D_g, D_v, D_s

The amount of a component in a specified amount of stationary phase, or in an amount of stationary phase of specified surface area, divided by the analytical concentration in the mobile phase. The subscripts g, v, s indicate as follows the way in which the stationary phase is specified:

$$D_g = \frac{\text{amount of component/gram of dry stationary phase}}{\text{amount of component/cm}^3\text{ of mobile phase}}$$

applicable in ion-exchange and gel chromatography, where swelling occurs, and in adsorption chromatography with adsorbents of unknown surface area,

$$D_v = \frac{\text{amount of component in the stationary phase/cm}^3\text{ of bed volume}}{\text{amount of component/cm}^3\text{ of mobile phase}}$$

applicable when it is not practicable to determine the weight of the solid phase, and

$$D_s = \frac{\text{amount of component/m}^2\text{ of surface}}{\text{amount of component/cm}^3\text{ of mobile phase}}$$

applicable in adsorption chromatography with a well-characterized adsorbent of known surface area.

9.35 Mass distribution ratio, D_m

The fraction $(1 - R)$ of a component in the stationary phase divided by the fraction (R) in the mobile phase:

$$D_m = \frac{\text{amount of component in stationary phase}}{\text{amount of component in mobile phase}}$$

This term is recommended in preference to the term *capacity factor* frequently used in gas chromatographic literature.

Note:

The subscripts in D_c, D_m, D_g, D_v, D_s may be omitted when there is no possibility of confusion of one term with another.

Values of these quantities, defined in Items 9.33–9.35, which allow the equilibrium between two phases to be specified, may be determined by static equilibrium measurements. They may also be related to retention volumes, and measurements of the latter frequently provide the most convenient experimental route for their determination.

9.36 Separation factor, $\alpha_{A/B}$
The ratio of the distribution ratios or coefficients, D_A/D_B for two substances A and B measured under identical conditions. By convention α is usually greater than unity.

9.37 Peak resolution, R_s
The separation of two peaks in terms of their average peak width (*Figure 2*):

$$R_s = 2y/(y_A + y_B)$$

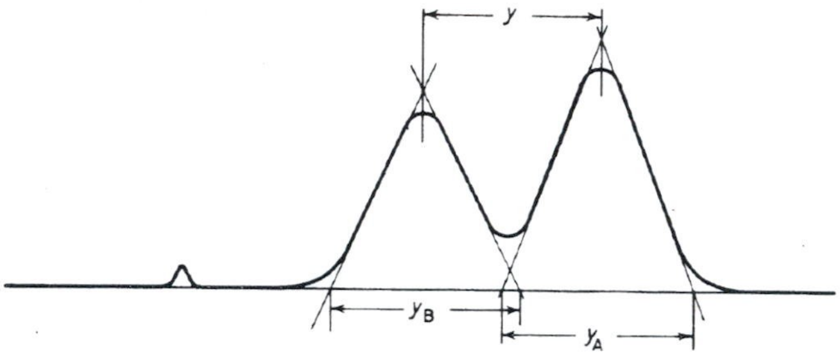

Figure 2

9.38 Theoretical plate number, n
A number indicative of column performance calculated from the equation

$$n = 16\ (\text{Peak elution volume/peak width})^2$$

In gas chromatography and some types of liquid chromatography the volumes of the sample injector and of the detector are negligible, and the expression for n can then be written as

$$n = 16\ (\text{Total retention volume/peak width})^2$$

In these expressions the units for the quantities inside the brackets must be consistent so that their ratio is dimensionless, i.e. if the numerator is a volume then peak width must be expressed in terms of volume also.

9.39 Effective theoretical plate number, N
A number indicative of column performance when resolution is taken into account:

$$N = 16\ R_s^2/(1 - \alpha)^2$$

9.40 Height equivalent to a theoretical plate, HETP, h
The column length divided by the theoretical plate number.

9.41 Height equivalent to an effective theoretical plate, HEETP, H
The column length divided by the effective theoretical plate number.

9.42 Retention index, I
A number, obtained by logarithmic interpolation, relating the adjusted retention volume of a component A to the adjusted retention volumes of the normal paraffins. Each n-paraffin is arbitrarily alotted by definition an index one hundred times its carbon number. The index I_A of substance A is then given by

$$I_A = 100N + 100n\left(\frac{\log V'_R(A) - \log V'_R(N)}{\log V'_R(N+n) - \log V'_R(N)}\right)$$

where $V'_R(N+n)$ and $V'_R(N)$ are the adjusted retention volumes of n-paraffins of carbon number N and $(N+n)$ that are respectively smaller and larger than $V'_R(A)$ the adjusted retention volume of A.

APPENDIX. LIST OF SYMBOLS

Symbol	*Item no.*	*Meaning*
A, B	9.28, 9.36	Components A and B
D_c	9.33	Concentration distribution ratio

Symbol	*Item no.*	*Meaning*
D_g	9.34	Distribution coefficient
D_m	9.35	Mass distribution ratio
D_s	9.34	Distribution coefficient
D_v	9.34	Distribution coefficient
F	9.19	Nominal linear flow
F_c	9.18	Volumetric flowrate
H	9.41	Height equivalent to an effective theoretical plate, HEETP
h	9.40	Height equivalent to a theoretical plate, HETP
I	9.42	Retention index
j	9.11	Pressure-gradient correction-factor
K_D	9.32	Distribution constant
N	9.39	Effective theoretical plate number
n	9.38	Theoretical plate number
p_i, p_o	9.11	Pressure of carrier gas at inlet and outlet of column
R	9.35	Fraction of component in the mobile phase
R_f	9.30	R_f value
R_B	9.31	R_B value
R_s	9.37	Peak resolution
$r_{A/B}$	9.28	Relative retention (of component A relative to component B)
u	9.20	Interstitial velocity
$\bar{u}$	9.21	Mean interstitial velocity
u_o	9.20	Interstitial velocity at the column outlet
$\overline{V}$	9.24	Peak elution volume
V_d	9.10	Dead volume
V_g	9.27	Specific retention volume
V_I	9.3	Interstitial volume
V_M	9.8, 9.9	Hold-up volume, gas hold-up volume
V_N	9.26	Net retention volume
V_R	9.23	Total retention volume
V'_R	9.25	Adjusted retention volume
V_S	9.5	Volume of the stationary phase
X	9.1	Column volume
$\alpha_{A/B}$	9.36	Separation factor
ε_I	9.4	Interstitial fraction
ε_S	9.6	Stationary-phase fraction

REFERENCES

[1] Recommendations on Nomenclature and Presentation of Data in Gas Chromatography. *Pure Appl. Chem.* **8**, 553 (1964).

[2] Recommended Nomenclature for Liquid–Liquid Distribution. *Pure Appl. Chem.* **21**, 111 (1970).

[3] Recommendations on Ion-Exchange Nomenclature. *Pure Appl. Chem.* **29**, 619 (1972).

[4] Recommended Practice for Gas Chromatography Terms and Relationships, *ASTM E355–68*. American Society for Testing and Materials: Philadelphia (1968); *J. Gas Chromatogr.* **6**, 1 (1968).

[5] E. Bayer *et al.*, *Chromatographia*, **1**, 153 (1968).

[6] E. Stahl, *Chromatographia*, **1**, 338 (1968).

[7] *British Standard 3282*, Glossary of Terms Relating to Gas Chromatography. British Standards Institution: London (1969).

[8] *Le Système International d'Unités*. Offilib; 48 rue Gay-Lussac, F 75 Paris 5 (1970); *SI The International System of Units*. HMSO: London (1970); *The International System of Units* (SI), *NBS Spec. Publ. 330*. National Bureau of Standards: Washington DC (1972).

[9] Manual of Symbols and Terminology for Physicochemical Quantities and Units. *Pure Appl. Chem.* **21**, 1 (1970).

RECOMMENDED TERMS, SYMBOLS, AND DEFINITIONS FOR ELECTROANALYTICAL CHEMISTRY

(Recommendations 1985)

COMMISSION ON ELECTROANALYTICAL CHEMISTRY*

L. MEITES[1], P. ZUMAN[2] and H. W. NÜRNBERG[3]†

[1]George Mason University, Fairfax, Virginia, USA
[2]Clarkson University, Potsdam, New York, USA
[3]Kernforschungsanlage Jülich GmbH, Jülich, FRG

INTRODUCTION

This document contains definitions and recommendations for terminology and the usage of symbols in electroanalytical chemistry. Its scope is the same as that of the 1975 rules on "Classification and Nomenclature of Electroanalytical Techniques," [Pure and Appl. Chem., 45 (1976) 81-97] and it supplements that document by giving explanations and definitions of the terms and symbols associated with the electroanalytical techniques there classified and named. The names given to specific techniques here are accompanied by numbers [e.g., "(4.21)"] that serve as cross-references to that document. Such numbers are always enclosed in parentheses.

Some of those techniques, including conductometry and potentiometry among others, are important to scientists in other areas as well as to electroanalytical chemists, and many terms, definitions, and symbols appropriate to such techniques may be found in Appendix III to the Manual of Symbols and Terminology for Physicochemical Quantities and Units [Pure and Appl. Chem., 37(4) (1974), 501] prepared by R. Parsons for Commission I.3 on Electrochemistry. Since material contained in Appendix III has not been duplicated here, both documents are needed to provide a complete list of the recommended terms, symbols, and definitions that are of importance to electroanalytical chemists. Some cross-references to Appendix III are given in the second column below and are always enclosed in square brackets; for example, the entry "[2.1]" under "n" below refers to the definition of "charge number of the cell reaction" in Appendix III.

*Membership of the Commission for 1977–83 during which period the recommendations were prepared was as follows:

Chairman: 1977–79 R. G. Bates (USA); 1979–81 J. F. Coetzee (USA); 1981–83 J. Jordan (USA); *Secretary*: 1977–79 J. F. Coetzee (USA); 1979–81 J. Jordan (USA); 1981–83 K. Izutsu (Japan); *Titular and Associate Members*: E. Bishop (UK; Associate 1977–79); J. F. Coetzee (USA; Associate 1981–83); A. K. Covington (UK; Titular 1977–83); W. Davison (UK; Associate 1981–83); R. A. Durst (USA; Associate 1979–83); T. Fujinaga (Japan; Titular 1977–79); Z. Galus (Poland; Titular 1977–79); L. Gierst (Belgium; Associate 1977–79); M. Gross (France; Associate 1979–83); K. Izutsu (Japan; Titular 1977–81); J. Jordan (USA; Titular 1977–79); J. Juillard (France; Associate 1977–79; Titular 1979–83); K. M. Kadish (USA; Associate 1979–83); R. Kalvoda (Czechoslovakia; Associate 1977–83); H. Kao (China; Associate 1981–83); R. C. Kapoor (India; Titular 1977–83); Y. Marcus (Israel; Associate 1979–83); L. Meites (USA; Associate 1977–81); T. Mussini (Italy; Associate 1979–83); H. W. Nürnberg† (FRG; Associate 1979–83); P. Papoff (Italy; Associate 1977–81); E. Pungor (Hungary; Titular 1979–83); M. Senda (Japan; Associate 1979–83); D. E. Smith (USA; Associate 1979–83); W. F. Smyth (Ireland; Associate 1977–79); O. Songina (USSR; Associate 1977–79); N. Tanaka (Japan; Associate 1979–83); *National Representatives*: D. D. Perrin (Australia; 1977–83); B. Gilbert (Belgium; 1981–83); W. C. Purdy (Canada; 1979–83); R. Neeb (FRG; 1977–83); G. Farsang (Hungary; 1977–79); K. Tóth (Hungary; 1979–83); H. V. K. Udupa (India; 1977–81); S. K. Rangarajan (India; 1981–83); W. F. Smyth (Ireland; 1981–83); E. Grushka (Israel; 1981–83); N. Tanaka (Japan; 1977–79); W. Kemula (Poland; 1977–79); Z. Galus (Poland; 1979–83); G. Johansson (Sweden; 1981–83); J. Buffle (Switzerland; 1981–83); B. Birch (UK; 1977–83); J. Osteryoung (USA; 1981–83); M. Branica (Yugoslavia; 1977–83).

†Deceased 12 May 1985

The terms listed in this document are given in alphabetical order, and each is given a number for ease of cross-reference. Many of the terms are accompanied by two or more symbols. The first of these is always the recommended one; any other not enclosed in parentheses is considered to be an acceptable alternative. Parentheses are used to denote symbols which have often been used in the past but of which the future use should be discouraged.

The use of the symbol $\underline{I}$ to denote electric current has been recommended by the Union and other international scientific organizations. The lower case $\underline{i}$ is preferred here, not only because it has been (and continues to be) far more widely used in the electroanalytical literature, but also to avoid confusion with the polarographic diffusion current constant (q.v.), which has always been denoted by the symbol $\underline{I}$. It may be hoped that electroanalytical chemists will eventually adopt the symbol $\underline{I}$ to denote electric current. This would be facilitated by using the symbol $\underline{\mathcal{I}}$ to denote the polarographic diffusion current constant and similarly using script capital letters to denote the fundamental response constants of other techniques — such as the chronocoulometric, chronopotentiometric, and voltammetric constants (q.v.).

For each recommendation the appropriate SI unit is given. Multiples and submultiples of these units are equally acceptable, and are often more convenient. For example, although the SI unit of concentration is mol m^{-3}, concentrations are frequently expressed in mol dm^{-3}, mol cm^{-3}, or mmol dm^{-3}. Similarly, although the SI unit of the chronopotentiometric constant defined below is A $s^{1/2}$ mol^{-1} m, nearly all of the chronopotentiometric constants reported in the literature employ the practical unit mA $s^{1/2}$ cm^{-2} (mol $dm^{-3})^{-1}$, or A $s^{1/2}$ $mmol^{-1}$ cm.

Recommended name and symbol; SI unit	Definition and remarks
1. Apex	See "12. Current, apex."
2. Area (of an electrode-solution interface) $\underline{A}$ m^2 [5.7]	In these definitions the area of an electrode-solution interface is understood to be the geometrical or projected area, and to ignore surface roughness.
3. Characteristic potential (no symbol recommended) V	An applied potential that is characteristic of a charge-transfer process and the experimental conditions (such as the composition of the solvent and supporting electrolyte and the temperature) under which it is investigated, and whose nature depends on the technique that is employed. Some typical characteristic potentials are the half-wave potential (cf. 60) in polarography (5.12), the quarter-transition-time potential (cf. 62) in chronopotentiometry (4.12), the peak potential (cf. 61) and the half-peak potential (cf. 59) in linear-sweep voltammetry (5.8), and the summit potential (cf. 63) in ac polarography (6.4).
4. Chronocoulometric constant $\mathcal{Q}$ (script capital Q) A $s^{1/2}$ mol^{-1} m	In chronocoulometry (4.12), the empirically evaluated quantity defined by the equation $\mathcal{Q} = \frac{1}{\underline{A}\,\underline{c}} \left(\frac{\Delta \underline{Q}}{\Delta \underline{t}^{1/2}}\right)$ where $\underline{A}$ = area of the electrode-solution interface (cf. 2), $\underline{c}$ = bulk concentration (cf. 6) of the electroactive substance, and $\Delta\underline{Q}/\Delta\underline{t}^{1/2}$ = the slope of a plot of $\underline{Q}$ against $\underline{t}^{1/2}$.

Note: An explanation has been provided in INTRODUCTION as to why, for the present, the symbol for electric current, lower case *i*, is preferred in electroanalytical chemistry rather than the symbol capital *I* recommended by IUPAC and other international scientific organizations.

Bibliography

Manual of Symbols and Terminology for Physicochemical Quantities and Units, Pure Appl. Chem., 51, No. 1 (1979), 1-41. Also published as a flexicover book (The Green Book), Pergamon Press, Oxford, 1979, 41 pages.

Electrochemical Nomenclature, Appendix III to the Manual, Pure Appl. Chem., 37, No. 4 (1974), 499-516.

Classification and Nomenclature of Electroanalytical Techniques, Pure Appl. Chem., 45, No. 2 (1976), pp. 81-97.

Compendium of Analytical Nomenclature, IUPAC Definitive Rules 1977, Pergamon Press, Oxford, 1978, 223 pages.

Recommended name and symbol; SI unit	Definition and remarks
5. Chronopotentiometric constant $\mathcal{T}$ (script capital T) $A\ s^{1/2}\ mol^{-1}\ m$	In chronopotentiometry (at constant current density) (4.12), the empirically evaluated quantity defined by the equation $\mathcal{T} = \underline{i}\ \tau^{1/2}/\underline{A}\ \underline{c}\ (= \underline{j}\ \tau^{1/2}/\underline{c})$ where $\underline{i}$ = electric current (cf. 7), τ = transition time (cf. 77), $\underline{A}$ = area of the electrode-solution interface (cf. 2), $\underline{c}$ = bulk concentration (cf. 6) of the electroactive substance, and $\underline{j}$ = current density = $\underline{i}/\underline{A}$.
6. Concentration, bulk $\underline{c}_B$ ([B]) $mol\ m^{-3}$	In any technique that involves the establishment of a concentration gradient, either within the material from which an electrode is made or in the solution that is in contact with an electrode, the bulk concentration of a substance B is the total or analytical concentration of B at points so remote from the electrode-solution interface that the concentration gradient for B is indistinguishable from zero at the instant under consideration. In common practice the bulk concentration of B is taken to be the total or analytical concentration of B that would be present throughout the electrode or solution if there were no current flowing through the cell and if the electrode and solution did not interact in any way. In the absence of any homogeneous reaction or other process that produces or consumes B, the bulk concentration of B is the total or analytical concentration of B that is present before the excitation signal is applied.
7. Current (electric) $\underline{i}$, $\underline{I}$ A [8.1]	Attention is called to this Commission's Recommendations for Sign Conventions and Plotting of Electrochemical Data [Pure and Appl. Chem., 45 (1976)131], which describes the consequences for electroanalytical chemistry of a recently adopted convention regarding the sign of the electric current.
8. Current, adsorption $\underline{i}_{ads}$, $\underline{I}_{ads}$ A	A faradaic current whose magnitude depends on the applied potential, and, at any particular applied potential, on the rate or extent of the adsorption of an electroactive substance (or the product obtained from the reduction or oxidation of an electroactive substance) onto the surface of the indicator or working electrode. See "9. Current, limiting adsorption."
9. Current, limiting adsorption $\underline{i}_{ads,\ell}$, $\underline{I}_{ads,\ell}$ A [8.3]	The potential-independent value that is approached by an adsorption current as the rate of reduction or oxidation of the electroactive substance is increased by varying the applied potential. The terms "adsorption current" and "limiting adsorption current" should not be applied to faradaic currents that have been increased or decreased by adding a non-electroactive surfactant to a solution containing an electroactive substance, nor to apparent waves resulting from the effect of adsorption or desorption on double-layer currents.
10. Current, alternating $\underline{i}_{ac}$, $\underline{I}_{ac}$ A	This term should be reserved for sinusoidal wave forms; all other wave forms should be termed "periodic."
11. Current, alternating, amplitude of $\underline{i}_{ac}$, $\underline{I}_{ac}$ A	This term should denote half of the peak-to-peak amplitude of the sinusoidal alternating current.
12. Current, apex $\underline{i}_{ap}$, $\underline{I}_{ap}$ A	In measurement of non-faradaic admittance (or tensammetry) (3.2), a plot of alternating current against applied potential shows a minimum or maximum when a non-electroactive

Note: **An explanation has been provided in INTRODUCTION as to why, for the present, the symbol for electric current, lower case i, is preferred in electroanalytical chemistry rather than the symbol capital I recommended by IUPAC and other international scientific organizations.**

Recommended name and symbol; SI unit | Definition and remarks

substance undergoes adsorption or desorption at the surface of the indicator electrode. Such a maximum or minimum may be called an "apex" to emphasize its non-faradaic origin and distinguish it from a "summit," which would result from a charge-transfer process. The highest value of the current on such an apex may be called an "apex current," and the corresponding applied potential may be called an "apex potential."

13. Current, capacity

See "20. Current, double-layer."

14. Current, catalytic

i_{cat}, I_{cat}

A

The faradaic current that is obtained with a solution containing two substances B and A may exceed the sum of the faradaic currents that would be obtained with B and A separately, but at the same concentrations and under the same experimental conditions. In either of the two following situations the increase is termed a "catalytic current."

14.1. B is reduced or oxidized at the electrode-solution interface to give a product B' that then reduces or oxidizes A chemically. The reaction of B' with A may yield either B or an intermediate in the overall half-reaction by which B' was obtained from B. In this situation the increase of current that results from the addition of A to a solution of B may be termed a "regeneration current."

14.2. The presence at the electrode-solution interface of one substance, which may be either A or the product A' of its reduction or oxidation, decreases the overpotential for the reduction or oxidation of B.

In either case the magnitude of the catalytic current depends on the applied potential.

14.3. If the current observed with a mixture of A and B is smaller than the sum of the separate currents, the term "non-additive current" should be used.

15. Current, limiting catalytic

$i_{cat,\ell}$, $I_{cat,\ell}$

A

[8.3]

The potential-independent value that is approached by a catalytic current as the rate of the charge-transfer process is increased by varying the applied potential.

16. Current, charging

See "20. Current, double-layer."

17. Current, diffusion (-controlled)

i_d, I_d

A

A faradaic current whose magnitude is controlled by the rate at which a reactant in an electrochemical process diffuses toward an electrode-solution interface (and, sometimes, by the rate at which a product diffuses away from that interface).

For the reaction mechanism

$$\mathrm{C} \underset{k_-}{\overset{k}{\rightleftharpoons}} \mathrm{B} \xrightarrow{\pm n e} \mathrm{B'}$$

there are two common situations in which a diffusion current can be observed. In one, the rate of formation of B from electroinactive C is small and the current is governed by the rate of diffusion of B toward the electrode surface. In the other, C predominates at equilibrium in the bulk of the solution, but its transformation into B is fast; C diffuses to the vicinity of the electrode surface and is there rapidly converted into B, which is reduced.

18. Current, limiting diffusion

$i_{d,\ell}$, $I_{d,\ell}$

A

[8.3]

The potential-independent value that is approached by a diffusion current as the rate of the charge-transfer process is increased by varying the applied potential.

19. Current, direct

i_{dc}, I_{dc}

A

This term and its symbol should be used (in preference to "current" alone) only to denote the steady (time-independent) component of a current that also has a periodic component.

Recommended name and symbol; SI unit	Definition and remarks
20. Current, double-layer i_{DL}, I_{DL} A	The non-faradaic current associated with the charging of the electrical double layer at an electrode-solution interface, given by $i_{DL} = d(\sigma A)/dt$ where σ = surface charge density of the double layer [5.2], A = area of the electrode-solution interface (cf. 2), and t = time. Capital letters should be used as subscripts to avoid the possibility of confusing this symbol with that for the limiting diffusion current (cf. 18).
21. Current, faradaic (no symbol recommended) A	A current corresponding to the reduction or oxidation of some chemical substance.
22. Current, net faradaic (no symbol recommended) A	The algebraic sum of all the faradaic currents flowing through an indicator or working electrode.
23. Current, faradaic demodulation i_{FD}, I_{FD} A	A component of the current that is due to the demodulation associated with an electrode reaction and that appears if an indicator or working electrode is subjected to the action of two intermodulated applied potentials of different frequency.
24. Current, faradaic rectification i_{FR}, I_{FR} A	A component of the current that is due to the rectifying properties of an electrode reaction and that appears if an indicator or working electrode is subjected to any periodically varying applied potential while the mean value of the applied potential is controlled.
25. Current, instantaneous i_t, I_t A	25.1. At a dropping electrode, the total current that flows at the instant when a time t has elapsed since the fall of the preceding drop. 25.2. At any other electrode, the total current that flows at the instant when a time t has elapsed since the beginning of an electrolysis. The instantaneous current is usually time-dependent and may have the character of an adsorption, catalytic, diffusion, double-layer, or kinetic current, and may include a migration current. A plot of the dependence of instantaneous current on time is commonly called an "i-t curve."
26. Current, kinetic i_k, I_k A	A faradaic current that corresponds to the reduction or oxidation of an electroactive substance B formed by a prior chemical reaction from another substance Y that is not electroactive, and that is partially or entirely controlled by the rate of that reaction. The reaction may be heterogeneous, occurring at an electrode-solution interface (surface reaction), or it may be homogeneous, occurring at some distance from the interface (volume reaction). See also "17. Current, diffusion."
27. Current, limiting kinetic $i_{k,\ell}$, $I_{k,\ell}$ A [8.3]	The potential-independent value that is approached by a kinetic current as the rate of the charge-transfer process is increased by varying the applied potential.
28. Current, limiting i_ℓ, I_ℓ A [8.3]	A limiting current is the limiting value of a faradaic current that is approached as the rate of the charge-transfer process is increased by varying the potential. It is independent of the applied potential over a finite range, and is usually evaluated by subtracting the appropriate residual current from the measured total current. A limiting current may have the character of an adsorption, catalytic, diffusion, or kinetic current, and may include a migration current.

Recommended name and symbol; SI unit	Definition and remarks

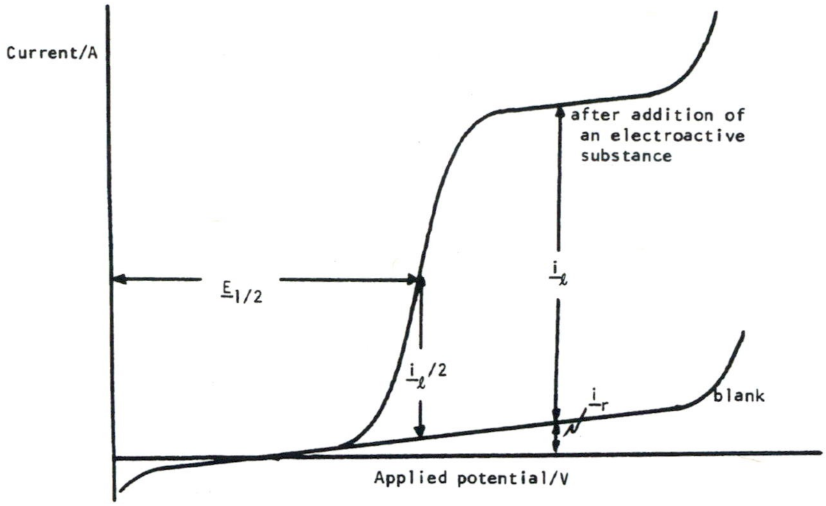

Fig. 1. Idealized polarogram, illustrating the limiting current $\underline{i}_\ell$, the residual current $\underline{i}_r$, and the half-wave potential $\underline{E}_{1/2}$.

29. Current, migration

$\underline{i}_m$, $\underline{I}_m$

A

The difference between the current that is actually obtained, at any particular value of the potential of the indicator or working electrode, for the reduction or oxidation of an ionic electroactive substance and the current that would be obtained, at the same potential, if there were no transport of that substance due to the electric field between the electrodes. The sign convention regarding current is such that the migration current is negative for the reduction of a cation or for the oxidation of an anion, and positive for the oxidation of a cation or the reduction of an anion. Hence the migration current may tend to either increase or decrease the total current observed. In any event the migration current approaches zero as the transport number of the electroactive substance is decreased by increasing the concentration of the supporting electrolyte, and hence the conductivity.

30. Current, limiting migration

$\underline{i}_{m,\ell}$, $\underline{I}_{m,\ell}$

A

[8.3]

The limiting value of a migration current, which is approached as the rate of the charge-transfer process is increased by varying the applied potential.

31. Current, non-additive

See "14(.3) Catalytic current."

32. Current, peak

$\underline{i}_p$, $\underline{I}_p$

A

In linear-sweep voltammetry (5.8), triangular-wave voltammetry (5.19), cyclic triangular-wave voltammetry (5.21), and similar techniques, the maximum value of the faradaic current due to the reduction or oxidation of a substance B during a single sweep. This maximum value is attained after an interval during which the concentration of B at the electrode-solution interface decreases monotonically, while the faradaic current due to the reduction or oxidation of B increases monotonically, with time. It is attained before an interval during which this current decreases monotonically with time because the rate of transport of B toward the electrode-solution interface is smaller than the rate at which it is removed from the interface by electrolysis.

Recommended name and symbol; SI unit	Definition and remarks

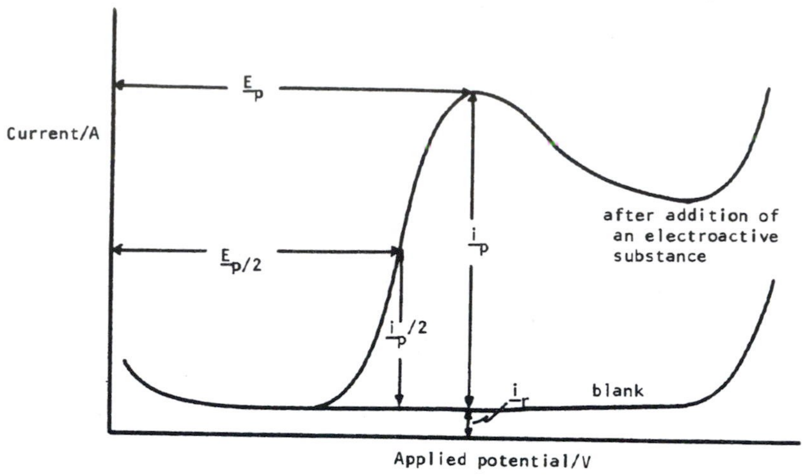

Fig. 2. Idealized linear-sweep voltammogram, illustrating the peak current i_p, the residual current i_r, the peak potential E_p, and the half-peak potential $E_{p/2}$.

The term "peak current" has also been used to denote the maximum value of the faradaic current attributable to the reduction or oxidation of an electroactive substance in techniques such as ac polarography (6.4), differential pulse polarography (6.3), and derivative polarography (5.13). However, these techniques give curves that arise in ways different from that cited above, and the terms "summit," "summit current," and "summit potential" are therefore recommended for use in connection with such techniques. See also "12. Current, apex."

Recommended name and symbol; SI unit	Definition and remarks
33. Current, periodic	See "10. Current, alternating."
34. Current, regeneration	See "14(.1) Current, catalytic."
35. Current, residual i_r, I_r A	The current that flows, at any particular value of the applied potential, in the absence of the substance whose behavior is being investigated (i.e., in a blank solution). See Figs. 1-3.
36. Current, square-wave i_{SW}, I_{SW} A	In square-wave polarography (6.5), the component of the current that is associated with the presence of a substance B. This component may be either faradaic (if B is electroactive) or non-faradaic (if B is surface-active).
37. Current, summit i_{su}, I_{su} A	In ac polarography (6.4), differential pulse polarography (6.3), derivative polarography (5.13), square-wave polarography (6.5), and similar techniques, the maximum value of the component of the current that is associated with the presence of a substance B. Normally this component of the current is faradaic, and the maximum arises because the rate of variation (with applied potential) of the rate of the charge-transfer process passes through a maximum. Similar maxima arise when this component is non-faradaic (and when B is surface-active rather than electroactive). In a case known to be of the latter type, the term "apex current" is recommended as being more specific.

Recommended name and symbol; SI unit	Definition and remarks

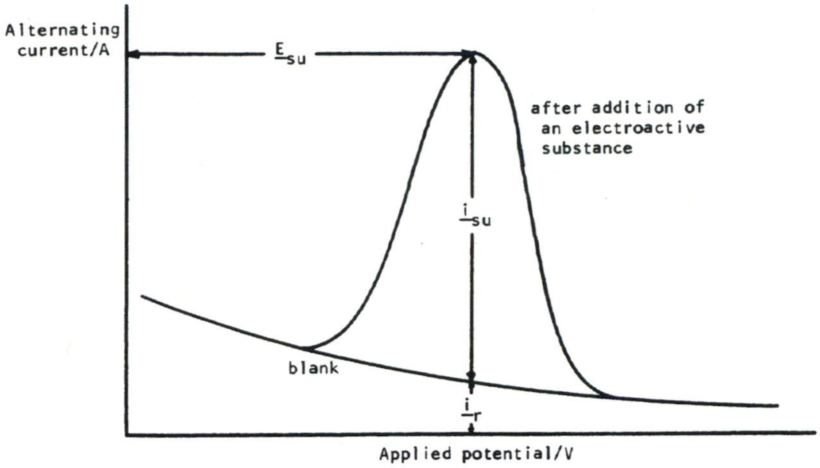

Fig. 3. Idealized ac polarogram, illustrating the summit current $\underline{i}_{su}$, the residual current $\underline{i}_r$, and the summit potential $\underline{E}_{su}$.

38. **Depolarizer**

The term "electroactive substance" (cf. 41) should be used in preference to "depolarizer."

39. **Diffusion current constant**

$\underline{I}$, $\mathcal{I}$ (script capital I)

A mol^{-1} m^3 kg$^{-2/3}$ s$^{1/2}$

In polarography (5.12), the empirical quantity defined by the equation

$$\underline{I} = \underline{i}_{d,\ell}/\underline{c}_B \underline{m}^{2/3} \underline{t}_1^{\ 1/6}$$

where $\underline{i}_{d,\ell}$ = limiting diffusion current (cf. 18), $\underline{c}_B$ = bulk concentration of the substance B whose reduction or oxidation is responsible for the wave in question (cf. 6), $\underline{m}$ = average rate of (mass) flow of mercury (or other liquid metal) (cf. 67) and $\underline{t}_1$ = drop time (cf. 40).

40. **Drop time**

$\underline{t}_1$, τ, $\underline{t}_d$

s

In polarography (5.12), the time that elapses between the instants at which two successive drops of liquid metal are detached from the tip of the capillary.

41. **Electroactive substance**

41.1 In voltammetry and related techniques, a substance that undergoes a change of oxidation state, or the breaking or formation of chemical bonds, in a charge-transfer step.

If an electroactive substance B is formed, in the solution or electrode, by a chemical reaction from another substance C, the substance C should be called the precursor of B.

41.2. In potentiometry with ion-selective electrodes, a material containing, or in ion-exchange equilibrium with, the sensed ion. The electroactive substance is often incorporated in an inert matrix such as poly(vinyl)-chloride or silicone rubber.

42. **Electrode, auxiliary**

Three-electrode cells comprise (1) an indicator (or test) electrode (cf. 44) or a working electrode (cf. 47), at the surface of which processes that are of interest may occur, (2) a reference electrode (cf. 45), and (3) a third electrode, the auxiliary or counter electrode, which serves merely to carry the current flowing through the cell, and at the surface of which no processes of interest occur.

If processes of interest occur at both the anode and the cathode of a cell [as in differential amperometry (4.17) or controlled-current potentiometric titration with two indicator electrodes (4.11)], the cell should be said to comprise two indicator (or test) or working electrodes.

Recommended name and symbol; SI unit	Definition and remarks
43. Electrode, counter	See "42. Electrode, auxiliary."
44. Electrode, indicator	An electrode that serves as a transducer responding to the excitation signal (if any) and to the composition of the solution being investigated, but that does not effect an appreciable change of bulk composition within the ordinary duration of a measurement.
45. Electrode, reference	An electrode that maintains a virtually invariant potential under the conditions prevailing in an electrochemical measurement, and that serves to permit the observation, measurement, or control of the potential of the indicator (or test) or working electrode.
46. Electrode, test	See "44. Electrode, indicator."
47. Electrode, working	An electrode that serves as a transducer responding to the excitation signal and the concentration of the substance of interest in the solution being investigated, and that permits the flow of current sufficiently large to effect appreciable changes of bulk composition within the ordinary duration of a measurement.
48. Electrolyte, base	See "50. Electrolyte, supporting."
49. Electrolyte, indifferent	See "50. Electrolyte, supporting."
50. Electrolyte, supporting	50.1. An electrolyte solution, whose constituents are not electroactive in the range of applied potentials being studied, and whose ionic strength (and, therefore, contribution to the conductivity) is usually much larger than the concentration of an electroactive substance to be dissolved in it. 50.2. The solutes that are present in such a solution.
51. Frequency Hz	It is essential to draw a careful distinction between the electrical frequency of an excitation signal or measured response and the rate of rotation of a rotating disc, wire, or other electrode.
52. Mass-transfer-controlled electrolyte rate constant $\underline{s}_B$ s^{-1}	In controlled-potential coulometry (4.27) and related techniques, the empirically evaluated constant of proportionality defined by the equation $\underline{s}_B = -(1/\underline{c}_B)(d\underline{c}_B/d\underline{t})$ where $\underline{c}_B$ is the bulk concentration (cf. 6) of the substance B, and $d\underline{c}_B/d\underline{t}$ is the rate of change of that concentration, resulting from the consumption of B by reduction or oxidation at the working electrode.
53. $\underline{n}$ dimensionless [1.2]	A stoichiometric ratio equal to the total number of electrons transferred between an electrode and a solution in the reduction or oxidation of one ion or molecule of an electroactive substance, whose identity must be specified. No other substance that is initially present may be reduced or oxidized during the process (see "14. Current, catalytic").
54. $\underline{n}_{app}$ dimensionless	An experimentally measured quantity equal to the total number of electrons transferred between an electrode and a solution in consequence of the oxidation or reduction of one ion or molecule of an electroactive substance, whose identity must be specified. When the reduction or oxidation of a substance B is accompanied by chemical processes such as the catalyzed or induced reduction of a second substance, or a side reaction that consumes B or an intermediate, the value of $\underline{n}_{app}$ will differ from that of $\underline{n}$ (cf. 53).

Recommended name and symbol; SI unit	Definition and remarks
55. Outflow velocity (of mercury or other liquid metal)	See "67 (and 68). Rate of flow."
56. Peak	See "32. Current, peak."
57. Potential, apex	See "12. Current, apex."
58. Potential, applied $\underline{E}_{app}$ V [2.2]	The difference of potential measured between identical metallic leads to two electrodes of a cell. The applied potential is divided into two electrode potentials, each of which is the difference of potential existing between the bulk of the solution and the interior of the conducting material of the electrode, an $\underline{i}\,\underline{R}$ or ohmic potential drop through the solution, and another ohmic potential drop through each electrode. In the electroanalytical literature this quantity has often been denoted by the term "voltage," whose continued use is not recommended.
59. Potential, half-peak $\underline{E}_{p/2}$ V	In linear-sweep voltammetry (5.8), triangular-wave voltammetry (5.19), cyclic triangular-wave voltammetry (5.21), and similar techniques, the potential of the indicator electrode at which the difference between the total current and the residual current is equal to one-half of the peak current (cf. 32). This potential is attained in the interval in which the rate of the charge-transfer process, and hence the (absolute value of the) current, increase monotonically with time. See Fig. 2.
60. Potential, half-wave $\underline{E}_{1/2}$ V	The potential of a polarographic or voltammetric indicator electrode at the point, on the rising part of a polarographic or voltammetric wave, where the difference between the total current and the residual current is equal to one-half of the limiting current (cf. 28). See Fig. 1. The quarter-wave potential $\underline{E}_{1/4}$, the three-quarter-wave potential $\underline{E}_{3/4}$, etc., may be similarly defined.
61. Potential, peak $\underline{E}_p$ V	In linear-sweep voltammetry (5.8), triangular-wave voltammetry (5.19), cyclic triangular-wave voltammetry (5.21), and similar techniques, the potential of the indicator electrode at which the peak current (cf. 32) is attained. See Fig. 2.
62. Potential, quarter-transition-time $\underline{E}_{\tau/4}$ V	In chronopotentiometry (at constant current density) (4.12), the potential of the indicator electrode at the instant when the time that has elapsed since the application of current is equal to one-fourth of the transition time (cf. 7). Appropriate correction for double-layer charging phenomena is needed in practice.

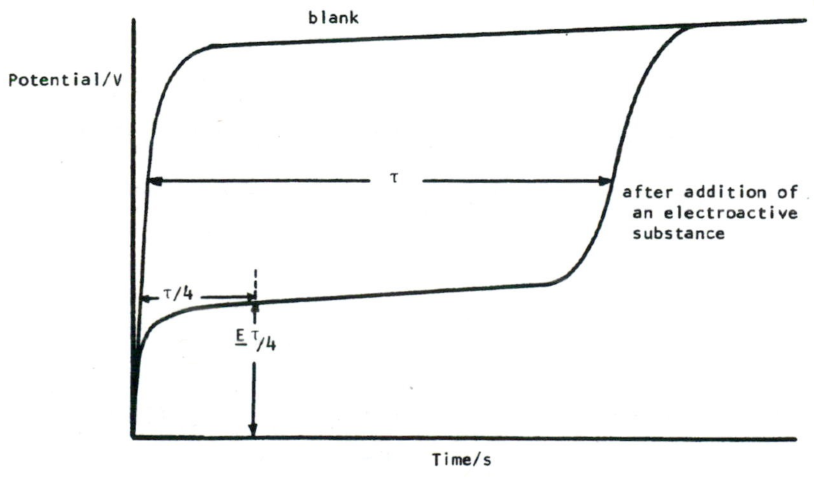

Fig. 4. Idealized chronopotentiogram, illustrating the quarter-transition-time potential $\underline{E}_{\tau/4}$ and the transition time τ.

Recommended name and symbol; SI unit	Definition and remarks
63. Potential, summit E_{su} V	In ac polarography (6.4), differential pulse polarography (6.3), derivative polarography (5.13), and similar techniques, the potential of the indicator electrode at which the summit current (*cf.* 37) is attained. See Fig. 3.
64. Precursor (of an electroactive substance)	See "41. Electroactive substance."
65. Pulse duration t_p s	In pulse polarography (5.22), differential pulse polarography (6.3), Kalousek polarography (5.24), and related techniques, the duration of an interval during which the excitation signal deviates from the base line. This interval includes the sampling interval (*cf.* 72).
66. Quantity of electricity Q C	The quantity of electricity that flows during the interval between t_1 and t_2 is given by $Q = \int_{t_1}^{t_2} i_t dt$ where i_t is the instantaneous current at any instant t during that interval. The reduction of an electroactive substance gives rise to negative values of Q; the oxidation of an electroactive substance gives rise to positive values of Q. The components of an overall value of Q should be named and given symbols similar to those of the corresponding currents, *e.g.*, Q_{DL} = double-layer quantity of electricity, Q_t = instantaneous quantity of electricity, etc.
67. Rate of flow (or mercury or other liquid metal), average m kg s^{-1}	In polarography (5.12), the ratio of the mass of a drop, at the instant when it is detached from the tip of the capillary, to the drop time t_1 (*cf.* 40); the average value of the instantaneous rate of flow over the entire life of the drop.
68. Rate of flow (or mercury or other liquid metal), instantaneous m_t kg s^{-1}	In polarography (5.12), the rate of increase of the mass of a drop at a particular instant t seconds after it has begun to form.
69. Reaction, surface	See "26. Current, kinetic."
70. Reaction, volume	See "26. Current, kinetic."
71. Response constant (no symbol recommended)	A quantity whose expression includes a current, whose value is characteristic of a charge-transfer process and the experimental conditions under which it is investigated, and whose nature depends on the technique that is employed. Some typical response constants are the diffusion current constant (*cf.* 39) in polarography (5.12), the voltammetric constant (*cf.* 82) in linear-sweep voltammetry (5.8), and the chronopotentiometric constant (*cf.* 5) in chronopotentiometry (4.12).
72. Sampling interval	In Tast polarography (5.15), square-wave polarography (6.5), and similar techniques, the interval during which the current is measured or recorded.
73. Sampling time t_m s	In Tast polarography (5.15), square-wave polarography (6.5), and similar techniques, the duration of the sampling interval (*cf.* 72).
74. Strobe interval	[In Tast polarography (5.15)] See "73. Sampling time."
75. Tast interval	[In Tast polarography (5.15)] See "73. Sampling time."
76. Thickness of the reaction layer μ m	When a kinetic current (*cf.* 26) flows, the concentrations of the electroactive substance B and its precursor C at very small distances from the electrode surface are influenced both by mass transfer and by the finite rate of establishment of the chemical

Recommended name and symbol; SI unit	Definition and remarks
	equilibrium. As the distance from the electrode surface increases, the chemical equilibrium is more and more nearly attained. The thickness of the reaction layer is the distance from the electrode surface beyond which deviations from the chemical equilibrium between C and B are taken to be negligibly small.
77. Transition time τ s	In chronopotentiometry (4.12) and related techniques, the time that elapses between the instant at which current is applied and the instant at which the concentration of an electroactive substance B at the electrode-solution interface becomes indistinguishable from zero. In experimental practice the latter time is often taken to be the instant at which the rate of variation of the potential of the indicator electrode attains a maximum value. See Fig. 4.
78. Voltage $\underline{e}$ V	The use of this term is discouraged, and the term "applied potential" should be used instead, for non-periodic signals. However, it is retained here for sinusoidal and other periodic signals because no suitable substitute for it has been proposed.
79. Voltage, alternating $\underline{e}_{ac}$ V	This term should be applied only to sinusoidal phenomena; the term "periodic voltage" should be used for other wave forms.
80. Voltage, alternating, amplitude $\underline{e}_{ac}$ V	This term should denote half of the peak-to-peak amplitude. Peak-to-peak and r.m.s. amplitudes should be so specified.
81. Voltage, periodic $\underline{e}_{pc}$ V	This general term is applicable to square, triangular, and other wave forms; the term "alternating voltage" should be reserved for sinusoidal wave forms.
82. Voltammetric constant $\mathcal{V}$ (script capital V) A mol^{-1} V$^{-1/2}$ m s$^{1/2}$	In linear-sweep voltammetry (5.8) and related techniques, the empirical quantity defined by the equation $$\mathcal{V} = \underline{i}_p/A\underline{v}^{1/2}\ \underline{c}_B\ (= \underline{j}_p/\underline{v}^{1/2}\underline{c}_B)$$ where $\underline{i}_p$ = peak current (cf. 32), A = area of the electrode-solution interface (cf. 2), $\underline{v}$ = rate of change of applied potential, and $\underline{c}_B$ = bulk concentration of the substance B whose reduction or oxidation is responsible for the peak in question (cf. 6).
83. Wave height	The limiting current (cf. 28) of an individual wave, frequently expressed in arbitrary units for convenience.

CONDITIONAL DIFFUSION COEFFICIENTS OF IONS AND MOLECULES IN SOLUTION:

AN APPRAISAL OF THE CONDITIONS AND METHODS OF MEASUREMENT

E. Bishop and Z. Galus

Chemistry Department, University of Exeter, Stocker Road, Exeter EX4 4QD U.K.

STATEMENT

In electroanalytical chemistry, many techniques involve the transport of electroactive species, charged or neutral, and among the migration processes, diffusion is the most widely encountered. A quantitative description of the processes and data on the relevant rate constants are necessary for the proper understanding and application of the technique. Many diffusion coefficients can be calculated from conductance data, but such values have only academic value in that they relate to ideal conditions not normally accessible in analytical practice.[1-4] This problem has been treated in detail from the experimental point of view by Turnham.[5] Tabulations of such values may, therefore, be more misleading than helpful, unless they be of conditional values,[6] the conditions being completely defined as below.

A conditional diffusion coefficient is of immediate use if the conditions under which it was determined are fully detailed and can be replicated. By the term conditional diffusion coefficient is meant the mass transfer rate under thermal diffusion conditions, i.e. correction has been made for electromigration where necessary, and is the proportionality constant between the flux and concentration gradient of the diffusing species. A thermodynamic diffusion coefficient is not of use unless the effects of all the matrix factors are fully and unambiguously correctable, and this is seldom possible in the complex matrices of high ionic strength encountered in typical analytical samples.

It appeared, therefore, that a critical compilation of such conditional values together with specifications of the conditions would be of use to electro-chemists. A brief examination of the requirements for such a compilation and a classification and brief appraisal of the principal methods of measurement are made in order to define the perspective and context.

CONDITIONS

Many factors affect the value of the conditional diffusion coefficient and would require specification. The value is know to vary, inter alia, with the solvent, the concentration of the diffusing species, the nature and concentration of other ionic and non-ionic species in the solution, the temperature, viscosity and density of the solution. Although the diffusion process is independent of the charge transfer kinetic parameters of an electrode process, it enters directly into the charge transfer overpotential, which is logarithmically dependent on the ratio of mass to charge transfer rate constants. The mass transfer rate is further dependent on current density and the potential of the working electrode, and, in many methods, on other transport processes such as convection, thermal and density gradients, electromigration, etc.

Diffusion coefficients in isolation can be determined by a variety of electrochemical methods wherein the whole system - electrodes and solution - is at perfect rest, free from convection arising from thermal effects at the electrode and from density gradients produced by mass transfer. They can likewise be determined when the solution and/or the electrode is moving and the hydrodynamic conditions are well-defined. When such conditions are not fulfilled, as with turbulently stirred solutions, the isolated diffusion coefficient ceases to be of individual value, and it is recommended that a conditional overall mass transfer rate constant be substituted in its stead. Non-electrical methods of both classical and recent origin may also be used, but values so obtained should be examined for their applicability in electrochemical contexts.

THE DETERMINATION OF CONDITIONAL DIFFUSION COEFFICIENTS

Certain general and cautionary points must be made before listing the methods individually.

1. The reliability and reproducibility of the selected method must be assessed by determination of the value of a diffusion coefficient for which good literature values are available, for example hexacyanoferrate(III)/(II) and dianisidine, and also by statistical evaluation of an adequate number of replicate determinations. A standard deviation of 2.5% of the quantity being determined is considered to be reasonable. Precisions of a few tenths of a per cent have been claimed for the trace-ion methods[7], and a highly sophisticated rotating electrode (disc, ring-disc, etc.) is also capable of attaining a relative standard deviation better than 1%.

2. Agreement between values obtained by different methods should be examined; more than one method should be used whenever possible. Again a standard deviation of 2-5% is considered acceptable.

3. The question, "can a value obtained by one method, e.g., chronoamperometry, be applied with confidence in a different technique, e.g., cyclic voltammetry?" must be considered with a good deal of care. Although fundamentally the diffusion coefficient should be independent of the method of measurement, provided that solution conditions are held constant, much depends on the care and insight with which the experiment is performed.

 0033–4545/79/0701–1575$02.00/0 © 1979 International Union of Pure and Applied Chemistry

A further question of interest concerns the relationship between values of diffusion coefficients determined by non-electrochemical methods and those determined by electrochemical methods. Of value in this respect is the work of Bearman[8] concerning the theoretical relationship between values of diffusion coefficients of the electroactive species determined polarographically and by tracer methods. These diffusion coefficients, strictly speaking, are not equal, the polarographic value being the greater. The difference, however, is of a sufficiently small magnitude to permit its being regarded as insignificant within the limit of the validity of the limiting current equations. Mills has reviewed the various theoretical approaches to the interpretation of diffusion process, including cross-term coefficients,[9] and has made a critical compilation of tracer diffusion coefficients.[10] The difference between polarographic and tracer coefficients can be summarized in the equations[9]

$$D_A^{p} = \frac{RT}{c_B\,\zeta_{AB} + c_R\,\zeta_{AR} + c_S\,\zeta_{AS} - c_A\,\zeta_{RA}} \qquad (1)$$

$$D_A = \frac{RT}{c_B\,\zeta_{AB} + c_R\,\zeta_{AR} + c_S\,\zeta_{AS}} \qquad (2)$$

As the concentration c_A becomes very small, the two coefficients tend to equality, the difference depending on the frictional factor, ζ_{RA}. Other theoretical treatments of diffusion in an electric field may be noted.[11,12] Comparison of the polarographic diffusion current with that calculated using tracer diffusion coefficients gave good agreement for monovalent thallium, but not for divalent ions such as lead and zinc.[13]

4. The question, "do values for a certain geometry, e.g., spherical diffusion, apply, with or without some conversion factor, to a different geometry, e.g., cylindrical diffusion?" must also be scrutinized in detail.[14]

5. The conditions of the determination must be fully spelt out in complete detail. This calls for a specification of which conditions must be so detailed, and the first paragraph under the heading "conditions" is no more than a partial listing of the more important factors.

Tamamushi of Commission I-3 has prepared a standard layout for presentation of results for later collation. Useful as this is, it does not include such important matters as the pretreatment of electrodes, potential scan rates, kinematic viscosity, angular velocity of rotating electrodes, extrapolation of Levich plots to determine the intercept, rate of deactivation of electrodes, established purity of solvent and dissolved substances, or ranges of conditions on which measurements have been made.

There now follows an annotated listing of a selection of the more useful methods.

I ELECTROCHEMICAL METHODS[14-18]

A. Stationary solution and electrodes

1. Conductance measurements, and ionic mobilities. Only when medium effects can be completely and specifically evaluated are results calculated from such measurements of value.[1-4,9]

2. Linear sweep voltammetry and polarography[6,14,19] offer excellent values when the limiting current region is well defined, but otherwise become approximate. Cyclic voltammetry, though used, is regarded as an unreliable technique for this particular application. There is a difference of opinion about the quality of values obtained by polarography: compilers of polarographic data in the Clarkson Project aver that the values are of little worth, probably because conditions are inadequately defined or the experimental techniques are not of the best. Critical compilation of electrode kinetic data by Tamamushi of Commission I-3 claim to contain a number of reliable values of diffusion coefficients.

3. Chronoamperometry affords a useful source of good values. Under defined conditions, chronoamperometry gives both reliable and precise results; it has been particularly successful in the determination of diffusion coefficients of metals in amalgams.[20,21]

4. Chronopotentiometry is also used, but gives less precise results because adequate precision in the measurement of transition times is not always attainable, and density gradients arise to disturb the mass transfer control.

5. Large perturbation methods, again provided that the limiting current regions, are well-defined, can give both individual values for D_{ox} and D_{red} and the useful ratio thereof, which in many cases is all that is required.

B. Moving electrodes or solutions under conditions of well-defined hydrodynamics

1. Spinning disc [14,15,17,18,22] electrode methods are frequently used and are capable of producing results of high accuracy and reliability, provided that the apparatus is soundly constructed. The strictest attention must be paid to avoidance of vibration, constancy of rotation speed, electrode centring, adequate ratio of vessel to electrode

assembly and electrode assembly to disc radii, adequate depth of a vessel, the design and disposition of counter and reference electrode and so on.

2. Spinning cylinder [14] electrode methods are considerably less precise than spinning disc methods, but have been used. The mass transfer pattern is ill-defined and the method is not at present considered to be reliable.

3. Tubular flowing systems, when the equipment is very carefully constructed so as to ensure near laminar flow, have virtues of their own which give them a useful flexibility, but in general the spinning disc method is best.

C. Undefined hydrodynamic conditions

Although Levich has shown, and many others have confirmed, that in turbulently stirred solutions mass transfer is proportional to the concentration of the active species, specific hydrodynamic equations cannot be derived for such a situation, and diffusion coefficients are neither measurable nor usable.

These conditions are, however, of considerable practical application, as in coulometry, and it is then convenient to replace the flux proportionality constant by a conditional overall mass transfer rate constant which incorporates all the factors which are difficult or impossible to evaluate separately.[23]

II NON-ELECTRICAL METHODS

Classical methods[1-4] need not be further elaborated here. Different vessels filled with solutions of the diffusing species, and diverse methods of determination of the concentration distribution of the test species, have been applied. The capillary method finds continuing use.[24,25] The tracer technique of more recent origin affords a considerable simplification, although it must be applied with caution to light elements. The method is particularly valuable for organic compounds labelled with tritium.[10,26,27] Gosting[28] has made use of the Gouy interferometric method, and the diaphragm cell has also been employed,[29,30] with considerable success, [31-5] and confirmed by radiotracer methods using ^{15}N.[34]

Conclusions

Isolated values of thermal diffusion coefficients are of merit only when the conditions, both hydrodynamic and experimental, are fully interpretable and fully detailed. On the other hand, in situations of well-defined hydrodynamics, a conditional diffusion coefficient in terms of a flux divided by a pulsatance can have a well recognized meaning and applications.

Fast perturbation methods are considered to be too limited by charge transfer phenomena to be of much assistance in the determination of diffusion coefficients. It is considered that linear sweep voltammetry and polarography, and especially chronoamperometry and spinning disc electrode methods, are the most reliable sources of electrochemical measurements. Non-electrical methods, particularly the capillary, diaphragm and tracer methods, offer useful alternatives.

In turbulently stirred solutions, a situation having special reference to coulometric analysis in stirred solutions with both solid and liquid electrodes and therefore of considerable practical importance, it is recommended that an overall conditional mass transfer rate constant be substituted for the diffusion coefficients.

This Report has been prepared for Commission V-5, Electroanalytical Chemistry, of IUPAC.

REFERENCES

1. MacInnes, D.A. "Principles of Electrochemistry", Reinhold, New York, 1939.

2. Glasstone, S., "Electrochemistry", McGraw Hill, N.Y., 1939.

3. Robinson, R.A. and Stokes, R.H., "Electrolyte Solutions", 2nd Ed., Butterworths, London, 1966.

4. Bockris, J.O'M. and Reddy, A.K., "Modern Electrochemistry", Volume I, MacDonald, London, 1970.

5. Turnham, D.S., J. Electroanal. Chem., 1965, 10, 19.

6. Meites, L., "Polarographic Techniques", 2nd Ed., Interscience, New York, 1965.

7. Mills, R., J. Phys. Chem., 1963, 67, 600.

8. Bearman, R.J., J. Phys. Chem., 1962, 66, 2072.

9. Mills, R., J. Electroanal. Chem., 1965, 9, 57.

10. Mills, R., Rev. Pure Appl. Chem., 1961, 11, 78.

11. Lopushanskaya, A.I., Pamfilov, A.V. and Tsisar, I.A., Russ. J. Phys. Chem., English Transl., 1963, 37, 1193.

12. Laity, R.W., J. Phys. Chem., 1963, 67, 671.

13. Sawada, S., Nishiyama, K., Yokochi, K. and Suzuki, M., Rev. Polarog. (Japan), 1972, 18, 62.

14. Adams, R.N., "Electrochemistry at Solid Electrodes", Dekker, New York, 1969.

15. Levich, V.J., "Physicochemical Hydrodynamics", Prenctice-Hall, Englewood Cliffs, New Jersey, 1962.

16. Vetter, K.J., "Electrochemical Kinetics", Academic Press, New York, 1967.

17. Galus, Z., "Theoretical basis of Electroanalytical Chemistry", P.W.N., Warsaw, 1971 (in Polish). English version "Fundamentals of Electrochemical Analysis", Ellis Horwood, Chichester, 1976.

18. Riddiford, A.C., Delahay, P. and Tobias, C.W., "Advances in Electrochemistry and Electrochemical Engineering", Vol. IV, p. 47. Interscience, New York, 1966.

19. Peters, D.G. and Lingane, J.J., J. Electroanal. Chem., 1961, 2, 1.

20. Stevens, W.J. and Shain, I., J. Phys. Chem., 1966, 70, 2276.

21. Baranski, A., Fitak, S. and Galus, Z., J. Electroanal. Chem., 1975.

22. Pleskov, Yu. V. and Filimovski, "Rotating Disc Electrode", Nauka, Moscow, 1972 (in Russian).

23. Bishop, E., Chem. Anal. (Warsaw), 1972, 17, 511.

24. Wang, J.H., J. Am. Chem. Soc., 1951, 73, 510; 4181.

25. Bacon, J. and Adams, R.N., Anal. Chem., 1970, 42, 524.

26. Miller, T.A., Prater, B., Lee, J.K. and Adams, R.N., J. Am. Chem. Soc., 1965, 87, 121.

27. Miller, T.A., Lamb, B., Prater, K., Lee, J.K. and Adams, R.N., Anal. Chem., 1964, 36, 418.

28. Wolf, L.A., Miller, D.G. and Gosting, L.J., J. Phys. Chem., 1962, 84, 317.

29. Kelly, F.J. and Stokes, R.H., "Electrolytes", ed. Pesce, B., Pergamon, London, 1962, p.96.

30. Henrion, P.N., Trans. Farad. Soc., 1964, 60, 75.

31. Hashitani, T. and Tamamusi, T., Trans. Farad. Soc., 1967, 63, 369.

32. Hashitani, T., Sci. Papers, Inst. of Phys. and Chem. Research, Tokyo, 1967, 61, 139.

33. Tanaka, K., Hashitani, T. and Tamamushi, R., Trans. Farad. Soc., 1967, 63, 74.

34. Tanaka, K. and Hashitani, T., Trans. Farad. Soc., 1971, 67, 2314.

35. Tanaka, K., J. Chem. Soc. Faraday Trans., 1975, 71, 1127.

GENERAL ASPECTS OF TRACE ANALYTICAL METHODS—IV. RECOMMENDATIONS FOR NOMENCLATURE, STANDARD PROCEDURES AND REPORTING OF EXPERIMENTAL DATA FOR SURFACE ANALYSIS TECHNIQUES

PROVISIONAL

COMMISSION ON MICROCHEMICAL TECHNIQUES AND TRACE ANALYSIS*

G. H. MORRISON, K. L. CHENG and M. GRASSERBAUER

The most important problem faced by a scientist at the start and at the conclusion of an experiment is communicating with other scientists. Since the type and quantity of information reported in scientific papers varies widely in the field of "surface analysis", it is the aim of these recommendations to suggest standard preparation procedures and methods of data reporting, and to clarify nomenclature for these various techniques.

It is first necessary to recommend standard technique names, then to clarify nomenclature, before suggesting a cleaning method and finally specifying the information needed to be included in a scientific paper describing an experiment using one of these techniques. (see Note a)

TECHNIQUE NOMENCLATURE

Auger Electron Spectroscopy (AES). The technique in which a sample is bombarded with keV-energy electrons or X-rays in a high vacuum apparatus, and the energy distribution of the electrons produced through radiationless de-excitation of the atoms in the sample is recorded; the derivative curve may also be recorded.

UV Photoelectron Spectroscopy (UPS). Any technique in which the sample is irradiated by monochromatic radiation of the ultraviolet region and the energy distribtuion of the photoelectrons emitted by the sample is measured. The spectra originate from excitation of valence electrons and particularly lead to the identification of interatomic groupings within molecules and to the electronic structure of clean surfaces.

X-Ray Photoelectron Spectroscopy (XPS). Any technique in which the sample is bombarded with X-rays and photoelectrons produced by the sample are detected as a function of energy. ESCA (Electron Spectroscopy for Chemical Analysis) refers to the use of this technique to identify elements, their concentrations, and their chemical state within the sample.

Rutherford Back Scattering (RBS) also referred to as Backscattering Spectrometry (BSS). Any technique using high energy particles directed toward a sample, in which the bombarding particles are detected and recorded as function of energy and/or angle. The technique is mostly used for determining depth distributions of elements based on the energy of the back-scattered particle. In general, He^+ or H^+ particles are used at energies in the order of 100 keV to some MeV.

Ion Scattering Spectrometry (ISS). Any technique using low energy ($\lesssim$ 10 keV) ions in which the bombarding particles scattered by the sample are detected and recorded as a function of energy and/or angle. This technique is used mainly for determining the composition and structure of the first few atomic layers of a sample.

Secondary Ion Mass Spectrometry (SIMS). Any technique in which the sample is bombarded with a stream of (primary) ions and the (secondary) ions ejected from the sample are detected after passage through a mass spectrometer.

Ion Probe Microanalysis refers to the use of SIMS for qualitative and quantitative elemental analysis with a spatial resolution of less than 10 µm.

Note a. The nomenclature given is provisional. Commission I.6 (COMMISSION ON COLLOID AND SURFACE CHEMISTRY) is engaged in the preparation of a manual of nomenclature in surface physics.

*Chairman: M. GRASSERBAUER (Austria); Secretary: M. PINTA (France); Titular Members: S. GOMIŠČEK (Yugoslavia); P. D. LA FLEUR (USA); A. MIZUIKE (Japan); E. A. TERENT'EVA (USSR); G. TÖLG (FRG); A. TOWNSHEND (UK); Associate Members: B. GRIEPINK (Netherlands); K. HEINRICH (USA); E. JACKWERTH (FRG); O. G. KOCH (FRG); Z. MARCZENKO (Poland); G. H. MORRISON (USA); J. M. OTTAWAY (UK); YU. A. ZOLOTOV (USSR); National Representatives: J. JANÁK (Czechoslovakia); A. D. CAMPBELL (New Zealand).

Reprinted with permission from *Pure and Applied Chemistry,* Volume 51, 1979, pages 2243–2250.

Ion Microscopy refers to the use of the SIMS technique to obtain micrographs of the elemental (or isotopic) distribution at the surface of a sample with a spatial resolution of 2 µm or better.

GENERAL NOMENCLATURE

In this section we attempt to define a number of terms used in the literature in connection with the above techniques.

Trace Element. Any element having an average concentration of less than about 100 parts per million atoms (ppma) or less than 100 µg per g.

Thin Film. A material which has been deposited or adhered to a substrate and has a uniform thickness within 20% of its average thickness, which is less than about 10 µm.

Interface. A boundary between two condensed phases. Experimentally, the portion of the sample through which the first derivative of any concentration vs. location plot has a measureable departure from zero. An interface between a solid phase and a gaseous phase is called a surface.

Surface. It is recommended that for the purpose of surface analysis a distinction be made between "surface" in general, "physical surface" and "experimental surface":

Surface - The "outer portion" of a sample of undefined depth; to be used in general discussions of the outside regions of the sample.

Physical Surface - That atomic layer of a sample which, if the sample were placed in a vacuum, is the layer "in contact with" the vacuum; the outermost atomic layer of a sample.

Experimental Surface - That portion of the sample with which there is significant interaction with the particles or radiation used for excitation. It is the volume of sample required for analysis or the volume corresponding to the escape for the emitted radiation or particle, whichever is larger.

Concentration in Experimental Surface (often called Surface Concentration). The amount of the material of interest divided by the total amount of substances in the volume of interest. Concentration may be defined in terms of numbers of atoms (particles) (ppma) or in terms of weight (µg/g).

Monolayer. Coverage of a substrate by one atomic or molecular layer of a species. The term monolayer expresses that all elementary units of the adsortive atoms or molecules are in contact with the surface as opposed to the term MULTILAYER which designates that more than one layer of the adsorptive species covers the surface and not all units are in contact with the surface layer of the substrate (1).

Monolayer Capacity: For chemisorption the amount of adsorbate which is needed to occupy all adsorption sites as determined by the structure of the adsorbent and by the chemical nature of the adsorptive. For physiosorption the amount needed to cover the surface with a complete monolayer of atoms or molecules in close packed array, the kind of close packing having to be stated explicitly when necessary.

Surface Coverage. For both monolayer and multilayer coverage defined as the ratio of the amount of substances covering the surface to the monolayer capacity (1). Coverage is unity when one complete monolayer is deposited onto the sample.

Surface Contamination. Material in the experimental surface which is either not characteristic of the sample or which would not be present if the sample had been prepared in an absolute vacuum by methods not contacting other substances to the sample.

Clean Surface. An experimental surface having no surface contamination observable by means of the used method.

Matrix Effects (2). Effects which cause changes in Auger-Electron, photoelectron, secondary ion yield, or scattered ion itensity, the energy or shape of the signal of an element in any environment as compared to these quantities in a pure element.

(a) Chemical Matrix Effects: Changes in the chemical composition of the solid which affect the signals as described above.

(b) Physical Matrix Effects: Topographical and/or crystalline properties which affect the signal as described above.

Chemical Shift. The displacement of photoelectron or Auger peak energies originating from changes in electron binding energies as a consequence of differences in the chemical environment of the atoms.

Auger Electron Yield. The fraction of the atoms having a vacancy in an inner orbital which relax by emission of an Auger electron.

Photo Electron Yield. The number of photoelectrons emitted by the sample per incident photon.

Secondary Ion Yield. The number of secondary ions generated at the surface of the sample per primary ion.

Sputter Yield (2). The number of particles sputtered from the surface of a target per primary ion.

Escape Depth. After Thomas (3), the distance into the sample measured from the physical surface from which all but a fraction 1/e of the particles or radiation detected have originated.

Depth Profile. Dependance of concentration on depth perpendicular to the surface in a solid sample. It can be obtained by a simultaneous or sequential process of erosion and surface analysis or by measurement of the energy loss of primary backscattered or particles produced by nuclear reactions.

Depth Resolution. The distance between the 84 and 16 percent level of the depth profile of an element in a perfect sandwich sample with an infinitesimally small overlap of the components (4). These limits corresponds to the 2 σ-value of the Gaussian distribution of the Gaussian distribution of the signal at the interface (Fig. 1).

Sputtering. Removal of surface material (atoms, clusters and molecules) by particle bombardment.

Absolute Detection Limit. Smallest detectable amount of an element or compound (3σ-criterion) on or in a particular sample - given in mass units or absolute number of atoms or molecules.

Relative Detection Limit (often incorrectly referred to as sensitivity). Smallest amount of material detectable (3σ-criterion) in a matrix relative to the amount of material analysed - given in atomic, mole or weight fractions.

Qualitative Elemental Specificity. Ability of a method to detect one element in the presence of another element.

PREPARATION OF SAMPLES FOR ELEMENTAL ANALYSIS OF SURFACES

Following whatever polishing may be deemed appropriate for the sample and method of analysis to be used, the following sample cleaning sequence (5) is suggested unless the desired analysis precludes or prevents the use of one or more steps; ultrasonic washing in each of the following solvents sequentially - trichlorethylene, acetone, deionized water, hot deionized water - at least one minute per wash. Heating in vacuum (preferable within the chamber in which the analysis is to be done to prevent contamination during transfer) to 150°C, with the vacuum conditions at least as good as the vacuum conditions to exist during the experiment.

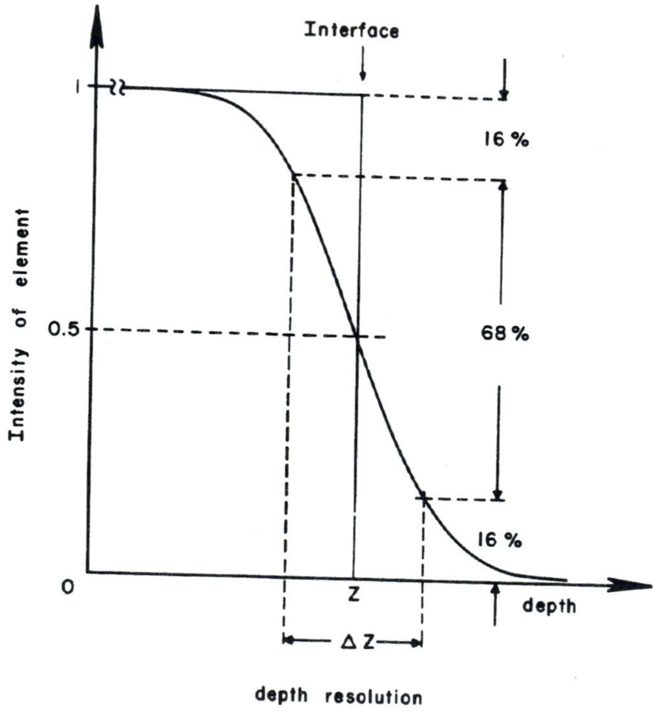

Fig. 1: Definition of Depth Resolution (according to Hofmann(4))

To clean residual adsorbed hydrocarbons and other contaminants clinging to the physical surfaces, very low energy Ar^+ bombarded with a diffuse beam is suggested. Farrell (6) has shown that at 0.1 to 1 eV and ion current density of $\lesssim$ 1 mA/cm^2 Ar^+ bombardment will remove impurities not bonded to the sample. In order to remove oxide layers, Ar^+ bombardment at 300-500 eV, at a current density of approximately 20 μA/cm^2 is suggested (7,8). It should be noted that composition changes in the experimental surface have been found for this method by Shimizu, et al. (9), and others (7,10). Morphology and coordination changes are also likely.

REPORTING OF EXPERIMENTS

The parameters and general information needed to define the experiment performed, such that the results may be properly interpreted and/or reproduced by other workers, constitute the minimum information which should be reported in a scientific paper.

For all of the techniques included in this discussion, the following information should be supplied in a paper:

1. The method of preparation of the sample, including cutting, polishing (material, size and methods), mounting and cleaning, indicating any special treatment such as annealing or fracture under ultra high vacuum.

2. Vacuum in the sample chamber and other chambers of importance under operating conditions (unit: N/m^2) and the method of achieving the vacuum and of determination the pressure.

3. Source of all materials, and their form.

4. Temperature at which the determinations were made if different from room temperature. If the sample is heated (either by external energy input or due to the experiment) the form, extent and characteristics of that heating should be described; e.g., resistive tungsten wire heating of the entire sample so as to raise the sample temperature from 150°C to 500°C at a rate of 10°/min.

5. Instrumental resolution for the measured signal. The stated, e.g. mass spectrometer m/e resolution should be stated for the m/e region in which the data were taken. If the resolution varies, the instrumental resolution for the various portions of a spectrum in which reported data lie should be reported. Mass, energy, spatial, and/or the spectrometer scanning rates (mass numbers per second or Vs^{-1}, etc.) and the integration time used for each peak should be indicated.

6. The instrumental "window", particularly in energy-measuring techniques. The window width and position should be reported (FWHM); if the width as full-width at half height maximum peak varies due to any experimental manipulation carried out during an experiment, the width vs. variable information should be included.

7. Whether a static or scanning method was used to deliver the primary particles or radiation, with the scanning information; e.g., area scanned, number of scans per area and scan rates.

8. Indicate the method of quantification (standards, calibration graphs, etc.) and all corrections made to the raw data and their magnitudes, e.g., charge-up corrections in XPS, secondary ion yield corrections in SIMS, etc.

9. The number of determinations made to obtain a reported value and the precision obtained.

10. Estimated or derived experimental error in all important derived quantities.

11. Incidence angle of the radiation or particles from the source, measured from the normal to the surface.

12. Takeoff angle of the detector; i.e., the angle from the surface normal at which the detection is accomplished. The solid angle of acceptance for the detected particle or radiation should be included.

<u>AES</u> -

a. The primary beam area and approximate shape, beam flux, nominal energy and FWHM of the energy distribution.

b. The instrumental value for the absolute energy of the predominant gold peaks determined on pure gold and the FWHM (full width at half maximum peak height) of the peaks.

c. Gold should be evaporated or sputter-deposited onto the sample when possible, to provide a reference in electrostatic equilibrium with the sample (11), and the predominant gold peaks should be used as a reference of each sample.

d. The numerical value of the modulation frequency.

e. For AES micrographs indicate magnification, spatial resolution, the mode (N(E) or dN/dE), the value of the modulation frequency, the time for a complete scan and if background normalization has been performed.

<u>RBS</u> -

a. The primary ion beam area, shape, flux, nominal energy and FWHM of the distribution.

b. The method used to calibrate the instrument for determining the areas of the peaks.

<u>ISS</u> -

The primary ion beam area, shape, flux, nominal energy and FWHM of the distribution.

SIMS -

a. The primary ion beam area, shape, flux, nominal energy and FWHM of the distribution.

b. The ions used in the experiment and the impurity level, including ions of the desired primary gas having a charge different from that of the majority used.

c. For ion micrographs indicate magnification, spatial resolution, thickness of sample sputtered off during exposure and changes in surface morphology.

XPS -

a. The primary X-ray beam area, shape, nominal energy, and FWHM line width.

b. The "impurity level" of other photon energies impinging on the sample.

c. Gold should be sputter - deposited on the sample when possible, to provide a reference which is in electrostatic equilibrium with the sample (11), and the gold 4f 7/2 peak should be used as a reference for each sample, particularly in making corrections for charge-up.

d. The instrumental value for the absolute energy of the gold 4f 7/2 peak and the FWHM of that peak, determined on pure gold.

Acknowledgements: This report has been critically evaluated by many renowned scientists in the field of surface analysis. The project leaders owe special thanks to all those colleagues who delivered contributions, proposals and criticism. We are especially grateful to R. Baetzold, R. D. Bauer, H. E. Beske, P. Chastagner, W. H. Christie, S. S. Cristy, H. Ebel, T. W. Haas, L. A. Harris, D. A. Harrison, K. F. J. Henrich, W. O. Hofer, S. Hofmann, A. Hubbard, D. N. Hume, A. Lodding, H. Malissa, G. Mason, D. Massignon, J. May, R. L. Park, C. J. Powell, B. Riggs, L. Rinderer, H. Shimizu, J. D. Winefordner and N. Winograd.

At the IUPAC General Assembly in Warsaw these problems have also been discussed with Commission I.6. The advice and suggestions given are gratefully acknowledged.

REFERENCES

1. These definitions have been adapted or taken without modification from the Definitions, Terminology and Symbols in Colloid and Surface Chemistry, Pure and Applied Chem. 31, 577 (1972).
2. These definitions have been adapted or taken without modification from the Tentative Nomenclature for Auger Electron Spectroscopy of ASTM Committee on Surface Analysis E. 42 (Chairman J. R. Cuthill, NBS).
3. J. M. Thomas, Prog. Surf. Memb. Sci., 8, 49-80 (1974).
4. S. Hofmann, Appl. Phys. 9, 59 (1976).
5. R. A. Langley, D. S. Sharp, J. Vac. Sci. Technol., 12(1), 155-59 (1975).
6. T. Farrell, J. Phys. E. (Sci. Instrum.), 6, 977-79 (1973).
7. M. L. Tarng, G. K. Wehner, J. Appl. Phys., 42, 2449-52 (1971).
8. H. S. Wildman, J. K. Howard, P. S. Ho, J. Vac. Sci. Technol., 12(1), 75-78 (1975).
9. H. Shimizu, M. Ono, K. Nakayama, Surf. Sci., 36, 817-21 (1973).
10. M. Ono, Y. Takasu, K. Nakayama, T. Yamashina, Surf. Sci., 26, 313 (1971).

NOMENCLATURE AND SPECTRAL PRESENTATION IN ELECTRON SPECTROSCOPY RESULTING FROM EXCITATION BY PHOTONS

(RECOMMENDATIONS 1975)

COMMISSION ON MOLECULAR STRUCTURE AND SPECTROSCOPY†

The Commission for Molecular Structure and Spectroscopy of IUPAC considers that it would be very valuable if agreement could be reached on nomenclature and conventions for spectral presentation in this rapidly developing and important field of spectroscopy. This type of spectroscopy involves the measurement of the kinetic energy of electrons emitted by chemical substances usually as a result of excitation by monochromatic X-rays (often termed ESCA—electron spectroscopy for chemical analysis) or ultraviolet radiation (often termed photoelectron spectroscopy). In many respects the two methods of excitation are complementary in their applications. In other cases, with which we shall not be concerned here, excitation of electron spectra can be by other processes such as electron bombardment.

AMBIGUITIES IN NOMENCLATURE

Examples of present ambiguities are as follows:

(a) The phrase "*photoelectron spectroscopy*"—commonly abbreviated to PES—is widely used to denote the analysis of the kinetic energies of electrons emitted after excitation by He I or He II photons in the UV region. But, logically, this description is equally applicable to the main features of spectra produced by X-ray photons (see later discussion).

(b) The phrase "*electron spectroscopy for chemical analysis*"—commonly abbreviated to ESCA—is widely used in connection with spectra produced by excitation with X-rays. But the spectra produced by ultraviolet irradiation also lead to applications to chemical analysis, particularly the identification of interatomic groupings within molecules.

It should be noted that other electron spectroscopic techniques, notably electron energy loss and Auger spectroscopy, also have important analytical uses.

A particular problem has arisen, leading to ungrammatical use of language, in connection with abbreviations which use "S" for "spectroscopy". For example, a phrase such as "NMR spectrometer"—meaning "nuclear magnetic resonance spectrometer"—is grammatically correct but, for example, the phrase "the ESCA spectrum" which literally means "the electron spectroscopy for chemical analysis spectrum" is incorrect. It is therefore recommended that as a general rule abbreviations or acronyms that incorporate S for "spectroscopy" or "spectra" should be discouraged. Exceptions are ESCA and PES which, for historical reasons, have been widely adopted. It is recommended that these abbreviations be limited to their respective contexts and that in each case care should be taken to see that the acronym or abbreviation is not used in grammatically incorrect situations, such as for example, in the phrases "ESCA spectroscopy" or "PES spectrometer". It is also recommended that any abbreviations within electron spectroscopy should include the letter E standing for "electron".

The widely adopted description of spectra excited by UV photons, such as from He I or He II, as "photoelectron spectra" is recommended for general use. With this type of excitation of electrons from valence orbitals virtually all the electrons of which kinetic energies are measured derive from photoionization processes. Higher energy excitation, such as by X-ray photons, commonly leads to the production of electrons derived from Auger processes as well as from photoionization processes. Such complete and uninterpreted spectra are therefore logically described by the more general phrase "electron spectra". However, the more specific description "photoelectron spectra" may logically be applied to the appropriately identified features of such spectra. Under such circumstances, the spectra should be reported, as appropriate, in the fashion "the He I photoelectron spectrum" (spoken—the helium one photoelectron spectrum) or the "Al K_α photoelectron spectrum" (spoken—the aluminium K alpha photoelectron spectrum). The use of the terms "photoemission spectroscopy" and "photoionization spectroscopy" in these contexts is to be discouraged.

THE ENERGY SCALE

Three designations of the energy scale—ionization energy, binding energy or electron kinetic energy—are in common usage in the presentation of diagrams of spectra derived from X-ray or UV excitation. The first two of these are alternative names for the same quantity but the electron kinetic energy, although linearly related to the others (it is the difference between the energy of the exciting photon and the ionization energy) increases in magnitude as the others decrease. Ionization potential (in volts) is also used as an alternative to ionization energy (in electron volts).

The general use of *ionization energy*, E_i, is recommended for the scale at the top of all spectral diagrams. It is also recommended that the experimentally measured parameter, electron kinetic energy, should be plotted along the bottom of the spectral diagram in the same units. The latter parameter is meaningful whether the electrons measured derive from photionization or Auger processes. However, those spectral features derived from Auger processes, such as occur when using high energy excitation by X-rays, should be indicated clearly by asterisks because the ionization energy scale is not applicable to them.

The customary unit for E_i or electron kinetic energy per molecule is the electron volt, eV. Alternatively, where there is need to relate these energies to other quantities expressed in SI units, the unit recommended for the *molar ionization energy* is joule per mole, J mol^{-1}.

The direction of the ionization energy scale is not well defined in the literature. For X-ray excitation, ionization (binding) energy is usually plotted so as to increase from right to left. For ultraviolet He I or He II excitation the literature shows a limited preference for ionization increasing from left to right. For consistency between the two related fields it is recommended that the spectra be plotted with increasing electron kinetic energy to the right, i.e. increasing ionization energy to the left. This presentation is consistent with other branches of electronic spectroscopy where spectral features involving higher levels of molecular excitation are normally found to the left of the spectral diagram.

OTHER RECOMMENDATIONS FOR THE PRESENTATION AND PUBLICATION OF SPECTRAL DATA

(i) The primary ordinate scale of the normal differential curve should be given as (dI/dV), i.e. ampere volt^{-1}, AV^{-1}, or as (counts s^{-1}).

(ii) Information about sweep rate (in Vs^{-1}) and/or integration time (in s) should be given where appropriate.

(iii) A statement should be given of the method of calibration and the precision achieved.

(iv) Because different types of apparatus exhibit different sensitivity as a function of electron kinetic energy, relevant information should be given about the performance of the equipment used.

† *Titular Members*: N. Sheppard (Chairman), M. A. Elyashévich (Vice-Chairman), F. A. Miller (Secretary), E. D. Becker, J. H. Beynon, E. Fluck, A. Hadni, G. Zerbi, *Associate Members*: G. Herzberg, B. Jeżowska-Trzebiatowska, Y. Morino, S. Nagakura, C. N. R. Rao, Sir Harold Thompson, D. W. Turner.

RECOMMENDATIONS FOR PUBLICATION OF PAPERS ON METHODS OF MOLECULAR ABSORPTION SPECTROPHOTOMETRY IN SOLUTION BETWEEN 200 AND 800 nm

COMMISSION ON ANALYTICAL NOMENCLATURE

Prepared for publication by

G. F. KIRKBRIGHT

Imperial College of Science and Technology, London, U.K.

INTRODUCTION

When new reagents for molecular absorption spectrophotometry in solution, or modified methods employing better known reagents, are presented in the literature at an increasing rate, it is inevitable that some confusion and contradiction should result with regard to the best reagent and optimum operating conditions for use in a particular spectrophotometric determination. Because of the non-critical nature of many publications concerning spectrophotometric methods, the advantages and disadvantages of the methods studied are often not made clear, and their comparison in terms of sensitivity and selectivity with well-established methods is not fully described. Such texts as those of Sandell[1], Boltz[2], Snell and Snell[3] and the IUPAC Commission on Optical Data[4] assist the critical assessment of various spectrophotometric reagents. The IUPAC publication provides the optical and chemical data necessary to evaluate the methods applicable to a particular problem. These texts are, however, largely concerned with well-established and proven reagents and methods; a newcomer to the field must be bewildered by the wealth of other information available in the literature for a variety of less well-known reagents. The capabilities of these reagents are less easily assessed, and the situation is aggravated by the different approaches employed for the examination of reagents and operating conditions for a particular determination. Different methods are used in the investigation of variables and interferences affecting the reaction which is to be made the basis of a spectrophotometric procedure. Different opinions and individual preferences prevail with regard to the form in which these data are published and concerning the expression of sensitivity, precision and accuracy.

The recommendations of the principal journals of analytical chemistry give a guide to the preferred method of presentation of precision and accuracy data, and most journals recommend that publications conform to a particular format. West[5] and Mellon[6] have made recommendations concerning the general approach to the development of a spectrophotometric procedure, and Yoe[7] presents a suggested outline for a comprehensive study of a new colour reaction and its development into a spectrophotometric method. Although it is obviously desirable, no agreement exists with regard to the presentation of data so that they are of maximum usefulness to other workers.

This report, which is based on the paper by Kirkbright[8] invited by the Editorial Board of Talanta as part of its series on publication of analytical methods, has been prepared for the Commission by Dr. G.F. Kirkbright.

NOTE ON NOMENCLATURE

In molecular absorption spectrophotometry the terms 'extinction', 'optical density' and 'absorbance' have been used to express the logarithm of the ratio of the reciprocal internal transmittance $\left[\log_{10}(T/T_0)\right]$.

The term 'absorbance' is preferred because it incorporates the root word which characterises the physical process involved.

RECOMMENDATIONS

The following recommendations are made concerning the <u>minimum</u> amount of data which should be incorporated into a paper which describes a method of analysis using molecular absorption spectrophotometry in solution.

1. Reagents

(a) Purity and stability of solid reagents used to form absorbing component
(b) Any procedure required for purification of solid reagents
(c) Data on preparation and properties of new reagents to allow their production and characterization by other workers
(d) Purity and stability of other reagents used (solvents, buffers, standard solutions, etc.)

(e) Suitable solvent(s) for any reagents used to form absorbing component and for the reaction to be studied. Reasons for choice of solvent (e.g., to give greatest sensitivity, highest stability, low cost, etc.)
(f) Stability of reagent solution under laboratory conditions (i.e., effect of daylight, oxygen, CO_2)

2. Spectral Characteristics

(a) Absorption spectra for reagent(s) alone and in presence of component to be determined. Statement of concentrations employed and composition of blank solution used to record spectra
(b) Wavelength chosen for determination. Reasons for choice (e.g., to give maximum sensitivity or precision)
(c) Effective spectral half-intensity band-width at chosen wavelength of the spectrometer employed
(d) Report of check on accuracy of wavelength (or frequency) scale

3. Effect of Concentration of Hydrogen Ion

(a) Details of composition of buffer system and ionic strength of solutions used for absorption measurements
(b) Method of calibration for 'pH' especially when partly non-aqueous solutions are used
(c) Effect of pH variation on absorbance at chosen wave-length (and on wavelength of maximum absorbance, if any) over range of pH and ionic strength expected in the examination of real samples

4. Effect of Reagent Concentration

(a) Effect on yield (i.e. final absorbance) of absorbing component of independent variation of concentration of each reagent used (at concentration of component to be determined and ionic strength similar to that expected in real sample solutions)

5. Order of Addition of Reagents and Rate of Reaction

(a) Effect of order in which reagents are mixed on rate of reaction and yield (i.e. final absorbance) of absorbing component (at optimum pH, ionic strength and reagent concentration)
(b) Time required to produce stable absorbance reading against reagent blank

6. Stability of Absorbing Component

(a) Stability of absorbance readings at wavelength chosen for analysis. Details of any precautions necessary to achieve stable absorbance reading (e.g., storage in dark, absence of dissolved oxygen, etc.)

7. Effect of Temperature

(a) Effect on rate of formation, yield (i.e., final absorbance) and stability of variation in temperature over range which might be encountered in laboratories (i.e., *ca.* 15 to 30 °C)
(b) Recommended temperature for operation of method

8. Calibration Graph and Optimum Concentration Range

(a) Description of analytical procedure used to establish calibration graph of absorbance *vs.* concentration of component to be determined
(b) Shape of graph (or equation for graph). Statement of range of concentration of test component over which linearity is obtained. Ringbom plot[5]. The publication of straight-line calibration graphs is not now the editorial policy of most journals
(c) Check on photometric accuracy of absorbance scale of spectrometer (with calibrated glass filters or liquid reference standards) (see e.g. ref. 10)

9. Sensitivity

(a) Statement of molar (linear) absorption coefficient, ε, of the absorbing component at wavelength of measurement and spectral half-intensity band-width, temperature and solvent employed. ε is defined by $\varepsilon = A/(l.c_B)$ where A = absorbance, l pathlength through the test system, and c_B = concentration of absorbing component. The unit employed for ε must be specified. $m^2\ mol^{-1}$ is preferred.

This definition of sensitivity is in accordance with that recommended for use in spectrochemical analysis[11], i.e. ε is the differential quotient(dA/dc) of the calibration function (analytical graph of absorbance *vs* concentration) for unit path-length through the test solution.

The absorptivity may be expressed with regard to one mole of the component to be determined per litre of solution instead of to one mole of the absorbing species. This may be termed the analytical molar (decadic) absorption coefficient and does not require knowledge of the molecular formula of the absorbing component or even the presence of the component determined in the solution whose absorbance is measured. This term may be applied readily to statement of sensitivity in combined solvent extraction-photometric and indirect methods of analysis.
(b) The characteristic concentration (sometimes known as the sensitivity index) i.e. number of micrograms of test component, converted to the absorbing component, which in a column of solution of cross-section 1 cm^2 shows an absorbance of 0.001, should not be used.

10. Interference

(a) Definition employed to establish presence or absence of interference from other components present with test component. Definition should be related to precision of method development (i.e., interference when error is greater than 2 or 3 times the standard deviation of method obtained in absence of foreign component)
(b) Effect of large excesses (e.g. 100 or 1000 times the substance concentration of test component) of other components on absorbance produced for test component at several concentration levels (e.g., near to top and bottom of recommended calibration range). Other components investigated must include those which show similar chemical properties to the component to be determined or which are commonly found with it in samples to be analysed. Results to be tabulated without concealment of data (e.g., as 'does not interfere' statement)
(c) Limitingpermissible concentrations of those components interfering at concentrations in (b) above
(d) Result of check that when two or more components which do not interfere individually, are present together with the component to be determined, that there is no slight interference from each which is additive
(e) Chemical or physical reasons for interferences observed (e.g., absorption at wavelength of measurement, precipitation, oxidation of reagent or test component).

11. Elimination of Interferences

(a) Recommendations for removal of effect of interfering components (e.g., subtraction of absorbance of extraneous component from that caused by component determined, use of masking agent) or preliminary separation of either the component to be determined or interfering component

12. Precision

(a) Examination of precision of complete analytical procedure in presence of any necessary masking agents or after preliminary separation procedures

Expression of relative standard deviation (as decimal fraction) for complete analytical procedure at low absorbance value and statement of number of measured values from which this is derived. The best straight line fit for the calibration data may be calculated by the method of least squares and the confidence limits for the slope of the calibration function established.

13. Accuracy

(a) Accuracy of method applied to standard or reference samples, or
(b) Comparison of analytical results with those obtained for identical samples by different established procedures.

14. Nature of Absorbing Component

(a) Report of investigation of nature of component
(b) Empirical formula of component in solution
(c) Conditional stability constant of complex (where applicable) under stated conditions of ionic strength, temperature, pH, etc., used in reaction

Many hundreds of methods have been proposed for the spectrophotometric determination of the elements copper and iron and anions such as phosphate and fluoride. Further methods should only be proposed for these and other ions if they are found to be markedly superior in several respects (sensitivity, selectivity, spectrophotometric stability, etc.) to the best of the existing methods. When a new method is proposed for any species its sensitivity and selectivity must be compared with other recommended methods for the spectrophotometric determination of the same species, and any particular advantages or disadvantages of the method proposed must be mentioned. It is again stressed that the above requirements represent the minimum acceptable amount of information which must be included in any report of a new spectrophotometric method of analysis.

REFERENCES

1. E.B. Sandell, Colorimetric Determination of Traces in Metals, Interscience, New York, 3rd Edn., (1959).
2. D.F. Boltz, Colorimetric Determination of Nonmetals, Interscience, New York, (1958).
3. F.D. Snell and C.T. Snell, Colorimetric Methods of Analysis, Vol II, Van Nostrand, Princeton, (1959).
4. International Union of Pure and Applied Chemistry, Commission on Optical Data, Spectrophotometric Data for Colorimetric Analysis, Butterworths, London, (1963).
5. T.S. West, Analyst, 87, 630, (1962).
6. M.G. Mellon, Analytical Absorption Spectroscopy, J. Wiley and Sons, New York, (1950).
7. J.H. Yoe, Analytical Chemistry, 29, 1246, (1957).
8. G.F. Kirkbright, Talanta, 13, 1, (1966).
9. A. Ringbom, Z. Anal Chem., 115, 332, (1939); A. Ringbom and F. Sundman, ibid, 115, 402, (1939); 116, 104, (1940).
10. R.W. Burke, E.R. Deardoff and O. Menis, Journal of Research, National Bureau of Standards, 76A, 469, (1972).
11. IUPAC Information Bulletin No 26, Nomenclature, Symbols, Units and their usage in Spectrochemical Analysis - II. Terms and Symbols related to Analytical Functions and their Figures of Merit, (1972).

Section 9

Crystallography and Electron Diffraction

Single-Crystal Data

This section treats data derived from X-ray and electron diffraction. Although no section of the Green Book is devoted directly to diffraction measurements, there is a section on the solid state that contains a few relevant definitions for lattices. The International Tables for Crystallography (*53*) contain more detailed information.

Crystallography is a highly organized discipline in which the final product is usually entered into standardized tables or automated databases to provide data on the unit cell, density, and symmetry. "Primary Crystallographic Data" (*54*) (reprinted here) provides detailed information on what should be presented in the publication of single-crystal data. Reference 55 contains similar information.

Papers and computer programs, which can assist the user in determining the unit cell and symmetry, are available. Because these papers and programs refer to the design of the measurement and to reporting it, they are to some extent supplementary to our primary purpose and are not reprinted here. Reference 56 is a computer program that can be used to calculate the reduced cell of a lattice, to calculate and reduce specified derivative supercells or subcells, and to calculate unit cell transformations and matrix inversions. In the important matter of determining symmetry, a matrix method (*57*) has been devised. An automated form is in development.

Reference 58, which considers Bravais lattice types and arithmetic classes, and reference 59, which handles polypeptide structures, cover nomenclature on crystal families. Reference 60 covers the definition of symmetry elements, and reference 61 studies statistical descriptions.

Powder Data

"Powder Data" (*62*) (reprinted here) contains general recommendations for the publication of powder diffraction data from the International Union on Crystallography. Reference 63 provides guidelines for publication of Rietveld analyses and pattern decomposition studies. "The Standard Data Form for Powder Diffraction Data" (*64*) (reprinted here) lists items that are considered essential documentation of data and information that is highly desirable.

Electron-Diffraction Data

"Guide for the Publication of Experimental Gas-Phase Electron Diffraction Data and Derived Structural Results in the Primary Literature" (*65*) is reprinted here.

Primary Crystallographic Data*

By O. Kennard†,
University Chemical Laboratory, Cambridge, England

J. C. Speakman
Department of Chemistry, University of Glasgow, Scotland

and J. D. H. Donnay
The Johns Hopkins University, Baltimore, Maryland, U.S.A.

(*Received* 18 *October* 1966)

A report containing a list of recommendations on the presentation of crystallographic data in primary publications relating particularly to single-crystal work. The more important items of information are discussed in detail with examples. Numerical values of certain constants in common use are recorded.

The increasing number of crystallographic publications, not only in *Acta Crystallographica*, but also in many other journals, has prompted the Commission on Crystallographic Data to draw up the following set of recommendations. The object of these recommendations is to ensure that essential information is given and to suggest a concise arrangement with standardized symbols. Also included are references to the latest numerical values of X-ray wavelengths and of Avogadro's number, for use with the ^{12}C scale of atomic weights. Flexibility in using the standard form is desirable and, in special circumstances, modifications of these recommendations may be necessary.

Previous recommendations were proposed by: Bernal, Ewald & Mauguin (1931); McCrone (1948, 1956); and The Chemical Society of London (1959). Relevant information is also contained in the three volumes of *International Tables for X-ray Crystallography* (1952, 1959, 1962) and in *Crystal Data* (Donnay, Nowacki & Donnay, 1954; Donnay, Donnay, Cox, Kennard & King, 1963). Attention is particularly directed to *Notes for Authors* (1965) and to the *Recommendations of the Commission on Crystallographic Computing* (1962).

For convenience the recommendations will be presented under three headings: (I) *Crystal data*, (II) *Structural data* and (III) *Structure factors.*

I. Crystal data

In this section we wish to suggest a convenient sequence for the essential items of information. Not all entries are applicable in every case; some may be omitted.

Standardized layout

Name(s) of substance. Chemical formula(e). Formula weight F.W. Melting point M.P. Provenance and size of crystals. Crystal system. Goniometric data. Crystal forms. Point group (but only if indicated by morphology). Habit, including twinning if observed. Cell parameters a, b, c, α, β, γ, as necessary (with standard deviations), at stated temperature. Cell volume V (with standard deviation). Measured density D_m in g.cm^{-3} (with method of measurement and limits of error). Number Z of formula units per cell. Calculated density D_x. Type(s) of X-rays used. Numerical value of wavelength used in calculating cell dimensions. Brief note on X-ray methods. Absorption coefficients for X-rays used for intensity measurements. Size of crystal used for such measurements. Total number of observed reflexions, and percentage of radiation sphere explored. Unusual correction factors. Space group. General position coordinates.* Any atoms in special positions. Any implications for molecular symmetry. Anomalous dispersion. Optical data (with wavelength of light used and temperature).

Notes

(1) *Name of substance.* For organic and inorganic compounds, as far as possible the IUPAC rules of chemical nomenclature should be used. Accepted trivial names, which facilitate identification, should also be given, as synonyms. In case of a new compound, the name originally used should be given, together with a reference to the chemical work. For minerals, the recommendations of the International Mineralogical Association are to be followed. Polymorphic forms should be distinguished. (Do not use the word *form* or *modification* to refer to a hydrate or to a twin).

(2) *Chemical formula.* For organic compounds a structural formula should accompany the name, except possibly with simple molecules. In addition, a formula that can be typed on a single line (such as CH_3CO_2H) is desirable. The alphabetized formula, in the form $C_xH_yA_pB_qD_r\ldots$, should also be given, as a check and to help the indexer. Any abbreviation should be explained (*e.g.*, φ stands for C_6H_5).

For inorganic compounds the formula should be written so as to reflect the conclusions derived from the crystal structure. Use of the dual formula, expressed in terms of oxides, is to be discouraged.

In case of significant incomplete site occupancy, give the non-stoichiometric formula.

For a phase of variable composition, give the formula of the specimen actually studied; for example, $(Fe_{1-x}Ni_x)_2O_3$ as $Fe_{1 \cdot 34}Ni_{0 \cdot 66}O_3$ or $(Fe_{0 \cdot 67}Ni_{0 \cdot 33})_2O_3$. Note that $(Fe_{1 \cdot 34}Ni_{0 \cdot 66})_2O_3$ is not acceptable.

(3) *Formula weight* (*or molecular weight*) F.W. Use the unified scale of atomic weights, based on ^{12}C = 12·0000. Make sure that F.W. refers to the formula actually given. For complicated mineral structures the formula weight of the whole cell is often preferable.

(4) *Point group.* When the point group has been determined from morphology, it should be given, in its oriented Hermann–Mauguin symbol. Usage of the Groth names, based on the general forms, to designate point groups is to be discouraged.

(5) *Habit.* The following adjectives are recommended: acicular, long prismatic, prismatic, short (or stout) prismatic, equant, thick tabular, tabular, thin tabular (or platy), leafy, flaky, pyramidal, dipyramidal and

* Recommendations prepared on behalf of the Commission on Crystallographic Data, International Union of Crystallography.

† External Scientific Staff, Medical Research Council.

* According to current English usage, it is cumbersome to differentiate between a point x,y,z ('position') and the collection of equivalent points ('set of positions'), for which a singular collective noun would be highly desirable. In this report we shall use *site* to designate the point itself and *position* to mean the point set; these terms correspond to the German *Punkt* and *Punktlage* respectively and to the French *point* and *position* respectively.

fibrous. The symbol for a face (···) or for an edge [···] follows where appropriate. Thus 'tabular (010)' means that (010) is the largest face, whilst 'prismatic *b*' means that the crystal is elongated in the *b* direction. Note that in this case of an axial direction, the designation *b* is preferred to [010]. 'Lath-shaped (010), elongated *c*' is self-explanatory.

If twinning is observed, it should always be described. For recommended nomenclature consult *International Tables* (1959, Vol. II, section 3). (Note especially that a twin is a heterogeneous edifice composed of two or more crystals – and not the other way round.)

(6) *Cell parameters.* Except for some good reason, the cell to be chosen should be the Bravais-reduced cell, *i.e.* the cell that has its edges along the shortest three lattice translations or lattice symmetry directions whenever such are available. [Donnay, 1943, 1952; see Preface to *Crystal Data* (Donnay, Donnay, Cox, Kennard & King, 1963); *cf.* also *International Tables* (1952, Vol. I, pp. 530–5) for the Delaunay cell]. Note that a triclinic Delaunay-reduced cell may or may not coincide with the Bravais-reduced cell. In every crystal system the set of coordinate axes should be right-handed. The cell can be uniquely oriented according to the following rules: In the triclinic system, choose $c<a<b$ with α and β non-acute.* In the monoclinic system, select the shortest two translations in the net perpendicular to the symmetry direction *b*, take $c<a$, β non-acute, and use appropriate centring. In the orthorhombic system, choose $c<a<b$ and appropriate centring if one-face-centered (*A*, *B*, or *C*). In both monoclinic and orthorhombic systems, this choice may result in a setting different from that used, for illustrating the space group, in *International Tables* (1952), in which case the coordinates of the equivalent sites in the necessary positions will have to be specified. Inasmuch as *International Tables* does not list the coordinates of equivalent points in all possible settings, authors of papers describing crystal structures may find it easier to follow the setting of *International Tables*.

In the high-symmetry systems, the choice of the shortest three lattice translations uniquely defines the cell. A tetragonal cell should be taken as *P* (not *C*) or *I* (not *F*); a hexagonal cell should be primitive; its symbol is now *P* (formerly *C*); the triple cell *H* is not used. In the cubic system, the cell is always a cube (*P*, *I*, or *F*).

If the crystal is better described by means of another cell, the conventional cell should also be given, together with the transformation matrix (from-unconventional-to-conventional). To save space the matrix rows may be given in linear form *uvw*/*u'v'w'*/*u''v''w''*. A program, written in FORTRAN, is available to perform this transformation (Takeda & Donnay, 1964).

(7) *Cell dimensions and standard deviations.* Lengths should be given in Å (not kX), angles in degrees and minutes or in degrees and decimal fractions. The cell volume should be given in Å^3.

Pending decision of the International Union of Crystallography, the numerical values of X-ray wavelengths remain those published in *International Tables* (1962).†

* In the low-symmetry systems the orientation of the cell is a matter of convention. (Note that, from the viewpoint of systematization, any cyclic permutation, such as $a<b<c$ with β and γ non-acute, is equally acceptable.)

† Note, however, the new values of Bearden (1964). They are expressed in Å*, a unit that is equal to the Å within a few parts per million (probable error), and is defined by the new primary standard for X-ray wavelengths, $\lambda(\mathrm{W}\ K\alpha_1)=0{\cdot}2090100$ Å*. Their order of accuracy cannot be attained in most crystallographic work.

All limits of error should be given in the form of a standard deviation σ; *e.g.*, $12{\cdot}431\pm11$, which means $\pm0{\cdot}011$ (*International Tables*, 1959, pp. 85–91). For the given uncertainty to qualify as a standard deviation, it must have been calculated by a least-squares treatment; for example, of the $\sin^2\theta$ values. If the uncertainty given differs from one standard deviation – if, for instance, it has been multiplied by three – this should be made clear: '$12{\cdot}431\pm33$ (three standard deviations)'. Optimistic guesses at the accuracy attained are to be shunned.

(8) *Number Z of formula units per cell and calculated density D_x.* The number *Z* must refer to the formula already specified, with F.W. expressed on the ^{12}C scale. For Avogadro's number, use $N_A=0{\cdot}602252\times10^{24}$; its reciprocal is $1{\cdot}660435\times10^{-24}$ (*National Bureau of Standards, Technical News Bulletin*, 1963). Note that the new values differ from those in *International Tables*, 1962, pp. 39–45).

(9) *Type(s) of X-rays.* The nature of the radiation used for intensity measurements should be specified (example: Ni-filtered Cu radiation). The numerical value of the wavelength of the radiation used to determine the cell dimensions should also be reported, to facilitate future corrections should the numerical values change.

(10) *Observed reflexions.* When the observed reflexions conform to the space-group criteria in the orientation of *International Tables*, no statement is needed. If the space group is presented in another orientation, the presence criteria should be stated. They should also be given when they show systematic anomalies.

(11) *Space group.* When the space group is unambiguously indicated by the observed reflexions and the choice of axes is that of the *International Tables*, simply give its oriented Hermann–Mauguin symbol.

When the choice of axes is not that of *International Tables*, the Schoenflies symbol should be added as a means of identifying the space group. Thus *Pnaa*, which appears in *International Tables* as *Pccn* (no. 56), should be followed by D_{2h}^{10}. The matrix of the transformation will also be useful. From *Pnaa* to *Pccn*=010/001/100.

When the observed absences are compatible with two or more space groups, the use of a condensed diffraction symbol (see Donnay & Kennard, 1964) is recommended. Thus the various possibilities *Pmmm*, *P2mm*, *Pm2m* and *Pmm2* are all covered by the symbol *P****. When one of the space groups has been chosen, give the reasons for the choice. For example: *Pmmm* was adopted because a statistical study of the intensities indicated centrosymmetry, which was subsequently confirmed by the structure analysis.

(12) *General position.* The coordinates of the equivalent sites should be given only when the labelling of axes differs from that used in *International Tables* (1959). They should be put into the condensed form. For example, for $P2_1/n$: $\pm(x,y,z;\ \tfrac{1}{2}+x,\tfrac{1}{2}-y,\tfrac{1}{2}+z)$.

(13) *Optical data.* For transparent substances the values of the indices of refraction and the optical orientation should be stated preferably for specified type of light and temperature.

Examples:

Cubic: *n* (Na, 20 °C) = $1{\cdot}650\pm2$.

Uniaxial: n_E $1{\cdot}650\pm2$, n_O $1{\cdot}595\pm2$ (Na, 25 °C).

If dichroic: n_E $1{\cdot}650\pm2$ (colourless), n_O $1{\cdot}595\pm2$ (blue).

Orthorhombic: Opt. neg. (5893 Å, 25 °C): 1·500 (*b*), 1·74 (*c*), 1·77 (*a*), 2*V* 39°; disp. $r>v$, weak.

Monoclinic: Opt. neg. (5893 Å, 25 °C): $1{\cdot}578\pm2$ (*b*), $1{\cdot}676\pm2$ (+41° to *c*), $1{\cdot}710\pm5$ (+31° to *a*), 2*V* 58°;

disp. $v > r$, strong. An extinction angle with c is taken as positive if in the obtuse angle β.

Triclinic: The optical orientation can best be presented by means of a stereographic projection. It may also be of interest to describe the optical properties of the cleavage planes and of the principal faces.

Pleochroism can be indicated after the values of the indices: Opt. neg.: 1·700 (*a*) (pale yellow), 1·750 (*b*) (golden yellow), 1·770 (*c*) (golden yellow).

Note that the given $2V$ (or $2E$) should be the *measured* value.

II. Structural data

(1) *Crystal-chemical unit.* To define the structure of a crystal, coordinates and other parameters must be stated for the atoms in one asymmetric unit (Schoenflies's *fundamental domain*). This asymmetric unit should be chosen so as to contain a chemical entity, even when this means giving, to some of the atoms, coordinates that are negative or improper fractions. Such a domain may be called the *crystal-chemical unit* (CCU). In a molecular compound, it may contain a single molecule, two or more molecules, half a molecule or less. In an ionic compound it may be only part of a cation and part of an anion (as in $PCl_4^+PCl_6^-$).

(2) *Numbering of atoms.* On a structural formula or diagram, the atoms may be numbered irrespective of chemical type [*e.g.*, O(1), O(2), C(3), C(4), C(5)] or according to chemical elements [*e.g.*, O(1), O(2), C(1), C(2), C(3)]. In either case the number should preferably be in parentheses, so as to avoid unwanted chemical implications (O_2, C_2).

(3) *Choice of origin.* The choice of the origin of coordinates should be clearly specified, particularly on the drawings.

(4) *Atomic coordinates.* Use fractional (or trimetric) coordinates x,y,z, defined by $X/a, Y/b, Z/c$, in which a,b,c are the cell edges and X,Y,Z the coordinates in Ångstroms (symbols of *Structure Reports*). Decimal points can be avoided by showing (say) 0·1907 as 1907, with $10^4\ X/a$ as column heading.

Ordinary coordinates (in Å) on orthogonal axes are useful when the symmetry is hexagonal, monoclinic or anorthic. Their symbols X', Y', and Z' must be defined; for example, in the monoclinic case: $X' = X \sin\beta$, $Y' = Y$, $Z' = Z + X\cos\beta$.

The published data should enable a reader to repeat the author's calculations: coordinates should, therefore, be given to one place (or possibly two places) beyond the last significant digit. Since the standard deviation will have been given, there will be no doubt as to which digits are significant.

(5) *Symmetry-related units.* It is often necessary to refer to atoms that lie outside the crystal-chemical unit. *Symmetry-related units*, including those repeated by lattice translations, are best designated by Roman numerals (preferably in lower case). Each of them contains one point equivalent to x,y,z. Example:

CCU	x,y,z
i	$x+\frac{1}{2}, y-\frac{1}{2}, z$
ii	$x+1, y, z$
iii	$x+\frac{3}{2}, y-\frac{1}{2}, z$

An intermolecular contact can be denoted as C(3) · · · O(2)(ii). Avoid using a continous line, which may suggest a chemical bond, between two non-bonded atoms.

(6) *Atomic scattering factors.* State the numerical values actually used. When the values given in *International Tables* (1962) are used, a statement to this effect is sufficient.

(7) *Thermal parameters.* If a single isotropic temperature factor e^{-M} has been used for all atoms, its value can be stated in the text. Individual Debye factors (B in $Å^2$) can be listed by adding a column to the table of coordinates. Individual anisotropic vibrational parameters require six parameters for each atom in a general position; they have been given in various forms. Definitions should be clearly stated. The exponential function in which the b_{ij} values appear, for instance, should be explicitly given. If the author lists the components U_{ij} of the tensor that expresses the ellipsoid representing the mean-square amplitude of vibration of the atom (Cruickshank, 1965), the axes of reference should be specified (a, b', c^*, for example, or some other orthogonal axes). The same remarks apply if librational or translational parameters are given for a molecule or part of a molecule acting as a rigid body.

(8) *Diagrams.* A good diagram is essential. It should show the whole, or a significant part, of the cell, with an indication of the symmetry elements. Such a diagram is a convenient place for giving the numbering of the atoms in the crystal-chemical unit and the numbering of the symmetry-related units.

The origin and the coordinate axes should be clearly marked and their positive senses shown by arrows. As in *International Tables* (1952), the axes should constitute a right-handed set. If feasible, place the origin of an orthogonal projection in the upper left-hand corner.

For example, if a triclinic cell is projected parallel to the c direction, onto a plane perpendicular to the c direction, the axes in the plane of the figure should be labelled b' and a', where $b' = b \sin\alpha$ is directed towards the right-hand side and $a' = a\cos\beta$ (approximately) towards the foot of the page. The positive c direction should be upwards (out of the paper); letting it point downwards (into the paper) has been a source of confusion. The three principal projections are illustrated in Fig. 1.

III. Structure factors

A paper reporting a successful structure analysis always includes a statement of the residual

$$R = \Sigma\ ||F_o| - |F_c|| / \Sigma\ |F_o|$$

(the treatment of non-observed reflexions should be specified). It is also customary to give a table of observed amplitudes $|F_o|$ and calculated structure factors F_c. There are strong arguments in favour of this procedure:

(*a*) The table is the ultimate evidence for the validity of the analysis.

(*b*) With these data, a future worker can always resume the analysis to check it, correct it, or carry out further refinement.

(*c*) The observed amplitudes constitute an important body of data, which it may not always be possible to remeasure. The crystal used may be the only single crystal in existence!

It may be argued that structure-factor tables occupy a good deal of space in a journal and are not of interest to the majority of readers. The problem is sometimes avoided by omitting the table of structure factors altogether, and substituting a statement that it may be obtained from the author or from the Library of Congress or other depository.

It seems desirable that structure factors should continue to be published until some better system of making data available is developed. Space may be saved by economical arrangement of the table. The table is clearer if only the currently changing index is listed in the first column with an indication whenever one of the other two indices is changed.

For a centrosymmetric crystal, the second and third columns will contain $|F_o|$ and F_c.

For a non-centrosymmetric structure, where phases have to be specified, the information can be given as $|F_o|$, followed by:

(1) A_c and B_c, the real and imaginary parts of F_c. (This carries the full information but has the disadvantage that $|F_c|$ is not explicitly given and cannot be compared with $|F_o|$).

(2) $|F_c|$, cos α and sin α. (This requires one additional column).

(3) $|F_c|$ and α.

Structurally absent reflexions may be omitted provided the calculated values are also small. Any discrepancies should be discussed in the text.

Some bolder suggestions should be included in this Report. It has been pointed out (Lipscomb 1963, 1964) that only $|F_o|$ needs to be recorded, since a reader with access to a computer can recalculate F_c from the structural parameters. Space can be further saved, though with sacrifice of elegance, by placing the $|F_o|$ values

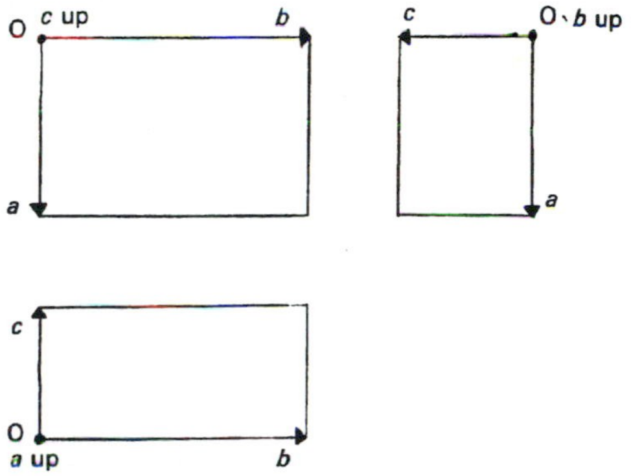

Fig. 1. The three principal projections with the conventional orientation of the axes.

serially along the line, with merely an indication of the changing indices.

In any case a large typescript table should be prepared (*e.g.*, as computer output) and considerably reduced to a photograph no larger than a full page in the journal. Such photographs (one or more) should be suitable for direct photographic reproduction. The editor of *Acta Crystallographica* reports that such a procedure greatly facilitates publication.

Our thanks are due to Professor Dame Kathleen Lonsdale, F.R.S., to Professor Caroline MacGillavry, to Professor P. P. Ewald, F.R.S., to Professor Frank-Kamenetskij, and to Professor V. B. Tatarskij for valuable suggestions.

References

BEARDEN, J. A. (1964). *X-Ray Wavelengths*. U.S. Atomic Energy Commission, Div. of Tech. Information Extension, Oak Ridge, Tennessee.

BERNAL, J. D., EWALD, P. P. & MAUGUIN, C. (1931). *Z. Kristallogr.* **79**, 498.

Chemical Society of London (1959). *Proc. Chem. Soc.* **5**.

COHEN, E. R. & DUMOND, J. W. M. (1964). *Proceedings of the Second International Conference on Nuclidic Masses and Related Constants*. Vienna: Springer Verlag.

CRUICKSHANK, D. W. J. (1965). *Acta Cryst.* **19**, 153.

DONNAY, J. D. H. (1943). *Amer. Min.* **28**, 507.

DONNAY, J. D. H. (1952). *Amer. Min.* **37**, 1063.

DONNAY, J. D. H., NOWACKI, W. & DONNAY, G. (1954). *Crystal Data*, Memoir **60**. New York: The Geological Society of America.

DONNAY, J. D. H., DONNAY, G., COX, E. G., KENNARD, O. & KING, M. V. (1963). *Crystal Data, Determinative Tables, 2nd Edition*, Monograph 5, American Crystallographic Association.

DONNAY, J. D. H. & KENNARD, O. (1964). *Acta Cryst.* **17**, 1337.

International Tables for X-ray Crystallography (1952, 1959, 1962). Vols. I–III. Birmingham: Kynoch Press.

LIPSCOMB, W. M. (1963). *J. Amer. Chem. Soc.* **85**, 846.

LIPSCOMB, W. M. (1964). *J. Chem. Phys.* **40**, 866.

MCCRONE, W. C. (1948). *Anal. Chem.* **20**, 274.

MCCRONE, W. C. (1956). *Anal. Chem.* **28**, 972.

National Bureau of Standards, Technical News Bulletin (1963). **47**, 39–45.

Notes for Authors (1965). *Acta Cryst.* **18**, 134.

Recommendations of the Commission on Crystallographic Computing (1963). *Acta Cryst.* **15**, 515.

TAKEDA, H. & DONNAY, J. D. H. (1964). IUCr *World List of Crystallographic Computer Programs, 2nd Edition*, p. 48. Transformation of crystal setting and space-group symbol (Program JHTRXL for IBM 7094).

Powder Data*

Commission on Crystallographic Data

By OLGA KENNARD,† *University Chemical Laboratory, Lensfield Road, Cambridge, England*, J.D. HANAWALT, *The University of Michigan, Ann Arbor, Michigan, U.S.A.*, A.J.C. WILSON,‡ *Department of Physics, University of Birmingham, P.O. Box 363, Birmingham 15, England*, P.M. DE WOLFF, *Laboratorium voor Technische Natuurkunde, Lorentzweg 1, Delft, The Netherlands*, and V.A. FRANK-KAMENETSKY, *Department of Geology, Leningrad State University, University Embankment 7/9, Leningrad, USSR*

The Commission on Crystallographic Data has published a set of recommendations on the publication of primary crystallographic data (Kennard, Speakman & Donnay, 1967). These recommendations, which were concerned almost entirely with single-crystal work, aroused considerable interest, and the Commission therefore took under consideration the preparation of recommendations on the publication of powder data. Publication at the present time is particularly appropriate, since the *Journal of Applied Crystallography* will carry considerably more powder data than *Acta Crystallographica* has done.

The purpose of obtaining powder data is usually identification, including phase analysis of metallurgical and ceramic systems. They are also used for structure determination, when the structure is comparatively simple or when single crystals are not available. The present recommendations are divided into six sections: (1) Presentation of data, (2) Experimental methods, (3) Derivation of spacings, (4) Indexing of patterns and derivation of cell parameters, (5) Reference intensities and related problems, and (6) Structure determination.

1. Presentation of data

The essential parts of a published powder pattern are set out in the following paragraphs. Portions printed in **heavy type** constitute the normal minimum requirements, but the requirements printed in ordinary type should also be fulfilled whenever possible.

1.1. *Name of substance*

The systematic chemical name should be given for a pure substance, together with the trivial name if this is better known. **If the substance is (or occurs as) a mineral, its mineral name should be given,** together with the idealized chemical name if the composition is reasonably simple and invariant.

1.2. *Formula*

The structural formula should be given for organic compounds whenever it has been established; it is convenient to have the empirical formula as well. **A full chemical analysis should be given for minerals and natural products of undetermined structural formula. The source of the substance (including locality for minerals) should always be given, the methods of preparation and treatment whenever relevant.**

1.3. *Standard of spacings*

The basis of the spacing scale should be stated explicitly. Depending on the technique employed, the spacings quoted may depend on the wavelength of the radiation used or on the spacings of an internal standard (§§ 2.2 and especially 3.2). If it is a wavelength, the numerical value assumed for the wavelength should be given, together with particulars of the filter or monochromator if used. *If an internal standard is used to establish the spacing scale, the standard should be named and its assumed parameter(s) given.*

1.4. *Temperature*

The temperature of the specimen should be given. 'Room temperature' may vary by 15°C or more, depending on the laboratory and the time of the year.

1.5. *Crystallographic data*

It has become conventional to list the spacings (d values) and intensities, either visually estimated, densitometric, or diffractometric. The methods of determining these are considered further in sections 2 and 3 below. Certain functions of d such as $Q(=d^{-2})$ have advantages over the spacings themselves (§ 4). If a unit cell is known, either from single-crystal data or from indexing the pattern (§ 4 below) the cell should be given together with the contents of the unit cell (2) and the lines indexed. The space group should be given if known. Any lines that cannot be accounted for by the unit cell and space group should be specially designated. *Observed spacings should not be replaced by spacings calculated from the unit cell.* 'Reference intensities' (§§ 2.3 and especially 5.2) should be given whenever possible.

1.6. *Other data*

It is convenient if other data relating to the material and its identification can be given. These will include the colour of the powder, optical data (if known) for crystals of the material, density, and melting point. Any available information about polymorphism of the substance should be given, and properties that influence its handling, such as deliquescence, efflorescence, toxicity and the like.

1.7. *References*

In many cases not all the above data will be determined by the authors of the publication. When data (such as the unit cell, density, chemical analysis, *etc.*) are not derived directly from observations made by the authors, **adequate literature references should be given.**

2. Experimental methods

2.1. *Apparatus*

The usual instruments used for obtaining powder data for identification are the Debye–Scherrer camera, the focusing camera, and the powder diffractometer. All are available in a range of radii, common values being between 28 and 100 mm for cameras and 100 and 200 mm for diffractometers. The focusing principle used in focusing cameras and diffractometers admits – without loss of intensity – a much better angular resolution than a Debye–Scherrer camera of corresponding size and, when used with a focusing monochromator (as in the Guinier camera and its modifications and in some diffractometer arrangements) a great measure of achromaticity is achieved as well. The resolution of a Guinier camera or diffractometer in practice may, therefore, be of the order of three to five times as good as that of a Debye–Scherrer camera. These figures should be taken only as a rough guide, since the resolution is affected by specimen size, specimen transparency, and collimation conditions, as well as instrument radius and wavelength. The advantage is particularly evident in the low-angle region, where the lines most important for identification occur. Debye–Scherrer patterns, however, can be made with much smaller quantities of the material than normal focusing-camera or diffractometer specimens, and occasionally this may be an overriding consideration. They are ordinarily sufficient for identification, but for standard file patterns

* Recommendations on the presentation of powder data for publication, prepared on behalf of the Commision on Crystallographic Data, International Union of Crystallography.

† External Staff, Medical Research Council.

‡ The final part of A.J.C.W.'s contribution was made while on leave at the Georgia Institute of Technology, Atlanta, Georgia 30332, U.S.A.

focusing-camera or diffractometer data are preferable, since it is always possible to deduce from a pattern obtained with good resolution what a pattern obtained with poorer resolution would be like, but not *vice versa*.

2.2. *Features of pattern important for identification*

The radiation used to obtain diffraction patterns is most commonly Cu $K\alpha$, but other wavelengths from Mo $K\alpha$ to Cr $K\alpha$ are also employed. An examination of the Powder Data File of the Joint Committee* which contains about 13000 inorganic materials indicates that the number of lines per pattern is between 10 and 100 with an average of about 35. The majority of the strongest lines of each pattern are found in a range of about 5·0 – 1·5 Å, corresponding to 18 – 60° in 2θ for Cu $K\alpha$ radiation. About ten per cent of the lines occur at 2θ less than 18°. The 5 or 10 strongest lines are of major importance and demand the highest accuracy of measurement.

(i) These lines represent the simpler (*hkl*) planes and are more useful for identification than the high-order lines, which become more numerous and weaker with increasing 2θ.

(ii) If the diffraction pattern is weak, either because the phase is only a minor constituent of a mixture or because the phase does not diffract X-rays strongly, then only the strongest lines of the pattern will be observable.

(iii) A special study (Rinn & Hanawalt, 1968) of the first 14 sets of Joint Committee Powder Data File (7017 inorganic patterns) showed that if the lines of each pattern are arranged in order of intensity, it requires relatively few lines to distinguish each pattern from all others in the file. The first count was made by assuming for simplicity that two lines coincide if their d values do not differ by more than 0·03 Å, corresponding to an assumed error of ±0·015Å throughout the range of d values. Under these conditions of the 7017 patterns listed all but 275 can be identified by their five strongest lines. Most of the 275 patterns could be distinguished if the accuracy of the measurements were increased to 0·1° 2θ. The remaining patterns included many from isostructural substances.

In the above studies intensity values were only used to determine the *order* of strongest lines of a pattern. If the intensity values were used as another parameter of the 'match', then the numbers of patterns identifiable by their three strongest lines would be greatly increased.

It is recommended that an *internal standard* be used for high-accuracy measurements of d values and intensities. For details see § 5 and Swanson, Morris & Evans (1966).

The weaker lines serve a useful purpose in detecting minor constituents and confirming identifications. The relative accuracy of measurement increases with 2θ. It is essential that the weaker lines, too, be stated as completely and accurately as possible.

3. Derivation of spacings

3.1. *Definition of spacing*

It has become a convention to express powder diffraction data as a list of spacings and intensities. The spacings could be defined with reference to the crystal geometry, but they are more often regarded as a wavelength-independent measure of the observed diffraction angles 2θ through the relation

$$d=\lambda/2\sin\theta \qquad (1)$$

It should be emphasized that from this standpoint (1) is to be regarded as a definition of d, not as a physical law.

3.2. *Standard of spacings*

It is recommended that authors state explicitly (i) **whether they have used an internal standard,** (ii) **if not, whether they have applied important corrections, such as the absorption correction for Debye–Scherrer cameras.**

The accuracy of 2θ measurements may be greatly enhanced by the use of an internal standard. This means an admixed standard substance, the lines of which are measured together with those of the specimen. Final 2θ values are obtained by interpolation between the standard lines. If the pattern contains lines at low θ, two exposures with different standards may have to be made. Some suitable standards are tungsten, corundum (α-Al_2O_3) silicon and KCl at medium angles and alum and $PbSO_4$ at low angles.*

3.3. *Listing of spacings*

The range of d or 2θ covered by the camera or diffractometer should be stated. It is recommended that not only d but also the value of $Q=1/d^2$ be given for each line.

The list of d has the drawback of giving a distorted image of the powder pattern. Other functions of $\sin\theta/\lambda$ can be chosen which yield a more faithful mapping. Among these, the quantity

$$Q=4\sin^2\theta/\lambda^2=1/d^2 \qquad (2)$$

has the additional advantage that it allows a quick review of indexing relations, since Q, because of Bragg's law, is equal to a homogeneous quadratic form in h, k, and l (§ 4). **The use of $\sin^2\theta$ (which is proportional to Q) should be avoided because it makes intercomparison of patterns taken with different wavelengths unnecessarily difficult.**

3.4. *Estimates of accuracy*

In Fig. 1 the errors in d and Q caused by an error of 0·01° in 2θ are plotted as a function of 2θ. The distortion of the d scale is apparent from the sharp drop of Δd, as well as from the Cu $K\alpha$ d scale drawn on the abscissa for easy reference. Up to $2\theta=90°$, the two errors do not deviate than 15% from

$$\Delta d=\text{constant}\times d^2\,, \qquad (3)$$

$$\Delta Q=\text{constant}\times Q^{1/2}\,. \qquad (4)$$

All considerations of accuracy and of discrepancies, when referred to d values, are hampered by this distortion, because each Δd is significant only in a very limited range of the pattern. For the same reason, the number of significant decimal places in d must be considered for each

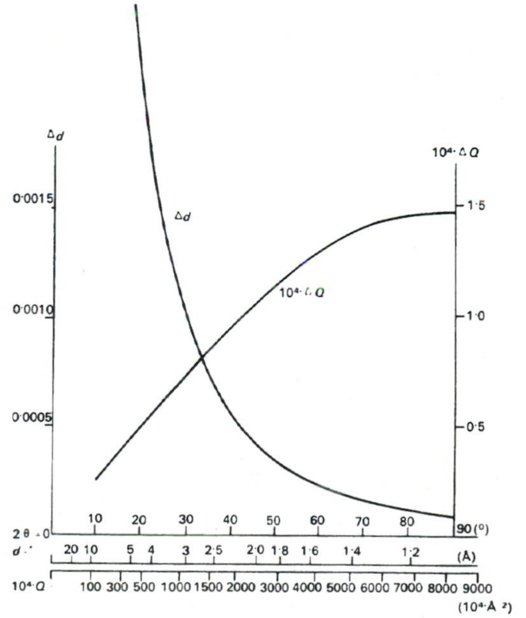

Fig. 1. Errors in d and Q, respectively, caused by an error of 0·01 in 2θ for Cu $K\alpha$ radiation, as a function of θ, d or Q.

* Most users of powder data will be familiar with the card index of powder patterns compiled and now also published by the Joint Committee on Powder Diffraction Standards, Inc. (formerly published by the American Society for Testing and Materials) 1845 Walnut St., Philadelphia, Pennsylvania 19103, U.S.A. It is often, though misleadingly, referred to as the 'ASTM file'. The term 'Joint Committee Powder Data File' is preferable.

* The development of a suitable material for use as internal standard for the determination of '*d*' spacings in powder diffraction is being undertaken by H.F. McMurdie & H. E. Swanson (Crystallography section, National Bureau of Standards, Washington D. C., U.S.A.). Details of these materials and how to obtain them will be published by McMurdie & Swanson in the *Journal of Applied Crystallography* as soon as they are available.

region of the pattern. Usually this number has to jump twice in the range of the d list. It is advisable to place the jumps on the safe side, so as to avoid unduly large rounding-off errors.

The estimated overall accuracy should, if possible, be stated as an average error in 2θ. For a rough indication, an average error in Q may also be useful, though the error in Q will usually vary more than that in 2θ. Indication of an error in d is impractical for the reasons discussed above.

3.5. *Values calculated from cell parameters*

Values of Q calculated from unit-cell data can be given for comparative purposes. They should be listed separately and in such a way that confusion with observed values is excluded. *Partial replacement of observed values (either of d or Q) by calculated ones is undesirable,* because the latter – though sometimes inherently much more accurate – depend on an indexing which is most often based largely on the lines in question (§ 4).

4. Indexing of patterns and derivation of cell parameters

4.1. *Basis of indexing*

Both indexing and unit-cell determination are based on the fact that, theoretically, $Q=1/d^2$ is a homogeneous quadratic form in h, k, and l:

$$Q=Ah^2+Bk^2+Cl^2+Dkl+Elh+Fhk, \quad (5)$$

with coefficients A to F depending on the reciprocal unit-cell parameters in the following way:

$$A=(A^*)^2,\ B=(b^*)^2,\ C=(c^*)^2,\ D=2b^*c^*\cos\alpha^*,$$
$$E=2a^*c^*\cos\beta^*,\ F=2a^*b^*\cos\gamma^*, \quad (6)$$

or

$$a^*=A^{1/2},\ b^*=B^{1/2},\ c^*=C^{1/2},\ \cos\alpha=D/2(BC)^{1/2},$$
$$\cos\beta^*=E/2(AC)^{1/2},\ \cos\gamma^*=F/2(AB)^{1/2}\,. \quad (7)$$

The complete relations between the reciprocal parameters a^* to γ^* and the actual cell parameters a to γ are given in *International Tables for X-ray Crystallography* (1959). Expressions for a to γ as functions of A to E are given by Visser (1969).

4.2. *Calculated values of Q*

If the unit cell is supposedly known, the constants A to F can be calculated and the resulting list of $Q(hkl)$, ordered in a sequence of increasing Q, can be compared with the observed values of Q. If the majority of discrepancies is not larger than the expected error (derived from the expected error in θ, as in § 3.4) both the indexing and the unit cell will be substantially correct. The cell parameters should be refined on the basis of those lines which have unambiguous indices. After one or more steps of such a least-squares refinement, a few lines may still have improbably large discrepancies. These should be marked in the list; the Powder Data File of the Joint Committee uses the indication nc (non cell).

4.3. *Determination and confirmation of unit cell*

If the unit cell is not known *a priori*, an indexing may be found by trial and error or by computer methods. These have been reviewed by de Wolff (1962). The further procedure is exactly as in § 4.2. In this case, however, a satisfactory agreement of calculated and observed of Q is not sufficient to make the unit cell convincing, because now the link with independent data from other sources is missing. The following quantity is recommended as an index of reliability (de Wolff, 1968):

$$M=Q_{20}/\{2(\Delta Q)\cdot N_{20}\}\,, \quad (8)$$

where Q_{20} = value of Q for the 20th observed line (not counting 'nc' lines); ΔQ = discrepancy, $|Q_{obs}-Q_{calc}|$ averaged over the first 20 observed and indexed lines; N_{20} = number of distinct Q values smaller then Q_{20}, calculated according to equation (1), not including any systematic absences. If the indexing has been performed as in § 4.2, ΔQ will be close to the average expected error in Q mentioned in § 3.4. The index M is the ratio of two average discrepancies (i) the expected discrepancy Q_{20}/N_{20} for an arbitrary unit cell yielding the same density of calculated lines as the cell proposed, and (ii) the actual discrepancy (ΔQ). Material gathered so far has indicated that incorrect indexings (disproved by single-crystal methods) may have values of M up to 5, and that correct indexings have M seldom less than 10 but usually between 10 and 40.

4.4. *Derivation of space group*

If the space group is not known from single-crystal work, it is desirable to state the information derived from the indexed powder pattern.

This information is less reliable than that from single-crystal investigations. Both the relatively high background and the restricted θ range of reliably indexed lines increase the chance that an apparently systematic absence is actually a fortuitous one. General (centring) extinctions are still quite reliable, because of the number of absent lines involved. However, glide planes and in particular screw axes often cannot be regarded as well established.

Quite apart from reliability, powders may yield less complete information than single crystals, because Laue groups such as $4/m$ and $4/mmm$ are in several cases indistinguishable. This makes it more difficult to state the findings in a compact form. For triclinic, monoclinic and orthorhombic crystals, the symbolism proposed by Donnay & Kennard (1964) may be used, since in these systems there is no fundamental difference between powder and single-crystal data.

In the systems of higher symmetry adoption of these symbols may be awkward; for example in a primitive hexagonal lattice five different symbols ($P6/{*}{*}{*}$, $P6/{*}$, $P3{*}1$, $P31{*}$, $\bar{3}P{*}$) correspond to a lack of systematic absences. For powder work the older aspect symbols (*International Tables for X-ray Crystallography*, 1962) are more suitable, but require the addition of the qualification 'trigonal', 'tetragonal', 'hexagonal', or 'cubic'.

5. Intensities

5.1. *Indication and measurement of intensities*

It is usual to give the intensities of the diffraction lines as percentages, so that the strongest one is tabulated as 100. The Joint Committee† qualifies this by the proviso that *the strongest line for this purpose must be chosen from those occurring with 2θ less than 90° for* Cu $K\alpha$ *radiation* ($d>1{\cdot}10$).

Ideally it would be preferable to use integrated intensities, but because of overlapping and other reasons this is impracticable. It is recommended, therefore, that the standard method of tabulation be peak diffractometric intensities expressed as percentages of the peak intensity of the strongest lines in the front-reflexion region for copper radiation.

Ordinarily the use of a different radiation will not change the observed intensity enough to disturb identification, but care must be taken to choose the strongest line from the d range corresponding to the θ range specified for copper. Line broadening may produce systematic variation of peak intensity with Bragg angle or with indices of reflexion, and specimens exhibiting it require special care. Noticeably broadened lines are indicated in the Joint Committee Powder Data File by the letter 'b': this normally refers to unresolved multiplets rather than diffraction broadening.

5.1.1. If a *diffactometer* is used, the diffraction intensity is usually recorded as a function of Bragg angle and the evaluation of line intensity is a matter of interpretation of the chart. Either peak values or integrated intensities (both corrected for background) can be stated. Peak intensities will, in general, yield a satisfactory characteristic of the pattern.

Special care should be taken to avoid five errors not easily detectable in a diffractometer record.

(i) *Statistical fluctuations due to large crystallites.* These can be detected by repeating the measurement on two or more specimens. If reduction of crystallite size is difficult, rotating the specimen in its own plane will greatly help to reduce the fluctuations.

† The Joint Committee now (1970) proposes the proviso $d>1{\cdot}000$ Å, as it is independent of wavelength.

(ii) *Unequal width of lines* due to anisotropic particle-size broadening. If this occurs, either integrated intensities should be measured or the particularly broad lines marked 'b' in the list of peak values.

(iii) *Preferred orientation.* This can severely influence the intensities, especially for platy and for needle-shaped crystallites. Diluting the specimen, for example with flour, or embedding it in a cement (such as Canada balsam), followed by regrinding of the hardened mixture, will reduce the effect. It is also advisable to fill the specimen holder from the side.

(iv) Cold work and grain disortion can adversely effect the quality of the diffraction pattern. Annealing the material at the proper temperature can reduce this distortion.

(v) Precautions should be taken against drift of the counter or apparatus.

5.1.2. *Films* obtained with Guinier or Debye–Scherrer cameras can be used as such, by visual estimation or comparison of the lines with a sequence of increasing known exposures of a single line. Visual estimates are, however, often unreliable. Therefore it is strongly recommended that films be measured quantitatively by densitometer or microphotometer. The possible errors (i) – (v) mentioned above are the same. Though they are more easily detected on a film, it is no easier to suppress the effects. The Debye–Scherrer pattern has the extra difficulty of a broad sometimes irregular line shape. The absorption factor is strongly dependent on the Bragg angle and difficult to correct for, so that the Debye–Scherrer technique is likely to yield less accurate intensities.

5.1.3. The absorption factor and certain other geometrical functions of angle differ for the diffractometer and the various types of camera, so that the relative intensities of the diffraction maxima are somewhat different for the different techniques. The differences are not normally great enough to interfere with identification, but may occasionally alter the order of the strongest lines (Hughes, Lewis & Wilson, 1956). *There is therefore, good reason for retaining both camera and diffractometric data in the Joint Committee Powder Data File, and for publishing new data obtained by one technique even when data obtained by another technique are already on record.*

5.2. *Reference intensities*

It is recommended to state, in addition to relative intensities, a reference intensity which is the intensity of the strongest line as a percentage of the intensity of the d= 2·085 Å line of α-alumina, measured in the diffraction pattern of a mixture of equal parts by weight of the pure substance and α-alumina. The experiment can be made with another ratio, say 1:R parts Al_2O_3; in that case the observed percentage must be multiplied by R. α-Alumina is obtainable as a fine powder, *e.g.* Linde A, Union Carbide Corporation, East Chicago, Illinois, U.S.A. Other suitable substances such as carbonyl nickel, magnesium oxide or potassium chloride can also be used. In this case, the results should be converted to the above-defined value by measuring the reference intensity (with respect to alumina) of the standard line used. For further details see Swanson, McMurdie, Morris & Evans (1968).

In order to allow a rough quantitative analysis of multiphase samples, it is desirable that the intensities of at least one line of the pure components be known also on an absolute scale. Both the quantitative analysis and the derivation of such absolute figures are easy if use is made of an internal standard admixed in a known proportion. One only has to measure the ratio of the intensity of a suitable line of the compound and of the standard substance. The unknown content then follows from the fact that this intensity ratio is equal to the product of the corresponding ratio of concentrations and constants factor (determined entirely by the choice of the two lines). Hence it is sufficient to state the intensity ratio for a mixture of known proportions of pure compound plus standard. Since this does not involve any really absolute measurement, the resulting figure is called a *reference intensity*. In principle, the choice of standard substance and of the line pair is arbitrary. The recommended choice is the procedure followed by the National Bureau of Standards of the U.S.A.

Neither the standard substance nor the specimen preparation for this measurement need be the same as that used in measuring diffraction angles (§ 3.2.). The measures taken to suppress preferred orientation, in particular, may well make the specimen unsuitable for precise angle measurement. Hence it is better to perform the reference-intensity determination on a separate specimen.

6. Structure determination

Except for simple substances, structure determination by means of powder data alone is unusual, but must be undertaken occasionally. The most satisfactory validation of a structure determination based on powder data is probably the publication of a comparison of the directly observed intensities with those calculated from the structure, allowance being made in the calculated intensities for the trigonometric factors, temperature factors, absorption, and other such effects. Since the number of observed powder lines is small in comparison with the number of reflexions obtainable in single-crystal work, publication of this comparison does not ordinarily present the same problems as the publication of F tables (III of Kennard, Speakman & Donnay, 1967). Overlapping and coincidence of lines makes a calculation of the usual R factor difficult, but a modified one based on directly observed intensities.

$$R_I = \frac{\sum |I_{obs} - I_{calc}|}{\sum I_{obs}}$$

may be useful.

Since in powder work more reflexions are unobserved than in single-crystal analysis, it is essential that the summation be extended over all (including unobserved) reflexions.

References

DONNAY, J. D. H. & KENNARD, O. (1964). *Acta Cryst.* **17**, 1337.

HUGHES, J. W., LEWIS, I. E. & WILSON, A. J. C. (1956). *Brit. J. Appl. Phys.* **7**, 80.

International Tables for X-ray Crystallography (1962). Vol. I, p. 347. Birmingham: Kynoch Press.

International Tables for X-ray Crystallography (1959). Vol. II. Birmingham: Kynoch Press.

KENNARD, O., SPEAKMAN, J. C. & DONNAY, J. D. H. (1967). *Acta Cryst.* **22**, 445.

SWANSON, H. E., MCMURDIE, H. F., MORRIS, M. C. & EVANS, E. H. (1968). National Bureau of Standards, U.S.A., Monograph 25, § 6.

SWANSON, H. E., MORRIS, M. C. & EVANS, E. H. (1966). National Bureau of Standards, U.S.A., Monograph 25, § 4.

THOMASSEN, L., RINN, H. W. & HANAWALT, J. D. (1968). Technical Publications N430, p. 204, American Society for Testing and Materials.

VISSER, J. W. (1969). *J. Appl. Cryst.* **2**, 142.

WOLFF, P. M. DE (1962), *Advanc. X-ray Anal.* Vol. 6. Proceedings of the 11th Annual Conference on the Application of X-ray Analysis, Denver).

WOLFF, P. M. DE (1968). *J. Appl. Cryst.* **1**, 108.

THE STANDARD DATA FORM FOR POWDER DIFFRACTION DATA

L. D. Calvert, J. L. Flippen-Anderson, C. R. Hubbard, Q. C. Johnson, P. G. Lenhert, M. C. Nichols, W. Parrish, D. K. Smith, G. S. Smith, R. L. Snyder, and R. A. Young

Subcommittee of the American Crystallographic Association

RECOMMENDATIONS

1. We recommend the establishment of publication standards for powder diffraction data through the use of a standard data-form. We have designed such a form to be used at the author-referee level to ensure that published data are of good quality.

An example of a filled-out form is shown in Appendix A. Suggestions for filling out the data-form are given in Appendix B.

The data-form itself is divided into five sections: Sample Characterization, Technique, Unit Cell Data, References, and the Powder Data. Items on the form that are considered essential documentation are indicated by bold-face print. The other items are highly desirable information; however, it is recognized that some experimental procedures will not yield data on some of these items. Hence, omissions of data for these items should not of themselves preclude publication of the powder data.

This list of items is not all-inclusive: an all-inclusive list might be longer than the paper accompanying it. But this list does include the data most important to today's instrumentation. It represents, we believe, what the referee and editor can reasonably expect of experimenters seeking to publish their data. However, we do not intend that the list be used to stifle either new techniques in powder diffraction or the publication of powder data obtained under difficult conditions, e.g., low and high temperatures, high pressure, ultra-small quantities of materials.

We have several specific recommendations pertaining to the reporting, dissemination, and archiving of powder diffraction data:

(a) The published powder pattern should be as complete as possible; for example, it should include weak as well as strong diffraction lines. An element of doubt often arises in phase identification when the observed strong diffraction lines match those of a particular reference pattern, but some weak lines are observed that are not listed in the reference pattern. Complete reference patterns would be a benefit in these cases and generally would make identification easier. Furthermore, the ease of computer-indexing of powder patterns depends, in most cases, on the accuracy and completeness of the data. Ordinarily, 40 lines permit ready characterization of the material.* In the absence of 40 lines, the data should extend to (at least) 100° 2θ (Cu Kα radiation).

*Although 40 lines are typically reported on a data card in the Powder Diffraction File, the author should not limit the published pattern to this number if more data is available. Patterns should be reported to at least 100°2θ. If no distinct lines can be detected beyond the high angle limit of the reported data, this point should be stated. Patterns with a small number of lines should be reported to the limit of the experimental method used.

(b) Researchers should report the experimentally observed 2θ values, in degrees and corrected for systematic instrumental errors. The d values are usually not an adequate substitute for the primary data, the 2θ values. Reporting the 2θ's does not distort information, but reporting the derived quantities, d's and Q's ($Q = 1/d^2$), usually does. For example, even the d values reported by NBS (for the example in Appendix A) do not match exactly their observed 2θ values. We had to add additional significant figures to the reported d values (in Appendix A) to accurately accomplish an otherwise simple transformation: 2θ's to d's. Hence, we believe that the magnetic-tape version of the PDF should contain 2θ values, which can be readily converted to d's, if necessary, by a user program. (Anyone using the magnetic-tape version, would have computer capability.) The PDF file cards, on the other hand, should probably continue to list d values.

(c) Intensities should be reported numerically, and on a scale from 0 to 100. This will avoid the non-quantitative notations $I < 1$ and $I \ll 1$ that sometimes occur with the present scale of 1-100. For example, 0.7 might be used instead of < 1. This new scale will require no additional digits. The recommended scale does not preclude the use of a scale from 1-10 or 1-100, when appropriate, nor should it be understood to imply an increased accuracy for $I < 1$.

(d) The reproducibility of the measured 2θ and I's should be indicated. These data should be obtained by multiple mountings of the sample material.

Reprinted with permission from an excerpt from the Subcommittee of the American Crystallographic Association's final report presented at the Symposium on Accuracy in Powder Diffraction in Washington, D.C., June 1979. Published by the International Centre for Diffraction Data: Swarthmore, PA, 1979.

(e) Indexing of the powder diffraction data should be required for all but the rarest and best-defended cases. Without indexing, there is little proof that the reported pattern is for a single phase. Patterns can often be automatically indexed with appropriate computer codes [4, 5, 16]. Authors should report a figure of merit based on accuracy of the 2θ measurements and completeness of their data. As mentioned above, a figure of merit has been proposed for this purpose.

(f) Information concerning line breadth of the sample should be supplied. This could be full-width at half-maximum (FWHM) of a resolved $K\alpha_1$ line in the region 30-60° 2θ.

(g) Additional information of value to future users should be supplied. Such data could be standard deviations, Chemical Abstracts Service Registry number, Crystal Data index number, etc.

We believe the use of a standard data-form has a number of benefits including:

- This data-form can be used as a checklist. Authors can use the form to ensure the necessary characterization of their data. Also, familiarity with the standard form will encourage authors to record complete information on their measurements.
- The completeness of the experimental data will be easy to evaluate. Because information pertaining to the experimentation will be collected in one place in the manuscript, referees or journal editors will be able to readily judge the future usefulness of the contribution. Moreover, an editor could send the powder data to an expert in powder diffraction for refereeing and send the rest of the manuscript to another reviewer.
- The data form could be recast into a camera-ready format for direct printing in the journals. In this way, data-base managers could receive the powder data exactly as sent by the authors.*

2. This subcommittee has also given some thought as to *what* constitutes publishable powder diffraction data. We believe that to justify being published, powder diffraction data must be potentially useful to someone else and must constitute in some way an original contribution to the literature. To constitute an original contribution, the data must be the first published for a well-characterized phase, must be a significant correction to or an improvement on published data, or must relate to the phase in a previously uncharacterized condition, e.g., at elevated temperatures or pressure. A powder pattern calculated from single-crystal structure data does not in itself meet the criterion of originality. Such data, we hope, will continue to be contributed directly to the data-base managers under guidelines being established by a JCPDS committee.

3. To help gain acceptance for publication standards within the world-wide powder diffraction community, we recommend that the ACA request the International Union of Crystallography (IUCr), through the Commission on Journals, to now consider implementation of these recommendations in all IUCr journals. As a first step in that process, the Commission on Journals may elect to refer these recommendations to the Commission on Crystallographic Data for review and approval. It is envisioned that a mutually acceptable version could be included in *Notes for Authors* and in the *Handbook for Co-Editors*. If the IUCr journals enact publication standards in the near future, other journals will in time follow that lead and the quality of published powder diffraction data will improve.

*JCPDS has offered to referee powder data and to receive powder data directly if they are not to be published by the journal.

APPENDIX A: EXAMPLE OF COMPLETED DATA FORM

POWDER DIFFRACTION DATA FOR PHASE CHARACTERIZATION

(Data from Swanson, H. E., et al. (1971). NBS Monograph No. 25, Section 9, p. 25)

BOLDFACE ITEMS ARE CONSIDERED ESSENTIAL

SAMPLE CHARACTERIZATION

NAME (CHEMICAL, MINERAL, Trivial) Magnesium Aluminum Oxide (Spinel)

EMPIRICAL FORMULA $MgAl_2O_4$

CHEMICAL ANALYSIS NO x **YES**

SOURCE/PREPARATION Synthetic; Fusion of binary oxides

CHEMICAL ABSTRACT REGISTRY NO. 12068-51-8 PEARSON PHASE DESIGNATION cF56

OTHER **Index of Refraction = 1.718 (Isotropic)**

TECHNIQUE

RADIATION TYPE, SOURCE X-rays, Cu **λ VALUE USED** 1.54056 Å K α_1

λ DISCRIM. (Filters Mono, Etc.) Diffracted beam, curved LiF mono.

λ DETECTOR (Film, Scint, Position Sensitive etc.) Geiger

INSTRUMENT DESCRIPTION (Type, Slits, etc.) 17 cm Vertical Diffractometer **DIV** 1° **REC** 0.003"

SOLLER Yes **No.** 1 **Position** Inc. **Aperture** q = 1.2

INSTRUMENTAL PROFILE BREADTH 0.10 °2θ **TEMP (°C)** 25 ± 1

SPECIMEN FORM/PARTICLE SIZE Edge loaded powder/< 10 μm particle size for I's, packed for **2θ's**

RANGE OF 2θ FROM 5 °2θ to 165.0 °2θ **SPECIMEN MOTION** None

INTERNAL/EXTERNAL 2θ STD (if any) Ag (internal) **LATTICE PARAMETER OF 2θ STD** 4.08641 Å

2θ ERROR CORRECTION PROCEDURE Linear interpolation from nearest **2θ**'s of std.

INTENSITY MEAS. TECHNIQUE Strip chart record **(peak heights)** **ERROR (~)** 5% **PEAK** x **INTEGRATED**

MINIMUM INTENSITY THRESHOLD (IN RELATIVE INTENSITY UNITS) 0.3

INTENSITY STD USED α - Al_2O_3 hkℓ's OF INTENSITY STD 113

INTENSITY RATIO I/I_c 1.70 (5) CONVERSION FACTOR IF CORUNDUM NOT USED

RESOLUTION (FWHM) FOR THIS MATERIAL: 0.10 °2θ AT 59.37 °2θ

2θ REPRODUCIBILITY FOR THIS MATERIAL: ± 0.02 °2θ AT All °2θ

UNIT CELL DATA

METHOD OF CELL DETN. Cell and structure known from Ref. 1

CELL REFINEMENT METHOD Least-squares. See Ref. 2

a = 8.0831 (1) Å ; b = __ ( ) Å ; c = __ ( ) Å

α = __°(°) ; β = __°(°) , γ = __°(°)

Z = 8 ; D_m = __ ( ) g cm^{-3} ; D_x = 3.578 g cm^{-3} ; V = 528.1 Å^3 ; Formula Wt = 142.25

CRYSTAL SYS. Cubic SPACE GROUP Fd3m (227) CRYSTAL DATA INDEX NO. 8.0831

FIGURE OF MERIT TYPE F_N. See Ref. 3 VALUE F_{29} = 58(0.015, 33)

REFERENCES

1. Bragg, W. H. (1915). Nature, 95, 561.
2. Appleman, D. E., Evans, H. T. (1973). NTIS Document No. PB-216188
3. Smith, G. S., and Snyder, R. L. (1979). J. Appl. Crystallography, 12, 60.

( ) INDICATES STANDARD DEVIATION IN LEAST SIGNIFICANT DIGIT(S)

POWDER DATA

ESSENTIAL		DESIRED		
2θ EXP (DEGREES)	I/I_0	d_{EXP} (Å)	$hk\ell$	$\Delta 2\theta$* (DEGREES)
19.02	35	4.66	111	+ .019
31.27	40	2.858	220	- .003
36.84	100	2.437	311	- .009
38.53	3	2.335	222	- .021
44.83	65	2.020	400	+ .016
55.64	9	1.650	422	- .020
59.37	45	1.5554	511	+ .008
65.24	55	1.4289	440	- .001
68.64	3	1.3662	531	+ .006
74.13	3	1.2780	620	+ .003
77.32	8	1.2330	533	- .029
78.40	1	1.2187	622	- .013
82.64	5	1.1666	444	+ .006
85.76	2	1.1320	711	- .012
90.97	5	1.0802	642	- .009
94.10	12	1.0524	731	- .005
99.34	7	1.0104	800	- .006
107.90	2	0.9527	822	- .020
111.22	8	0.93343	751	- .014
112.32	1	0.92738	662	- .035
116.91	6	0.90384	840	- .025
120.50	1	0.88722	911	+ .004
121.69	0.9	0.88203	842	- .021
126.76	0.8	0.86161	664	+ .013
130.74	8	0.84737	931	- .011
138.07	17	0.82488	844	+ .033
142.97	0.4	0.81232	933	+ .024
152.70	2	0.79266	10.2.0	- .033
160.65	11	0.78139	951	+ .025

* $2\theta_{EXP} - 2\theta_{CALC}$

Appendix B. Information to Aid in the Completion of the Powder Diffraction Data Form

The underlined items below correspond to each of the items on the Powder Diffraction Data-Form. Authors may not be able to complete every item, but are urged to be as complete as possible.

SAMPLE CHARACTERIZATION

Chemical Name. Names should be consistent with the conventions of the journal in which the pattern is to be published. Such nomenclature is often in accord with IUPAC conventions [1-3]. This name should include polymorphic phase identification (e.g., β-manganese dioxide).

Mineral Name. The mineral name should be included for all minerals or synthetic compounds known to have mineral counterparts (e.g., lead sulfide, galena).

Trivial. Common names like Rochelle salt or methylene blue, which may be of value, should be included.

Empirical Formula. The empirical or stoichiometric formula should, if possible, connote structural information such as functional groups (e.g., $(PO_4)_2$). Such stoichiometric formulas can be readily converted to strictly empirical formulas by computers. The stoichiometric formula should, however, be consistent with the conventions of the journal in which the pattern is to be published.

Chemical Analysis. The results of a partial or full chemical analysis should be communicated either on this form or in the journal article.

Source. List the source of the material. If it is a natural mineral, the location must be given. If the material is commercially obtained, state the supplier.

Preparation. State method and pertinent conditions of preparation.

Other Data. Give any other data that will help to assure that the specimen is well characterized apart from the diffraction pattern. (A high-quality diffraction pattern of an improperly identified material is of less than no use.) Such data might include: melting point or transformation temperatures, the fact that single-crystal studies were done; structure type (e.g., NaCl type), color, indices of refraction, etc.

Chemical Abstracts Service Registry Number. This number is uniquely assigned to each compound by Chemical Abstracts. It is very useful in cross correlating between various computer data bases. This number should be included it if is known to the author. For further information see reference [4].

Pearson Phase Designation. A method for classifying structures of metals and alloys is described by W. B. Pearson [5]. The "Pearson Symbol" is quite useful and, if known, should be included (e.g., α-Se has the symbol mP32, Al_3Zn has the symbol tI16). The three parts of the Pearson symbol are: the first, lower-case letter (a, m, o, t, h, c) designating crystal system, the second, capital letter (P, C, F, I, R) designating Bravais lattice type, and a number indicating the number of atoms in the conventional unit cell: NOTE: This is not the number of formula units.

TECHNIQUE

Radiation Type. X-rays, neutrons, electrons, etc.

Radiation Source. X-ray target material, neutron source, electron accelerating voltage, etc.

λ Value Used. The numerical value of the wavelength used in calculating d values. List $\lambda(K\alpha_1)$ if the α_1 component was fully resolved throughout the pattern or if analysis leads to 2θ's for the α_1 component. List $\lambda(K\alpha)$ if the $K\alpha$ doublet was not resolved. If both $\lambda(K\alpha)$ and $\lambda(K\alpha_1)$ were used, indicate the dividing line and list both wavelengths.

λ Discriminator (Filters, Mono, etc.). Give the method used to monochromatize the beam. State whether the monochromator or filter was used in the incident or the diffracted beam.

λ Detector (Film, Scint., Position Sensitive, etc.). The type of radiation detector used and any unusual electronic processing should be given. (e.g., pulse-height discrimination for a scintillation or proportional detector is conventional and need not be stated.)

Instrument Description (Type, Slits, etc.). State the type and size of instrument used (e.g., 114.6-mm Debye-Scherrer camera, 17-cm diffractometer, 200-mm Guinier camera, etc.) and the conditions for the experiment. For diffractometers, state the divergence angle (Div) of the incident beam, receiving-slit width (Rec), and whether or not different or continuously varying divergence slits were used for various 2θ ranges.

Soller (Number, Position, Apertures). For diffractometers, give the number of sets of Soller slits, their positions (e.g., in incident or diffracted beam or in both) and aperture(s), because axial divergence can be a source of large profile displacement or broadening. Aperture is best characterized by $q = R\Delta/h$, where R is the radius of the diffractometer; h is 1/2 the axial extension of the sample, and

$$\Delta = \frac{\text{spacing between Soller foils}}{\text{length of Soller foils}}$$

Instrumental Profile Breadth. This parameter is the full width at half maximum (FWHM) of a reference sample that has minimal intrinsic broadening. It allows the instrument used to be compared with others and indicates the resolving ability. A recommended procedure is to measure the 311 reflection ($d_{311} = 1.638$ Å) of a well-annealed Si specimen whose particle size is between 1 and 20 µm. The FWHM of only the $K\alpha_1$ peak should be reported (for Cu $K\alpha_1$ radiation, this occurs at 56.12° 2θ). If another material or peak is chosen, it should be stated. If the α_1 and α_2 peaks are not resolved, report that fact along with the FWHM for the peak observed.

Temperature. State the temperature of the sample when the pattern was obtained.

Specimen Form. Indicate if the sample is a loose powder, sintered compact, metallographic mount, etc. State how the specimen was prepared for diffraction analysis (e.g., side loaded or vertically packed into a diffractometer holder, dusted onto a substrate, packed into a capillary). If different preparations were used in the measurement of d values and intensities, indicate both methods.

Particle Size. Give the average or maximum particle size in the specimen (e.g., 10 μm, <325 mesh, <20 μm).

Range of 2θ. Indicate the range of 2θ that was examined. All peaks above the stated intensity threshold (given below) in the reported 2θ range should be included under Powder Data. Include a minimum of 40 peaks or all data up to 2θ = 100°.

Specimen Motion. State type of sample motion during the diffraction experiment (e.g., sample spinner, rotated cylinder, Gandolfi motion, none).

Internal/External 2θ Standard (if any). If the instrument was calibrated with a known standard before or after the pattern was obtained, state "external--(material of standard)." If an internal standard was mixed with the sample to check for a specimen displacement and/or 2θ zero error, state "internal--(material of standard)." A standard material recommended for both procedures is National Bureau of Standards SRM silicon [6].

Lattice Parameter of 2θ Standard. State the numerical value of the lattice parameter used in calculating the expected 2θ positions of the lines of the standard.

2θ Error Correction Procedure. State the method used to correct the observed 2θ values for systematic instrumental error. A recommended procedure is to use one or more external standards such as Si (ref. [6]) and determine the Δ2θ ($2\theta_{exp} - 2\theta_{calc}$) vs 2θ curve. These data may be fitted using a least squares polynominal regression (e.g., a third-degree equation: $\Delta 2\theta = a_0 + a_1 2\theta + a_2 2\theta^2 + a_3 2\theta^3$). If this procedure is used, state the coefficient (a_i's) of the polynominal. A cubic standard having a large cell dimension (>10 Å) is recommended for patterns having many lines at low angles.

Intensity Measuring Technique. Indicate the method used to determine diffracted intensities: peak heights or integrated areas from a strip chart, or densitometer or step-scan data. Indicate if monochromator polarization effects have been removed from the I/I_0 values reported under Powder Data.

Minimum Intensity Threshold (in Relative Intensity Units). Indicate the relative intensity (see I/I_0 below) not considered to be a real peak. Typically this threshold is taken as the background reading plus 2.5 times $\sqrt{\text{background}}$. NOTE: Because this value will differ in regions of high and low background, limiting values should be reported.

Intensity Standard Used. Indicate the material used in determining the reference intensity ratio. Intensity standards allow for the direct comparison of intensities in different materials. Methods involving the use of corundum as an intensity standard for direct quantitative phase analysis have been described [7-9].

hkl's of Intensity Standard. Give the hkl(s) of the line(s) used for determining the I/I_c standard value.

Intensity Ratio I/I_c. Give the value of the ratio of the strongest line in the pattern to the strongest line of corundum (d_{113} = 2.085 Å) in a 50-50 wt percent mixture of the two [7-9], and give the reproducibility, $\alpha(I/I_c)$.

Conversion Factor if Corundum not used. If a-Al_2O_3 is not used as the intensity standard, then the conversion factor between the standard used and corundum should be listed. This conversion factor can be used to calculate the I/I_c for the specimen.

Resolution (FWHM) for this Material. The full width at half maximum, in degrees 2θ, for a well-resolved α_1 line (if possible) should be given along with the 2θ of this line. This peak should be chosen between 30° and 60° 2θ to avoid wavelength broadening effects. This value, of course, includes the instrumental profile breadth given above. If α_1 and α_2 are not resolved, this fact should be stated.

2θ Reproducibility for this Material. State any measured reproducibility of the positions of the diffraction maxima; if measured for one line, state the 2θ.

UNIT CELL DATA

Method of Cell Determination. If the unit cell was obtained from the literature, give the reference. If the unit cell was derived from the powder data, state the method or program used.

Cell Refinement Method. If the cell was refined, state the method and computer program used (if any).

Cell Parameters a, b, c, α, β, and γ. These values should be reported to the proper number of significant digits; put the standard deviations, in terms of the least significant digit, in parentheses (parameters in angstroms, angles in degrees).

D_m: experimentally determined density in g/cm^3

D_x: calculated density; from $D_x = Z\ FW/N_oV$, where Z is the number of formula units in each unit cell, FW is formula weight, N_o = Avogadro's number (0.60225 x 10^{24}), and V is the volume of the unit cell in cm^3

Space Group. If the space group has been determined by single-crystal studies, give the Hermann-Mauguin symbol followed in brackets by the space-group number as listed in reference [10]. For space group determinations that are not your own, a reference should be given below. For unit cells determined solely from powder diffraction data, give the diffraction aspect as defined in reference [11] (pp. S-8 to S-18).

Crystal Data Index Number(s). List the axial ratios, derived from the cell parameters, that would be used to locate this material in Crystal Data Determinative Tables [11]. Warning: these axial ratios are based on a unit cell defined by Crystal Data, which may not be the cell for which the parameters are given above. Reference [10] describes how the Crystal Data cell is obtained. These rules have been incorporated in the powder pattern evaluation program AIDS [12].

Figure of Merit Type and Value. List the type of figure of merit used (e.g., M_{20} or F_N--see references [13] and [14]) and its value. The figure M_{20} is defined as:

$$M_{20} = \left(\frac{Q_{20}}{2\,|\overline{\Delta Q}|}\right)\left(\frac{1}{N_{poss}}\right),$$

where Q_{20} is the Q value (= $1/d^2$) of the 20th observed line, $|\overline{\Delta Q}|$ is the average absolute discrepancy between Q_{obs} and Q_{calc}, and N_{poss} is the number of independent diffraction lines possible up to the 20th observed line. The figure F_N is defined as:

$$F_N = \left(\frac{1}{|\overline{\Delta 2\theta}|}\right)\left(\frac{N}{N_{poss}}\right),$$

where $|\overline{\Delta 2\theta}|$ is the average absolute discrepancy between observed and calculated 2θ values; and N_{poss} is the number of independent diffraction lines possible up to the Nth observed line.

With regard to the figures of merit, some guidelines for the counting of possible independent diffraction lines are in order:

Systematic absences caused by symmetry elements and lattice type are excluded in the tallying of N_{poss}.

Only one plane from the complete set of planes related by crystal symmetry is counted in N_{poss}. For example, in the cubic system, the 100 line is counted as one independent line although it is composed of diffracted intensities from all six planes of that crystallographic form.

Some forms, though not related by symmetry, have exactly the same spacing and would give rise to the same line in the powder pattern (e.g., 333 and 511 in the cubic system). Forms of this kind are also counted as one independent line. Note that this rule means that the higher-symmetry Laue group is always assumed. When a lower-symmetry Laue group is definitely known from single-crystal studies, this rule is not strictly correct for some of the diffraction lines. However, for most of these cases, the effect on the value of the figure of merit for the overall pattern should be insignificant.

For the case of accidental degeneracy (i.e., nonequivalent forms which have spacings so nearly identical that the individual lines would not be experimentally resolved), all lines in such a cluster are counted as possible independent lines; the line having the smallest Δ2θ is used in the calculation of $|\overline{\Delta 2\theta}|$ for the pattern, and the other line or lines are listed as not observed, thereby increasing N_{poss}.

The function is reported in the form

$$F_N = \text{Value } (|\overline{\Delta 2\theta}|, N_{poss}).$$

For example,

$$F_{30} = 73.5(0.012, 34).$$

We recommend that N in F_N be the 30th observed line or the last line if there are fewer than 30 lines in the pattern.

Note that if the space group or the diffraction aspect is not determined, N_{poss} can only be reported as a maximum value; i.e., no allowance would have been made for systematic absence, a fact which could appreciably reduce M_{20} or F_N.

REFERENCES

Cite pertinent literature references for previous x-ray or preparative studies.

POWDER DATA LISTING

$2\theta_{exp}$. Report the experimentally observed 2θ values, in degrees and corrected for systematic instrumental errors. If multiple determinations of each peak were made, then the standard deviation of each 2θ value should follow in parentheses.

I/I_0. List the relative intensities I/I_0 of the diffraction lines on a numeric scale; $I/I_0 = I/I_{max} \times$ scale, where I_{max} = the numeric value chosen for the most intense refection. The scale value should be chosen such that the maximum intensity value is not greater than 100. Intensities less than 1 are reported as decimal fractions. If the standard deviation of each intensity was computed, place the value in parenthesis after each I/I_0 value.

d_{exp}. List the d values in angstroms, derived from the observed 2θ values using Bragg's law: $d = \lambda/[2 \sin (2\theta/2)]$, where λ is the value of the wavelength stated under Technique. In reporting d's, the number of significant figures given should be sufficient to allow recomputation of the experimental 2θ's to their measured accuracy, i.e., the number of significant figures should be in accordance with $\sigma(d) = (d/2) \cot \theta\sigma(2\theta)$.

hkl. List the Miller indices of the diffraction lines determined from the known unit cell. If several peaks are overlapped such that separate peak position measurements cannot be made, the hkl's should be grouped together and given for the single intensity value. The hkl's of all peaks can be listed, or a "+" can be post-scripted to the last one given to indicate others not listed.

Δ2θ. List the difference, with sign, between the experimental and calculated 2θ values: $\Delta 2\theta = 2\theta_{exp} - 2\theta_{calc}$. A useful rule of thumb is that, if systematic errors have been removed completely, the absolute value of each Δ2θ should be less than 3σ(2θ). Alternatively, a small bar-graph can be presented showing Δ2θ (with sign) versus line number. The reader, if interested, could reconstruct more accurate values of Δ2θ from the experimental 2θ's and the reported lattice parameters.

References, Appendix B

[1] International Union of Pure and Applied Chemistry, Inorganic Chemistry Section, Definitive Rules for Nomenclature of Inorganic Chemistry 1957 (Butterworths, London, 1959).

[2] Pure and Applied Chem. 28, 1 (1971).

[3] Fletcher, J. H., Dermer, O. C., and Fox, R. B., Nomenclature of Organic Compounds, Principles, and Practice, Advances in Chemistry Series No. 126 (American Chemical Society, Washington, DC, 1974).

[4] Chemical Abstracts Service Chemical Registry Structure Conventions (Chemical Abstracts Service Publishers, Columbus, Ohio, 1968).

[5] Pearson, W. B., Handbook of Lattice Spacings and Structures of Metals and Alloys, Vol. 2 (Pergamon Press, Oxford, 1967).

[6] Hubbard, C. R., Swanson, H. E., and Mauer, F. A., A silicon powder diffraction standard reference material, J. Appl. Cryst. 8, 45 (1975).

[7] Visser, J. W. and de Wolff, P. M., Absolute intensities, Report No. 641.109, Technisch Physische Dienst., Delft, Netherlands (1964).

[8] Chung, F. H., Quantitative Interpretation of x-ray diffraction patterns. III. Simultaneous determination of a set of reference intensities, J. Appl. Cryst. 8, 17 (1975).

[9] Hubbard, C. R., Evans, E. H., and Smith, D. K., The reference intensity ratio, I/I_c, for computer simulated powder patterns, J. Appl. Cryst. 9, 169-174 (1976).

[10] International Tables for X-ray Crystallography, N. F. M. Henry and K. Lonsdale, eds., Vol. 1, 2nd edition (Kynoch Press, Birmingham, England, 1965).

[11] Donnay, J. D. H. and Ondik, H. M. (eds.), Crystal Data Determinative Tables, Third Ed., Vols. 1-4 (National Bureau of Standards and Joint Committee on Powder Diffraction Standards, Publishers, 1972, 1973, 1978).

[12] Mighell, A., et al., Single crystal and powder data evaluation and standardization program, U.S. National Bureau of Standards, to be published.

[13] de Wolff, P. M., A simplified criterion for the reliability of a powder pattern indexing, J. Appl. Cryst. 1, 108 (1968).

[14] Smith, G. S. and Snyder, R. L., F_N: a criterion for rating powder diffraction patterns and evaluating the reliability of powder pattern indexing, J. Appl. Cryst. 12, 60 (1979).

Discussion

Question (Zwell): Many analysts have asked for publication of all diffraction patterns. Wouldn't your proposal decrease the number of patterns being published and thereby reduce the information available? (Would the improvement in quality of pattern compensate for the loss of data?)

Comments (Calvert, et al.): It would be hard to predict the effect with any degree of confidence. A similar set of publication recommendations made earlier by the single-crystal community resulted in a marked improvement in the average quality of papers published in Acta and many other journals. Some journals continue to accept lower quality data but the prestige of publication in the lead crystallographic journals is such that authors generally strive to meet these standards. That result can't be bad.

Question (Zwell): It appears that it is implied that 2θ would be for Cu radiation. Wouldn't the use of such angles reduce the value of data for those persons using other radiations chromium, cobalt, molybdenum? (d's are constant and independent of radiation.)

Comments (Calvert, et al.): Two theta would be recorded as measured. These data would not be transformed to a different wavelength since that process could potentially introduce noise.

POWDER DIFFRACTION DATA FOR PHASE CHARACTERIZATION

BOLD FACE ITEMS ARE CONSIDERED ESSENTIAL

SAMPLE CHARACTERIZATION

NAME (CHEMICAL MINERAL, Trivial) ____________

EMPIRICAL FORMULA ____________

CHEMICAL ANALYSIS. NO ____ **YES** ____________

SOURCE/PREPARATION ____________

____

CHEMICAL ABSTRACT REGISTRY NUMBER ____________ PEARSON ALLOY DESIGNATION ____________

TECHNIQUE

RADIATION TYPE, SOURCE ____________ **λ VALUE USED** ____________

λ DISCRIM. (Filters Mono, Etc.) ____________

λ DETECTOR (Film, Scint, Position Sensitive, etc.) ____________

INSTRUMENT DESCRIPTION (Type, Slits, etc.) ____________ **DIV =** ______ **REC =** ______

INSTRUMENTAL PROFILE BREADTH ____________ °2θ ______ **TEMPERATURE (°C)** ______

SPECIMEN FORM/PARTICLE SIZE ____________

RANGE OF 2θ FROM ______ °2θ **TO** ______ °2θ, **SPECIMEN MOTION** ____________

INTERNAL/EXTERNAL 2θ STD (if any) ____________ **LATTICE PARAMETER OF 2θ STD** ____________

2θ ERROR CORRECTION PROCEDURE ____________

INTENSITY MEAS. TECHNIQUE ____________ **PEAK** ____ **INTEGRATED** ____

MINIMUM INTENSITY THRESHOLD (IN RELATIVE INTENSITY UNITS) ____________

INTENSITY STD USED ____________ 2θ's OF INTENSITY STD ____________

INTENSITY RATIO I/I_c ______ CONVERSION FACTOR IF CORUNDUM NOT USED ____________

RESOLUTION (FWHM) FOR THIS MATERIAL. ____________ °2θ AT ____________ °2θ

2θ REPRODUCIBILITY FOR THIS MATERIAL: ____________ °2θ AT ____________ °2θ

CELL DATA

METHOD OF CELL DETN. ____________

CELL REFINEMENT METHOD ____________

a = ____________ ( ) Å; b = ____________ ( ) Å; c = ____________ ( ) Å

α = ____________ ° (°); β = ____________ ° (°); γ = ____________ ° (°)

Z = ______; D_m = ______ ( ) g cm^{-3}, D_x = ______ g cm^{-3}, V = ______ Å^3; Formula Wt = ______

CRYSTAL SYS. ____________; SPACE GROUP ____________; CRYSTAL DATA INDEX No. ____________

FIGURE OF MERIT TYPE ____________ VALUE ____________

REFERENCES

____

( ) INDICATES STANDARD DEVIATION IN LEAST SIGNIFICANT DIGIT/S

POWDER DATA

ESSENTIAL		DESIRED		
2θ EXP (DEGREES)	I/I_o (1-999)	d_{EXP} (Å)	$hk\ell$	$\Delta 2\theta$* (DEGREES)

$^*2\theta_{EXP} - 2\theta_{CALC}$

RECOMMENDATIONS FOR THE PRESENTATION OF INFRARED ABSORPTION SPECTRA IN DATA COLLECTIONS—A. CONDENSED PHASES

COMMISSION ON MOLECULAR STRUCTURE AND SPECTROSCOPY

E. D. BECKER

US Department of Health, Education and Welfare,
Bethesda, Maryland, U.S.A.

PREAMBLE

These recommendations relate to the infrared spectra of condensed phase materials that are intended for permanent retention in data collections. They are revised from "Tentative Specifications for the Measurement and Evaluation of Infrared Spectra for Documentation Purposes," published in IUPAC Information Bulletin No. 34 in 1969. Further action concerning that report has been held in abeyance pending the assessment of new developments in the technique of infrared spectrophotometry resulting from the emergence of interferometric techniques of spectra measurement. These recommendations are based on two reports published by the Coblentz Society (1,2) of which the latter takes account of the interferometric method of measuring the spectra.

Although these recommendations are directed toward infrared spectra prepared for reference collections, many of them are also pertinent to spectra presented in journals.

The Coblentz Society has designated three levels of quality evaluation for infrared spectra:

I. Critically Defined Physical Data. This is a quality level in which all instrumental sources of error have been assessed and their tolerances specified. This applies specifically to the absolute band intensities and shapes; these are factors that are not considered in II and III. It is unlikely that substantial collections of spectra of this category will be amassed. No specifications have yet been written but it can be anticipated that small collections of spectra in this category covering limited wavenumber ranges will be obtained to serve as reference standards for infrared intensity measurements.

II. Research Quality Analytical Spectra. This category represents a quality level which can be achieved by good infrared spectrophotometers operated by competent technicians under conditions appropriate to a research laboratory.

III. Approved Analytical Spectra. This represents a minimal quality level for spectra that are to be retained for permanent cataloging.

There are, of course, many other spectra of high quality that cannot be classified into the above categories because certain information is lacking. These may be referred to as Unevaluated Spectra.

The recommendations in this report are based on the Coblentz Society's Research Quality Analytical Spectra category but the lower tolerances acceptable for the Approval Analytical Spectra are noted parenthetically where the tolerance differences are purely numerical. Other differences between the two quality characteristics mainly concern the documentation of the chemical structure and related matters. These are important where the spectra are deposited in atlases, but are less relevant where the spectra are used for other purposes, e.g. in a chemical journal in which it may be presumed that the documentation of the sample origin and structure will be referenced in the same article.

It is recommended that the Coblentz categorization of spectral quality be retained in the IUPAC specifications.

SPECIFICATIONS

These recommendations relate to both dispersively and interferometrically measured spectra unless otherwise indicated. When used without qualification the term spectrophotometer designates either a dispersive or an interferometric instrument. For interferometrically measured spectra all evaluation is based on the spectrogram, not on the interferogram. Tests to evaluate these specifications based on measurements on standard reference materials are described in the Coblentz Society's report (2) and the additional references cited therein.

(1) Anal. Chem. 38, No. 9, 27A (1966).
(2) Anal. Chem. 47, No. 11, 945A (1975).

I. SPECTROPHOTOMETER OPERATION

A. Resolution

For dispersively measured spectra the spectral slit width should not exceed 2 cm^{-1} (5 cm^{-1})* through at least 80% of the wavenumber range and at no place should it exceed 5 cm^{-1} (10 cm^{-1}). For interferometrically measured spectra the optical retardation must be at least 0.5 cm (0.2 cm) and the apodization function must be stated.

B. Wavenumber Accuracy

The abscissa, as read from the chart, should be accurate to ±5 cm^{-1} (15 cm^{-1}) at wavenumbers greater than 2000 cm^{-1} and to ±3 cm^{-1} (5 cm^{-1}) at wavenumbers less than 2000 cm^{-1}. Fiduciary marks should be recorded on each chart shortly after the beginning and near the end of each uninterrupted scanned segment of the spectrum. These marks are required to guard against errors from paper shrinkage and from mismatch between the printed chart grid and the spectrophotometer.

C. Noise Level

The noise level should not exceed 1% (2%) average peak-to-peak or 0.25% (0.5%) root mean square of a full-scale deflection in internal transmittance.

D. Energy

The spectrophotometer should be purged with dry gas or evacuated to ensure that at least 50% of the source energy is available at all wavenumbers (except in the regions of carbon dioxide absorption near 2350 and 670 cm^{-1}; if less than 25% of the original energy is available, these narrow regions of the spectrum should be deleted).

E. Other Performance Criteria

1. False radiation. Apparent stray radiation should be less than 2% (5%) transmittance at wavenumbers greater than 500 cm^{-1}.

2. Servo System and Recording System. Spectra should not exhibit evidence of dead spots or of excessive recorder overshoot. The spectrophotometer and recorder time constant should be compatible with the scan rate.

3. Temperature. It is to be assumed that the spectrum is measured at the ambient temperature unless stated otherwise.

II. PRESENTATION

A. Information to Appear on the Chart

Both the structural and the molecular formulae of the compound should appear on the chart.

The make and model of the spectrophotometer should be recorded as well as the date on which the spectrum was measured. For spectra obtained dispersively all changes of gratings and filters should be indicated, including the wavenumber at which they occur. A bar or arrow would be appropriate for this purpose. No mechanical attenuator should be placed in the reference beam additional to the optical attenuator, which is integral to some infrared spectrophotometers. For interferometrically measured spectra, the apodization function must be specified (See I. A). For all spectrophotometers the wavenumber positions at which cell changes occur should be specified.

The physical condition of the sample should be stated (e.g. solution, liquid, mull, halide pellet matrix, etc.). For measurements on solutions the solvent used in each region of the spectrum should be recorded. The concentration and nominal path length should be given for both solutions and pellets. The nominal path lengths of liquid samples should be indicated; a very thin layer may be described as a capillary film (See III B.1.c). In all cases the cell window or support window should be stated.

B. Spectral Range

The chart should cover the range 3800 (3700) to 450 (700) cm^{-1} without gaps, except under the conditions defined in I.D. Extensions above and below this range are strongly encouraged. For such extended-range spectra the wavenumber accuracy, false radiation, atmospheric absorption and resolution should be stated.

For special purposes, such as spectra specifically categorized as "far infrared spectra" or "near infrared spectra" the inclusion of the range 3800 (3700) to 450 (700) cm^{-1} need not be mandatory.

C. Intensity Scale

It is preferred that the intensity ordinate values be expressed in absorbance units. Where the charts are plotted on a linear transmittance scale, a logarithmic ordinate grid is preferable so that the absorbance can be interpolated directly. Spectra plotted on a linear absorbance scale or in linear transmittance on a linear percent transmission ordinate grid are acceptable.

---

*Values in parentheses refer to Approved Analytical Spectra.

Any band over 1.5 absorbance units should be reproduced on a less absorbing sample. A significant fraction of the useful bands should have absorbance greater than 0.2. At least one band in the spectrum should have absorbance exceeding 0.7 (0.6). When multiple traces are required on the same chart their number should be kept to a minimum.

D. Wavenumber Readability

Sharp peaks should be readable to within ±3 cm^{-1} (±8 cm^{-1}) at wavenumbers greater than 2000 cm^{-1} and to ±1 cm^{-1} (±2 cm^{-1}) at wavenumbers less than 2000 cm^{-1}. Only spectra recorded with the abscissal scale linear in wavenumber are acceptable, but scale changes at designated abscissal positions are allowed. Presentation in a smaller format is allowable provided the absorption maxima are labelled with their wavenumber positions, in figures at least 1 mm high in the final reproduction.

E. Recording

Recording should be continuous with no gaps in wavenumber subject to the exception described in I.D. It is, however, permissible for spectra to extend over more than one chart. Discontinuities in ordinate (absorbance), if present, should not exceed 0.01 absorbance unit. Hand retraced spectra are not acceptable.

F. Atmospheric Absorption

Atmospheric absorption should not exceed the allowable noise level when observed in the double beam mode (See I.D.).

III. SAMPLE IDENTIFICATION AND PREPARATION

A. Compound Identification and Purity

Spectra should show no inconsistencies with the postulated structure. Some relaxation of this requirement may be permitted in the case of isotopically labelled substances in which complete isotopic exchange cannot reasonably be achieved. In such cases, the bands associated with the minor isotopic species should be indicated on the chart.

B. Sample Preparation

1. Liquid State Samples. (a) For analytical purposes it is preferable that the sample be run in liquid solution, normally at concentrations in the range 5% to 10% weight (g) per volume (cm^3). Solvent bands should be compensated, but not more than 75% of the energy should be removed from the beam by such compensation and then only over a short region of the spectrum. Any solvent bands resulting from incomplete compensation should be indicated on the chart. Suitable solvents which between them cover the range 3800-450 cm^{-1} are carbon tetrachloride (3800-1335, 650-450 cm^{-1}) and carbon disulfide (1350-450 cm^{-1}); both solvents should be used at path lengths in the range 0.03 - 0.3 mm.

Cases may arise that require the use of other solvents, and solubility limitations or other concentration-dependent factors may necessitate the use of cells of longer path length. These conditions are acceptable provided the reference beam energy is not attenuated by more than 75%.

(b) For documentation purposes it is desirable that the spectrum of the liquid be recorded. Solution spectra and liquid spectra are to be regarded as complementary and not as substitutes for one another.

(c) The spectra of liquids not soluble in transparent solvents should be measured as capillary films (See II. A).

2. Solid State Samples. (a) For most analytical purposes solution spectra in transparent solvent are preferred, provided the solvents and path lengths can be chosen to leave no significant gaps due to solvent obscuration. (See III.B.1.a). This does not apply where it is required to characterize the specific crystalline structure as with many natural minerals and biological samples.

(b) Solid state spectra may be recorded from samples incorporated in mulls or pressed disks, but care should be taken to ensure that no interaction occurs between the sample and the supporting medium. The possibility of thermal decomposition or other changes in the crystalline structure must be recognized, as must the possibility of preferred orientation of platy or fibrous crystals.

Solid state spectra should meet the following criteria:

(i) Isotropic materials. The background absorbance should be less than 0.20 near 3800 cm^{-1} and less than 0.10 near 2000 cm^{-1}. No gross abnormalities should be evident in the background. Compensation in the reference beam by a blank mull or pellet should be indicated, and in no case should it reduce the reference beam intensity by more than 50%. Only minor distortion resulting from the Christiansen effect may be permitted. Interference fringes should not be apparent. Pellets should exhibit no bands due to absorbed water greater than

0.03 absorbance unit. Mulls should be made with perhalogenated oils (or equivalent) for the range 3800 - 1335 cm^{-1} and the intensity of the overtone band near 2300 cm^{-1} should not exceed 0.02 absorbance unit. Liquid paraffin (Nujol or equivalent) should be used below 1350 cm^{-1}, and the intensity of the band near 720 cm^{-1} should not exceed 0.05 absorbance unit. If there are suitable bands present, they should be overlapped in the two mull spectra and one of the bands should be identified in both mull spectra; this aids in establishing the intensity ratio between the two sections of the spectrum.

(ii) Non-isotropic materials. Spectra of non-isotropic materials, such as single crystals or oriented polymers, should be accompanied by a record of the orientation of the sample with respect to the radiation beam and, in the case of dispersively measured spectra, the orientation with respect to the grating rulings should also be indicated.

RECOMMENDATIONS FOR THE PRESENTATION OF NMR DATA FOR PUBLICATION IN CHEMICAL JOURNALS

COMMISSION ON MOLECULAR STRUCTURE AND SPECTROSCOPY†

The extensive use of nuclear magnetic resonance spectroscopy in chemical research makes it desirable to encourage the presentation of n.m.r. data in a uniform manner. Commission I.5 (Molecular Structure and Spectroscopy) therefore recommends that the following conventions should be followed in the graphical presentation of n.m.r. data in chemical journals.

A. CONVENTIONS RELATING TO PROTON SPECTRA

(1) The dimensionless scale factor for chemical shifts should be 10^6, i.e. parts per million, for which p.p.m. is a convenient abbreviation.

(2) The unit for measured data should be hertz (cycles per second), for which Hz is the appropriate abbreviation, and the frequency scale should run in the same direction as the dimensionless scale.

(3) The unit for spin–spin coupling constants should be hertz (cycles per second).

(4) The graphical presentation of spectra should show the frequency decreasing to the right (applied field increasing to the right), absorption increasing upwards, and the standard sweep direction should be from high to low frequency (low to high field).

(5) Whenever possible the dimensionless scale should be tied to an internal reference, which should normally be tetramethylsilane. The proton resonance of tetramethylsilane should be taken as zero; if some other internal reference is used that reference, and the conversion shift used to convert the measured shifts to the tetramethylsilane reference scale, should be explicitly stated. The dimensionless scale should be defined as positive in the high frequency (low field) direction. The scale in parts per million based on zero for tetramethylsilane should be termed the δ scale. A shift measured on this scale should be given as, for example, $\delta = 5.00$, not $\delta = 5.00$ p.p.m. The symbol δ_H may be used if there is ambiguity about the nucleus under investigation.

(6) When the spectra are submitted for publication, additional information should include:

(a) The name of the solvent used.
(b) The concentration of the solute.
(c) The name and concentration of the internal reference.
(d) The name of the external reference if one is used. Water should not normally be so used because of the temperature dependence of its resonance.
(e) The temperature of the sample.
(f) The procedure used to measure the peak positions.
(g) The radio-frequency at which the measurements were made; alternatively the magnetic field should be stated if the spectrum was obtained by a frequency sweep method.

† *Titular Members*: R. N. Jones (Canada) (Chairman); A. R. H. Cole (Australia) (Vice-Chairman); F. A. Miller (USA) (Secretary); *Members*: M. A. Elyashevich (USSR), Th. Förster (Germany), A. Hadni (France), Y. Morino (Japan), N. Sheppard (UK); *Associate Members*: E. Fluck (Germany), E. R. Lippincott (USA), R. C. Lord (USA), S. Nagakura (Japan), J. Plivá (Czechoslovakia), Sir Harold Thompson (UK), D. W. Turner (UK), *Advisory Counsellor*: G. Herzberg (Canada); *National Representative*: T. Urbanski (Poland).

Reprinted with permission from *Pure and Applied Chemistry,* Volume 29, Number 4, 1972, pages 627–628.

Other information should be added where appropriate or necessary, e.g. the sweep rate, the magnitude of the H_1 fields, data pertinent to the use of spin decoupling, and whether oxygen has been removed from the sample. Solvent and impurity bands, carbon-13 satellites, or spinning sidebands should be indicated as such.

If single resonances or part spectra are presented as diagrams there should be a graphical indication of the distance corresponding to a suitable range of Hz so that fine-structure spacings, or widths of broad resonances, can be estimated.

B. CONVENTIONS RELATING TO SPECTRA FROM OTHER NUCLEI

At this stage the Commission does not wish to make definite recommendations relating to n.m.r. spectra of other nuclei. However, increasing interest in the use of double resonance techniques to relate chemical shifts of other nuclei to the proton resonance of tetramethylsilane suggests that it may become advantageous to adopt the same conventions for other nuclei.

Standard Practice for Data Presentation Relating to High-Resolution Nuclear Magnetic Resonance (NMR) Spectroscopy[1]

This standard is issued under the fixed designation E 386; the number immediately following the designation indicates the year of original adoption or, in the case of revision, the year of last revision. A number in parentheses indicates the year of last reapproval. A superscript epsilon (ϵ) indicates an editorial change since the last revision or reapproval.

1. Scope

1.1 This standard contains definitions of basic terms, conventions, and recommended practices for data presentation in the area of high-resolution NMR spectroscopy. Some of the basic definitions apply to wide-line NMR or to NMR of metals, but in general it is not intended to cover these latter areas of NMR in this standard. This version does not include definitions pertaining to double resonance nor to rotating frame experiments.

2. Nomenclature and Basic Definitions

2.1 *nuclear magnetic resonance (NMR) spectroscopy*—that form of spectroscopy concerned with radio-frequency-induced transitions between magnetic energy levels of atomic nuclei.

2.2 *NMR apparatus; NMR equipment*—an instrument comprising a magnet, radio-frequency oscillator, sample holder, and a detector that is capable of producing an electrical signal suitable for display on a recorder or an oscilloscope, or which is suitable for input to a computer.

2.3 *high-resolution NMR spectrometer*—an NMR apparatus that is capable of producing, for a given isotope, line widths that are less than the majority of the chemical shifts and coupling constants for that isotope.

NOTE —By this definition, a given spectrometer may be classed as a high-resolution instrument for isotopes with large chemical shifts, but may not be classed as a high-resolution instrument for isotopes with smaller chemical shifts.

2.4 *basic NMR frequency,* ν_o—the frequency, measured in hertz (Hz), of the oscillating magnetic field applied to induce transitions between nuclear magnetic energy levels. The static magnetic field at which the system operates is called H_o (Note 1) and its recommended unit of measurement is the tesla (T) (1 T = 10^4 gauss).

2.4.1 The foregoing quantities are approximately connected by the following relation:

$$\nu_o = \frac{\gamma}{2\pi} H_o$$

where γ = the magnetogyric ratio, a constant for a given nuclide (Note 2). The amplitude of the magnetic component of the radio-frequency field is called H_1. Recommended units are millitesla and microtesla.

NOTE 1—This quantity is normally referred to as B by physicists. The usage of H to refer to magnetic field strength in chemical applications is so widely accepted that there appears to be no point in attempting to reach a totally consistent nomenclature now.

NOTE 2—This expression is correct only for bare nuclei and will be only approximately true for nuclei in chemical compounds, since the field at the nucleus is in general different from the static magnetic field. The discrepancy amounts to a few parts in 10^6 for protons, but may be of magnitude 1×10^{-3} for the heaviest nuclei.

2.5 *NMR absorption line*—a single transition or a set of degenerate transitions is referred to as a line.

2.6 *NMR absorption band; NMR band*—a region of the spectrum in which a detectable signal exists and passes through one or more maxima.

2.7 *reference compound (NMR)*—a selected material to whose signal the spectrum of a sample may be referred for the measurement of chemical shift (see 2.9).

2.7.1 *internal reference (NMR)*—a reference compound that is dissolved in the same phase as the sample.

2.7.2 *external reference (NMR)*—a reference compound that is not dissolved in the same phase as the sample.

2.8 *lock signal*—the NMR signal used to control the field-frequency ratio of the spectrometer. It may or may not be the same as the reference signal.

2.8.1 *internal lock*—a lock signal which is obtained from a material that is physically within the confines of the sample tube, whether or not the material is in the same phase as the sample (an annulus for the purpose of this definition is considered to be within the sample tube).

2.8.2 *external lock*—a lock signal which is obtained from a material that is physically outside the sample tube. The material supplying the lock signal is usually built into the probe.

NOTE —An external lock, if also used as a reference, is necessarily an external reference. An internal lock, if used as a reference, may be either an internal or an external reference, depending upon the experimental configuration.

2.8.3 *homonuclear lock*—a lock signal which is obtained from the same nuclide that is being observed.

2.8.4 *heteronuclear lock*—a lock signal which is obtained from a different nuclide than the one being observed.

2.9 *chemical shift,* δ—the defining equation for δ is the following:

$$\delta = \frac{\Delta\nu}{\nu_R} \times 10^6$$

where ν_R is the frequency with which the reference substance is in resonance at the magnetic field used in the experiment

---

[1] These definitions are under the jurisdiction of ASTM Committee E-13 on Molecular Spectroscopy and are the direct responsibility of Subcommittee E13.07 on Nuclear Magnetic Resonance Spectroscopy.

Current edition approved June 29, 1990. Published August 1990. Originally published as E 386 - 69 T. Last previous edition E 386 - 78.

and $\Delta\nu$ is the frequency of the subject line minus the frequency of the reference line at constant field. The sign of $\Delta\nu$ is to be chosen such that shifts to the high frequency side of the reference shall be positive.

2.9.1 If the experiment is done at constant frequency (field sweep) the defining equation becomes

$$\delta = \frac{\Delta\nu}{\nu_R} \times \left(1 - \frac{\Delta\nu}{\nu_R}\right) \times 10^6$$

2.9.2 In case the experiment is done by observation of a modulation sideband, the audio upper or lower sideband frequency must be added to or subtracted from the radio frequency.

2.10 *spinning sidebands*—bands, paired symmetrically about a principal band, arising from spinning of the sample in a field (dc or rf) that is inhomogeneous at the sample position. Spinning sidebands occur at frequencies separated from the principal band by integral multiples of the spinning rate. The intensities of bands which are equally spaced above and below the principal band are not necessarily equal.

2.11 *satellites*—additional bands spaced nearly symmetrically about a principal band, arising from the presence of an isotope of non-zero spin which is coupled to the nucleus being observed. An isotope shift is normally observed which causes the center of the satellites to be chemically shifted from the principal band. The intensity of the satellite signal increases with the abundance of the isotope responsible.

2.12 *NMR line width*—the full width, expressed in hertz (Hz), of an observed NMR line at one-half maximum height (FWHM).

2.13 *spin-spin coupling constant (NMR), J*—a measure, expressed in hertz (Hz), of the indirect spin-spin interaction of different magnetic nuclei in a given molecule.

NOTE —The notation $^nJ_{AB}$ is used to represent a coupling over n bonds between nuclei A and B. When it is necessary to specify a particular isotope, a modified notation may be used, such as, $^3J\,(^{15}NH)$.

3. Types of High-Resolution NMR Spectroscopy

3.1 *sequential excitation NMR; continuous wave (CW) NMR*—a form of high-resolution NMR in which nuclei of different field/frequency ratio at resonance are successively excited by sweeping the magnetic field or the radio frequency.

3.1.1 *rapid scan Fourier transform NMR; correlation spectroscopy*— a form of sequential excitation NMR in which the response of a spin system to a rapid passage excitation is obtained and is converted to a slow-passage spectrum by mathematical correlation with a reference line, or by suitable mathematical procedures including Fourier transformations.

3.2 *broad-band excitation NMR*—a form of high-resolution NMR in which nuclei of the same isotope but possibly different chemical shifts are excited simultaneously rather than sequentially.

3.2.1 *pulse Fourier transform NMR*—a form of broad-band excitation NMR in which the sample is irradiated with one or more pulse sequences of radio-frequency power spaced at uniform time intervals, and the averaged free induction decay following the pulse sequences is converted to a frequency domain spectrum by a Fourier transformation.

3.2.1.1 *pulse Fourier difference NMR*—a form of pulse Fourier transform NMR in which the difference frequencies between the sample signals and a strong reference signal are extracted from the sample response prior to Fourier transformation.

3.2.1.2 *synthesized excitation Fourier NMR*—a form of pulse Fourier NMR in which a desired frequency spectrum for the exciting signal is Fourier synthesized and used to modulate the exciting radio frequency.

3.2.2 *stochastic excitation NMR*—a form of broad band excitation NMR in which the nuclei are excited by a range of frequencies produced by random or pseudorandom noise modulation of the carrier, and the frequency spectrum is obtained by Fourier transforming the correlation function between the input and output signals.

3.2.3 *Hadamard transform NMR*—a form of broad band excitation NMR in which the phase of the excitation signal is switched according to a binary pseudorandom sequence, and the correlation of the input and output signals by a Hadamard matrix yields an interference pattern which is then Fourier-transformed.

4. Operational Definitions

4.1 *Definitions Applying to Sequential Excitation (CW) NMR:*

4.1.1 *field sweeping (NMR)*—systematically varying the magnetic field strength, at constant applied radio-frequency field, to bring NMR transitions of different energies successively into resonance, thereby making available an NMR spectrum consisting of signal intensity versus magnetic field strength.

4.1.2 *frequency sweeping (NMR)*—systematically varying the frequency of the applied radio frequency field (or of a modulation sideband, see 4.1.4), at constant magnetic field strength, to bring NMR transitions of different energies successively into resonance, thereby making available an NMR spectrum consisting of signal intensity versus applied radio frequency.

4.1.3 *sweep rate*—the rate, in hertz (Hz) per second at which the applied radio frequency is varied to produce an NMR spectrum. In the case of field sweep, the actual sweep rate in microtesla per second is customarily converted to the equivalent in hertz per second, using the following equation:

$$\frac{\Delta\nu}{\Delta t} = \frac{\gamma}{2\pi} \cdot \frac{\Delta H}{\Delta t}$$

4.1.4 *modulation sidebands*—bands introduced into the NMR spectrum by, for example, modulation of the resonance signals. This may be accomplished by modulation of the static magnetic field, or by either amplitude modulation or frequency modulation of the basic radio frequency.

4.1.5 *NMR spectral resolution*—the width of a single line in the spectrum which is known to be sharp, such as, TMS or benzene (^{1}H). This definition includes sample factors as well as instrumental factors.

4.1.6 *NMR integral (analog)*—a quantitative measure of the relative intensities of NMR signals, defined by the areas of the spectral lines and usually displayed as a step function in which the heights of the steps are proportional to the areas (intensities) of the resonances.

4.2 *Definitions Applying to Multifrequency Excitation (Pulse) NMR:*

4.2.1 *pulse (v)*—to apply for a specified period of time a perturbation (for example, a radio frequency field) whose amplitude envelope is nominally rectangular.

4.2.2 *pulse (n)*—a perturbation applied as described above.

4.2.3 *pulse width*—the duration of a pulse.

4.2.4 *pulse flip angle*—the angle (in degrees or radians) through which the magnetization is rotated by a pulse (such as a 90-deg pulse or $\pi/2$ pulse).

4.2.5 *pulse amplitude*—the radio frequency field, H_1, in tesla.

NOTE —This may be specified indirectly, as described in 8.3.2.

4.2.6 *pulse phase*—the phase of the radio frequency field as measured relative to chosen axes in the rotating coordinate system.[2]

NOTE —The phase may be designated by a subscript, such as, $90°_x$ or $(\pi/2)_x$.

4.2.7 *free induction decay (FID)*—the time response signal following application of an r-f pulse.

4.2.8 *homogeneity spoiling pulse; homo-spoil pulse; inhomogenizing pulse*—a deliberately introduced temporary deterioration of the homogeneity of the magnetic field *H*.

4.2.9 *filter bandwidth; filter passband*—the frequency range, in hertz, transmitted with less than 3 dB (50 %) attenuation in power by a low-pass filter.

NOTE 1—On some commercial instruments, filter bandwidth is defined in a slightly different manner.

NOTE 2—Other parameters, such as rate of roll-off, width of passband, or width and rejection of center frequency in case of a notch filter, may be required to define filter characteristics adequately.

4.2.10 *data acquisition rate; sampling rate; digitizing rate*—the number of data points recorded per second.

4.2.11 *dwell time*—the time between the beginning of sampling of one data point and the beginning of sampling of the next successive point in the FID.

4.2.11.1 *aperture time*—the time interval during which the sample-and-hold device is receptive to signal information. In most applications of pulse NMR, the aperture time is a small fraction of the dwell time.

NOTE —*Sampling Time* has been used with both of the above meanings. Since the use of this term may be ambiguous, it is to be discouraged.

4.2.12 *detection method*—a specification of the method of detection.

4.2.12.1 *single-phase detection*—a method of operation in which a single phase-sensitive detector is used to extract signal information from a FID.

4.2.12.2 *quadrature detection*—a method of operation in which dual phase-sensitive detection is used to extract a pair of FID's which differ in phase by 90°.

4.2.13 *spectral width*—the frequency range represented without foldover. (Spectral width is equal to one half the data acquisition rate in the case of single-phase detection; but is equal to the full data acquisition rate if quadrature detection is used.)

4.2.14 *foldover; foldback*—the appearance of spurious lines in the spectrum arising from either *(a)* limitations in data acquisition rate or *(b)* the inability of the spectrometer detector to distinguish frequencies above the carrier frequency from those below it.

NOTE —These two meanings of *foldover* are in common use. Type *(a)* is often termed "aliasing." Type *(b)* foldover is obviated by the use of quadrature detection.

4.2.15 *data acquisition time*—the period of time during which data are acquired and digitized; equal numerically to the product of the dwell time and the number of data points acquired.

4.2.16 *computer-limited spectral resolution*—the spectral width divided by the number of data points.

NOTE—This will be a measure of the observed line width only when it is much greater than the spectral resolution defined in 4.1.5.

4.2.17 *pulse sequence*—a set of defined pulses and time spacings between these pulses.

NOTE —There may be more than one way of expressing a sequence, for example, a series $(90°, \tau)_n$ may be one sequence of *n* pulses or *n* sequences each of the form $(90°, \tau)$.

4.2.18 *pulse interval*—the time between two pulses of a sequence.

4.2.19 *waiting time*—the time between the end of data acquisition after the last pulse of a sequence and the initiation of a new sequence.

NOTE —To ensure equilibrium at the beginning of the first sequence, the software in some NMR systems places the waiting time prior to the initiation of the first pulse of the sequence.

4.2.20 *acquisition delay time*—the time between the end of a pulse and the beginning of data acquisition.

4.2.21 *sequence delay time; recovery interval*—the time between the last pulse of a pulse sequence and the beginning of the succeeding (identical) pulse sequence. It is the time allowed for the nuclear spin system to recover its magnetization, and it is equal to the sum of the acquisition delay time, data acquisition time, and the waiting time.

4.2.22 *sequence repetition time*—the period of time between the beginning of a pulse sequence and the beginning of the succeeding (identical) pulse sequence.

4.2.23 *pulse repetition time*—the period of time between one r-f pulse and the succeeding (identical) pulse; used instead of *sequence repetition time* when the "sequence" consists of a single pulse.

4.2.24 *inversion-recovery sequence*—a sequence that inverts the nuclear magnetization and monitors its recovery, such as $(180°, \tau, 90°)$, where τ is the pulse interval.

4.2.25 *saturation-recovery sequence*—a sequence that saturates the nuclear magnetization and monitors its recovery, such as the sequence (90°, homogeneity-spoiling pulse, τ, 90°, *T*, homogeneity-spoiling pulse) or the sequence $(90°)_n$, τ, 90°, *T*, where $(90°)_n$ represents a rapid burst of 90° pulses.

4.2.26 *progressive saturation sequence*—the sequence 90°, $(\tau, 90°)_n$, where *n* may be a large number, and data acquisition normally occurs after each pulse (except possibly the first three or four pulses).

4.2.27 *spin-echo sequence*—the sequence 90°, τ, 180°

[2] For a discussion of the rotating coordinate system, see Abragam, "Principles of Nuclear Magnetism," Oxford, 1961, pp. 19ff.

4.2.28 *Carr-Purcell (CP) sequence*—the sequence 90°, τ, 180°, $(2\tau, 180°)_n$, where n can be a large number.

4.2.29 *Carr-Purcell time*—the pulse interval 2τ between successive 180° pulses in the Carr-Purcell sequence.

4.2.30 *Meiboom-Gill sequence; CPMG sequence*—the sequence $90°_x$, τ, $180°_y$, $(2\tau, 180°_y)_n$.

4.2.31 *spin-locking sequence*—the sequence $90°_x$, $(SL)_y$, where SL denotes a "long" pulse (often measured in milliseconds or seconds, rather than microseconds) and H (lock) $\gg$ H (local).

4.2.32 *zero filling*—supplementing the number of data points in the time response signal with trailing zeroes before Fourier transformation.

4.2.33 *partially relaxed Fourier transform (PRFT) NMR*—a set of multiline FT spectra obtained from an inversion-recovery sequence and designed to provide information on spin-lattice relaxation times.

4.2.34 *NMR integral (digital)*—the integrals (see 4.1.6) of pulse-Fourier transform spectra or of digitized CW spectra, obtained by summing the amplitudes of the digital data points that define the envelope of each NMR band. The results of these summations are usually displayed either as a normalized total number of digital counts for each band, or as a step function (running total of digital counts) superimposed on the spectrum.

5. NMR Conventions

5.1 The dimensionless scale used for chemical shifts for any nucleus shall be termed the δ scale. The correct usage is $\delta = 5.00$ or δ 5.00. Alternative forms, such as $\delta = 5.00$ ppm or shift = 5.00 δ shall not be used.

5.2 The unit used for line positions should be hertz.

5.3 The dimensionless and frequency scales should have a common origin.

5.4 The standard sweep direction should be from high to low radio frequency (low to high applied magnetic field).

5.5 The standard orientation of spectra should be with low radio frequency (high field) to the right.

5.6 Absorption mode peaks should point up.

6. Referencing Procedures and Substances

6.1 *General:*

6.1.1 Whenever possible, in the case of proton and carbon-13 spectra, the chemical shift scale should be tied to an *internal* reference.

6.1.2 In case an external reference is used, either a coaxial tube or a capillary tube is generally adequate.

6.1.3 For nuclei other than protons or ^{13}C, for which generally agreed-upon reference substances do not yet exist, it is particularly important to report the reference material and referencing procedure fully, including separations in hertz and the spectrometer radio frequency when it is known.

6.2 *NMR Reference Substances for Proton Spectra:*

6.2.1 The primary internal reference for proton spectra in nonaqueous solution shall be tetramethylsilane (TMS). A concentration of 1 % or less is preferred.

6.2.2 The position of the tetramethylsilane resonance is defined as exactly zero.

6.2.3 The recommended internal reference for proton spectra in aqueous solutions is the sodium salt of 2,2,3,3-tetradeutero-4,4-dimethyl-4-silapentanoic acid (TSP-d_4). Its chemical shift is assigned the value zero.

6.2.4 The numbers on the dimensionless (shift) scale to high frequency (low field) of TMS shall be regarded as positive.

6.3 *NMR Reference Substances for Nuclei Other than Protons:*

6.3.1 For all nuclei the numbers on the dimensionless (shift) scale to high frequency (low field) from the reference substance shall be positive. In the interim, until this proposal has been fully adopted, the sign convention used should be explicitly given.

NOTE —The existing literature on NMR contains examples of both the sign convention given above and its opposite. It seems desirable to adopt a uniform convention for all nuclei, and the convention recommended herein is already widely used in both proton and ^{13}C NMR. The recommended convention will result in assigning the most positive numerical value to the transition of highest energy.

6.3.2 The primary internal reference for ^{13}C spectra of nonaqueous solutions shall be tetramethylsilane (TMS). For aqueous solutions, secondary standards such as dioxane have been found satisfactory. When such standards are used the line positions and chemical shifts should be reported with reference to TMS, and the conversion factor should be stated explicitly.

6.3.3 The primary external reference for boron spectra (^{10}B and ^{11}B) shall be boron trifluoride-diethyletherate [$(C_2H_5)_2O:BF_3$].

6.3.4 The primary external reference for ^{31}P spectra shall be phosphorus trioxide (P_4O_6).

6.3.5 Specific recommendations for nuclei other than those mentioned above are not offered here. The following guidelines should be used: If previous work on the nucleus under study exists, any earlier reference should be used unless there are compelling reasons to choose a new reference. A reference substance should have a sharp line spectrum if possible. A singlet spectrum is preferred. A reference substance should be chosen to have a resonance at low frequency (high field) so far as possible, in order that the majority of chemical shifts will be of positive sign. Internal references should be avoided unless it is possible to include a study of solvent effects on chemical shift.

7. Recommended Practice for Signal-to-Noise Determination in Fourier Transform NMR

7.1 *General*—This section gives the recommended practice for signal-to-noise ratio (S/N) determination in three specific situations: () proton single pulse mode; (*b*) carbon-13 single pulse mode; and (*c*) carbon-13 multiple pulse mode.

NOTE 1—Some of the materials recommended for use in this section are known to present health hazards if used improperly. Anyone making up solutions containing benzene, dioxane, or chloroform should consult and abide by OSHA regulations 29CFR 1910.1000 (solvents) and 29CFR 1910.1028 (benzene).

7.2 *Proton Single Pulse Mode:*

7.2.1 *Sample*—Dilute ethylbenzene in $CDCl_3$.

7.2.2 *Measurement*—Proton signal-to-noise ratio is measured using a single pulse of radio-frequency power applied to a dilute solution of ethylbenzene in $CDCl_3$. Choose the concentration of ethylbenzene appropriate to the sensitivity of the instrument under test, such that the S/N as measured on the methylene quartet is 25 : 1. State the determined S/N as "equivalent one percent ethylbenzene sensitivity." Carry out the measurement using the following conditions:

Spectral width	0 to 10 ppm ($\delta^{1H}_{TMS} \equiv 0$)
Data acquisition time	≥0.4 s
Flip angle	90°
Analog filter	appropriate for method of detection
Detection method	specify (for example, single phase, SSB, QPD)
Equilibration delay	60 s

Following the data acquisition, multiply the data by a decaying exponential function of the form $e^{-t/A}$, where A is equivalent to a T_2 contribution. A may be expressed as a time constant in units of seconds, or, alternatively, the line broadening (LB) resulting from the exponential multiplication may be expressed in units of hertz (Hz). For the measurement, A = 0.3 or LB = 1 Hz. Perform no data smoothing after transformation. Plot the resulting absorption mode spectrum over the full 0 to 10 ppm. Measure S/N on a plot expansion covering the range of 2 to 6 ppm, in which the methylene quartet is plotted to fill the chart paper as closely as practical. Use sufficient vertical amplitude to obtain a peak-to-peak noise measurement greater than 2 cm. Measure peak-to-peak noise over the 4 to 6 ppm region on the same trace or calculate rms noise by computer (see Note 2). The S/N is then calculated on the strongest line in the quartet as follows (see Fig. 1):

$$[(\text{signal intensity})/(\text{peak-to-peak noise})] \times 2.5 = \text{S/N}$$

NOTE 2—The true rms noise can be calculated by computer and used in the S/N determination. Since peak-to-peak noise is approximately five times rms noise, rather than 2.5 times, the rms noise must be doubled to obtain a comparable S/N. When this is done, it is felt that the S/N determined by computer should be reliable and less subject to human error than the alternate method of estimating peak-to-peak noise from a chart recording. The computer program should do the following:

(*a*) Select the region in which noise is to be measured as specified in the above test.

(*b*) Obtain the algebraic mean of all the observed points in this region, and subtract the mean from each point (zero-order correction).

(*c*) If the base line slopes, a first order correction may be made by using a standard least-squares method to obtain the slope and intercept of the baseline, then subtracting each calculated point from the corresponding observed point.

(*d*) Corrections calculated on the noise in the specified region of the spectrum should be applied to that region and also to the spectral region containing the signal.

(*e*) Form the sum of the squares of each amplitude (point), corrected as described previously, divide by one less than the number of points in the region, and take the square root. This is the rms noise.

$$\text{rms noise} = [(\Sigma[\text{amplitude}]^2)/(N - 1)]^{1/2}$$

No other processing should be done; in particular, points that appear to be extreme should not be deleted. S/N becomes simply (signal intensity/2)/(rms noise).

7.2.3 *Discussion*—The 1 % ethylbenzene S/N measurement is a widely used method for ^{1}H S/N both in CW and FT NMR. Although presenting few difficulties in CW work, the typical samples used in FT NMR do present some problems which we hope to avoid using this procedure.

7.2.3.1 The 1 % concentration traditionally employed generates a very high S/N on modern FT spectrometers, particularly at very high magnetic field strengths.

7.2.3.2 TMS is usually present in standard samples at the 1 % level. This causes a very strong signal which can lead to an erroneous S/N measurement.

7.2.3.3 The variety of sample tube sizes and S/N values has made it inconvenient to use a uniform concentration. The solution(s) should be made up by volume composition at 25°C using good volumetric practice. Suggested solutions:

No.	*Ethylbenzene*, %	*TMS*, % (Note 3)
1	3.0	0.3
2	1.0	0.1
3	1.0	1.0 (also valuable for CW TMS-locked spectrometers)
4	0.33	0.03
5	0.10	0.01
6	0.033	0.003
7	0.010	0.001

NOTE 3—The TMS is added for a reference material.

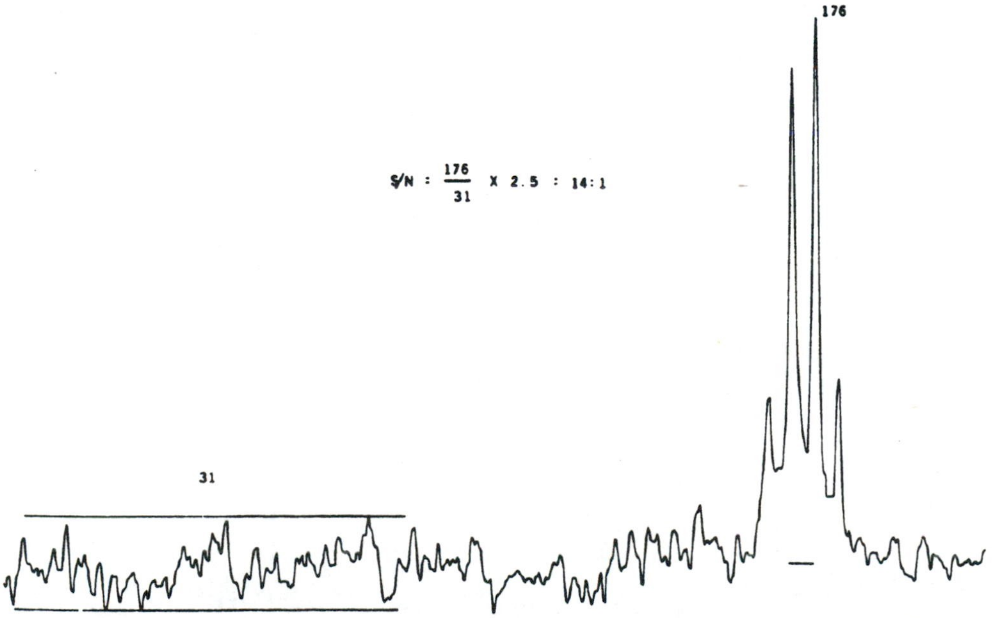

FIG. 1 Typical S/N Measurement on the Proton Signal in Dilute Ethylbenzene

7.3 *Carbon-13 Single Pulse Mode:*

7.3.1 *Sample*—60 %C_6D_6(>98atom %D),40 % *p*-dioxane (v/v).

7.3.2 *Measurement*—Measure carbon-13 signal-to-noise ratio on the benzene carbon signal in a solution of 60 % perdeuterobenzene - 40 % *p*-dioxane, with the spectrometer locked to the deuterium in the sample, using the following conditions:

Spectral width	0 to 200 ppm ($\delta\ C^{13}_{TMS} \equiv 0$ ppm)
Data acquisition time	≥0.4 s
Flip angle	90°
Analog filter	appropriate for method of detection
Detection method	specify (for example, single phase, SSB, QPD)
Equilibration delay	300 s
Decoupler	off

Following the data acquisition, multiply the data by a decaying exponential function of the form $e^{-t/A}$, where A is equivalent to a T_2 contribution. A may be expressed as a time constant in units of seconds, or, alternatively, the line broadening (LB) resulting from exponential multiplication may be expressed in units of hertz (Hz). For the measurement, A = 0.3 or LB = 1 Hz. Perform no data smoothing after transformation. Plot the resulting absorption mode spectrum over the full 0 to 200 ppm chemical shift range. Plot the C_6D_6 triplet to fill the vertical range of the chart paper as closely as practical. Use sufficient vertical amplitude to obtain a peak-to-peak noise measurement greater than 2 cm. Signal-to-noise is to be measured as:

[(average triplet intensity)/(peak-to-peak noise)]× 2.5 = S/N

Measure the peak-to-peak noise between the C_6D_6 and dioxane triplets, specifically between and inclusive of 80 and 120 ppm on the ^{13}C chemical shift scale, or calculate rms noise by computer (see Note 2 and Fig. 2).

7.3.3 *Characteristics of the Proposed Standard:*

7.3.3.1 The S/N of the C_6D_6 triplet is low enough to permit a plot from which both signal and noise may be measured. For a full scale vertical display of the C_6D_6 triplet, the peak-to-peak noise amplitude should be adequately measured and have two significant figures. (For those spectrometers with very high sensitivity, noise would still have to be blown up to at least 2 cm peak-to-peak in a separate trace of the same transformed data.)

7.3.3.2 The C_6D_6 triplet has linewidth of 14 Hz under these conditions, reasonably independent of magnet resolution, permitting easy tune up and small 4 K data table for the measurement.

7.3.3.3 The C_6D_6 S/N can be measured in the presence of or absence of high power proton decoupling facilitating servicing diagnostic procedures. It is particularly valuable in diagnosing decoupler-caused noise contributions.

7.3.3.4 The broad lines of the C_6D_6 result from long-range ^{13}C-^2H coupling and thus the linewidth is not field-dependent.

7.3.3.5 C_6D_6 has no nuclear Overhauser enhancement (NOE).

7.3.3.6 The reference material is widely available and can serve as an internal ^2H lock.

7.3.3.7 The C_6D_6 S/N is independent of applied lock power in normal locking power range up to and beyond saturation of the deuterium signal.

7.3.3.8 The C_6D_6 S/N is temperature independent over normal working temperatures.

7.3.3.9 The dioxane serves several purposes: ready reference to prior data; a conveniently short T_1 (<10 s); under decoupled conditions it possesses a strong signal serving for $\gamma H_1/2\pi$ measurement by means of a 90° pulse determination; under off-resonance conditions its residual ^{13}C-^1H coupling can serve to measure $\gamma H_2/2\pi$; the decoupled singlet can be used to measure resolution in terms of full linewidth at half-height, also line shape and spinning sidebands; and under coupled conditions and longer acquisition times, it can provide a coupled spectrum with long-range couplings. The strong signal available from decoupled dioxane permits facile tests of decoupler gating through measurement of the NOE via "Suppressed Overhauser" gating schemes vs use of coupled dioxane as the base point for calculating the NOE. The short T_1 of dioxane allows routine check of automatic T_1 programs and calculations.

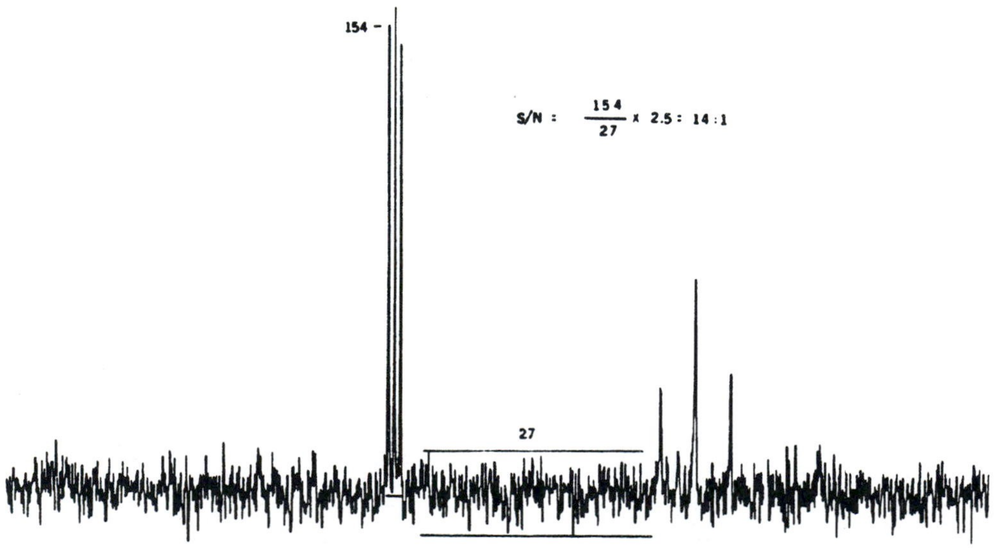

FIG. 2 Typical S/N Measurement on Single Pulse ^{13}C Spectrum of C_6D_6-Dioxane Mixture

7.3.4 *Discussion*—The proposed measurement is possible and convenient on any modern FT instrument. This method ensures that the maximum available S/N is obtained, thus preventing confusion in parameter choice, particularly in the case of the exponential weighting. A new standard is necessary in view of the difficulty in widespread reliable use of the 90 % ethylbenzene sample previously used. The natural linewidths of the ethylbenzene lines are less than 0.1 Hz requiring exacting field homogeneity to obtain maximum resolution. The narrow lines also demand long data acquisition times in each FID to define the lines adequately. Since ethylbenzene S/N is measured on a decoupled protonated carbon signal, decoupler power, modulation efficiency, and offset are all factors in determining S/N. The S/N for most spectrometers is >100 : 1 for 90 % ethylbenzene making noise measurements the primary factor in the derived S/N.

7.3.4.1 Dioxane has been proposed for the S/N sample but it has some serious drawbacks in addition to several advantages shared with deuterobenzene. Its T_1 is dipole-dipole dominated and has full NOE in the decoupled experiment. It is easily possible to have residual NOE in a *coupled* spectrum by not waiting long enough for the NOE to decay away prior to the sampling pulse. Although deuterobenzene has the common requirement of sufficient equilibration delay the error is *always* on the side of lower S/N, whereas dioxane's apparent S/N can be up to a factor of three greater than that assumed by simple inspection of the spectrum. This makes comparison of intrinsic S/N susceptible to error. The addition of dioxane to the 40 % level provides all the advantages listed above for routine tuning up and quick S/N checking, while the C_6D_6 permits an absolute measurement. The other major disadvantage of dioxane is the dependence of the character of the spectrum on acquisition time and weighting function. If more than 0.5-s acquisition is used with a less severe weighting function than above, the fine structure from the long-range coupling becomes visible. While no problem for the experienced spectroscopist, this can be and has been confusing to inexperienced users.

7.3.4.2 In summary, the sample in 7.3 for S/N measurement is recommended particularly when comparing instruments in different laboratories. For use within a laboratory by knowledgeable operators, ethylbenzene still offers a practical sample for simultaneous checking of S/N, resolution and decoupling efficiency. The adoption of an intrinsic S/N sample such as that described above also identifies the need for separate measurement of resolution and $\gamma H_2/2\pi$ to more completely characterize the performance of an FT spectrometer on ^{13}C. In addition, this measurement is understood to measure only intrinsic sensitivity and not the sensitivity of a time-averaged spectrum on a "routine" sample.

7.4 *Carbon-13 Multiple Pulse Mode:*

7.4.1 *Sample*—0.1 *M* Sucrose in D_2O equilibrated with toluene. Dissolve 3.423 g of sucrose (stored at a relative humidity of 50 % or less; NBS SRM sucrose is satisfactory) in about 90 cc of D_2O in a 100-cc volumetric flask, then dilute to the mark at 25°C with D_2O after all the sucrose is dissolved. Add 0.05 ml of toluene as a preservative.

7.4.2 *Measurement*—Carry out the measurement in the multiple-pulsed mode locked to the internal D_2O using the following conditions:

Spectral width	0 to 200 ppm ($\delta^{13C}_{TMS} \equiv 0$)
Data acquisition time	≥0.4 s
Flip angle	90°
Analog filter	appropriate for method of detection
Detection method	specify (for example, single phase, SSB, QPD)
Pulse repetition rate	1 pulse/s
^1H decoupler	broadband
^1H decoupler frequency	centered at 5 ± 1 ppm in the ^1H spectrum
^1H decoupler modulation mode	specify (for example, noise, square wave, etc.)
^1H decoupler modulation frequency	specify
Number of transients	4000 for 5-mm sample size 1000 for 10 to 12-mm sample size 100 for >12-mm sample size
Operating temperature	specify

Following the data acquisition, multiply the data by a decaying exponential function of the form $e^{-t/A}$, where A is equivalent to a T_2 contribution. A may be expressed as a time constant in units of seconds, or, alternatively, the line broadening (LB) resulting from the exponential multiplication may be expressed in units of Hz. For the measurement, $A = 0.3$ or LB = 1.0 Hz. Perform no data smoothing after transformation. Plot the resulting absorption mode spectrum over the full 200 ppm chemical shift range. Plot the spectrum to fill the vertical range of the chart paper as closely as practical. Measure the peak-to-peak noise between 120 and 140 ppm of the spectral window or calculate rms noise by computer (see Note 2). For those spectrometers with very high sensitivity, noise may have to be blown up to at least 2 cm peak-to-peak in a separate trace of the same transformed data. Measure signals Nos. 2, 3, 9, and 12 (identified on Fig. 3) and calculate S/N as follows:

$$[(2 + 3 + 9 + 12)/(\text{peak-to-peak noise})] \times 0.625 = S/N$$

7.4.3 *Discussion*—This measurement permits evaluation of sensitivity under "typical" conditions; that is, the decoupler is on and many transients are obtained. In addition to a knowledge of the basic, or intrinsic, ^{13}C sensitivity as measured in the C_6D_6 test, it is extremely important to evaluate the long term sensitivity as reflected in a proton-decoupled, time-averaged spectrum. The type and quality of the decoupling, as well as long term and short term instabilities in any instrument element, can profoundly affect sensitivity. This test is designed to monitor this performance.

7.4.3.1 Sucrose is chosen because of its widespread availability, purity, low cost, stability (in toluene equilibrated water) and spectral characteristics. Among these are the reasonable (1 Hz) linewidths, short T_1s, and full NOE. The number of transients is chosen to provide a reasonable total experimental time, typically 20 min, while still running long enough to simulate normal experiments adequately.

7.4.3.2 Decoupling efficiency is another highly variable element in "routine sensitivity." It certainly determines the ultimate sensitivity in the 90 % ethylbenzene sensitivity test (magnet homogeneity permitting). For this reason ethylbenzene is unsuitable for an absolute sensitivity determination. Yet, it is necessary to include the decoupler in sensitivity considerations since a poorly operating decoupler can be the main determinant in apparent sensitivity. Thus, proper consideration must be given not only to intrinsic sensitivity but also to "routine" sensitivity in characterizing spectrometer performance.

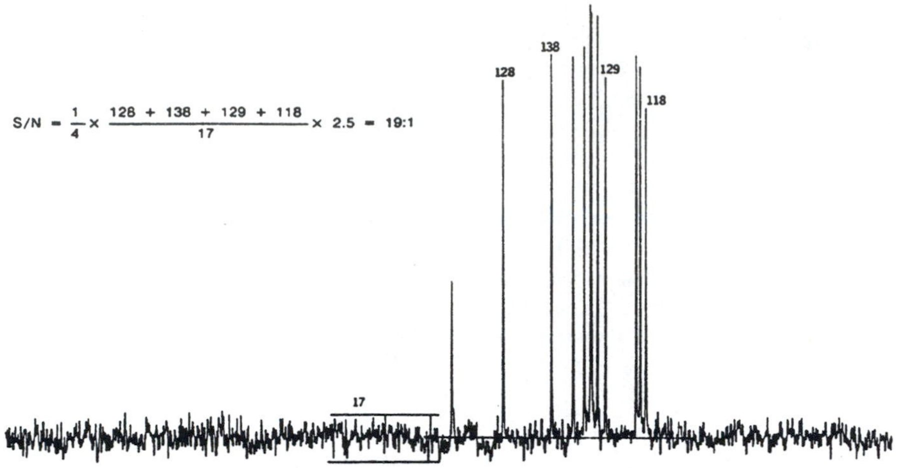

FIG. 3 Typical S/N Measurement on Accumulated ^{13}C Spectrum of 0.1 *M* Sucrose in D_2O

8. Presentation of NMR Data and Spectrometer Parameters

8.1 *General*—The following should be specified whenever NMR data are published:

8.1.1 Nucleus observed. In cases where possible ambiguity exists, the isotope must be specified, for example, ^{14}N, ^{11}B. In other cases the isotope may be specified, even though superfluous, such as, ^{19}F, ^{31}P.

8.1.2 Name of solvent and concentration of solution.

8.1.3 Name of external reference, or name and concentration of internal reference, as applicable.

8.1.4 Temperature of sample and how measured.

8.1.5 Procedure used for measuring peak positions.

8.1.6 Radio frequency at which measurements were made.

8.1.7 Magnitude of radio frequency field (see 2.4), or assurance that saturation of the signal has not occurred (in the case of CW spectra), or both.

8.1.8 Mathematical operations used to analyze the spectra. In cases where a computer program has been used to assist in the analysis of the spectrum, the following information should be included: Identification/source of program, number of lines fitted, identity of parameters varied, rms deviation of all lines, estimated precision of fitted parameters, and maximum deviation of worst line.

8.1.9 Numbers on the frequency scale (if used). They should increase from low to high frequency (high to low applied field if field sweep is used).

8.2 When CW spectra are published the following information should be included:

8.2.1 Sweep rate.

8.2.2 Values of both r-f fields when spin decoupling or double resonance is employed.

8.2.3 The shifts and couplings obtained from the spectra should be reported when available, the former in dimensionless units (ppm) and the latter in frequency units (hertz).

8.3 *Pulse-Fourier Transform Spectra*—For high-resolution pulse-Fourier transform experiments, all of the following that are applicable should be specified:

8.3.1 Pulse flip angle used.

8.3.2 90° pulse width, or pulse amplitude.

NOTE —Both 8.3.1 and 8.3.2 must always be specified. They may be given indirectly, for example, as pulse width used *and* as pulse width for a 90° pulse for the nucleus being studied.

8.3.3 Bandwidth and rolloff characteristics of all limiting filters (low-pass and crystal filters). Usually given as bandwidth (see 4.2.9) and type (such as, a 4-pole Butterworth).

8.3.4 Spectral width (or data acquisition rate or dwell time).

8.3.5 Data acquisition time (and acquisition delay time if relevant).

8.3.6 Pulse repetition time and number of pulses if the "sequence" consists of a single pulse.

8.3.7 Description of pulse sequence including *(a)* common name or details of pulses and phases, *(b)* sequence repetition time, *(c)* pulse intervals, *(d)* waiting time, *(e)* number of sequences, and *(f)* the specific pulse intervals during which data are acquired.

8.3.8 Quadrature phase detection, if used.

8.3.9 Number of data points Fourier transformed (it is desirable to indicate specifically whether zero filling is used).

8.3.10 The time constant of exponential weighting function (exponential filter), if used.

8.3.11 Details of apodization or other weighting of the time response signal.

8.3.12 Details of any other data processing such as spectral smoothing, baseline corrections, etc.

8.3.13 Details of systematic noise reduction, if used.

8.3.14 Relation of pulse frequency to observed frequencies.

PRESENTATION OF NMR DATA FOR PUBLICATION IN CHEMICAL JOURNALS—B. CONVENTIONS RELATING TO SPECTRA FROM NUCLEI OTHER THAN PROTONS

(RECOMMENDATIONS 1975)

COMMISSION ON MOLECULAR STRUCTURE AND SPECTROSCOPY†

Section B constitutes the second part of the "Recommendations for the Presentation of NMR Data for Publication in Chemical Journals". Section A dealt with proton NMR spectra (*Pure Appl. Chem.* **29**, 627 (1972)). Although the present recommendations are directed toward nuclei other than protons, they are equally applicable to proton NMR.

(1) The nucleus giving rise to the spectrum should always be explicitly stated in full or in abbreviation, e.g. ^{10}B NMR spectrum (spoken: boron 10 NMR spectrum). The isotopic mass number should be stated except in cases where a single abundant isotope leads to a situation without ambiguity, e.g. NMR spectra from ^{19}F or ^{31}P.

Abbreviations such as PMR (for 'proton NMR' or 'phosphorus NMR') or CMR (for 'carbon NMR') are strongly discouraged.

(2) The dimensionless scale factor for chemical shifts should be 10^{-6}, i.e. parts per million, for which ppm is a convenient abbreviation. When large chemical shifts are given exactly, the radiofrequency of the standard substance should be reported with sufficient accuracy.

(3) The unit for spin–spin coupling constants should be hertz (cycles per second). The symbol for coupling constants is *J*. The coupling between two nuclei separated by *n* chemical bonds can be indicated by the left *super*script *n*, e.g. ^4J denotes the coupling constant between two nuclei separated by 4 chemical bonds. Right *sub*scripts may be used to give the symbols of the coupling nuclei, e.g. the coupling constant between the phosphorus nucleus and protons in trimethylphosphite, $P(OCH_3)_3$, would bear the symbol $^3J_{PH}$. Alternatively a notation of the type ^3J(PH) or ^3J(^{11}BH) may be used.

(4) The graphical presentation of spectra should show the frequency decreasing to the right (applied field increasing to the right), absorption increasing upwards, and the standard sweep direction should be from high to low frequency (low to high field). Solvent and impurity bands, and spinning side-bands, should be indicated as such.

(5) Whenever possible the dimensionless scale should be tied to an internal reference, which should be explicitly stated. The dimensionless scale should be defined as positive in the high frequency (low field) direction. The scale in parts per million should be termed the δ scale. A shift measured on this scale should be given as, for example, $\delta = 5.00$, not $\delta = 5.00$ ppm. If data from more than one nucleus are reported, the symbol δ should be used with the corresponding symbol of the element given in brackets, e.g. δ(C) or δ(^{11}B). The position of the nucleus in the structural formula could be denoted by an additional number following this symbol for the nucleus, e.g. δ(C-5) or δ(^{11}B-5).

(6) When the spectra are submitted for publication, additional information should include:

(a) A statement of how the spectrum was recorded, e.g. using the continuous wave (CW), pulse Fourier transform (pulse FT), or other technique. The number of spectra accumulated should be stated.

When a pulse FT technique is employed, both the duration of the pulse and the duration of a 90° pulse should be stated. These two pieces of data may be expressed in other forms, e.g. as the angle of reorientation of the magnetic vector and the duration of a 90° pulse.

(b) The name of the solvent used:

(c) The concentration of the solute.

(d) The name and concentration of the internal reference.

(e) The name of the external reference, if one is used.

(f) The diameter of the sample tube and whether or not it was rotated.

(g) The temperature of the sample.

(h) The approximate radio-frequency or magnetic field at which the measurements were made.

(i) If spectra are presented as diagrams there should be a graphical indication of the distance corresponding to a suitable range of Hz, so that fine structure spacings or widths of broad resonances can be estimated.

(k) Where relevant it should be stated whether oxygen has been removed from the sample.

(l) Other experimental information should be added where appropriate or necessary, e.g.

(i) in CW experiments the sweep rate and magnitude of the B_1 fields;‡

(ii) in double resonance experiments the magnitude of the irradiating field,‡ B_2, whether monochromatic or noise decoupling is used, and whether CW or pulsed operation is carried out;

(iii) in cases where the experimental data are processed by a computer, all relevant information about spectral width, filtering, apodization, deconvolution processes, number of data points, etc.

†*Titular Members*: N. Sheppard (Chairman), M. A. Elyashévich (Vice-Chairman), F. A. Miller (Secretary), E. D. Becker, J. H. Beynon, E. Fluck, A. Hadni, G. Zerbi, *Associate Members*: G. Herzberg, B. Jeżowska-Trzebiatowska, Y. Morino, S. Nagakura, C. N. R. Rao, Sir Harold Thompson, D. W. Turner.

‡In contrast to the previous recommendations concerned with proton spectra the more recently adopted symbol *B* is used for the magnetic induction (field).

Recommendations for EPR/ESR NOMENCLATURE AND CONVENTIONS FOR PRESENTING EXPERIMENTAL DATA IN PUBLICATIONS

(Recommendations 1989)

COMMISSION ON MOLECULAR STRUCTURE AND SPECTROSCOPY*

Prepared for publication by

HIDEO KON

Laboratory of Chemical Physics, Room B1–14, Building 2,
NIDDK/National Institutes of Health, Bethesda, Maryland 20892, USA

COMMENTS

The very first version of the Recommendations on EPR Nomenclature was drafted in 1977 by James R. Bolton, University of Western Ontario, and was the subject of discussion in a group meeting chaired by Dr. Bolton in the VIth ISMAR in Banff, Canada in 1977.

The second version, which was sent out for reviewing by a number of EPR experts in several countries in 1984, was based on the Bolton draft and was revised by Hideo Kon, National Institutes of Health.

The present proposal has been re-written by Kon, taking into account the suggestions and criticisms by twenty-five reviewers during 1984-1985 in consultation with James Vincent of the University of Maryland.

Rationales for the choice in some of the items are inserted in the text (indented and italicized paragraphs).

In view of the extensive use of electron paramagnetic resonance (EPR) spectroscopy (electron spin resonance (ESR) spectroscopy) in chemistry, physics and biology, it is desirable to encourage the use of consistent nomenclature and the presentation of experimental data in a uniform manner. The Commission on Molecular Structure and Spectroscopy recommends the following:

*Membership of the Commission for varying periods during which the report was prepared (1983–1987) was as follows:

Chairman: J. R. Durig (USA); *Secretary*: H. A. Willis (UK); *Titular Members*: A. M. Bradshaw (FRG); B. G. Derendjaev (USSR); S. Forsen (Sweden); J. G. Graselli (USA); E. Hirota (Japan); J. F. J. Todd (UK); *Associate Members*: R. D. Brown (Australia); R. K. Harris (UK); H. Kon (USA); G. Martin (France); Yu. N. Molin (USSR); W. B. Person (USA); H. Ratajczak (Poland); C. J. H. Schutte (R.S. Africa); M. Tasumi (Japan); *National Representatives*: R. Colin (Belgium); J. E. Collin (Belgium); Y. Kawano (Brazil); G. Herzberg (Canada); J. Lu (Chinese Chemical Society); Z. Luz (Israel); S. Ng (Malaysia); B. Jezowska-Trzebiatowska (Poland); G. Varsanyi (Hungary).

SCOPE

These recommendations contain definitions of basic terms, conventions and practices for data presentation in the area of EPR/ESR spectroscopy. This part A includes those pertaining to spectra of systems with $S = \frac{1}{2}$. A version for systems with $S > \frac{1}{2}$ may follow. This part also does not include the areas pertaining to saturation transfer, double resonance and time domain techniques.

NOMENCLATURE AND BASIC DEFINITIONS

2.1 Electron paramagnetic resonance (EPR) and/or electron spin resonance (ESR) is defined as the form of spectroscopy concerned with microwave-induced transitions between magnetic energy levels of electrons having a net spin and orbital angular momentum. In the present part, the magnetic field scanning method is assumed. Other methods, however, are also conceivable. The term electron paramagnetic resonance and the symbol EPR are preferred and should be used for primary indexing.

2.1 Use of the upper case, with no punctuation, "EPR" as opposed to "e.p.r.", is adopted, since it is consistent with the existing IUPAC NMR nomenclature. Also "paramagnetic" should be used, since it comprises other than 'spin only' systems. On the other hand, "ESR" has been so widely used that it is not practical to exclude it completely.

2.2 The frequency ($\underline{\nu}$) of the oscillating magnetic field applied to induce transitions between the magnetic energy levels of electrons is measured in gigahertz (GHz) or megahertz (MHz).

2.3 The static magnetic field at which the EPR spectrometer operates is measured by the magnetic flux density B and the recommended unit is the tesla (T) (1 T = 10^4 gauss).

2.3 tesla vs. gauss: both tesla and gauss are units of magnetic induction (magnetic flux density) for which the symbol B *has been used. The magnetic field strength* H *has been recorded in ampere·turn/meter or oersted, and not by gauss. One may argue that "the abscissa of every EPR spectrum is recorded in units of* H, *namely the field applied to the sample achieved by passing current through coils". However, the calibration of the field strength is inevitably based upon the magnetic induction in matter, whether one uses a standard sample, a proton probe, or the Hall effect. For that reason, it is more appropriate to use* B *(the magnetic flux density) and specify the method of calibration. The "tesla" rather than "Tesla" is used in SI convention.*

2.4 The amplitude of the oscillating magnetic field is designated by B_1. The recommended unit is the millitesla (mT).

2.5 EPR absorption and dispersion. A single transition and a set of degenerate or unresolved transitions are referred to as a line. The line shape is often described to be Lorentzian, Gaussian, or a mixture of the two. Absorption or dispersion lines are commonly presented in the first or the second derivative mode.

Symbols for the modes are U_1 and U_2 for dispersion first and second derivatives, respectively, and V_1 and V_2 for absorption. Spectra recorded out-of-phase with respect to the Zeeman field modulation are indicated by adding primes to the previous symbols (e.g. V_2' for second derivative out-of-phase absorption).

2.5 Although this version of the recommendation does not include the saturation transfer technique (ST-EPR), it seems appropriate to standardize the mode designations here for future extension.

2.6 In the absence of nuclear hyperfine interactions (vide infra), B and $\underline{\nu}$ are related by

$$h\underline{\nu} = g\mu_B B$$

where h is the Planck constant, μ_B is the Bohr magneton $eh/(4\pi m_e)$, and the dimensionless scalar g is called g-factor. Use of the term g-value is discouraged.

2.6 "In the absence of nuclear hyperfine interaction" the nuclei involved have no nuclear spin, and therefore, there will be no nuclear Zeeman term or nuclear electric quadrupole term. Thus the relation is rigorously correct for S = ½ systems. IUPAC Manual of Symbols & Terminology for Physicochemical Quantities and Units adopts μ_B *(Pure & Appl. Chem. vol. 51, pp. 1-41, 1979) as the Bohr magneton.*

2.7 When the paramagnetic species exhibits an anisotropy, the spatial dependency of the g-factor is represented by a 3×3 matrix $\underline{g}$. The matrix representation is referred to as $\underline{g}$-matrix. In a general coordinate system, such as (x,y,z), the components may be designated as g_{xx}, g_{xy}, ..., etc. In cases where a principal axis system can be assigned, in which the off-diagonal terms are zero, the three principal values of the g-matrix will be expressed by g with a single subscript identical to the principal axis designation adopted for the g-matrix. A recommended example is: g_X, g_Y, g_Z for the principal axes (X,Y,Z).

2.7 That g is not in general a tensor (neither is the hyperfine coupling constant A (ref. 2.8)) was shown by Abragam and Bleaney ("Electron Paramagnetic Resonance of Transition Ions", Clarendon Press, Oxford 1970, pp. 166, 170, 651), while the nuclear electric quadrupole coupling P (ref. 2.11) is a tensor in the strict sense. Some authors clearly distinguish them by calling g and A a "matrix" and P a "tensor" when all three are involved.

Use of the double subscripts for a general coordinate system and a single subscript for the principal axes has a definite merit of clearly distinguishing the principal components.

The matrices (g and A) and tensor (P) quoted here are defined through the Hamiltonian expression

$$\hat{H} = +\underline{B} \cdot \underline{g} \cdot \underline{S}\mu_B - \Sigma_a \underline{S} \cdot \underline{A}_a \cdot \underline{I}_a - \Sigma_a \underline{I}_a \cdot \underline{P}_a \cdot \underline{I}_a$$

where $\hat{H}$ is the Hamiltonian operator, B is the magnetic flux density, S and I are vector spin operators, the summation index a covers all nuclear species (except that I = 0 nucleus can be omitted and I = ½ nuclei have no quadrupole term) and the direct interaction of nuclear spin with the magnetic flux density is omitted.

Use of the double subscripts for a general coordinate system and a single subscript for the principal axes has a definite merit of clearly distinguishing the principal components.

In general, the square roots of the principal values of a symmetric tensor G represent the principal values of g.

$$G_{ik} = \Sigma_j g_{ji} g_{jk}$$

A similar condition applies to A; in this case the correct signs of the square roots may be uncertain.

It is too restrictive to choose a specific set of nomenclature for the principal components of g, another set for A, and another for P, in addition to the molecular coordinate axis system which normally is dictated by orbital designations such as $d_{x^2-y^2}$, $2p_z$, etc. The recommendation must allow for the most general case, assuming that all these axis systems occur and do not coincide, and yet, has to make certain that there is no conflict under any circumstances. While this could be done, it would make the convention too complicated and cumbersome for authors to remember. In most writing situations, however, not all of the above mentioned principal axis systems occur and some of them may coincide. Thus authors can choose the principal axis designations for only the necessary matrices.

Considering all these complications, it seems much more practical to leave designation of principal axis systems, to a certain extent, to individual authors. One conceivable drawback in so doing, would be that two authors, describing the same compound, may adopt two different principal axis designations so that one author's g_X, for example, may correspond to g_C of the other. However, this kind of inconsistency can not be completely avoided anyway, even if the two authors adopt the same principal axis designation, unless the convention dictates also the order of g-factors such as, e.g., $g_X > g_Y > g_Z$. In fact, one might suggest a convention for setting the ordering of the principal values by "taking the average and designating the farthest one from the average as g_Z and the next as g_Y, etc." However, ordering can not be determined a priori, because it depends on the individual compound. For example, in Cu^{2+}-porphyrin, the g-factor measured with the magnetic field parallel to the four-fold axis is the largest, whereas in low-spin Co^{2+}-porphyrin, the g-factor measured similarly is in the middle, but one would not call the four-fold axis Z in one compound and Y in another.

Thus, forcing too many details of the principal component designations tends to run into conflict with the orbital designations.

For a powder spectrum, if a specific assignment is not made, the following conventions are recommended:

(a) In a spectrum having the characteristics of lower than axial symmetry with three distinct lines, g_1, g_2, and g_3 are used for the low, middle and high field line in that order.

> *2.7a There are, even in powder spectra, cases in which one can make a specific assignment of principal axes. The recommended convention 2.7(a) is strictly for the cases where no such assignment can be made. Designation of unassigned g-factors in terms of (x, y, z) subscripts should be discouraged, because it may turn out to be in conflict with molecular axis system designation as explained above. Non-committal (1,2,3) is much preferred for unassigned cases.*

(b) In a spectrum having the characteristics of apparently uniaxial symmetry, exhibiting a parallel and a perpendicular feature, the lines are designated as $g_{\parallel}$ and $g_{\perp}$, respectively.

(c) In a spectrum representing more than one paramagnetic species, designations of g-factors must include some species identification in parenthesis, e.g., g_1(radical 1) or $g_1(1)$.

2.8 Hyperfine interactions. The interaction energy between the electron spin and a magnetic nucleus is characterized by the hyperfine coupling constant A with units in joules. A/h and $A/(hc)$ may be reported in MHz and cm^{-1}, respectively. Expressing A in units of tesla, millitesla, or gauss is rejected. When the paramagnetic species has magnetic anisotropy the hyperfine coupling is expressed by a 3×3 matrix called a hyperfine coupling matrix $\underline{A}$. $\underline{A}$ is often divided into an anisotropic and an isotropic term as follows:

$$\underline{A} = \underline{T} + [\mathrm{Tr}(\underline{A})/3]\,\underline{1}$$

$\underline{T}$ is a traceless 3×3 matrix (sum of the diagonal elements being equal to zero), $\underline{1}$ is a unit matrix of the same dimension. The principal components of $\underline{A}$, when resolved, are denoted by A with the principal axis designation added as the subscript (e.g., A_a, A_b, A_c, if (a,b,c) is chosen as the principal axes for $\underline{A}$). If the absolute sign of a principal component is deduced theoretically, it should be given in parentheses as e.g. A_a/h = (+) 70 MHz. The principal axis systems for matrix $\underline{A}$ and $\underline{g}$ may or may not coincide with each other. In cases where the principal components of $\underline{A}$ are not resolved, the hyperfine interaction in a line is described in terms of A' with the same subscript adopted for the g-factor of the line in which the hyperfine interaction is observed.

2.9 Hyperfine interaction usually results in splitting of lines in an EPR spectrum. The splitting (a) is measured in units of millitesla (mT). The relation between the hyperfine splitting and hyperfine coupling constant must be derived for each system, e.g. by computer simulation, depending upon the accuracy desired. For cases where higher-order terms can be neglected and the effects of the nuclear Zeeman term need not be taken into account, the splitting a is related to the absolute value of the hyperfine coupling constant A by

$$A = g\mu_B a$$

2.10 The nuclear species giving rise to the hyperfine interaction should be explicitly stated, e.g. "the hyperfine splitting due to ^{65}Cu". When additional hyperfine splittings due to other nuclear species are resolved, the nomenclature should include the designation of the nucleus, and the isotopic number, e.g. $a(^{14}N)$. If the splittings are assigned to more than one nucleus of the same nuclear species, they may be distinguished by adding subscripts such as $a(^{15}N_1)$ and $a(^{15}N_2)$. The same conventions apply to the nuclear hyperfine coupling constant A.

2.11.a When the nucleus has an electric quadrupole moment (i.e., $I > \frac{1}{2}$), its interaction with the surrounding molecular electric field-gradient is expressed by a second rank tensor $\underline{P}$ called the nuclear electric quadrupole coupling tensor. The principal components of $\underline{P}$, if resolved by analysis of a hyperfine spectrum, are denoted by P with the principal axis designation as the subscript (e.g. P_1, P_2, P_3 for the principal axes (1,2,3)).

In the presence of axial symmetry, the axial component of $\underline{P}$ tensor, e.g. P_3 is defined by

$$P_3 = eQq/[2I(2I - 1)]$$

where eQ is the nuclear quadrupole moment, q is the axial electric field gradient at the nucleus and I is the nuclear spin. Deviation from axial symmetry is expressed by the asymmetry factor $\eta = (P_1 - P_2)/P_3$. The nuclear species involved must be specified following the conventions for hyperfine interactions, e.g. $P_1(^{75}As)$.

2.11.b Values of P are expressed in joules; P/h and $P/(hc)$ may be reported in MHz and cm^{-1}, respectively.

2.11 Nomenclature for nuclear quadrupole interaction is limited to the minimum necessary for reporting EPR data. Use of notations such as DQ (= $3P_{zz}/2$), which appear in some EPR articles, are not included. However, use of these and other parameters derived from P should be considered optional, as long as they are clearly defined in terms of the present notation. Numerous different designations are also in use for nuclear quadrupole coupling tensors in other fields of spectroscopy. The relationship of P to them should be derived from the definitions given in this recommendation.

PRESENTATION OF EPR DATA AND EXPERIMENTAL CONDITIONS

Reference back to previous publications for experimental method is permitted only if the required specific information is present there.

3.1 At least the following items should be specified in presentation of EPR data to facilitate transfer of spectral information:

(a) If in solution, name of the solvent (or matrix) and concentration of the solution; for solid materials, methods of sample preparation and mounting
(b) Temperature of the sample and how it is controlled
(c) Type of sample cell (e.g. aqueous flat cell)
(d) Type of resonator used, microwave frequency (GHz or MHz), power level (mW) incident on the resonator and loading information (e.g. dewar insert used); whether the frequency and power level are calibrated or taken from the spectrometer settings; power saturation distorting the spectrum, if any, must be so stated.

3.1(d) The commonly used term "cavity resonator" is replaced by a generic term "resonator", since various types of non-cavity resonators are now in use.

(e) If Zeeman field modulation is used, frequency (MHz or Hz) and amplitude (mT) of modulation, and whether they are calibrated or from the spectrometer settings

3.1(e) "If Zeeman field modulation is used ..."; in some EPR measurements, Zeeman field modulation and/or analog field scan are not used.

(f) Type of standard sample, if used for field calibration and/or quantitation of the magnetic species
(g) Method of g-factor measurement and experimental uncertainties

3.2 When EPR spectra are graphically presented, the following information for abscissa and ordinate should be supplied in addition to the above items:

(a) If the spectra are obtained by analog field scan, the field scan rate, otherwise, the field step size
(b) Total field extent and/or field calibration marker and method of field calibration
(c) Markers indicating the point in a line where the g-factor is measured; use of g-factors in place of field scale is discouraged.
(d) Ordinate information, if the ordinate is not the direct spectrometer output (e.g. computer processed normalization)
(e) Presentation mode, e.g. "the first derivative of absorption"
(f) Filtering information (e.g. analog time constant, digital smoothing specifications, etc.)

3.3 Other conventions in graphical presentation of the spectra:

(a) The Zeeman field increases to the right.
(b) The phase should be adjusted so that the start of the first low field line in V_1 mode has a positive excursion.

NOMENCLATURE AND CONVENTIONS FOR REPORTING MÖSSBAUER SPECTROSCOPIC DATA

(RECOMMENDATIONS 1975)

COMMISSION ON MOLECULAR STRUCTURE AND SPECTROSCOPY†

INTRODUCTION

These Recommendations are in considerable measure based on a report of the *ad hoc* Panel on Mössbauer Data of the Numerical Data Advisory Board of the Division of Chemistry and Chemical Technology of the National Research Council, U.S.A. (Chairman, Professor J. J. Zuckerman), which took into account several earlier documents, especially the National Bureau of Standards (U.S.A.) Special Publication 260-13 and the report of the Mössbauer Spectroscopy Task Group of Committee E-4 (Metallography) of the American Society for Testing and Materials (Chairman, Professor R. H. Herber).

The Recommendations incorporate modifications suggested by the present and immediate past Chairman of Commission I.1 (Professor D. H. Whiffen and Dr. M. A. Paul), the Chairman of Commission I.5 (Professor N. Sheppard), Professor N. N. Greenwood (consulted by Commission 1.5) and a number of other scientists who wrote to make specific suggestions.

A. PROPOSED CONVENTIONS FOR THE REPORTING OF MÖSSBAUER DATA

I. *Text*

The text should include information about:

(a) the method of sample mounting, sample thickness, sample confinement, and appropriate composition data for alloys, solid solutions or frozen solution samples;

(b) the form of the absorber (single crystal, polycrystalline powder, inert matrix if used, evaporated film, rolled foil, isotopic enrichment, etc.);

(c) the apparatus and detector used and comments about the associated electronics (e.g. single channel window, escape peak measurements, solid-state detector characteristics, etc.) if unusual; data acquisition time if unusual;

(d) the geometry of the experiment (transmission, scattering, in-beam, angular dependence, etc.);

(e) the critical absorbers or filters, if used;

(f) the method of data reduction (e.g. visual, by computer, etc.) and curve-fitting procedure: (See Notes A-1 and A-2);

(g) the isomer shift convention used or the isomer shift of a standard (reference) absorber. Positive velocities are defined as source approaching absorber. Sufficient details concerning the isomer shift standard should be included to facilitate interlaboratory comparison of data (See Note A-3 and Table); and

(h) an estimate of systematic and statistical errors of the quoted parameters.

II. *Numerical or tabulated data*

Information collected and summarized in tabular form should include:

(a) the chemical state of source matrix and absorber;

(b) the temperature of source and absorber and the constancy of these parameters over the length of the data acquisition period;

(c) values of the parameters required to characterize the features in the Mössbauer spectrum (given in mm/s, cm/s or other appropriate units (See IIIB)) with estimated errors;

(d) the isomer shift reference point with respect to which the position parameters are reported;

(e) the observed line-widths defined as the full-width at half maximum peak-height;

(f) the line intensities or (relative) areas of each component of the hyperfine interaction spectrum observed, when pertinent.

III. *Figures illustrating spectra*

Scientific communications in which Mössbauer effect measurements constitute a primary or significant source of experimental information should include an illustration of at least one spectrum (i.e. % transmission or absorption or counting rate vs an energy parameter) to indicate the quality of the data. Such figures should include the following information features:

(a) a horizontal axis normally scaled in velocity or frequency units (e.g. mm/s or MHz. Channel number or analyzer address-values should not be used for this purpose); (see Note A-4)

(b) a vertical axis normally scaled in counts per channel or related units; (See Note A-5)

(c) an indication, for at least one data point, of the statistical counting error limits; (See Note A-6)

(d) individual data points (rather than a smoothed curve alone) should be shown. Computed fits should be indicated in such a way that they are clearly distinguishable from the experimental points.

B. MANUAL OF TERMINOLOGY, SYMBOLS, AND UNITS FOR MÖSSBAUER SPECTROSCOPY

The symbols for physical quantities are in italics. If the units selected by the experimenter are not SI units, they should be defined in the text.

---

†*Titular Members*: N. Sheppard (Chairman), M. A. Elyashévich (Vice-Chairman), F. A. Miller (Secretary), E. D. Becker, J. H. Beynon, E. Fluck, A. Hadni, G. Zerbi, *Associate Members*: G. Herzberg, B. Jeżowska-Trzebiatowska, Y. Morino, S. Nagakura, C. N. R. Rao, Sir Harold Thompson, D. W. Turner.

Reprinted with permission from *Pure and Applied Chemistry*, Volume 45, 1976, pages 211–216.

Name	Symbol	SI unit	Suggested decimal multiple or sub-multiple SI units for Mössbauer data	Definition and comment
Isomer shift	δ	m/s	mm/s ($=10^{-3}$ m/s)	Measure of the energy difference between the source (E_s) and the absorber (E_a) transition. The measured Doppler velocity shift, δ, is related to the energy difference by $E_a - E_s = \delta E_\gamma / c$ (where E_γ is the Mössbauer gamma energy and c is the speed of light in vacuum) (See Note B-1a, B-1b).
Nuclear quadrupole moment (spectroscopic)	eQ	C m^2	C cm^2 ($=10^{-4}$ C m^2)	A parameter which describes the effective shape of the equivalent ellipsoid of the nuclear charge distribution, $Q>0$ for prolate (e.g. ^{57}Fe, ^{197}Au); $Q<0$ for oblate (e.g. ^{119}Sn, ^{129}I) nuclei
Electric Field Gradient (EFG) tensor		V/m^2	V/cm^2 ($=10^4$ V/m^2)	A second rank tensor describing the electric field gradient specified by η and V_{zz}; in addition the Euler angles may be required specifying the orientation of the tensor principal axes. (See below) (See Note B-3)
Principal component of EFG	$-V_{zz}$	V/m^2	V/cm^2 ($=10^{-4}$ V/m^2)	$(\partial^2 V/\partial z^2) = eq$ (e is the proton charge, $\lvert V_{zz}\rvert$ is the largest component of the diagonalized EFG)
Quadrupole coupling constant	e^2qQ/h	Hz	MHz ($=10^6$ Hz)	Product of V_{zz}/h and the nuclear quadrupole moment, eQ. (See Note B-1b)
Quadrupole splitting	Δ	m/s	mm/s ($=10^{-3}$ m/s)	The measured Doppler velocity difference between the two peaks seen in quadrupole split spectra from nuclides such as ^{57}Fe and ^{119}Sn. Its value is related to the quadrupole coupling constant as $\Delta = 1/2(c/E_\gamma)e^2qQ\sqrt{(1+\eta^2/3)}$ (See Note B-2)
Asymmetry parameter	η			$=(V_{xx} - V_{yy})/V_{zz}$
Line width	Γ_{exp}	m/s	mm/s ($=10^{-3}$ m/s)	Full width at half maximum of the observed resonance line(s)
Natural line width	Γ_{nat}	m/s	mm/s ($=10^{-3}$ m/s)	Theoretical value of the full width at half maximum of the nuclear transition, usually calculated from lifetime data
Resonance effect magnitude	ϵ			Ratio of the difference in the transmitted or scattered intensity at resonance maximum and off-resonance, relative to the intensity off-resonance. (See Note B-4)
Recoil-free fraction	f			The fraction of all Mössbauer gamma rays of the transition which are emitted (f_s) or absorbed (f_a) without significant recoil energy loss. (See Note B-5)
Mossbauer thickness	t			The effective thickness of a source (t_s) or absorber (t_a) in the optical path. (See Note B-6)
Resonance cross-section	σ_0	m^2	barn ($=10^{-28}$ m^2)	The cross-section for resonant absorption of the Mössbauer transition gamma ray (See Note B-7)
Magnetic flux density	B	T = Wb/m^2		Magnetic flux density at the nucleus (from experiment) in those cases in which the magnetic hyperfine interaction can be described by an effective field. In other cases the vector components of the magnetic hyperfine interaction should be reported if possible. (See Note B-8)
Vibrational anisotropy	ϵ_m			When the vibrational anisotropy tensor $(\langle x_{ij}^2\rangle)$ is axially symmetric $\epsilon_m = (1/\lambda\!\!\!^{-2})(\langle x_\parallel^2\rangle - \langle x_\perp^2\rangle)$ where $\langle x_\parallel^2\rangle$ and $\langle x_\perp^2\rangle$ are the mean square vibrational amplitudes of the Mössbauer nucleus parallel and perpendicular to the cylindrical symmetry axis through the Mössbauer atom, and $\lambda\!\!\!^{-}$ is the wavelength of the Mössbauer radiation divided by 2π
Intensity asymmetry (Gol'danskii–Karyagin asymmetry)	A			For a randomly oriented sample with $\eta = 0$ and for $1/2 \rightarrow 3/2$ magnetic dipole transitions, this is the ratio of the area under the π transition peak to the area under the σ transition peak $A = \dfrac{\int_0^\pi e^{-\epsilon_m \cos^2\theta}(1+\cos^2\theta)\sin\theta\, d\theta}{\int_0^\pi e^{-\epsilon_m \cos^2\theta}\left(\frac{5}{3} - \cos^2\theta\right)\sin\theta\, d\theta}$ and is independent of sample orientation. (See Note B-9)

NOTES

(A-1). If data are analyzed by computer, a brief description of the program should be given to identify the algorithm used. The number of constraints should be specified, (e.g. equal line-widths or intensities, etc.) and a measure of the goodness of fit should be indicated.

(A-2). If measurements of very high accuracy are reported and the discussion of the reality of small effects is an important part of the work, then the following items should be included:

1. the functional form and all parameters used in fitting (i.e. the constraints should be clearly stated);
2. the treatment of the background (e.g. assumed energy independent, experimentally subtracted, etc.);
3. the relative weighting of abscissa and ordinate (e.g. equal weighting);
4. a measure of the statistical reliability;
5. the number of replications and the agreement between these if applicable;
6. an estimate of systematic errors as primary results.

(A-3). The table in the adjacent column contains a list of materials which have been proposed and are being used for isomer shift standards. The proposals are partly based on the tabulation by R. L. Cohen and G. M. Kalvius [*Nucl. Instr. Methods* **86**, 209 (1970)]. That article also discusses some criteria in the choice of materials for isomer shift standards, and provides an extensive list of references.

(A-4). Constant acceleration spectrometers to be used for work in the mm/s range can be calibrated with respect to velocity using either metallic iron foil of at least 99.99% purity (e.g. N.B.S. Standard Reference Material SRM 1541 or equivalent) or an optical method based on interferometric or Moiré pattern techniques. The ground state (g_0) and excited state (g_1) splitting in metallic iron have been reported. [See, for example, the values quoted in J. G. Stevens and R. S. Preston, *Mössbauer Effect Data Index, Covering the* 1970 *Literature*, (edited by J. G. Stevens and V. E. Stevens), p. 16: IFI/Plenum, New York (1972), and in other references cited therein. (See also J. J. Spijkerman, J. R. DeVoe and J. C. Travis, N.B.S. Spec. Pub. 260-20, Washington, D.C. 1972 and C. E. Violet and D. N. Pipkorn, *J. Appl. Phys.* **42**, 4339 (1971)]. The temperature dependence of g_0 can be estimated from data quoted in the J. I. Budnick, L. J. Bruner, P. J. Blume and B. L. Boyd, *J. Appl. Phys.* **32**, 1205 (1961), and R. S. Preston, S. S. Hanna and J. Heberle, *Phys. Rev.* **128**, 2207 (1962). [See R. H. Herber in *Mössbauer Effect Methodology*. Vol. 6, (Edited by I. Gruverman) Plenum Press, New York (1971)]. In the absence of independent linearity measurements, a quadrupole split doublet should not be used to effect velocity calibration of spectrometers.

(A-5). It has become customary to display data obtained in transmission geometry with the resonance maximum 'down' and scattering data with the resonance maximum 'up'. In either case sufficient data should be shown far enough from the resonance peaks to establish the non-resonant base line.

(A-6). In most instances (where the data are uncorrected counting results), the standard deviation (i.e.: the square root of the second moment of the distribution) is given by $N^{1/2}$, where N is the number of counts scaled per velocity point. For corrected data (i.e. when background or other non-resonant effects are subtracted from the raw data), the error propagated should be computed by normal statistical methods which are briefly described in the text or figure legend. Fiducial marks bracketing the data point to show the magnitude of the standard deviation are often useful in indicating the spread of the data.

Element	Material	Element	Material
K	KCl	Eu	EuS EuF_3
Fe	αFe[a]	Gd	$GdAl_2$
	$Na_2[Fe(CN)_5NO]\cdot 2H_2O$[a]	Dy	Dy
Zn	ZnS	Er	Er
Ge	Ge	Yb	$YbAl_2$
Kr	Kr (Solid)	W	W
Ru	$K_4[Ru(CN)_6]\cdot 3H_2O$[b]	Os	$K_4[Os(CN)_6]$
Sn	SnO_2[c]	Ir	Ir
	$BaSnO_3$[c]	Pt	Pt
Sb	InSb	Au	Au
Te	SnTe[d]	Np	$NpAl_2$
I	CsI		
Xe	Xe (Solid)		
Sm	SmF_2		

[a]Both of these materials are in wide use. The spectrum of disodium pentacyanonitrosoferrate(2-)dihydrate does not have unresolved hyperfine structure, footnote (b) of R. L. Cohen and G. M. Kalvius [*Nucl. Instr. Methods* **86**, 209 (1970)] misinterpreted the EFG results of R. W. Grant, R. M. Housley and U. Gonser [*Phys. Rev.* **178**, 523 (1969)]. In view of the reported line spacing anomaly in Fe metal, [J. J. Spijkerman, J. C. Travis, D. N. Pipkorn and C. E. Violet, *Phys. Rev. Lett.* **26**, 323 (1971)], very precise results reported with respect to Fe should specify which lines are being used as a references.

[b]The small size of the effect in this compound may make ruthenium metal also useful as a secondary, or working, standard despite its unresolved quadrupole splitting.

[c]These compounds present minor materials problems which should not interfere with their effective use. Barium stannate has been well established to be a cubic perovskite at room temperature [A. J. Smith and A. J. E. Welch, *Acta Cryst.* **13**, 653 (1970)], but must be made by the ceramic process, and may very well undergo a crystallographic distortion at low temperatures. Much of the older Mössbauer literature reports ^{119}Sn isomer shift data with respect to SnO_2 which has a small quadrupole splitting usually observed as line broadening. Such data are directly comparable to those reported with respect to $BaSnO_3$ (both at room temperature) within an experimental uncertainty to ±0.02 mm/s.

[d]SnTe of exactly 1 : 1 stoichiometry is difficult to prepare. At low temperatures there is evidence for line broadening in the ^{119}Sn resonance spectrum due to unresolved hyperfine interactions. The transition temperature varies with composition. This material should only be used as a standard at temperatures where no unresolved hfs is present.

(B-1a). The centre of a Mössbauer spectrum is defined as the Doppler-velocity at which the resonance maximum is (or would be) observed when all magnetic dipole, electric quadrupole, etc. hyperfine interactions are (or would be) absent. The contribution of the second order Doppler shift (δ_T) should be indicated, if possible. The isomer shift (δ) is the sum of this term and the chemical isomer shift (δ_C).

(B-1b). The SI unit of energy for both isomer shift and quadrupole coupling constant should be J, as also for Δ. The measured quantity is the velocity (m/s) which can be converted to energy as appropriate.

(B-2). Quadrupole splittings are frequently reported in megahertz, especially when direct comparison with NMR or NQR data can be effected (e.g. in the case of ^{129}I). If such units are used in conjunction with data derived from Doppler shift measurements, the conversion factors should be stated.

(B-3). The sum $(V_{xx} + V_{yy} + V_{zz}) = 0$ regardless of the choice of axes. In the absence of magnetic hyperfine interaction, principal axes are chosen so that the off-diagonal matrix elements vanish, $V_{ij} = 0$ $(i, j = x, y, z;$ $i \neq j)$ and are defined such that

$$|V_{zz}| \geqslant |V_{yy}| \geqslant |V_{xx}|, \quad \text{so that} \quad 0 \leqslant \eta \leqslant 1.$$

$$(EFG)_{ij} = -(\partial^2 V / \partial x_i \partial x_j) \quad x_i, x_j = x, y, z.$$

(B-4). This parameter is calculated from the relationship $\epsilon = [I(\infty) - I(0)]/I(\infty)$, where $I(0)$ is the counting rate (or transmission or scattering intensity) at the resonance maximum, and $I(\infty)$ is the corresponding rate at a velocity at which the resonance effect is negligible. If corrections for non-resonant gamma- or X-rays, or other base-line corrections, have been made in evaluating I, these should be stated.

(B-5). The recoil-free fraction can be related to the expectation value of the mean square displacement of the Mössbauer atom by the relationship

$$f = \exp(-k^2\langle x^2 \rangle)$$

where k is the wave number of the Mössbauer gamma ray and x is the displacement taken along the optical axis.

(B-6). The t parameter is usually calculated for a thin absorber from the relationship $t = n\sigma_0 af$, in which n is the number of Mössbauer element atoms per unit area in the optical path, σ_0 is the cross-section for recoilless scattering, a is the fractional abundance of the Mössbauer transition active nuclides, and f is the recoil-free fraction (*vide supra*).

(B-7). This parameter is usually calculated from the relationship

$$\sigma_0 = (h^2c^2/2\pi)E_\gamma^{-2}(1 + \alpha_T)^{-1}(1 + 2I_e)(1 + 2I_g)^{-1}.$$

Where E_γ is the transition energy; I_e and I_g are the excited and ground state spins, respectively; and α_T is the total internal conversion coefficient of the Mössbauer transition.

(B-8). 1 Tesla (10 kG) equals 1 Wb/m^2.

(B-9). For ^{57}Fe and ^{119}Sn the following table summarizes the conventions which relate the shape of the equivalent ellipsoid of electronic charge surrounding the nucleus, the sign of V_{zz}, the sign of the excited state nuclear quadrupole moment, $Q_{3/2}$, the sign of the quadrupole coupling constant e^2qQ/h, the doublet intensity ratio R (defined as the area ratio of the more positive velocity peak divided by that of the more negative velocity peak in the absence of a magnetic hyperfine interaction), and the angular dependence ratio A for the σ ($\Delta m = 0, \pm 1$) and π ($\Delta m = \pm 1$) transitions for the axially symmetric case:

	^{57}Fe		^{119}Sn	
shape		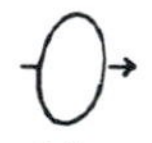		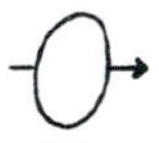
	prolate	oblate	prolate	oblate
V_{zz}	−	+	−	+
$Q_{3/2}$	+	+	−	−
e^2qQ	−	+	+	−
	$E_{3/2} < E_{1/2}$	$E_{3/2} > E_{1/2}$	$E_{3/2} > E_{1/2}$	$E_{3/2} < E_{1/2}$
R	$\sigma/\pi\ (=A^{-1})$	$\pi/\sigma\ (=A)$	$\pi/\sigma\ (=A)$	$\sigma/\pi\ (=A^{-1})$

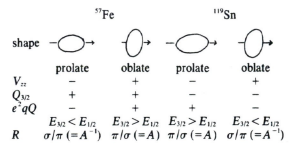

Guide for the Publication of Experimental Gas-Phase Electron Diffraction Data and Derived Structural Results in the Primary Literature*

International Union of Crystallography

Commission on Electron Diffraction.

By L. S. BARTELL, *Department of Chemistry, The University of Michigan, Ann Arbor, Michigan* 48109, *U.S.A*, KOZO KUCHITSU, *Department of Chemistry, Faculty of Science, The University of Tokyo*, 3–1 *Hongo 7-chome, Bunkyo-ku, Tokyo* 113, *Japan* and H. M. SEIP, *Department of Chemistry, University of Oslo, Blindern, Oslo* 3, *Norway*

Introduction

This report is concerned with the presentation of structural investigations by electron diffraction. Its aim is to make results obtained by specialists more accessible to those in other disciplines and, at the same time, to increase the potential value of the original data to other specialists if subsequent events warrant reinvestigation. The needs of compilers and correlators of information will also be benefited by attention to these considerations. Only if enough information is provided to allow readers to appraise the precision and accuracy of the work, and only if reasonably uniform standards of reporting the results are adhered to, can all ends be met.

This report is an abbreviated version of a document approved by the Commission on Electron Diffraction of the International Union of Crystallography in August, 1975. For its general recommendations the original document quoted extensively from the *Guide for Publication of Experimental Data and Derived Numerical Results in the Primary Literature* prepared for CODATA and UNESCO in 1973 by the CODATA Task Group on Publication in the Primary Literature (1973). The present report concentrates on specialized problems encountered in the field of gas-phase electron diffraction, certain aspects of which have been referred to in review papers in the field (Akishin, Rambidi & Spiridonov, 1967; Bartell, 1971; Bastiansen, Seip & Boggs, 1971; Bauer, 1970; Beagley, 1973; Davis, 1971; Haaland, Vilkov, Khaikin, Yokozeki & Bauer, 1975; Hilderbrandt & Bonham, 1971; Karle, 1973; Kuchitsu, 1972*a*, *b*; Robiette, 1973; Seip, 1973). Literature citations are illustrative, not exhaustive.

While an adequate documentation of experiment and interpretation is vital, so also, in the avalanche of scientific literature we must contend with, is brevity and conciseness. Possible ways to achieve both ends are as follows. It would be desirable to develop a compact style of reporting essential details that vary from analysis to analysis. Standard equipment and procedures in a given laboratory that have been described clearly in readily available journals or accessible depository services may be documented simply by citing the appropriate references. Workers in every laboratory have an obligation to provide this information, as outlined in the following sections and to revise it every few years when substantial changes are made. Special procedures and certain data that are important for a critical evaluation of results but not of general interest to readers should be summarized and placed in a suitable depository service or published as microfilm together with the article.

More and more highly specialized computer routines are being used to process data, to convert it to molecular parameters, to compute the effects of a host of influences such as electron density shifts, distortions of diffracted waves, molecular vibrations, *etc.*, and to interpret derived structures in terms of quantum-chemical or other models. References to important computer packages employed and their sources should be given.

Structural investigations by gas-phase electron diffraction differ so much in complexity and in aim that it is impractical to recommend rigid rules for the reporting of procedures and presentation of results. Investigations in which low precision suffices need not be documented as minutely as those in which high precision is claimed. In the following are presented recommendations intended to be helpful in the preparation of a full paper. This guide may not fit all cases, and future developments may necessitate modifications. However, it is our hope that authors will deviate from the recommendations only after careful consideration.

I. Experimental apparatus and procedures

An adequate description of the experimental procedures used to obtain the numerical results should be made available. The major points to be considered are:

A. *Sample*

The source, verification of purity, and relevant handling procedures should be stated. When temperatures or reactivities are such that several species may be present, information concerning the vapor phase composition should be cited. The sample pressure should always be recorded when possible.

B. *Reference to the diffraction unit used*

If the detail of the unit has been published previously, a simple citation may suffice. If not, all relevant details must be given.

C. *Uncertainty in the s-scale*

The accuracy of measurement and means by which that accuracy is checked and maintained should be described.

D. *Nozzle and sample temperature*

The nozzle and reservoir temperatures are straightforward and should be reported. There is as yet no consensus on the effective temperatures of various internal molecular motions after the free expansion of the gas jet to the electron beam. Suitable caution in expressing the sample temperature seems advisable. The sample temperature at the electron beam depends upon (and, in principle, can be calculated from) the sample pressure, the nozzle dimensions, the distance of the electron beam from the nozzle lip, and the electron beam diameter. In all studies in which temperature is important, it is imperative that the above experimental quantities be given. Routinely the nozzle-beam dimensions should be made available.

* This report is based on a draft written in 1973 by the Gas-Diffraction subcommittee of the IUCr Commission on Electron Diffraction in consultation with workers in a majority of the existing laboratories of electron diffraction. The present version incorporates suggestions received during discussions of the *Guide* in scheduled open sessions at the Austin Symposium on Molecular Structure, Austin, Texas, March, 1974; the Second European Crystallographic Meeting in Keszthely, Hungary, August, 1974; and the Tenth International Congress of Crystallography in Amsterdam, August, 1975; and by correspondence from interested IUCr members.

E. Sector calibration (if used)

Some laboratories do and some do not calibrate the shapes of the sectors they use. It is important that an explicit statement be given about whether a calibration has been made, what means have been used, and what accuracy is achieved. A purely optical measurement with a standard traveling microscope may not be sufficient for the calibration of the inner range of a sector. For example, if the traveling microscope has a precision of 2×10^{-4} cm and if a two parts per thousand precision in the sector opening is desired, the smallest radius of an r^3 sector that can be measured optically with the requisite accuracy is 1·1 cm for a sector with $r_{max} = 4{\cdot}4$ cm and 1·8 cm for a sector with $r_{max} = 8{\cdot}8$ cm.

F. Intensity measurement

The precision but not accuracy of the intensity measurements will be revealed, in part, in the least-squares residuals to be discussed later. Reference should be made to the calibration of the measuring device, both with respect to scattering angle and with respect to intensity value. Some laboratories favor establishing intensity measurement with reference to some standard molecule such as benzene. If photographic recording is used, the assumption of a linear density–exposure relation may distort derived amplitudes of vibration (or, in some cases, it may even interfere with the determination of structure). Whether plates are spun, oscillated (over what amplitude), or read with a linear scan should be stated.

G. Number of measurements

The number of independent data points used should be given (but see II*D* and III*D* below). If photographic recording is used, the numbers of plates for each camera geometry should be stated.

II. Treatment of diffraction data

In operations to transform observed intensity values into a form convenient for comparison with theoretical expressions, the major points to be enumerated are:

A. Leveling procedure

If intensities are leveled, converted to an $s^4I(s)$ basis, or otherwise modified prior to determination of the background function, the requisite scattering factors or assumptions about electron distributions, polarization corrections, *etc.*, should be identified.

B. Extraneous scattering corrections

Significant excursions of the background of leveled intensities from a flat, horizontal line should be noted. They may signify extraneous scattering (an additive effect), variation of plate sensitivity (a multiplicative effect) or inaccurate scattering factors. The manner of compensation can influence derived molecular parameters and needs to be known if later reanalyses of data are made.

C. Background function

The derivation of this important function (analytical or hand drawn) should be stated. It would be good practice to report the effective number of shape parameters implicit in the background if it has non-uniform derivatives.

D. Interpolation of data

The procedures used to interpolate and/or smooth data points and the means of determining data correlations should be referred to (see III*D* below).

III. Derivation of structural parameters

(Akishin, Rambidi & Spiridonov, 1967; Bartell, 1971; Bastiansen, Fritsch & Hedberg, 1964; Bauer, 1970; Beagley, Cruickshank, Hewitt & Haaland, 1967; Corbet, Dallinga, Oltmans & Toneman, 1964; Harshbarger, Lee, Porter & Bauer, 1969; Hedberg & Iwasaki, 1964; Iwasaki, Fritsch & Hedberg, 1964; Karle, 1973).

The principal points requiring attention are:

A. Equations relating intensities to molecular quantities

An explicit reference should be made to the electron scattering formulae used including scattering factors or electron distribution functions, polarization corrections, dynamic corrections, *etc.* Note that the functions adopted in structure refinements may not be the same as those employed in the leveling of data.

B. Auxiliary information

The values of force constants, rotational constants, or related quantities used in analyzing data should be given and their origin should be cited. Such quantities may enter the analysis in (1) computation of non-varied amplitudes of vibration; (2) computation of shrinkage corrections; (3) estimation of asymmetry parameters in internuclear distribution peaks; (4) searching for plausible models of refinement *via* the Westheimer–Hendrickson 'molecular mechanics' approach; (5) adopting spectroscopic quantities such as rotational constants as constraints or merging them with diffraction intensities in the matrix of observations; (6) correcting spectroscopic quantities in (5) to be compatible with vibrational averages derived by diffraction.

C. Geometry adjustments

A discussion of the range of structural models tried must be given with statements: (1) whether symmetry constraints or other simplifications were imposed (*e.g.*, assumptions of local C_{3v} symmetry for a CF_3 group). All such constraints should be stated explicitly. (2) Whether multiple minima in least-squares solutions were searched for or encountered (false, deceptive minima are encountered in molecules as simple as SeO_2F_2). (3) Whether a static model with broadened peaks or a superposition of models distributed along various internal coordinates was taken.

D. Analysis of uncertainties

Estimates of the precision and probable accuracy should be given. It is essential to describe the various sources of uncertainty with a clear separation between measurement imprecisions, numerical analysis deviations, and possible systematic biases. The methods and assumptions made in the statistical analyses should be indicated including the weighting scheme and the inference of random errors and data correlations. Discussions of sources of experimental errors are given in several references, including Akishin, Rambidi & Spiridonov (1967); Bartell (1971); Bastiansen, Seip & Boggs (1971); Bauer (1970); Beagley (1973); Davis (1971); Harshbarger, Lee, Porter & Bauer (1969); Hilderbrandt & Bonham (1971); Karle (1973); Kuchitsu (1972*a*, *b*); Robiette (1973); Seip (1973); Vilkov (1964). Statistical analyses are outlined in several places including Bartell (1971); Bartell & Yow (1973); Bastiansen, Fritsch & Hedberg (1964); Hamilton (1964); Hedberg & Iwasaki (1964); Iwasaki, Fritsch & Hedberg (1964); MacGregor & Bohn (1971); Morino, Kuchitsu & Murata (1965); Murata & Morino (1966); Seip & Stølevik (1972); Seip, Strand & Stølevik (1969); Vilkov & Sadova (1967). One useful indicator that should *always* be determined and reported in diffraction studies is the 'index of pattern contrast' or 'index of resolution' defined as the ratio of $[I_{molec}(s)/I_{backgr}(s)]_{obs}$ to $[I_{molec}(s)/I_{backgr}(s)]_{calc}$ best characterizing the adjustable scale factor for the molecular intensity in the strong part of the pattern. Values much lower than unity indicate a washed-out pattern or imperfect intensity calibration. In complex molecules with overlapping internuclear distances it may be prudent to investigate couplings between systematic errors in intensities and derived parameters over and above simple scale-factor errors (Bartell & Yow, 1973).

E. The meaning of the parameters determined

The physical significance of the lengths, angles, and amplitudes of vibration deduced is implicit in the form of the equation adopted to relate observations to derived quantities. The complexity of the possible range of corrections (see III*A*, III*B*, and IV*B*) makes it necessary to identify explicitly the meaning of the final values reported insofar as possible. Many of the problems encountered are discussed in Iijima (1972); Kuchitsu (1968); Kuchitsu & Cyvin (1972); Rambidi & Ezhov (1968); and Sutton (1965).

IV. Presentation of results

The most important considerations are:

A. Conventional symbols, terminology, and units

It is suggested that authors follow wherever feasible the recommendations of the international system of units (SI) (IUPAP Commission for Symbols, Units, and Nomenclature 1965; *Le Systéme International d'Unités (SI)*, 1970; McGlashan, 1970; *Rules for the Use of Units of the International Systems...*, 1969) and the symbols and nomenclature approved by the various international unions (e.g., CODATA, 1973; International Union of Crystallography, 1973; IUPAC Commission on Thermodynamics and Thermochemistry, 1972; Triple Commission for Spectroscopy, 1963). The Commission on Electron Diffraction, however, accepts the IUCr recommendation to crystallographers (IUCr, 1973) that, in reporting structures, the ångstrom unit is preferred to an SI unit. If SI units are not adopted, the CODATA Task Group (CODATA, 1973) recommends footnotes such as 'Throughout this paper Torr = (101·325/760) kPa, and Å = 100 pm'.

B. Specialized symbols

In the field of electron diffraction no universally accepted notation has emerged, and the diversity of (continually evolving) procedures extant makes it unprofitable to try to impose a standard notation in such quantities as reduced intensity functions and radial distribution functions. It is desirable, however, to have a consensus on meanings of symbols for well defined molecular parameters since so many different types of averages are of concern in structural chemistry that the risk of confusion is great in the absence of standardization. Most of the following symbols are in widespread use and seem clear and concise enough to be adopted (Kuchitsu, 1968; Kuchitsu & Cyvin, 1972; Sutton, 1965).

(1) *Internuclear distances of special importance in electron diffraction*:

r_g Thermal average value of internuclear distance.

r_a Thermal average value of internuclear distance evaluated with an r^{-1} weight factor in the averaging. See IV*B*(4) below for relation to diffraction pattern.

r_z, r_α^0 Distance between average nuclear positions in ground vibrational state (r_z often refers to spectroscopic and r_α^0 to diffraction determinations).

r_α Distance between average nuclear positions at thermal equilibrium.

r_e Distance between equilibrium nuclear positions (structure at minimum potential energy).

If the above set of alternative parameters seems redundant and needlessly esoteric it should be noted that each member serves a useful role and, unless the associated distinctions are made and understood, it is impossible to publish fundamentally significant and precise structure results based on electron diffraction (or any other) studies. The parameters r_e, r_α, and r_z enjoy the property of corresponding to distances between points representing atoms in a geometrically self-consistent structure capable of representation by Cartesian coordinates. The true mean distance r_g and the natural diffraction distance r_a do not share this property, in general, owing to vibrational effects. But r_e, the theorists' preferred quantity, is seldom accessible (though plausible estimates can often be made), and r_α (and r_z) are misleadingly far from physical average internuclear distances to be optimally useful for consumption by nonspecialists who may seek to relate distances to interatomic forces, *etc.* One possible compromise is a structure (sometimes called an r_γ structure) in which *bond* lengths are assigned their r_g values and bond angles and torsion angles are given the values in an r_α structure. In such an 'r_γ' structure, which can always be expressed in Cartesian coordinates in the case of acyclic molecules, internuclear distances differ from true mean distances only by relatively small 'Bastiansen–Morino shrinkage corrections'. A disadvantage of the r_γ convention is that its basis is less fundamental in that it requires an arbitrary distinction between bonded and nonbonded distances. Furthermore, in the case of those cyclic molecules whose ring angles can be altered by a totally symmetric stretching deformation, the r_γ structure cannot be precisely self-consistent (except by accident).

(2) *Bond angles*:

No special notation has arisen and none seems necessary at this time. It is well to mention, however, that angles deduced solely from r_g or r_a distances without shrinkage corrections in general cannot correspond exactly to angles in a geometrically self-consistent structure, and the difference between r_g-based angles and the self-consistent r_α and r_e angles may far exceed experimental uncertainties. When amplitudes of vibration are very large, the physical meaning of bond angles may be obscure, particularly in the case of quasilinear or quasiplanar molecules. Due caution in reporting should be exercised in these cases.

(3) *Amplitudes of vibration*:

These are commonly designated by the symbols l or u, either of which is acceptable. For purposes of illustration we follow Cyvin (1968) and write l:

l_g represents $[\int(r-r_g)^2\ P(r)dr]^{1/2}$

l_e represents $[\int(r-r_e)^2\ P(r)dr]^{1/2}$

l_m represents the effective amplitude found by identifying the vibrational modulation of the molecular intensity curve with the damping factor $\exp(-l_m^2 s^2/2)$.

(4) *Interference patterns*:

s The scattering variable $(4\pi/\lambda)\sin(\varphi/2)$ where φ is the scattering angle.

κ A parameter, characteristic of the asymmetry in an internuclear distribution peak, showing up in the argument of the associated sinusoidal interference features as $\sin\{s[r_a-\kappa s^2+O(s^4)]\}$.

$f(s)$, $\eta(s)$ Complex atomic scattering factor for electrons where $f(s)=|f(s)|\exp i\eta(s)$, assuming that atoms are spherical. Notation for corrections for atoms in molecular environments is not yet standardized.

C. Mode of presentation of the results

The CODATA Task Group (CODATA, 1973) recommends, as a general principle, that results be reported in a form as free from interpretation as possible (*i.e.* as close as is practical to experimentally observed quantities). These results should be reported in such a manner that the degree of experimental randomness can be assessed. The reader should be able to recover enough of the experimental data so that he can reanalyze them in terms of different hypotheses. Graphical and analytical representations of important results, although convenient for the reader, are not acceptable substitutes for tabular presentation of accurate experimental results.

D. Quantities to be reported in standard structure analyses

Publication of the following tabular and graphical information is recommended:

(1) *Tables*

(*a*) *Digital values of leveled total intensities (molecular plus background)* **and** *the background used by the author.* Alternatively, the molecular intensities might be substituted for the leveled total intensities (while retaining a report of the background). Molecular intensities are less 'primary' than total intensities but are more convenient to analyze. These vital data should *always* be made available in work meriting publication in standard professional journals but may be deposited as supplementary material rather than as a tabulation in the journal article itself. It would be desirable to report indices of resolution for the various camera geometries here *as well as in the text.*

(*b*) *Bond distances, bond angles, and their uncertainties.* The meaning of the uncertainties must be specified in the tables as well as in the abstract and it is recommended that 2σ or 3σ be reported since they are more appropriate quantities than σ for consumption by non-specialists. Present methods of estimating σ are unreliable because errors are not statistically distributed in conventional diffraction analyses. It is to be hoped that progress will be made in error analyses.

The table should refer to a description of the principal systematic errors as well as random errors, either by direct inclusion, in a footnote of the table, or by an explicit reference in the table to the part of the text or other publication where the errors are discussed. The discussion should include how *known* systematic errors are corrected and how the magnitudes of poorly known systematic errors are estimated and included in the final uncertainties.

(*c*) *Mean square amplitudes.* The temperature should be specified insofar as is possible.

(*d*) *Error matrix or correlation matrix for derived parameters* (Hamilton, 1964). The correlation matrix with elements ϱ_{ij} is more immediately diagnostic of potential troubles in the analysis (if correlations are high) than is the error matrix. The full error matrix with elements $\varrho_{ij}\sigma_i\sigma_j$ is needed to calculate regression lines. The regression slope $\varrho_{ij}\sigma_i/\sigma_j$ may be valuable in allowing a quick estimation of $\Delta\theta_i/\Delta\theta_j$ where $\Delta\theta_i$ is the expected change in parameter θ_i from its least-squares value if subsequent information indicates that parameter θ_j should be changed from its least-squares value by $\Delta\theta_j$.

A good compromise between convenience and space is to tabulate the row matrix with elements σ_i and to tabulate immediately beneath it the correlation matrix. In some cases the correlation matrix is too large to warrant publication and a useful compromise might be to list only the elements with values exceeding, say, 0·5.

Some authors may wish to present a table of atomic coordinates to make it convenient for readers to calculate nonbonded distances, dependent angles, moments of inertia, *etc.* Interpretational difficulties (see IV*B*), however, mar the utility of such a table.

(2) *Figures*

(*a*) *Radial distribution curve.* In many cases these curves and their residuals (the differences between experimental and calculated curves) provide insights not readily apparent in intensity curves. They should be labeled with clear indications of the assignments of the prominent features.

(*b*) *Molecular intensity curves.* It may be useful, particularly when the tabulated intensities are not published in the article itself, to illustrate the molecular intensity curves, including the residuals. These should never be considered as a replacement of the tabular information in IV*D*(1*a*) above, however, for such figures are not easily subjected to reanalysis.

Some authors may wish to present figures showing experimental and theoretical intensities for separate plates or separate camera geometries. In addition, figures may compare residuals with experimental uncertainties obtained by computing values from each plate separately. Such plots may often be helpful.

Abbreviated check-list for electron diffraction publications

I. *Experimental apparatus and procedures*
   - *A*. Sample (source, purity, pressure)
   - *B*. Apparatus (description)
   - *C*. Uncertainty in the *s*-scale
   - *D*. Nozzle specifications and sample temperature
   - *E*. Sector calibration
   - *F*. Intensity measurement
   - *G*. Number of independent measurements

II. *Treatment of diffraction data*
   - *A*. Leveling scheme (description if used)
   - *B*. Extraneous scattering corrections (description if used)
   - *C*. Background function (method of derivation, number of implicit shape parameters)
   - *D*. Interpolation of data (procedures, smoothing involved, correlations in interpolated data)

III. *Derivation of structural parameters*
   - *A*. Equations (scattering formulae, scattering factors)
   - *B*. Auxiliary information (force constants, rotational constants, asymmetry parameters, *etc.*)
   - *C*. Geometry adjustments
     1. Symmetry constraints imposed
     2. Whether multiple solutions searched for or encountered
     3. Method to represent peak broadenings
   - *D*. Analysis of uncertainties (how done)

References

AKISHIN, P. A., RAMBIDI, N. G. & SPIRIDONOV, V. P. (1967). *The Characterization of High Temperature Vapors*, Edited by J. L. MARGRAVE, Chap. 12. New York: John Wiley.

BARTELL, L. S. (1971). *Physical Methods of Chemistry*, Edited by A. WEISSBERGER & B. W. ROSSITER, 4th ed., Vol. 1, Pt. III*D*, Chap. 2, pp. 125–158. New York: Wiley–Interscience.

BARTELL, L. S. & YOW, M. (1973). *J. Mol. Struct.* **15**, 173–188.

BASTIANSEN, O., FRITSCH, F. N. & HEDBERG, K. (1964). *Acta Cryst.* **17**, 538–543.

BASTIANSEN, O., SEIP, H. M. & BOGGS, J. E. (1971). *Perspectives in Structural Chemistry*. Edited by J. D. DUNITZ & J. A. IBERS, Vol. 4, pp. 60–165. New York: John Wiley.

BAUER, S. H. (1970). *Physical Chemistry, An Advanced Treatise*. Edited by D. HENDERSON, Vol. IV, Chap. 14. New York: Academic Press.

BEAGLEY, B. (1973). *Molecular Structures by Diffraction Methods (Specialist Periodical Report)*, Edited by G. A. SIM & L. E. SUTTON, Vol. 1, Part 1, Chap. 1. London: The Chemical Society.

BEAGLEY, B., CRUICKSHANK, D. W. J., HEWITT, T. G. & HAALAND, A. (1967). *Trans. Faraday Soc.* **63**, 836–845.

CODATA TASK GROUP ON PUBLICATION IN THE PRIMARY LITERATURE. (1973). D. GARVIN, T. GOLASHVILI, H. V. KEHIAIAN, N. KURTI & E. F. WESTRUM (Chairman). *Guide for Publication of Experimental Data and Derived Numerical Results in the Primary Literature*. Guide prepared under the auspices of UNESCO. *CODATA Bulletin* No. 9, December 1973.

CORBET, H. C., DALLINGA, G., OLTMANS, F. & TONEMAN, L. H. (1964). *Rec. Trav. Chim. Pay-Bas*, **83**, 789–798.

CYVIN, S. J. (1968). *Molecular Vibrations and Mean Square Amplitudes*. Amsterdam: Elsevier.

DAVIS, M. (1971). *Electron Diffraction in Gases*. New York: Marcel Dekker.

HAALAND, A., VILKOV, L., KHAIKIN, L. S., YOKOZEKI, A. & BAUER, S. H. (1973). *Topics in Current Chemistry*, Vol. 53. Berlin: Springer Verlag.

HAMILTON, W. C. (1964). *Statistics in Physical Science*. New York: Ronald Press.

HARSHBARGER, W., LEE, G., PORTER, R. F. & BAUER, S. H. (1969). *Inorg. Chem.* **8**, 1683–1689.

HEDBERG, K. & IWASAKI, M. (1964). *Acta Cryst.* **17**, 529–533.

HILDERBRANDT, R. L. & BONHAM, R. A. (1971). *Ann. Rev. Phys. Chem.* **22**, 279–312.

IIJIMA, T. (1972). *Bull. Chem. Soc. Japan*, **45**, 3526–3530.

INTERNATIONAL UNION OF CRYSTALLOGRAPHY (1973). *Notes for Authors. Acta Cryst.* **A29**, 97–108, or **B29**, 145–156.

IUPAC COMMISSION ON THERMODYNAMICS AND THERMOCHEMISTRY (1972). Definitive publication of *A Guide to Procedures for the Publication of Thermodynamic Data. Pure Appl. Chem.* **29**, 397–408.

IUPAP COMMISSION FOR SYMBOLS, UNITS, AND NOMENCLATURE (1965). *Symbols, Units, and Nomenclature in Physics*. Document U.I.P. 11 (S.U.N. 65-3).

IWASAKI, M., FRITSCH, F. N. & HEDBERG, K. (1964). *Acta Cryst.* **17**, 533–537.

KARLE, J. (1973). *Determination of Organic Structures by Physical Methods*. Edited by F. C. NACHOD & J. J. ZUCKERMAN, Vol. V., Part I, pp. 1–74. New York: Academic Press.

KUCHITSU, K. (1968). *J. Chem. Phys.* **49**, 4456–4462.

KUCHITSU, K. (1972*a*). *Molecular Structures and Vibrations*, Edited by S. J. CYVIN, Chap. 10, pp. 148–170. Amsterdam: Elsevier.

KUCHITSU, K. (1972*b*). *MTP International Review of Science* (*Phys. Chem., Ser.* 1, Vol. 2). Edited by G.

ALLEN, Chap. 6, pp. 203–240. Oxford: Medical & Technical Publ. Co.

KUCHITSU, K. & CYVIN, S. J. (1972). *Molecular Structures and Vibrations.* Edited by S. J. CYVIN, Chap. 12, pp. 183–211. Amsterdam: Elsevier.

Le Système International d'Unités (SI) (1970). Sèvres, France: International Bureau of Weights and Measures. Authorized English translations are available: (*a*) London: Her Majesty's Stationery Office; and (*b*) *National Bureau of Standards Special Publication* 330 (1970). Edited by C. H. PAGE & P. VIGOREUX. Washington, D.C.: U.S. Government Printing Office.

McGLASHAN, M. L. (Editor) (1970). *Manual of Symbols and Terminology for Physicochemical Quantities and Units*, prepared for the International Union of Pure and Applied Chemistry. *Pure Appl. Chem.* **21**, 3–44.

MACGREGOR, M. A. & BOHN, R. K. (1971). *Chem. Phys. Lett.* **11**, 29–34.

MORINO, Y., KUCHITSU, K. & MURATA, Y. (1965). *Acta Cryst.* **18**, 549–557.

MURATA, Y. & MORINO, Y. (1966). *Acta Cryst.* **20**, 605–609.

RAMBIDI, N. G. & EZHOV, Y. S. (1968). *Zh. Strukt. Khim.* **9**, 363–371.

ROBIETTE, A. G. (1973). *Molecular Structures by Diffraction Methods (Specialist Periodical Report).* Edited by G. A. SIM & L. E. SUTTON, Vol. 1, Part 1, Chap. 4. London: The Chemical Society.

Rules for the Use of Units of the International Systems of Units and a Selection of the Decimal Multiples and Sub-Multiples of the SI Units (1969). 1st ed. ISO Recommendation R 1000 (February, 1967). New York: American National Standards Institute.

SEIP, H. M. (1973). *Molecular Structures by Diffraction Methods (Specialist Periodical Report).* Edited by G. A. SIM & L. E. SUTTON, Vol. 1, Part 1, Chap. 1. London: The Chemical Society.

SEIP, H. M. & STØLEVIK, R. (1972). *Molecular Structures and Vibrations.* Edited by S. J. CYVIN, Chap. 11, pp. 171–182. Amsterdam: Elsevier.

SEIP, H. M., STRAND, T. G. & STØLEVIK, R. (1969). *Chem. Phys. Lett.* **3**, 617–623.

SUTTON, L. E. (Editor) (1965). *Tables of Interatomic Distances and Configuration in Molecules and Ions, Supplement* 1956–1959, pp. 6, 7. Special Publication No. 18. London: The Chemical Society.

TRIPLE COMMISSION FOR SPECTROSCOPY. (1963). *Report of Subcommittee on Units and Terminology*, Appendix A to Minutes of the Meeting for the Triple Commission for Spectroscopy (IAU, IUPAP, IUPAC) at Tokyo, 9 September 1962. *J. Opt. Soc. Amer.* **53**, 883–893.

VILKOV, L. V. (1964). *Zh. Strukt. Khim.* **5**, 809–813.

VILKOV, L. V. & SADOVA, N. I. (1967). *Zh. Strukt. Khim.* **8**, 1088–1092.

Section 10

Spectroscopies

The topics in this section all involve excitation of a sample and separation of the resulting signal into a characteristic spectrum. However, the modes of excitation and the types of signal widely vary. Placing each type into a separate section would produce a number of small sections that could also be subdivided.

Spectroscopic techniques are primarily used for studies of atomic or molecular structure or for chemical analysis. The relationship with parts of Section 8 is close.

Table 10.1 presents the spectroscopic data from the Green Book, which is primarily concerned with the structural aspects of spectroscopies. Table 7.1 on electromagnetic radiation may also be relevant in some cases.

Mass Spectroscopy

"Critical Evaluation of Class II and Class III Electron Impact Mass Spectra" (*66*) is reprinted here. Reference 67 presents further information on symbolism and nomenclature.

Infrared Spectroscopy

Much effort has gone into developing specifications for publication of infrared data. The papers reprinted here are "Specifications for Infrared Reference Spectra of Molecules in the Vapor Phase" (*68*) and "Recommendations for the Presentation of Infrared Absorption Spectra in Data Collections—A. Condensed Phases" (*69*). References 68 and 69 replace and supersede earlier provisional versions, including references 70 and 71.

"Presentation of Molecular Parameter Values for Infrared and Raman Intensity Measurements" (*72*) is also included in this book.

Raman Spectroscopy

"Presentation of Raman Spectra in Data Collections" (*73*) is reprinted. Reference 72 should also be noted.

Nuclear Magnetic Resonance Spectroscopy

Two reprinted papers, "Recommendations for the Presentation of NMR Data for Publication in Chemical Journals" (*74*) and "Standard Definitions of Terms, Symbols, Conventions, and References Relating to High-Resolution Nuclear Magnetic Resonance (NMR) Spectroscopy" (*75*) cover presentation of NMR data for measurements made by using protons. "Presentation of NMR Data for Publication in Chemical Journals—B. Conventions Relating to Spectra from Nuclei Other Than Protons" (*76*) is also reprinted.

Electron Paramagnetic Resonance Spectroscopy

"Recommendations for EPR/ESR Nomenclature and Conventions for Presenting Experimental Data in Publications" (*77*) is reprinted here.

Mössbauer Spectroscopy

"Nomenclature and Conventions for Reporting Mössbauer Spectroscopic Data" (*78*) is reprinted here.

Molecular Spectroscopy

Reference 79 discusses standard notation for diatomic molecules, reference 80 gives notation for polyatomic molecules, reference 81 covers the labeling of parity doublet levels in linear molecules, and reference 82 discusses the vibrational numbering of bands in the spectra of polyatomic molecules.

"Definitions and Symbolism of Molecular Force Constants" (*83*) is reprinted.

Atomic Spectroscopy

Atomic spectroscopy, because of its nature, is usually of more interest to physicists and astrophysicists than to chemists. It is included here for completeness and because several papers in the field cover terminology, symbolism, and the preparation of papers for publication.

Monographs presenting evaluated data on the lighter elements and the rare earth elements have been published. The sections covering nomenclature and spectroscopic notation are cited in reference 84 for the lighter elements and in reference 85 for the rare earth elements.

Reference 86 provides information on atomic transition probabilities. The authors (*86*) also outline the procedures they used to evaluate data from the literature, and clearly indicate the items that measurers should cover when preparing their papers.

Reference 87 deals with the accuracy of transition probabilities and reference 88 provides critical assessment of data from atomic spectroscopy.

Table 10.1. Spectroscopy. Continued on next page.

This section has been considerably extended compared with the previous edition [1.c] and with the corresponding section in the IUPAP document [3]. It is based on the recommendations of the ICSU Joint Commission for Spectroscopy [36, 37] and current practice in the field which is well represented in the books by Herzberg [38]. The IUPAC Commission on Molecular Structure and Spectroscopy has also published various recommendations which have been taken into account [9–15].

Name	*Symbol*	*Definition*	*SI unit*	*Notes*
total term	T	$T = E_{tot}/hc$	m^{-1}	1, 2
transition wavenumber	$\tilde{\nu}$, (ν)	$\tilde{\nu} = T' - T''$	m^{-1}	1
transition frequency	ν	$\nu = (E' - E'')/h$	Hz	
electronic term	T_e	$T_e = E_e/hc$	m^{-1}	1, 2
vibrational term	G	$G = E_{vib}/hc$	m^{-1}	1, 2
rotational term	F	$F = E_{rot}/hc$	m^{-1}	1, 2
spin orbit coupling constant	A	$T_{s.o.} = A\langle \hat{L} \cdot \hat{S} \rangle$	m^{-1}	1
principal moments of inertia	$I_A; I_B; I_C$	$I_A \leqslant I_B \leqslant I_C$	$kg\ m^2$	
rotational constants,				
in wavenumber	$\tilde{A}; \tilde{B}; \tilde{C}$	$\tilde{A} = h/8\pi^2 c I_A$	m^{-1}	1, 2
in frequency	A; B; C	$A = h/8\pi^2 I_A$	Hz	
inertial defect	Δ	$\Delta = I_C - I_A - I_B$	$kg\ m^2$	
asymmetry parameter	κ	$\kappa = \frac{(2B - A - C)}{(A - C)}$	1	3

(1) In spectroscopy the unit cm^{-1} is almost always used for wavenumber, and term values and wavenumbers always refer to the reciprocal wavelength of the equivalent radiation in vacuum. The symbol c in the definition E/hc refers to the speed of light in vacuum.

(2) Term values and rotational constants are sometimes defined in wavenumber units (e.g. $T = E/hc$), and sometimes in frequency units (e.g. $T = E/h$). When the symbol is otherwise the same, it is convenient to distinguish wavenumber quantities with a tilde (e.g. $\tilde{\nu}$, $\tilde{T}$, $\tilde{A}$, $\tilde{B}$, $\tilde{C}$ for quantities defined in wavenumber units), although this is not a universal practice.

(3) The Wang asymmetry parameters are also used: for a near prolate top $b_p = (C - B)/(2A - B - C)$, and for a near oblate top $b_o = (A - B)/(2C - A - B)$.

Table 10.1. Continued. Continued on next page.

Name	*Symbol*	*Definition*	*SI unit*	*Notes*
centrifugal distortion constants,				
S reduction	$D_J; D_{JK}; D_K; d_1; d_2$		m^{-1}	4
A reduction	$\Delta_J; \Delta_{JK}; \Delta_K; \delta_J; \delta_K$		m^{-1}	4
harmonic vibration wavenumber	ω_e; ω_r		m^{-1}	5
vibrational anharmonicity constant	$\omega_e x_e$; x_{rs}; $g_{tt'}$		m^{-1}	5
vibrational quantum numbers	v_r; l_t		1	5
Coriolis zeta constant	ζ_{rs}^{α}		1	
angular momentum quantum numbers	see additional information below			
degeneracy, statistical weight	g, d, β		1	6
electric dipole moment of a molecule	$\boldsymbol{p}$, $\boldsymbol{\mu}$	$E_p = -\boldsymbol{p}\cdot\boldsymbol{E}$	C m	7
transition dipole moment of a molecule	$\boldsymbol{M}$, $\boldsymbol{R}$	$\boldsymbol{M} = \int \psi' \boldsymbol{p} \psi'' \mathrm{d}\tau$	C m	7
molecular geometry, interatomic distances,				8, 9
equilibrium distance	r_e		m	
zero-point average distance	r_z		m	
ground state distance	r_0		m	
substitution structure distance	r_s		m	
vibrational coordinates,				8
internal coordinates	R_i, r_i, θ_j, etc.		(varies)	
symmetry coordinates	S_i		(varies)	
normal coordinates				
mass adjusted	Q_r		$kg^{\frac{1}{2}}$ m	
dimensionless	q_r		1	
vibrational force constants,				10
diatomic	f, (k)	$f = \partial^2 V/\partial r^2$	$J\ m^{-2}$	

(4) S and A stand for the symmetric and asymmetric reductions of the rotational hamiltonian respectively; see [39] for more details on the various possible representations of the centrifugal distortion constants.

(5) For a diatomic: $G(v) = \omega_e(v+\frac{1}{2}) - \omega_e x_e(v+\frac{1}{2})^2 + \dots$ For a polyatomic molecule the $3N-6$ vibrational modes ($3N-5$ if linear) are labelled by the indices $r, s, t, \dots$, or $i, j, k, \dots$. The index r is usually assigned in descending wavenumber order, symmetry species by symmetry species. The index t is kept for degenerate modes. The vibrational term formula is

$$G(v) = \sum_r \omega_r (v_r + d_r/2) + \sum_{r \leq s} x_{rs}(v_r + d_r/2)(v_s + d_s/2) + \sum_{t \leq t'} g_{tt'} l_t l_{t'} + \dots$$

(6) d is usually used for vibrational degeneracy, and β for nuclear spin degeneracy.

(7) Molecular dipole moments are often expressed in the non-SI unit debye, where $D \approx 3.335\,64 \times 10^{-30}$ C m.

(8) Interatomic (internuclear) distances and vibrational displacements are often expressed in the non-SI unit ångström, where $Å = 10^{-10}$ m $= 0.1$ nm $= 100$ pm.

(9) The various slightly different ways of representing interatomic distances, distinguished by subscripts, involve different vibrational averaging contributions; they are discussed in [40], where the geometrical structure of many free molecules is listed. Only the equilibrium distance r_e is isotopically invariant. The effective distance parameter r_0 is estimated from the rotational constants for the ground vibrational state and has only approximate physical significance for polyatomic molecules.

(10) Force constants are often expressed in mdyn $Å^{-1}$ = aJ $Å^{-2}$ for stretching coordinates, mdyn Å = aJ for bending coordinates, and mdyn = aJ $Å^{-1}$ for stretch–bend interactions. See [16] for further details on definitions and notation for force constants.

Table 10.1. Continued. Continued on next page.

Name	*Symbol*	*Definition*	*SI unit*	*Notes*
polyatomic,				
internal coordinates	f_{ij}	$f_{ij}=\partial^2 V/\partial r_i \partial r_j$	(varies)	
symmetry coordinates	F_{ij}	$F_{ij}=\partial^2 V/\partial S_i \partial S_j$	(varies)	
dimensionless normal coordinates	$\phi_{rst\ldots}$, $k_{rst\ldots}$		m^{-1}	11
nuclear magnetic resonance (NMR),				
magnetogyric ratio	γ	$\gamma=\mu/I\hbar$	C kg^{-1}	
shielding constant	σ_A	$B_A=(1-\sigma_A)B$	1	12
chemical shift, δ scale	δ	$\delta=10^6(\nu-\nu_0)/\nu_0$	1	13
(indirect) spin–spin coupling constant	J_{AB}	$\hat{H}/h=J_{AB}\hat{I}_A\cdot\hat{I}_B$	Hz	14
direct (dipolar) coupling constant	D_{AB}		Hz	15
longitudinal relaxation time	T_1		s	16
transverse relaxation time	T_2		s	16
electron spin resonance, electron paramagnetic resonance (ESR, EPR),				
magnetogyric ratio	γ	$\gamma=\mu/s\hbar$	C kg^{-1}	
g factor	g	$h\nu=g\mu_B B$	1	
hyperfine coupling constant,				
in liquids	a, A	$\hat{H}_{hfs}/h=a\hat{S}\cdot\hat{I}$	Hz	17
in solids	T	$\hat{H}_{hfs}/h=\hat{S}\cdot T\cdot\hat{I}$	Hz	17

(11) The force constants in dimensionless normal coordinates are usually defined in wavenumber units by the equation $V/hc=\Sigma\phi_{rst\ldots}\, q_r q_s q_t \ldots$, where the summation over the normal coordinate indices $r, s, t, \ldots$ is unrestricted.

(12) σ_A and B_A denote the shielding constant and the local magnetic field at nucleus A.

(13) ν_0 is the resonance frequency of a reference molecule, usually tetramethylsilane for proton and for ^{13}C resonance spectra [11]. In some of the older literature proton chemical shifts are expressed on the τ scale, where $\tau=10-\delta$, but this is no longer used.

(14) $\hat{H}$ in the definition is the spin–spin coupling hamiltonian between nuclei A and B.

(15) Direct dipolar coupling occurs in solids; the definition of the coupling constant is $D_{AB}=(\mu_0/4\pi)r_{AB}^{-3}\gamma_A\gamma_B(\hbar/2\pi)$.

(16) The longitudinal relaxation time is associated with spin–lattice relaxation, and the transverse relaxation time with spin–spin relaxation. The definitions are

$$dM_z/dt=-(M_z-M_{z,e})/T_1,$$

and

$$dM_x/dt=-M_x/T_2,$$

where M_z and M_x are the components of magnetization parallel and perpendicular to the static field B, and $M_{z,e}$ is the equilibrium value of M_z.

(17) $\hat{H}_{hfs}$ is the hyperfine coupling hamiltonian. The coupling constants a are usually quoted in MHz, but they are sometimes quoted in magnetic induction units (G or T) obtained by dividing by the conversion factor $g\mu_B/h$, which has the SI unit Hz/T; $g_e\mu_B/h \approx 2.8025$ MHz T^{-1}, where g_e is the g factor for a free electron. In liquids the hyperfine coupling is isotropic, and the coupling constant is a scalar a. In solids the coupling is anisotropic, and the coupling constant is a 3×3 tensor T.

Symbols for angular momentum operators and quantum numbers

In the following table, all of the symbols denote (*angular momentum*)/$\hbar$, and are dimensionless. (Although this is a universal practice for the quantum numbers, some authors use the operator symbols to denote *angular momentum*, in which case the operators would have SI units: J s.) The column heading 'Z-axis' denotes the space-fixed component, and the heading 'z-axis' denotes the molecule fixed component along the symmetry axis (linear or symmetric top molecules), or the axis of quantization.

Table 10.1. Continued. Continued on next page.

Angular momentum[1]	*Operator symbol*	*Quantum number symbol Total*	*Z-axis*	*z-axis*	*Notes*
electron orbital	$\hat{L}$	L	M_L	Λ	2
one electron only	$\hat{l}$	l	m_l	λ	2
electron spin	$\hat{S}$	S	M_S	Σ	
one electron only	$\hat{s}$	s	m_s	σ	
electron orbital + spin	$\hat{L}+\hat{S}$			$\Omega=\Lambda+\Sigma$	2
nuclear orbital (rotational)	$\hat{R}$	R		K_R, k_R	
nuclear spin	$\hat{I}$	I	M_I		
internal vibrational					
spherical top	$\hat{l}$	$l(l\zeta)$		K_l	
other	$\hat{j}, \hat{\pi}$			$l(l\zeta)$	2
sum of $R+L(+j)$	$\hat{N}$	N		K, k	2
sum of $N+S$	$\hat{J}$	J	M_J	K, k	2, 4
sum of $J+I$	$\hat{F}$	F	M_F		

(1) In all cases the vector operator and its components are related to the quantum numbers by eigenvalue equations analogous to:

$$\hat{J}^2\psi = J(J+1)\psi,\ \hat{J}_Z\psi = M_J\psi,\ \text{and}\ \hat{J}_z\psi = K\psi,$$

where the component quantum numbers M_J and K take integral or half-odd values in the range $-J \leqslant M_J \leqslant +J$, $-J \leqslant K \leqslant +J$. (If the operator symbols are taken to represent *angular momentum*, rather than *(angular momentum)/ħ*, the eigenvalue equations should read $\hat{J}^2\psi = J(J+1)\hbar^2\psi$, $\hat{J}_Z\psi = M_J\hbar\psi$, and $\hat{J}_z\psi = K\hbar\psi$.)
(2) Some authors, notably Herzberg [38], treat the component quantum numbers Λ, Ω, l and K as taking positive or zero values only, so that each non-zero value of the quantum number labels two wavefunctions with opposite signs for the appropriate angular momentum component. When this is done, lower-case k is often regarded as a signed quantum number, related to K by $K=|k|$. However, in theoretical discussions all component quantum numbers are usually treated as signed, taking both positive and negative values.
(3) There is no uniform convention for denoting the internal vibrational angular momentum; j, π, p and G have all been used. For symmetric top and linear molecules the component of j in the symmetry axis is always denoted by the quantum number l, where l takes values in the range $-v \leqslant l \leqslant +v$ in steps of 2. The corresponding component of angular momentum is actually $l\zeta\hbar$, rather than $l\hbar$, where ζ is a Coriolis coupling constant.
(4) Asymmetric top rotational states are labelled by the value of J (or N if $S\neq 0$), with subscripts K_a, K_c, where the latter correlate with the $K=|k|$ quantum number about the a and c axes in the prolate and oblate symmetric top limits respectively.

Example $J_{Ka,Kc} = 5_{2,3}$ for a particular rotational level.

Symbols for symmetry operators and labels for symmetry species

(i) *Symmetry operators in space-fixed coordinates* [41]

identity	E
permutation	P
space-fixed inversion	E^*
permutation-inversion	P^* $(=PE^*)$

The permutation operation P permutes the labels of identical nuclei.

Example In the NH_3 molecule, if the hydrogen nuclei are labelled 1, 2 and 3, then $P=(123)$ would symbolize the permutation 1 is replaced by 2, 2 by 3, and 3 by 1.

The inversion operation E^* reverses the sign of all particle coordinates in the space-fixed origin, or in the molecule-fixed centre of mass if translation has been separated. It is also called the parity operator; in field-free space, wavefunctions are either parity + (unchanged) or parity − (change sign) under E^*. The label is used to distinguish between the two nearly degenerate components formed by K-, Λ-, or l-doubling; it is also used in conjunction with the total angular momentum J to define the symmetry labels e and f on parity doublets: see [42].

Table 10.1. Continued. Continued on next page.

(ii) *Symmetry operators in molecule fixed coordinates* [38]

identity	E
rotation by $2\pi/n$	C_n
reflection	σ, σ_v, σ_d, σ_h
inversion	i
rotation-reflection	S_n $(=C_n\sigma_h)$

If C_n is the primary axis of symmetry, wavefunctions that are unchanged or change sign under C_n are given species labels A or B respectively, and wavefunctions that are multiplied by $\exp(\pm 2\pi i\ s/n)$ are given the species label E_s. Wavefunctions that are unchanged or change sign under i are labelled g (gerade) or u (ungerade) respectively. Wavefunctions that are unchanged or change sign under σ_h have species labels with a ′ or ″ respectively. For more detailed rules see [37, 38].

Other symbols and conventions in optical spectroscopy

(i) *Term symbols for atomic states*
The electronic states of atoms are labelled by the value of the quantum number L for the state. The value of L is indicated by an upright capital letter: S, P, D, F, G, H, I, and K, . . . , are used for $L=0$, 1, 2, 3, 4, 5, 6, and 7, . . . , respectively. The corresponding lower-case letters are used for the orbital angular momentum of a single electron. For a many-electron atom, the electron spin multiplicity $(2S+1)$ may be indicated as a left-hand superscript to the letter, and the value of the total angular momentum J as a right-hand subscript. If either L or S is zero only one value of J is possible, and the subscript is then usually suppressed. Finally, the electron configuration of an atom is indicated by giving the occupation of each one-electron orbital as in the examples below.

Examples B: $(1s)^2(2s)^2(2p)^1$, $^2P_{1/2}$
C: $(1s)^2(2s)^2(2p)^2$, $^3P_0$
N: $(1s)^2(2s)^2(2p)^3$, ^4S

(ii) *Term symbols for molecular states*
The electronic states of molecules are labelled by the symmetry species label of the wavefunction in the molecular point group. These should be Latin or Greek upright capital letters. As for atoms, the spin multiplicity $(2S+1)$ may be indicated by a left superscript. For linear molecules the value of Ω $(=\Lambda+\Sigma)$ may be added as a right subscript (analogous to J for atoms). If the value of Ω is not specified, the term symbol is taken to refer to all component states, and a right subscript r or i may be added to indicate that the components are regular (energy increases with Ω) or inverted (energy decreases with Ω) respectively.

The electronic states of molecules are also given empirical single letter labels as follows. The ground electronic state is labelled X, excited states of the same multiplicity are labelled A, B, C, . . . , in ascending order of energy, and excited states of different multiplicity are labelled with lower-case letters a, b, c, . . . In polyatomic molecules (but not diatomic molecules) it is customary to add a tilde (e.g. $\tilde{X}$) to these empirical labels to prevent possible confusion with the symmetry species label.

Finally the one-electron orbitals are labelled by the corresponding lower-case letters, and the electron configuration is indicated in a manner analogous to that for atoms.

Examples The ground state of CH is $(1\sigma)^2(2\sigma)^2(3\sigma)^2(1\pi)^1$, X $^2\Pi_r$, in which the $^2\Pi_{1/2}$ component lies below the $^2\Pi_{3/2}$ component, as indicated by the subscript r for regular.

The ground state of OH is $(1\sigma)^2(2\sigma)^2(3\sigma)^2(1\pi)^3$, X $^2\Pi_i$, in which the $^2\Pi_{3/2}$ component lies below the $^2\Pi_{1/2}$ component, as indicated by the subscript i for inverted.

The two lowest electronic states of CH_2 are . . . $(2a_1)^2(1b_2)^2(3a_1)^2$, $\tilde{a}$ $^1A_1$,
. . . $(2a_1)^2(1b_2)^2(3a_1)^1(1b_1)^1$, $\tilde{X}$ $^3B_1$

The ground state of C_6H_6 (benzene) is . . . $(a_{2u})^2$ $(e_{1g})^4$, $\tilde{X}$ $^1A_{1g}$.

The vibrational states of molecules are usually indicated by giving the vibrational quantum numbers for each normal mode.

Table 10.1. Continued.

Examples for a bent triatomic molecule,
(0, 0, 0) denotes the ground state,
(1, 0, 0) denotes the ν_1 state, i.e. $v_1 = 1$, and
(1, 2, 0) denotes the $\nu_1 + 2\nu_2$ state, etc.

(iii) *Notation for spectroscopic transitions*
The upper and lower levels of a spectroscopic transition are indicated by a prime ′ and double-prime ″ respectively.

Example $h\nu = E' - E''$

Transitions are generally indicated by giving the excited state label, followed by the ground state label, separated by a dash or an arrow to indicate the direction of the transition (emission to the right, absorption to the left).

Examples B–A indicates a transition between a higher energy state B and a lower energy state A;
B→A indicates emission from B to A;
B←A indicates absorption from A to B.
(0, 2, 1)←(0, 0, 1) labels the $2\nu_2 + \nu_3 - \nu_3$ hot band in a bent triatomic molecule.

A more compact notation [43] may be used to label vibronic (or vibrational) transitions in polyatomic molecules with many normal modes, in which each vibration index r is given a superscript v_r' and a subscript v_r'' indicating the upper and lower state values of the quantum number. When $v_r' = v_r'' = 0$ the corresponding index is suppressed.

Examples 1_0^1 denotes the transition (1, 0, 0)–(0, 0, 0);
$2_0^2\ 3_1^1$ denotes the transition (0, 2, 1)–(0, 0, 1).

For rotational transitions, the value of $\Delta J = J' - J''$ is indicated by a letter labelling the branches of a rotational band: $\Delta J = -2$, -1, 0, 1, and 2 are labelled as the O-branch, P-branch, Q-branch, R-branch, and S-branch respectively. The changes in other quantum numbers (such as K for a symmetric top, or K_a and K_c for an asymmetric top) may be indicated by adding lower-case letters as a left superscript according to the same rule.

Example ^{p}Q labels a 'p-type Q-branch' in a symmetric top molecule, i.e. $\Delta K = -1$, $\Delta J = 0$.

(iv) *Presentation of spectra*
It is recommended to plot both infra-red and visible/ultraviolet spectra against wavenumber, usually in cm^{-1}, with decreasing wavenumber to the right (note the mnemonic 'red to the right') [9, 20]. (Visible/ultraviolet spectra are also sometimes plotted against wavelength, usually in nm, with increasing wavelength to the right.) It is recommended to plot Raman spectra with increasing wavenumber shift to the left [10].

It is recommended to plot both electron spin resonance (ESR) spectra and nuclear magnetic resonance (NMR) spectra with increasing magnetic induction (loosely called magnetic field) to the right for fixed frequency, or with increasing frequency to the left for fixed magnetic field [11, 12].

It is recommended to plot photoelectron spectra with increasing ionization energy to the left, i.e. with increasing photoelectron kinetic energy to the right [13].

Dear Sir

Critical Evaluation of Class II and Class III Electron Impact Mass Spectra. Operating Parameters and Reporting Mass Spectra

The accepted use of mass spectrometry for qualitative analysis in various areas of science, including component identification of fossil fuels, drugs, pesticides, metabolites, organometallic compounds and air and water pollutants, emphasizes the importance and need for reference mass spectra. A systematic collection, evaluation, and publication of reliable reference mass spectra is indispensable in carrying out such analyses. To aid in attaining the most useful quality of spectra, a set of specifications has been developed for guiding laboratory procedures and for documentation of the resulting spectra. The specifications may also be used in the evaluation of existing reference spectra. In addition, it is hoped that these specifications will be used in establishing standards for individual laboratories.

Proposed specifications

Specifications are proposed for mass spectra obtained using electron impact ion sources. The specifications are intended as a guide for the collection, reporting and evaluation of mass spectra. As such the specifications represent the manner in which new mass spectra should be collected, reported and evaluated for the ideal reference file, recognizing the desirability that such a file contain spectra of many compounds of diverse types.

In the draft of these specifications it was assumed with the present state of the art of mass spectrometry that it is not feasible to obtain mass spectra which may be regarded as Standard Reference Spectra (Class I) i.e. spectra that represent physical constants of the material, independent of instrumentation. (Instruments which can be controlled to reproduce designated reference spectra within close quantitative tolerances, such as the CEC 21-103 used for early API reference data,[1] could conceivably measure Class I spectra.)[2] Therefore, specifications are recommended for two levels of operation and reporting. Spectra of high quality measured on pure compounds in accord with the established instrument operational guidelines and data reporting, and reproducible to within the specified limits are classified as Research Reference Spectra (Class II). Spectra on compounds of established purity in accord with the particular instrument operational guidelines and data reporting, and evaluated to eliminate obvious discrepancies and errors are classified as Analytical Reference Spectra (Class III).

Specifications for Research Spectra (Class II) are designed to provide a set of preferred procedures so that determinations of spectra in different laboratories, on similar instruments, may yield spectra which agree within the error limits designated. The information obtained from the comparison of the mass spectrum for an unknown compound, determined under the experimental conditions for Class II, with the spectrum for a known pure compound, may be used as evidence to establish the identity of a component in an unknown sample, and to indicate the presence of other components. The information obtained from the comparison of the mass spectrum of an unknown compound, determined in accord with Class III specifications, with that of a known pure compound may be used in a qualitative fashion to establish the identity and probable structure of the unknown compound.

The specifications first establish guidelines to assure proper spectrometer and inlet system operation, and adequate and systematic reporting of spectral data. A second portion of the specifications is designed to establish procedures and criteria for evaluating mass spectra.

Class II Research reference spectra

(A) Mass spectrometer. The model and type of mass spectrometer should be identified. Resolution of the mass spectrometer must correspond to unit resolution at the highest mass observed on the 10% valley definition or an equivalent definition. To assure proper operation of the mass spectrometer, the mass spectrum of a standard reference compound must be measured under the same instrumental and data reduction conditions. Peak intensities and pattern data for suitable standard compounds must be obtained to assure proper mass spectrometer operation. The electron energy should be 70 eV, but spectra measured in the range 50–100 eV will be acceptable if such conditions give data in agreement with those of the reference compound. The type of inlet system (gas chromatograph, solid, liquid, gas, etc.) should be given. Performance of the inlet system should be evaluated by measuring the spectrum of a standard compound. Inlet systems, other than those of conventional design, should be described in detail and their performance evaluated. The temperature of the ion source should be given.

(B) Spectra. Spectra should be determined and reported from m/z 10 to m/z (M+40), where M is the mass of the molecular ion. M/z values whose relative intensities are between 0.1% and 100% (base peak: most intense m/z in the spectrum), should be tabulated. Multiply charged ion masses should be tabulated to the nearest 0.1 u.

Presentation of spectra should be cast in the form of m/z versus relative intensity. For research quality reference spectra, the reproducibility of individual relative ion intensities should be the larger of ±20% relative or ±5% absolute.

(C) Compound purity and identity. The purity of the compound entering the ion source should be equal to or greater than 99.0%. Information on either the source or the preparation and purification procedures for the sample should be provided. Additional data (boiling point, gas chromatography, infrared, etc.) which support the assigned structure and purity may be presented. (Spectra of materials for which insufficient data are supplied to authenticate the chemical structure are assigned to Class III). The operational reporting specifications for Class II Research Reference Spectra are summarized below.

(D) Operation-reporting

(I) *Instrumentation*

(a) *Type of mass spectrometer.* Make and model identified.

(b) *Performance.* The performance should be checked by measurement and reporting of peak intensities and pattern data for a reference compound. The preferred reference compound is PFTBA (perfluorotributylamine), but others may be used including DFTPP (decafluorotriphenylphosphine).[2] The performance check shall be made within ±one week of the measurement of the mass spectra submitted for evaluation. The reference spectrum must meet the criteria set forth in IIa: the larger of ±20% relative error or ±5% abolute error.

(c) *Inlet system.* The type of inlet system and the source temperature should be specified. Performance of the inlet system shall be tested by measuring the mass spectrum of cholesterol as a reference compound. Cholesterol is the preferred reference but other compounds may be used. The reference spectrum must meet the criteria set forth in IIa: the larger of ±20% relative error or ±5% absolute error.

(d) *Resolution.* Unit resolution at the highest m/z recorded; 10% valley definition or equivalent.

(e) *Electron energy.* The electron energy should be between 50 and 100 eV with 70 eV preferred.

(II) *Data Presentation*

(a) *Ion intensity and mass values.* Tabular presentation of m/z values versus relative intensity, reported for singly and multiply charged ions of intensities from 0.1% to 100% (base peak).

(b) *Mass values.* *M/z* values of peaks should be based on the sum of the whole number values of the isotopic masses of the constituent elements; *m/z* values of multiply charged ions should be reported to one decimal place. A value for the relative intensity of the molecular ion should be given. Reproducibility of ion intensities for a separate measurement of the spectrum shall be the larger of ±20% relative or ±5% absolute error.

(c) *Spectra.* The complete spectrum with relative intensities for *m/z* values from 10 to (M+40) should be reported. (M = molecular ion mass.)

(d) *Background spectrum.* A background spectrum should be measured before the standard spectrum is measured and appropriate corrections should be made.

(III) *Compound Purity*

(a) The purity of the compound entering the ion source shall be equal to or greater than 99.0%.

(b) The means of establishing the purity of the compound shall be specified; i.e. gas chromatography, melting point, etc.

(c) Impurities and the level (percentage) of such impurities shall be specified.

Class III. Analytical reference spectra

Specifications for Class III spectra center on compliance with accepted operation of the mass spectrometer as outlined below. Spectra in this class are for qualitative use only since no limits of reproducibility are designated.

Operating-reporting. Class II criteria, unless as noted below.

(I) *Instrumentation*

(a) *Type of mass spectrometer.* Make and model identified (Class II criteria).

(b) *Performance.* Performance checked by measurement of peak intensities for a standard reference compound. Reasonable assurance of accurate mass calibration and abundance measurements for the reference compound.

(c) *Inlet system.* Type of inlet system specified. Ion source temperature specified, if available.

(d) *Resolution.* Unit resolution at the highest *m/z* recorded; 10% valley definition or equivalent. (Class II criteria.)

(d) *Electron energy.* The electron energy should be between 50 and 100 eV, preferably 70 eV. (Class II criteria.)

(II) *Data Presentation*

(a) *Ion intensity.* Tabular presentation of *m/z* versus relative intensity, reported for the intensities from 1.0% to 100.0% (base peak). A value must be given for the molecular ion abundance.

(b) *Mass values.* *M/z* values of peaks should be based on the whole number values of the isotopic masses of the constituent elements.

(c) *Spectra.* Complete spectrum with relative intensities for *m/z* values from 29 to (M+40). (M = molecular ion).

(d) *Background.* Appropriate corrections for background peaks should be made.

(III) *Purity*

The purity of the compound entering the mass spectrometer shall be greater than 90%.

Evaluation plan

Evaluation of spectra should be carried out by individuals familiar with spectra of similar compounds according to the following guidelines. The 'Quality Index' described by Speck[3] is used, except that full credit is given for the 'source of spectrum'.

The evaluation process for Classes II and III focuses on presentation of spectra and accompanying data in accord with the specifications, and preliminary examination for obvious discrepancies and errors. For Class II certification a further critical evaluation of the spectrum shall be made for each of the points summarized below.

(I) Sufficient evidence of compound purity has been presented.

(II) The specifications for measuring spectra have been followed.

(III) The important peaks have been formed by mechanistically reasonable processes.

(IV) If the spectrum has a Quality Index[3] less than 0.80, it will not be certified unless detailed examination shows that this low value results from the unusual mass spectral behavior of the compound, i.e. that the mass spectrum of this compound of >99% purity measured carefully under Class II conditions would still yield a Quality Index <0.80.

Further modifications

These are being published as a suggested procedure by the Mass Spectrometry Subcommittee of the Joint Committee on Atomic and Molecular Physical Properties (FWM, Chairman). Further suggestions are invited from the mass spectrometry community.

Acknowledgements

Valuable suggestions were given by W. L. Budde, L. H. Gevantman, H. S. Hertz, R. L. Kiser, D. P. Martinsen, J. M. McGuire and A. G. Sharkey Jr. Sponsorship of the Joint Committee on Atomic and Molecular Physical Data, the American Society for Mass Spectrometry and The Office of Standard Reference Data of the US National Bureau of Standards is gratefully acknowledged. The basic document was prepared by J. G. D. under a grant from the US National Bureau of Standards.

Yours

J. G. DILLARD
Department of Chemistry,
Virginia Polytechnic Institute,
Blacksburg,
Virginia 24061,
USA

S. R. HELLER
Environmental Protection Agency,
401 M Street, SW,
Washington DC 20460,
USA

F. W. McLAFFERTY (to whom correspondence should be addressed)
Department of Chemistry,
Cornell University,
Ithaca,
New York 14853,
USA

G. W. A. MILNE
Building 10,
The National Institutes of Health,
Bethesda,
Maryland 20205,
USA

R. VENKATARAGHAVAN
Lederle Laboratories,
Pearl River,
New York 10965,
USA

August 1980

References

1. American Petroleum Institute Project 44, Thermochemical Research Center, Texas A & M University, College Station, Texas.
2. J. W. Eichelberger, L. E. Harris and W. L. Budde, *Anal. Chem.* **47**, 995 (1975).
3. D. D. Speck, R. Venkataraghavan and F. W. McLafferty, *Org. Mass Spectrom.* **13**, 209 (1978).

SPECIFICATIONS FOR INFRARED REFERENCE SPECTRA OF MOLECULES IN THE VAPOR PHASE

(Recommendations 1987)

PHYSICAL CHEMISTRY DIVISION
COMMISSION ON MOLECULAR STRUCTURE AND SPECTROSCOPY*

in conjunction with

VAPOR PHASE SUBCOMMITTEE OF THE
COBLENTZ SOCIETY SPECTRAL EVALUATION COMMITTEE†

Prepared for publication by
JEANETTE G. GRASSELLI
Standard Oil R & D, Cleveland, Ohio 44128, USA

Abstract - The document supplies specifications for the measurement, collection, presentation and storage of high quality infrared reference spectra of materials in the vapor phase. The specifications include details of instrument conditions to be used, sampling procedures, hard copy presentation of data, classification of data to facilitate later spectral searching, and recommendations for digital storage. A summary of the parameters for storing spectra on magnetic tape is also presented.

PREAMBLE

Introduction

Infrared spectrometry is being used to an increasing extent for the qualitative and quantitative analysis of trace components of the atmosphere. Several collections of reference spectra of samples in the vapor phase at ambient temperature are now available (1-6). The spectra in these collections are measured at medium resolution ($\Delta\tilde{\nu}$ = 1 - 4 cm^{-1}); in some the samples are neat at low pressure while in others an atmosphere of air or nitrogen has been added. Although these reference data are certainly useful to the many users of medium resolution spectrometers (both grating and Fourier transform), the spectra have been measured at too high resolution to provide even semi-quantitative calibration factors for users of tunable diode laser (TDL) spectrometers. There is now a very definite need for collections of spectra of samples in the vapor phase, measured at a variety of resolutions, with and without air broadening, in as uniform a format as possible.

Many infrared spectrometers are now mini- and micro-computer controlled, and can readily generate data in digital form for archival storage. In order for digitally stored data to be readily transferable from one site to another, it is necessary to ensure that the format of the stored information is as uniform as possible. In this document we are not only proposing specifications for the measurement and presentation, as hard copy, of infrared reference spectra of molecules in the vapor phase, but also for formatting digitally stored reference data on magnetic tape.

*Membership of the Commission during the preparation of this report (1983–85) was as follows:

Chairman: J. R. Durig (USA); *Secretary*: H. A. Willis (UK); *Titular Members*: A. M. Bradshaw (FRG); B. G. Derendjaev (USSR); S. Forsen (Sweden); J. F. J. Todd (UK); *Associate Members*: R. D. Brown (Australia); J. G. Grasselli (USA); R. K. Harris (UK); H. Kon (USA); Yu. N. Molin (USSR); W. B. Person (USA); H. Ratajczak (Poland); C. J. H. Schutte (R.S. Africa); M. Tasumi (Japan); *National Representatives*: R. Colin (Belgium); Y. Kawano (Brazil); G. Herzberg (Canada); J. Lu (Chinese Chemical Society); Z. Luz (Israel); S. Ng (Malaysia); B. Jezowska-Trzebiatowska (Poland).

†Membership of the Committees is as follows:

Chairman, Vapor Phase Subcommittee: P. R. Griffiths (USA); *Members*: A. R. H. Cole (Australia); P. L. Hanst (USA); W. J. Lafferty (USA); R. J. Obremski (USA); J. H. Shaw (USA).

Chairman, Spectral Evaluation Committee: R. Norman Jones (Canada).

Reprinted with permission from *Pure and Applied Chemistry*, Volume 59, Number 5, 1987, pages 673–681.

Instrumental conditions

Unlike the criteria previously proposed by the Coblentz Society (7), and subsequently approved by the International Union of Pure and Applied Chemistry, for medium resolution reference spectra of condensed phase samples (8), equally firm specifications cannot be proposed which are rigidly applicable to the many different types of instruments used for measuring vapor phase spectra. For example, a reference spectrum measured using a selective wavelength analyzer is of little or no use to users of TDL spectrometers, and vice versa. Thus, we are proposing that reference spectra measured on any type of infrared spectrometer should not be rejected simply on the grounds of resolution, but rather that the resolution should be recorded with the data. In theory, at least, reference spectra measured at very high resolution may be degraded to simulate spectra measured on a lower resolution spectrometer by convolution of the high resolution spectrum with the instrument line shape function of the lower resolution spectrometer. Thus, primary emphasis should be placed on the acquisition of high resolution data ($\Delta\tilde{\nu} < 1$ cm^{-1}), over as wide a spectral range as possible.

Spectra should be plotted over a "useful" wavenumber range, without extensive regions where little or no information on the sample of interest is present. For spectra measured at medium or low resolution ($\Delta\tilde{\nu} > 1$ cm^{-1}), no regions of the spectrum should be left blank. For spectra measured at higher resolution, spectra should only be plotted in regions where "significant" spectral information is to be found. Obviously what is "significant" will vary from application to application and from laboratory to laboratory, but as a default condition we define "significant" information as an absorption band containing at least one feature with a peak absorbance greater than 0.05.

Specification of the low wavenumber cutoff also presents a problem. In many cases the lower wavenumber limit of spectrometers with KBr optics (400 cm^{-1}) may not be low enough to enable the lowest frequency fundamental modes of some heavy molecules to be measured. On the other hand, the increasingly popular use of the mercury cadmium telluride photodetector on many spectrometers may necessitate a cutoff as high as 700 cm^{-1}. Thus we have specified as an upper limit a low wavenumber cutoff for reference spectra as 700 cm^{-1}, with the recommendation that spectra be plotted to lower wavenumbers than this if possible.

Sampling

Ideally vapor phase spectra should be measured under at least two conditions, both neat at low pressure (self-broadening) and with an atmosphere of dry air free of CO_2 added (air-broadening). As an alternative to air, dry nitrogen can be added (nitrogen-broadening). The (partial) pressure of the sample should be great enough to cause the strongest features to absorb between about 70% and 90% of the incident radiation. If important weak bands are present, the spectra should be rerun at higher pressure or at longer pathlength to yield an interpretable spectrum in this region. Samples should be held at ambient temperature unless an increase in temperature is needed to prevent sample condensation in the cell.

Since quantitative data are required, an accurate knowledge of the pathlength of the cell and (partial) pressure of the sample is necessary. The product of these two parameters should be accurate to better than ± 10%. If the error limits are believed to lie outside this range, they should be estimated and recorded.

Presentation of data

Spectra should be plotted linearly in transmittance or percent transmission on charts which are linear in transmittance or percent transmission, since this is the format in which they are usually measured originally. Nonlinear absorbance grids are acceptable but not recommended. Spectra plotted linearly in absorbance are also acceptable but again not recommended.

Spectra are to be plotted with high wavenumber to the left, and in such a format as to permit clear identification of each feature. To this end, the abscissal scale expansion and pen width should provide a trace in which the width of the plotted line is at least four times narrower than the full width at half maximum absorbance of the narrowest spectral feature.

It is recognized that it is often difficult to prepare absolutely pure samples. If impurity bands are apparent on a reference spectrum, they should be indicated and preferably identified.

Classification of data

Since the spectra of vapor phase samples are so strongly dependent on instrumental resolution and the presence or absence of a broadening gas, spectral search routines using vapor phase reference spectra must be applied with great care. To simplify the programming for spectral search and to provide a simple classification for vapor phase reference data, we propose the following classification:

a. Instrumental Resolution: A number will be assigned to a spectrum to denote that the resolution of the instrument fell in a certain range. This number will change each time the resolution changes by a factor of $\sqrt{10}$, and will be equal to $2[1-\log_{10}(R/\text{cm}^{-1})]$, where R is the minimum numerical resolution (cm^{-1}) in the range. The code will be as follows(Table I):

b. Broadening Gas: A letter will be assigned to the spectrum to denote the nature of the broadening gas. This letter will immediately follow the numerical code above. The letter code will be as follows:

- S self-broadened;
- A broadened with one atmosphere of air, free of water and carbon dioxide;
- N broadened with one atmosphere of dry nitrogen;
- X broadened by some other gas.

Table 1

Numerical Code	Resolution (cm^{-1})	
	Greater than	Less than
0	10	30
1	3	10
2	1	3
3	0.3	1
4	0.1	0.3
5	0.03	0.1
6	0.01	0.03
7	0.003	0.01
8	0.001	0.003
9		0.001

Thus a reference spectrum of a gas broadened by nitrogen and measured at 2 cm^{-1} resolution will be given the classification of 2N. In this way it is a simple matter to program a search. If, for example, a spectrum of an unknown in air is measured at 4 cm^{-1}, it is probable that only reference spectra with the classification 1N, 1A, 2N and 2A would be searched initially. If no matching spectrum is found, it may also be desirable to subsequently search the 1S and 2S data base.

Digital storage

Perhaps the most universal medium for digital storage of data is magnetic tape. We are therefore proposing a format for data stored on magnetic tape and we hope that this format may be easily modified for data stored on other media, e.g. punched cards or discs. The format which we propose is similar to the format proposed earlier by another subcommittee for GC-IR reference data (8).

Digitally stored reference data must contain at least two sets of data, a general name and data file (file A) and the digitized spectrum itself (file B). The absorbance spectrum is used in preference to the transmittance spectrum since it is more closely related to the fundamental parameter, the absorption coefficient and within certain limitations line or band intensities are directly proportional to the (partial) pressure of the sample.

The wavenumber interval at which spectra are digitized is a matter of some controversy. In theory, if the instrument line shape function (slit function) of the spectrometer is known and used as an interpolation function, only one data point per resolution element need be stored in order for the original analog spectrum to be completely reproduced. In practice, if at least two data points per resolution element have been sampled, many commonly used interpolation functions can be applied in order to accurately reproduce the analog trace of the spectrometer. As progressively smaller sampling intervals are used, the digital spectrum plotted using only linear interpolation more closely approximates the analog spectrum, but the storage required gets unnecessarily high. Therefore, the specification of at least two data points per resolution element and a known instrument line shape function has been adopted for all spectra measured on spectrometers other than TDL spectrometers. For laser spectrometers the half width of the laser line is considerably less than the Doppler width of each line in the spectrum. Therefore, in this case, we would recommend a minimum sampling interval of 1×10^{-3} cm^{-1} or one half the Doppler width, with a large interval for air-broadened spectra.

In an attempt to produce an efficient storage format, the data base will be stored in FORTRAN readable unformatted magnetic tape records (although File A may be considered formatted since the character data cannot be stored more efficiently in an unformatted mode). Since most computer systems can read IBM compatible formats, this format mode is considered the most convenient. Since the data records are unformatted, the use of ASC11 Variable Block Size record formats is required. Some computer systems have low limits on the maximum record length, so the record length has been limited to 2058 8-bit bytes, the first eight bytes of which are header information bytes used by the tape handlers for assessing the actual record length. Thus, for a spectrum recorded from 4000 to 400 cm^{-1} with a sampling interval of exactly 2 cm^{-1}, there are 1801 data points which can be recorded in 3602 bytes. This results in two records, one 2050 (2048 + 2) bytes in length, and the other 1556 (1554 + 2) bytes. All records have two more types than required by the number of data points to allow the serial number of the spectrum to be written in the first two bytes of each record.

A typical schematic for a tape is shown in Fig. 1 of reference (9). The first two bytes are the total byte length of the physical record (including those two bytes). For file A of the digitally stored data described in the section on RECORDING ON MAGNETIC TAPE, File A - Information Record and Table 2, vide infra, the data length is 624_{10} bytes, plus eight bytes for the header. Thus the total record length is 632_{10} bytes which, expressed as a hexadecimal number, is 278_{16}. The next two bytes must be null, or zero. Bytes five and six are the logical record length, which, in this case, is the entire remainder of the record; i.e., 628_{10} bytes, or 274_{16}. Bytes seven and eight must also be null in this case. The header blocks for file B are generated in an analogous fashion. Employing the correct Job Control Language Statements, IBM systems will automatically write and read these header blocks; however, many other computer systems will not read them. To generate the header block for file A on such systems, eight should be added to the data record length, and that number expressed hexadecimally, should be written in the first 2-byte work of the record. A null word follows; i.e., a word of zeros, then the data record length plus four, another null word, and finally the data. Upon reading these records on non-IBM systems, the first eight bytes may simply be ignored.

SPECTROMETER OPERATION

Spectral range

For spectra measured at a resolution of 1 cm^{-1} or lower resolution ($\Delta\tilde{\nu} > 1$ cm^{-1}), spectra should be recorded from 4000 cm^{-1} to 700 cm^{-1} or below without gaps; reduction of the lower limit to 400 cm^{-1} is encouraged. For spectra measured at higher resolution ($\Delta\tilde{\nu} < 1$ cm^{-1}), spectral regions in which no absorption feature of the sample shows an absorbance greater than 0.05 need not be recorded. For very high resolution spectra ($\Delta\tilde{\nu} < 0.01$ cm^{-1}) reference spectra of single bands are acceptable.

Resolution

The resolution, given as the full width at half maximum of the instrument line shape function, should appear on the chart. No restrictions on the resolution are specified since reference spectra are needed at a variety of resolutions. Every effort, however, should be made to ensure that the instrumental resolution does not change dramatically across the spectrum.

Abscissa

The abscissal scale should be linear in wavenumber. Scale changes on low resolution spectra (e.g., a 2:1 scale change at 2000 cm^{-1}) are acceptable but not desirable. Spectra should be plotted from high to low wavenumber with high wavenumber to the left.

Wavenumber accuracy

The frequencies of sharp absorption features as read from the chart should be accurate to one half the resolution for all spectra except those measured using a laser spectrometer. The wavenumbers of spectra measured using a laser spectrometer should be accurate to 2×10^{-3} cm^{-1}.

Noise level

The peak-to-peak noise level in the spectrum should not exceed 2% of full scale, except over very short regions where no absorption bands of the sample are observable (such as the 2347 cm^{-1} band due to carbon dioxide). It is preferable that digital smoothing routines are not applied to the spectra after acquisition except for the purpose of reducing the resolution of a reference spectrum in order to simulate data measured at a lower resolution on a different instrument. If a smoothing function is used, it must be specifically defined.

Ordinate expansion

If, for some reason, the absorbance of the strongest feature in the spectrum does not exceed 0.50, the spectrum may be measured or replotted using ordinate scale expansion (zero suppression), provided that the ratio of the peak absorbance of the strongest feature in the spectrum to the maximum peak-to-peak noise level in the spectrum is at least 40:1. When a spectrum is measured or plotted using ordinate expansion it must be clearly indicated on the chart, and the measured peak absorbance of the most intense spectral feature must be given.

Baseline flatness

The baseline (I_0 line, 100% line) must be flat to better than 5% of full scale across the recorded spectrum.

Recording

It is permissible for spectra to appear on more than one chart, even though for low or medium resolution spectra ($\Delta\nu > 1$ cm^{-1}) no spectral regions should be omitted. Discontinuities in the ordinate scale should not exceed 2%. Hand retraced spectra are unacceptable.

Atmospheric interferences

The effects of atmospheric water vapor lines should not exceed the allowable noise level (see SPECTROMETER OPERATION, Wavenumber Accuracy). The effect of CO_2 absorption lines can exceed this level provided that the sample has no absorption features around the 2347 or 668 cm^{-1} bands of CO_2.

Ordinate scale

It is preferred that the intensity ordinate values be plotted linearly in transmittance or percent transmission on charts with a linear transmittance or percent transmission grid; a logarithmic ordinate grid or spectra plotted linearly in absorbance are acceptable. <u>Note</u>: This specification does <u>not</u> refer to digitally recorded data on magnetic tape, but <u>only</u> to the original hard-copy record.

Pen width

The width of the plotted line should be at least four times less than the resolution for all spectra other than those recorded using TDL spectrometers. For spectra measured using a laser spectrometer, the width of the plotted line should be at least four times less than the width of the narrowest absorption feature in the spectrum.

Apodization function

Spectra measured using a Fourier transform spectrometer should be computed from the interferogram using a triangular apodization function.

SAMPLE HANDLING

Peak absorbance

All samples must be present in the cell so that the combination of partial pressure of the sample and pathlength of the cell causes the strongest feature in each plotted range to have a peak absorbance of between 1.0 and 0.50.

Pressure

Samples may either be present in the cell with no foreign gas present or, with an atmosphere of dry air (free of CO_2), nitrogen, or some other gas added. In all cases, the (partial) pressure of the sample and (where applicable) the partial pressure and identity of the added gas should be recorded on the chart.

Temperature

All spectra should be measured with the sample at ambient temperature, unless an increased temperature is needed to permit a sufficiently high sample vapor pressure for the peak absorbance criterion to be obeyed (see SPECTROMETER OPERATION, Ordinate Expansion and SAMPLING HANDLING, Peak Absorbance). The temperature should be recorded on the chart for all spectra.

Cell pathlength

No restriction is placed on the cell pathlength. The pathlength should be recorded on the chart for all spectra.

Quantitative accuracy

The product of the pathlength of the cell and the (partial) pressure of the sample should be accurate to better than ± 10%. If the error limits are believed to lie outside this range, the estimated limits should be recorded on the chart.

INFORMATION TO APPEAR WITH THE SPECTRUM (HARD COPY)

Sample

Both the structural and the molecular formula should appear on the chart. It is also recommended that the compound name be included, and that the name should conform with the nomenclature used by IUPAC. Common or proprietary names and the CAS Registry Number can be included if desired. The source of the sample should also be included. The (partial) pressure of the sample (Pa) should be recorded along with the pressure and identity of the added gas (where applicable) and the temperature of the sample (K).

Instrumental

The make and model of the spectrometer should be recorded, as well as the date on which the spectrum was measured. For spectra measured on dispersive spectrometers, all changes of gratings and filters should be recorded, together with the wavenumber at which they occur and the slit program used. The resolution at which the spectrum was measured should be recorded, together with the alphanumeric classification (see PREAMBLE, Classification of Data).

Peak absorbance

The peak absorbance of the strongest band in the spectrum should be given. If a second spectral plot was necessary, the conditions under which the second spectrum was measured or plotted should appear on the chart. If ordinate expansion was required, an indication should also appear with the spectrum.

RECORDING ON MAGNETIC TAPE

General

For spectra measured on spectrophotometers with a digital data system, data should also be output on magnetic tape, both for archival purposes and for later modification to computer searchable files. The aim for magnetic tape storage should be to store data in a uniform manner no matter on what type of spectrometer or data system the spectra have been measured. To this end, it is recommended that data be written on 9-track magnetic tape at 800 bpi in FORTRAN readable, IBM compatible, unformatted records. Since the data records are unformatted the use of ASC11 Variable Block Size record formats is required. The record length will be limited to 2058 8-bit bytes (see PREAMBLE, Digital Storage).

Two files are required for each spectrum. File A will contain the sample information and instrument parameters. File B will contain the absorbance spectrum of the sample. Only a single spectrum will be stored in this file; spectra run under different conditions (e.g., sample pressure, air-broadening, resolution) will be stored in a separate file with a separate header. At least two points per resolution element will be stored for spectra measured at resolutions down to 0.003 cm^{-1}. For spectra measured at higher resolution, a sampling interval of 0.001 cm^{-1} is acceptable.

Table 2. Information Record

Field No.	No. of bytes required	Description	Suggested Format	Degree of necessity
1	10	Serial number, contributor assigned	I10	Mandatory
2	32	Molecular formula <u>e.g.</u>, C.2.F.5.Cl.1	32A1	Mandatory
3	12	Connectivity matrix[a] and/or CAS Registry Number	12A1	Mandatory
4	128	Common or proprietary name(s), <u>e.g.</u>, FREON 115; use a semi-colon and a blank to separate each name.	128A1	Mandatory, if applicable
5	256	IUPAC name, <u>e.g.</u>, CHLOROPENTAFLUOROETHANE	256A1	Mandatory
6	7	Molecular weight	F7.2	Desirable
7	10	Starting spectrum wavenumber	F10.4	Mandatory
8	10	Final spectrum wavenumber	F10.4	Mandatory
9	8	Sampling interval, cm^{-1}	F8.5	Mandatory
10	7	Number of spectrum points	I7	Mandatory
11	8	(Partial) pressure of sample, kPa (see footnote)	F8.5	Mandatory
12	4	Partial pressure of broadening gas kPa	I4	Mandatory
13	7	Cell length, mm	I7	Mandatory
14	6	Cell temperature, °C	F6.1	Mandatory
15	32	Manufacturer and model number of spectrometer	32A1	Mandatory
16	5	Nominal resolution at 750 cm^{-1}, cm^{-1}	F5.3	Mandatory
17	5	Nominal resolution at 1500 cm^{-1}, cm^{-1}	F5.3	Mandatory
18	5	Nominal resolution at 3000 cm^{-1}, cm^{-1}	F5.3	Mandatory
19	32	Grating changes, cm^{-1}	32A1	Desirable, if applicable
20	32	Filter changes, cm^{-1}	32A1	Desirable, if applicable
21	32	Infrared detector type	32A1	Very desirable
22	2	Spectrum classification code from PREAMBLE, Classification of Data	2A1	Mandatory
23	32	Operator's name	32A1	Mandatory
24	10	Date spectrum was measured, Month-Date-Year, <u>e.g.</u>, 02-09-1981	10A1	Mandatory
25	32	Laboratory name	32A1	Mandatory

[a]Any matrix that describes structural features and is transferable.

File A—information record

The information stored in file A concerns compound identification and properties, parameters for the spectrum stored in file B, sampling parameters, instrumental details, and contributor information. A schematic of the nature, format and degree of necessity for File A is shown in Table 2. It may be summarized as follows:

Field number 1 through 6: Compound identification and properties. These fields include a contributor-assigned serial number, molecular formula, connectivity matrix or Chemical Abstracts Service registry number, common and IUPAC or CAS name, and molecular weight. The connectivity matrix is one which describes the structure of the compound.

Field number 7 through 10: Parameters for the absorbance spectrum stored in file B. These fields include the starting and final wavenumber, sampling interval and number of data points. The starting wavenumber should <u>always</u> be greater than the final wavenumber.

Field number 11 through 14: Sampling information, such as sample pressure, broadening gas pressure, cell pathlength and temperature.

Field number 15 through 21: Instrumental information, such as the manufacturer and model number of the spectrometer, grating and filter changes and detector. The resolution is specified at three different wavenumbers (750, 1500 and 3000 cm^{-1}), since the spectral resolution is not constant on many types of spectrometers.

Field number 22: Alphanumeric classification code for spectrum (see PREAMBLE, Classification of Data).

Field number 23 through 25: Contributor information, such as the date, and the operator's name and laboratory.

File B—absorbance spectrum

After the IBM header data, the first two bytes of each data record will contain an integral serial number corresponding to the serial number in field No. 1 of file A. Each spectrum will be divided up into records of 1024 data points (2048 bytes) as discussed in the PREAMBLE, Digital Storage. The final record of each spectrum may be shorter than 2048 bytes.

The ordinates will be expressed to 0.002 absorbance units from 0.000 to 1.998. In order to save space, the absorbance units will be expressed as integers (0 to 1998) by multiplying each ordinate by 1000.

FUTURE NEEDS

To a certain extent this may be considered to be a preliminary document in that, whereas specifications as to, for example, the wavenumber accuracy and spectrometer resolution have been set down, no universally accepted method of testing these specifications has been found. For example, the wavenumber calibration data published in the two IUPAC-sponsored books (10,11) are sufficient for many users of fairly high resolution spectrometers but not good enough at all wavenumbers for users of laser spectrometers. If spectra are measured at lower resolution, lines begin to blend and shift and so these data once again cannot be used. Obtaining accurate, reliable data for wavenumber calibration across the complete mid infrared spectrum for spectrometers operating at all the resolution settings listed in the PREAMBLE, Classification of Data, is a gigantic task, and far beyond the scope of this document.

The verification of resolution is also by no means an easy task. We favor the definition of resolution as the full width at half height of the instrument line shape function, but for many types of spectrometers the width of this function does not remain constant over the entire spectrum. Measurement of the function is not a trivial task, and a spectroscopic determination requires the use of many different compounds for all the resolution settings listed in Table 1 over the entire spectrum. Again the specification of resolution standards is considered to be outside the scope of this document.

Similarly, we have not discussed procedures for sample handling, and good sampling is critical to obtaining accurate quantitative reference spectra. Another parameter necessary for obtaining good quantitative data is cell pathlength. While the pathlength of single-pass cells may be measured fairly accurately, pathlength determination of multi-pass cells is not at all easy, since it is possible that different parts of the beam take different paths. Again, we have not addressed specifications for pathlength determination.

In spite of these limitations, we believe that the specifications above will enable high quality infrared reference spectra of materials in the vapor phase to be measured, collected and used.

REFERENCES

1. Dow Chemical Company, "Infrared Spectra of Gases and Vapors," (Vols. 1 and 2); available from Foxboro/Wilks Analytical, South Norwalk, CT.
2. The Perkin-Elmer Corporation, "Reference Spectra of Gases," Technical Bulletin IRB-36, Norwalk, CT.
3. The Perkin-Elmer Corporation, "IR Reference Spectra of Vapors at the OSHA Concentration Limit," Technical Bulletin IRB-36, Norwalk, CT.
4. Sadtler Research Laboratories, Inc., "Gases and Vapors" (150 spectra), Philadelphia, PA.
5. Beckman Instruments, Inc., "Infrared Spectra of Hazardous Gases and Vapors at OSHA Maximum Tolerance Levels," Industrial Technical Report TR-590, Fullerton, CA.
6. D. G. Murcray and A. Goldman, "Handbook of High Resolution Infrared Spectra of Atmospheric Interest," CRC Press, Boca Raton, FL.
7. "The Coblentz Society Specifications for Evaluation of Research Quality Analytical Infrared Spectra (Class II)," Anal. Chem., 47, 945A (1975).
8. "Recommendations for the Presentation of Infrared Absorption Spectra in Data Collections - A. Condensed Phases," Commission on Molecular Structure and Spectroscopy, International Union of Pure and Applied Chemistry, Pure and Appl. Chem., 50, 321 (1978).
9. "Specifications for Infrared Reference Spectra in the Vapor Phase above Ambient Temperature," The Coblentz Society, Appl. Spectrosc., 33, 543 (1979).
10. "Tables of Wavenumbers for the Calibration of Infrared Spectrometers," Prepared by the Commission on Molecular Structure and Spectroscopy of IUPAC, Butterworth, London (1961).
11. A. R. H. Cole, "Tables of Wavenumbers for the Calibration of Infrared Spectrometers," 2nd Edition, Pergamon Press, Oxford (1977).

Recommendations for the Presentation of Raman Spectra in Data Collections

(Recommendations 1981)

COMMISSION ON MOLECULAR STRUCTURE AND SPECTROSCOPY*

Prepared for publication by
E. D. BECKER, J. R. DURIG, W. C. HARRIS
and G. J. ROSASCO

INTRODUCTION

These recommendations relate to the Raman spectra of isotropic materials that are intended for permanent retention in data collections. They are a more comprehensive version of the "Recommendations for the Presentation of Raman Spectra for Cataloging and Documentation in Permanent Data Collections" which were published in *Pure and Applied Chemistry* 36, 277 (1973). The current recommendations are based on a report published by the Ad Hoc Panel on Raman Spectral Data, convened by the Numerical Data Advisory Board of the National Academy of Sciences-National Research Council of the United States (1). A provisional version of these Recommendations was published (2) and has been modified slightly as a result of comments received.

There is a recognized need to establish a set of guidelines for the presentation of Raman spectral data that deal with the format for data presentation and the experimental parameters required to define the spectrum properly. Although these recommendations are directed toward Raman spectra prepared for permanent collections, many of them are also pertinent to spectra presented in journals.

There are a number of different types of phenomena which come under the general category of Raman scattering. These include "normal" Raman scattering, resonance Raman scattering, coherent anti-Stokes Raman scattering, hyper-Raman effect, etc. The present recommendations are not intended to encompass all these aspects of the field of Raman scattering, but rather will deal with the presentation of data primarily representing "normal" Raman scattering from isotropic materials. The guidelines define a minimum set of parameters which should be specified. These parameters, or analogous parameters adapted to the particular field of interest, will be necessary for adequate data presentation in many fields of Raman spectroscopy. For example, the parameters presented below should be regarded as a minimal subset of those necessary for the specification of resonance Raman spectra in isotropic materials.

It seems useful to define different classes for spectral data which allow variation in data quality. These classes are defined in analogy to similar infrared data classifications which have been proposed by the Coblentz Society (3,4) and adapted by IUPAC (5).

Critically defined physical data. Spectra in this category are of such high quality that they are acceptable as physical constants of the substances under precisely defined conditions. Specifications for this class are not discussed in this document.

Research quality analytical spectra. Data in this category are those in which the sample and spectrum conform to the best current and commonly practiced Raman spectrometry procedures.

Approved analytical spectra. Data in this category are those for which the sample and spectrum are of sufficient quality for use in the identification of materials.

There are, of course, many other spectra of high quality that cannot be classified into the above categories because certain information is lacking. These may be referred to as Unevaluated Spectra.

*Membership of the Commission during the preparation of these recommendations (1977-1979) was as follows:

Chairman: E. D. BECKER (USA); *Secretary:* G. ZERBI (Italy); *Titular Members:* J. H. BEYNON (UK); V. A. KOPTYUG (USSR); C. N. R. RAO (India); C. SANDORFY (Canada); T. SHIMANOUCHI (Japan); D. W. TURNER (UK); *Associate Members:* P. DIEHL (Switzerland); F. DÖRR (FRG); J. R. DURIG (USA); J. G. GRASSELLI (USA); B. JEŽOWSKA-TRZEBIATOWSKA (Poland); S. LEACH (France); P. M. RENTZEPIS (USA); H. A. WILLIS (UK); *National Representatives:* G. HERZBERG (Canada); H. W. THOMPSON (UK).

Reprinted with permission from *Pure and Applied Chemistry,* Volume 53, 1981, pages 1879–1885.

SPECIFICATIONS

The recommendations for parameter specifications are listed in Table I under four categories: (A) Sample, (B) Excitation Source, (C) Raman Spectrometer and (D) Experimental Configuration. The guideline used in the selection and specification of parameters is that all information which is significant to the intended physical or chemical interpretation of the data should be presented. In addition, for research quality analytical data, it is suggested that the information presented should be sufficient to allow some evaluation of the data.

The list of parameters under each category in Table I is generally self-explanatory, but additional discussion of the categories is presented below.

DISCUSSION

A. Sample

The information presented in this category should include sample characteristics which either affect the observed spectrum or make the data more useful. It is extremely important that the structure and purity of the sample be established by accepted independent techniques. In the case of a sample in which there is resonance Raman scattering the spectrum can be very different in appearance from that observed in a normal Raman experiment; an acknowledgement of this resonance and possibly the presentation of the electronic absorption spectrum for the material would be very useful.

It is important to recognize that sample heating caused by the exciting laser beam may be significant. Thus, when a temperature is quoted for the spectrum, a description of the method of measurement should be given (i.e., the temperature of a surrounding heat sink, an internal probe temperature, or a spectroscopically determined temperature), and an estimate of the temperature uncertainty would be useful.

B. Excitation source

It is not normally necessary to present details of the measurement of laser power or the transmission function of the filters. The intent of specifications B3 and B4 is to allow a reliable judgement of the power and/or irradiance levels used to perform the experiment.

C. Raman spectrometer

Selection of the optimum value of spectral slit width in terms of signal level and resolution is left to the discretion of the scientist. As a general rule the scan rate (cm^{-1}/s) should not exceed the value of the ratio of the spectral slit width (cm^{-1}) to 4 times the value of the time constant (sec), i.e.,

$$\text{Scan rate } \left(\frac{cm^{-1}}{s}\right) \leqq \frac{\text{Spectral slit width } (cm^{-1})}{4 \times \text{time constant (s)}}$$

In this context, the time constant is defined as the 1/e-time of the electronics, i.e., the time for the recording system to reach a value of (1 - 1/e) for a unit step function input. The spectral slit width is operationally defined as the full width (in cm^{-1}) at half height of the peak observed as the spectrum of a "narrow line source" (e.g., a laser plasma line).

For Research Quality Analytical spectra, the spectral slit width should be determined by actual measurement. For Approved Analytical spectra, this information may be obtained from a calculation utilizing the linear dispersion and mechanical slit width of the instrument or from the manufacturer's specifications. Any significant variations of the spectral slit width over the range of the Raman spectrum should also be specified.

The parameters describing response presented in C6 are important in determining the relative "intensity" of a series of bands covering a fairly broad frequency range in the Raman spectrum of a material. The term "system" is meant to include the effects of both the spectrometer and the detector. "Relative response function" can be defined in the context of a calibration utilizing a broad band, continuous standard light source. It is quite feasible to extract useful information on the response of the system by recording the Raman spectrum of a defined standard material.

D. Experimental configuration

It is recommended (D1) that the scattering geometry and the sample cell orientation be described by reference to some conveniently defined orthogonal axis system. This axis system in many cases might be defined by axes parallel to the optical axis of the fore-optics and/or spectrometer, the long direction of the entrance slit, and a third direction perpendicular to these two. This form of specification (D1) is preferred to traditional descriptions (i.e., back scattering, transverse illumination, etc.)

Item 4a recommends that the experimental geometry [X(Y,Y)Z] employed to determine mode symmetry by the measured depolarization ratios can be more generally specified by a symbol $\mathbf{k}_i(\mathbf{e}_i,\mathbf{e}_s)\mathbf{k}_s$ (6). The vector $\mathbf{k}_i(\mathbf{k}_s)$ refers to the direction of propagation of the incident (scattered) light, and the vector $\mathbf{e}_i(\mathbf{e}_s)$ refers to the direction of electric field polarization of the incident (scattered) light. These vectors should be referred to a convenient orthogonal axis system. This system will generally be identical to the system defined in D1, but in any case should be easily relatable to the system defined in D1. This form of specification (D4a) is preferred to traditional names such as parallel and perpendicular or horizontal and vertical, etc.

For the purposes of this document the depolarization ratio will be defined in a restricted, operational sense. Two separate spectra are measured, corresponding to the following geometries (see Fig. 1):

TABLE I. Parameters recommended for the specification of Raman spectral data.

Type of parameter	Research Quality Analytical Spectra	Approved Analytical Spectra
A. Sample		
1. Name and structural formula	Specify	Specify
2. a. State, e.g., gas, liquid, solid (powder, etc.), solution (solvent and concentration)	Specify	Specify
b. Color (if any), reference to an absorption spectrum	Specify absorption spectrum	Specify color
3. Impurities	No impurity bands evident	Impurity bands acceptable if identified
4. Temperature (kelvin, K)	Specify	Specify
5. Pressure, if other than ambient	Specify	Specify
B. Excitation source		
1. Type of laser	Specify	Specify
2. Wavelength (nanometers, nm)	Specify	Specify
3. Laser power (watts, W)		
a. at laser	Specify	Specify
b. at sample	Specify	...
4. Optical elements in beam		
a. Narrow band filter, attenuators, etc.	Specify	Specify
b. Effective aperture and focal length of laser focusing lens	Specify	...
C. Raman spectrometer		
1. Spectrometer manufacturer and model (or equivalent description)	Specify	Specify
2. Detector characteristics, e.g., response type or manufacturer and model	Specify	Specify
3. Spectral slit width at the exciting wavelength (cm^{-1})	Specify	Specify
4. Scan rate (cm^{-1}/s)	Specify	Specify
5. Time constant of recording electronics (s)	Specify	Specify
6. System response as a function of Raman shift		
a. Relative response function	Specify either a or b	...
b. Demonstrate by reference to a "standard" compound		Specify
D. Experimental configuration		
1. Scattering geometry specified with respect to a set of orthogonal axes	Specify	Specify
2. Sample cell	Specify	Specify
3. Special arrangements; e.g., spinning sample, multipass irradiation (approximate enhancement)	Specify	Specify
4. Polarization measurements		
a. Specify X(Y,Y)Z for experiments used to determine mode symmetry or measure depolarization ratio (see Fig. 1)	Specify	Specify
b. Optical elements; e.g., polarizer, scrambler	Specify	Specify
c. Solid angle of collection lens	Specify	...

$$k_i(e_i, e_s)k_s \mathrel{\hat{=}} X(Y,Y)Z$$

$$k_i(e_i, e_s)k_s \mathrel{\hat{=}} X(Y,X)Z$$

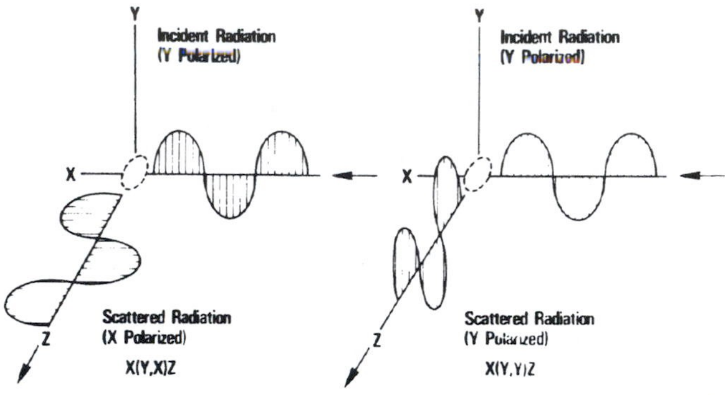

Figure 1. Geometry involved in the measurement of depolarization ratios. (A left-handed coordinate system is used here for ease of visualization but is not recommended for general use.)

This experiment implies that the incident light is linearly polarized and that the scattered light is polarization-analyzed to select the necessary components. Further, the polarization dependence of the system response (C6) must be eliminated by a suitable optical technique (e.g., use of a scrambler plate) or a correction applied to the recorded spectra. The ratio of the suitably corrected peak height or the integrated area of a band observed in experiment X(Y,X)Z to that observed in experiment X(Y,Y)Z is defined as the depolarization ratio, i.e.,

$$\rho = \frac{I[X(Y,X)Z]}{I[X(Y,Y)Z]}$$

Depolarization ratios may be measured from peak heights or integrated areas, and both may be useful in different applications. It should be realized that these measurements may be different, since the depolarization ratio varies significantly across the band. The basis of the measurement (peak height or area) should always be stated.

This quantity is useful for assigning mode symmetries and for verification of structural formula and chemical identification. The technique is recommended for routine measurement of depolarization ratios for analytical purposes.

SPECTRAL FORMAT

Within the context of these recommendations the most important spectral information involves the frequency shifts, relative "intensity," and depolarization ratio (related to the vibrational symmetry type) or the Raman bands of the material.

A. Graphical presentation

It is necessary that a graphical presentation of the spectrum be made for both Research Quality and Approved Analytical Spectra. The format for this presentation is as follows:

Abscissa - This should be linear in wavenumber shift, zero at the right, with the wavenumber shift increasing from right to left. The accuracy of the wavenumber scale should be not less than $\pm$ 5 cm^{-1} for Approved Analytical data and $\pm$ 2 cm^{-1} for Research Quality Analytical data. Note that the wavenumber shifts should refer to free space values; however, the accuracy required is not so stringent, in general, as to require a differentiation between reciprocal wavelength in vacuo and in air.

Ordinate - This should be linear and proportional to the intensity (power) or the Raman signal, increasing upwards. The units of the intensity scale should be actually observed photocurrent or count rate or an equivalent measurement. This intensity scale will be more significant when a "standard" spectrum is defined. The amount of any suppression of the "zero" light level should be stated.

B. Depolarization ratio

Research Quality Analytical data should present the two spectra utilized in the determination of the depolarization ratio such that the two spectra may be easily compared and the approximate values of the ratio are graphically evident. It is suggested that the two spectra be superimposed on a single chart if this does not detract from the clarity of the presentation. This presentation is also urged for Approved Analytical data.

C. Tables of Raman data

It is required for Research Quality Analytical data and urged for Approved Analytical data that the information on wavenumber shift, intensity relative to a specified feature of the spectrum or to some other reference signal, and depolarization ratio be presented for the principal bands of the spectrum in a table accompanying the spectral chart. For Approved Analytical Data the numerical values of the depolarization ratio of the principal bands should be presented on the spectral chart if not included in a table. The method of calculation of the depolarization ratio, i.e., peak height or band area ratios, and any dependence on exciting wavelength should be specified.

D. Spectral data catalogs

It is urged that the format for the presentation of spectral data catalogs be in reasonable and practical agreement with the format previously summarized in part A. There are additional specific recommendations applying to such catalogs as follows:

1. It is desirable that the spectrum should show the date of measurement and, if the spectrum is a part of a compilation, the name of the contributing laboratory.

2. The Raman format should agree with the infrared format; i.e., a 2:1 scale compression in the abscissa above 2000 cm^{-1}.

3. A minimum ratio of the maximum of the strongest band in the spectrum to the peak-to-peak noise at the "baseline" should be 100:1 or better.

4. The intensity unit of the ordinate may be selected such that the strongest bands are off scale by a factor not exceeding 1000 if this is necessary to obtain a satisfactory record of the weaker significant bands. The strongest bands should then be rescanned at an appropriately reduced intensity unit with the specific reduction ratio indicated in the chart.

REFERENCES

(1) *Applied Spectroscopy* 30, 20 (1976).

(2) *IUPAC Inf. Bull.*, No. 2 (1978).

(3) *Anal. Chem.* 38, No. 9, 27A (1966).

(4) *Anal. Chem.* 47, No. 11, 945A (1975).

(5) *IUPAC Inf. Bull.*, No. 50 (1976).

(6) T. C. Damen, S. P. S. Porto and B. Tell, *Phys. Rev.* 142, 570 (1966).

PRESENTATION OF MOLECULAR PARAMETER VALUES FOR INFRARED AND RAMAN INTENSITY MEASUREMENTS

(Recommendations 1988)

COMMISSION ON MOLECULAR STRUCTURE AND SPECTROSCOPY*

Prepared for publication by

M. GUSSONI[1], G. ZERBI[1] AND W. PERSON[2]

[1]Dipartmento di Chimica Industriale e Ingegneria Chimica, Politecnico di Milano, Piazza Leonardo da Vinci 32, Milano, Italy
[2]Department of Chemistry, University of Florida, Gainesville, Florida 32611, USA

Abstract - The field of basic and applied studies of infrared and Raman intensities is growing, and more experimental data as well as theoretical works are appearing in the scientific literature. The aim of the document is to point out where lack of common definitions and common use of dynamical and spectroscopic quantities may give rise to confusion among various authors. Recommendations are made as to the correct description of the sign of the various vibrational coordinates used, and of the eigenvectors derived from the dynamical matrix.

Studies of vibrational intensities in infrared and Raman spectroscopy have recently attracted the attention of an increasing number of workers (1). Knowledge of the wavenumber dependence of the absorption coefficient in the infrared or scattering power of molecules in the Raman spectrum is increasingly important for applications which range from ordinary analytical chemistry through solid state physics all the way to astrophysics. At present such studies are conducted in laboratories all over the world, and the interest is both experimental and theoretical.

The theory attempts to correlate and organize the experimental results by interpretation based on quantum mechanical calculations or on models which try to find parameters to characterize experimental intensities in the same way that force constants have parametrized potential fields for the vibrational frequencies.

It is sometimes difficult to compare experimental data or theoretical predictions in the literature because of a lack of proper definition of the quantities used. Some of these quantities are considered here.

Infrared or Raman intensities of the ith fundamental mode of vibration (designated I_i^{IR} or I_i^{Raman}, respectively) can be measured by the peak height of the absorption or scattering band, but are usually and more meaningfully measured experimentally by the integrated absorbance (taken to base e, either $\int \ln(I_0/I)d\nu$ or $\int \ln(I_0/I)d(\ln\nu)$, where the integral is

*Membership of the Commission for varying periods during which the report was prepared (1981–85) was as follows:

Chairman: 1981–83 C. N. R. Rao (India); 1983–85 J. R. Durig (USA); *Secretary*: 1981–83 J. R. Durig (USA); 1983–85 H. A. Willis (UK); *Titular Members*: A. M. Bradshaw (FRG); B. G. Derendjaev (USSR); S. Forsen (Sweden); V. A. Koptyug (USSR); C. Sandorfy (Canada); J. F. J. Todd (UK); D. W. Turner (UK); *Associate Members*: R. D. Brown (Australia); J. G. Grasselli (USA); R. K. Harris (UK); H. Kon (USA); J. P. Maier (Switzerland); Yu N. Molin (USSR); W. B. Person (USA); H. Ratajczak (Poland); C. J. H. Schutte (R. South Africa); M. Tasumi (Japan); *National Representatives*: R. Colin (Belgium); Y. Kawano (Brazil); G. Herzberg (Canada); J. Lu (Chinese Chemical Society); Z. Luz (Israel); B. Jezowska-Trzebiatowska (Poland); S. Ng (Malaysia); H. Thompson† (UK).

†Deceased

Reprinted with permission from *Pure and Applied Chemistry,* Volume 60, Number 9, 1988, pages 1385–1388. Copyright 1988 by the International Union of Pure and Applied Chemistry—All rights reserved.

taken over the ith band (2)) or the corresponding expression of the scattering power I_i/I_0 (where I_i is the integrated radiant intensity for the ith Raman band and I_0 is the incident radiance) for the Raman scattering. Whatever the experimental definition of the infrared intensity, it is proportional to the square of the theoretical matrix element for the transition moment $\langle 1|\mathbf{M}|0\rangle$ obtained from theory by integrating the product of the excited state vibrational wavefunction, ψ_{1i}, times the result of operating on the vibrational ground state wavefunction ψ_{0i} by the vector dipole moment operator, **M**. The integration is made with respect to the vibrational coordinate Q_i, over all space. Similarly the intensity of the Raman scattering is proportional to $|\langle 1|\boldsymbol{\alpha}|0\rangle|^2$, where α is the polarizability operator.

If the vibration is both electrically and mechanically harmonic, (the usual simplifying assumption for interpretation of data),

$$I_i^{IR} \sim \sum_u^3 |\partial M_u/\partial Q_i|^2 \qquad \text{and} \qquad I_i^{Raman} \sim \sum_{u,v}^3 |\partial\alpha_{u,v}/\partial Q_i|^2.$$

Here $\mathbf{M} = M_x\mathbf{i} + M_y\mathbf{j} + M_z\mathbf{k}$ is the molecular dipole moment, α is the molecular polarizability tensor, Q_i is the ith normal coordinate and the summation is over all the components ($u = x, y, z$; $v = x, y, z$). The most important point to be made here is that the experimentally measured intensity is proportional to the square of the theoretically important intensity parameter $(\partial M_u/\partial Q_i)$ or $(\partial\alpha_{u,v}/\partial Q_i)$. Hence the experimental measured intensity does not determine the sign of these parameters. In the theoretical interpretation of intensities, the next step is usually to derive new parameters, such as those relating to bond or group motions, from the values of $\partial M_u/\partial Q_i$ by linear combinations of the latter. As a result, the values of these new parameters depend strongly on the choice of sign for each of the $3N$-6 experimental values of $\partial M_u/\partial Q_i$ (or $\partial\alpha_{u,v}/\partial Q_i$). The determination of these signs is the primary problem in the interpretation of experimental vibrational intensities.

There are several methods for the determination of these signs, both by experimental studies and by approximate quantum mechanical calculations of these derivatives. However, the results from studies by different authors often cannot be compared because there had been no definition by the author either of the reference system of cartesian coordinates in which **M** or $\boldsymbol{\alpha}$ are described or of the phase of the normal coordinate Q_i. For this reason it is hereby recommended that the following conventions be adopted.

1. <u>The coordinate systems used in the interpretation of the data must be clearly and explicitly defined</u>.

The orientation of the cartesian axes system with respect to the molecule should be shown or described explicitly, as should the orientation of any permanent dipole moment.

In addition to the cartesian axes system, there are several other coordinate systems that must be defined. These include (3,4) the column vector of cartesian displacement coordinates **X**, defined so that a displacement of atom i in the positive x direction Δx_i, for example, is positive; the column vector of internal displacement coordinates **R**, the column vector of symmetry displacement coordinates **S** and the column vector of the normal coordinates **Q**. Here the internal displacement coordinates must be clearly defined. The sign convention for the out-of-plane bending and for the torsional coordinates must be specified explicitly, as pointed out in the IUPAC Recommendations on Definition and Symbolism of Molecular Force Constants (5).

The definition of the S_i elements as linear combinations of R_j elements, and the sign of S_i, must be reported explicitly. When two **S** coordinates are degenerate, the definitions of the degenerate coordinates in **S** or in a symmetrized **X** matrix should be iso-oriented with cartesian axes which belong to the same irreducible representation.

When all of these coordinate definitions have been given, the phase of the normal coordinate is then defined by giving the eigenvector matrix **L**:

$$\mathbf{R} = \mathbf{L}\mathbf{Q}$$

$$\mathbf{S} = \mathbf{L}^S\mathbf{Q}$$

or

$$\mathbf{X} = \mathbf{L}^X\mathbf{Q}$$

2. All studies which deal with sign determinations of $\partial M_u/\partial Q_i$ or of $\partial\alpha_{u,v}/\partial Q_i$ should report the **L** matrix completely, in addition to the coordinate definition recommended in 1.

Notice that the ambiguities in the definition of the phase of Q_i arise from the fact that both L_i and $-L_i$ are column eigenvectors that solve the same secular equation with the same eigenvalues. Two solutions of the same problem may be obtained with opposite phases from two computing programs, even when exactly the same input data are used. From the viewpoint of dynamics the phase of Q_i is irrelevant, but this is not the case in intensity calculations.

3. If quantum mechanical calculations of intensity parameters are made by finite difference methods (for example $\partial M_u/\partial R_j \sim [M_u(\Delta R_j) - M_u^e]/\Delta R_j$), then it is very important to describe carefully exactly how the displacement was made.

These displacements (ΔR_j or ΔS_j) in a finite difference calculation should in principle be displacements corresponding to a purely vibrational motion, to give a displacement coordinate that is orthogonal to the pure translations and rotations of the molecule. The vibrational displacements satisfy the Eckart-Saytzev conditions (3,4). In that case the ith element of the column vector of symmetrical cartesian displacements for each atom, $\mathbf{X}_i$, is related to the column vector of symmetry displacement coordinates **S** by:

$$\mathbf{X}_i = (\mathbf{A}^S)_i\,\mathbf{S}.$$

Here $(\mathbf{A}^S)_j$ is the ith row of the $\mathbf{A}^S$ matrix defined by:

$$\mathbf{A}^S = \mathbf{m}^{-1}\,(\mathbf{B}^S)^{\dagger}\,(\mathbf{G}^S)^{-1};$$

$\mathbf{m}^{-1}$ is the diagonal matrix of inverse atomic masses and the $\mathbf{B}^S$ and $\mathbf{G}^S$ matrices are defined in Ref. 3. (The "†" designates the transpose of the $\mathbf{B}^S$ matrix.)

If S_i in a finite difference calculation is not defined in this way (or by some other equivalent method for satisfying the Eckart-Saytzev conditions) then the displacement coordinate mixes pure vibrational displacements with translational and rotational displacements and the so called "rotational corrections" (6) must be introduced in the interpretation of the calculated dipole derivative for infrared intensity calculations or of the calculated polarizability derivatives for Raman intensity calculations in the case of a non-spherical molecular polarizability tensor.

REFERENCES

1. "Vibrational Intensities in Infrared and Raman Spectroscopy", (W. B. Person and G. Zerbi, eds.), Elsevier (1982).
2. B. Crawford, Jr., J. Chem. Phys., 29, 1042 (1958).
3. E. B. Wilson, Jr., J. C. Decius and P. C. Cross, "Molecular Vibrations", McGraw-Hill, N.Y. (1955).
4. M. V. Volkenstein, M. A. Elyashevich and B. I. Stepanov, "Kolebaniya Molekul", Gosteknizdat, Moscow (1949).
5. Definition and Symbolism of Molecular Force Constants (Recommendations 1978), Pure Appl. Chem., 50, 1707 (1978).
6. B. Crawford, Jr., J. Chem. Phys., 20, 977 (1952).

DEFINITION AND SYMBOLISM OF MOLECULAR FORCE CONSTANTS

(RECOMMENDATIONS, 1978)

COMMISSION ON MOLECULAR STRUCTURE AND SPECTROSCOPY†

Definitions and notations for force constants often vary between different authors. This arises from the fact that force constant problems have wide varieties and different aims in their applications. At the same time the whole subject is still constantly changing. Therefore it seems unwise to attempt to fix all minor details at the present stage. For these reasons, some conventions of fundamental importance will be given below as general guidelines, minor details being left to each person, who should fit them to the specific problem to be studied in conformity with the suggested guidelines as far as possible. This document deals principally with harmonic force constants. Anharmonic force constants are mentioned in an Appendix.

Force constants are the coefficients in an expansion of the intramolecular potential function in terms of a definite set of coordinates the values of which define the deformation of the molecule away from its equilibrium configuration. In order to define force constants, one has to indicate both the definition of each coordinate and the expansion of the potential.

1. *The definitions and symbols of force constants used should always be clearly defined in each paper.*

This self-evident comment must be the first recommendation, because confusion always results from the unspecified use of definitions and notations.

Often the same notation is used with different meanings by different authors, and hence the particular notation used in a paper should be clearly defined even when the variation is small, perhaps only in the coefficients or in the numbering.

2. *Among various kinds of force field, the most important is the general force field, GFF, which is expressed in terms of 3N-6 basis coordinates:*

$$V = \frac{1}{2}\sum_{ij} f_{ij}\ (\text{or } F_{ij})\cdot S_i\cdot S_j \qquad (1)$$

We recommend that the force constants in this force field be denoted by f or F (preferably by f). The basis coordinates S (or sometimes s) may be internal symmetry coordinates, local symmetry coordinates or any others most suitable to the problem, but the number of the coordinates has to be reduced to 3N-6 (3N-5 for linear molecules), N being the number of atoms in the molecule.

Since many kinds of basis coordinate systems are possible for the GFF of a molecule, the distinction should be made clear by the use of subscripts or superscripts.

The choice of coordinate in the GFF strongly depends on the symmetry of the molecule, but it is not uniquely given by the symmetry alone. This is easily illustrated by the simple example that the most suitable coordinates are not always the same for CH_3D and CH_3I or for CH_2D_2 and CH_2I_2, although the two molecules belong to the same symmetry group.

3. • *The other extreme and fundamental force field is the internal valence force field, IVFF, which is expressed in terms of bond-stretching, angle-bending, torsional and other displacements directly connected to the structural parameters of the molecule. We recommend that the force constants of this molecular field be denoted by k or K (preferably by k) in the form of*

$$V = \frac{1}{2}\sum_{ij} k_{ij}\ (\text{or } K_{ij})\cdot R_i\cdot R_j\ , \qquad (2)$$

where R denotes the internal valence coordinates. When the internal valence force field itself is the GFF, either k (or K) or f (or F) can be used for the force constant.

The notations R may be replaced by other specified symbols, such as Δr, $\Delta\phi$ or others. Subscripts or superscripts are used for defining minor details of the coordinates and the force constants, in the same manner as for f or F described in paragraph 2.

The most conspicuous difference between the IVFF and the GFF lies in the fact that the former suffers from redundancy such as among the bending coordinates of the bond angles around an atom, or among the stretching and bending coordinates in a ring structure. Therefore not all the force constants of the IVFF can be determined, unless assumptions are made about the force constants related to the redundancy (an example is shown in Appendix 3).

---

†Report prepared by Y. Morino and T. Shimanouchi

For molecules with no angle redundancies (e.g. H_2O, NH_3) the IVFF representation is a possible choice for the GFF representation. In such cases it is open to the author to use either the symbol k or the symbol f for his force constants.

4. Between the two extremes of the IVFF and the GFF described above, there might be a number of intermediate systems which consist of more than 3N-6 coordinates. The force constants in these systems should be expressed by k or K with appropriate subscripts or superscripts, in the sense that it is a modified IVFF which is not yet reduced to a GFF.

The intermediate systems described above can be derived from, or can be correlated with, the IVFF by an orthogonal transformation. The number of coordinates in the system is reduced from that of the IVFF by the number of redundant coordinates considered.

5. We recommend that the force constants for the Urey-Bradley force field, UBFF, be denoted by upper-case letters, K, H, F and Y for bond-stretching, angle-bending, nonbonded repulsive and torsional motions, respectively.

When there is danger of confusion between the force constants in the UBFF and those in other force fields, it is advisable to use lower-case letters, *f* or *k*, for the latter. When *F* or *K* is used, the definition should be clearly stated in the paper.

6. We recommend that the geometry of the molecule used in the calculation of force constants be clearly shown in the paper.

Some of the force constants are independent of the geometry, whereas others are clearly related with the geometry and their numerical values have meanings only when the geometry is clearly specified. Bond lengths, bond angles and other molecular parameters used in the calculation should therefore be mentioned in the paper, when necessary. When the value of the equilibrium distance is not known, an estimated value may be used. In any case, the value used in the calculation should clearly be mentioned.

7. The force constants for angle variation depend upon the choice of the coordinates, for which the following definitions are recommended:

(i) For the coordinate of angle-bending, the deformation of bond angle itself is taken, namely $\Delta\phi$ but not $r\Delta\phi$, though the latter has often been used in the past.

(ii) For the coordinate of out-of-plane bending:

$$\Delta\theta_{i-jkl} = \frac{\Delta z_j}{r_{eij}} \sin\Phi_{kil} \qquad (3)$$

instead of

$$\Delta\theta_{i-jkl} = \frac{\Delta z_j}{r_{eij}} \qquad (4)$$

where the numbering of the atoms is given in Fig. 1. Φ_{kil} denotes the angle between the bonds ik and il, Δz_j the perpendicular distance of the atom j from the instantaneous plane ikl and r_{eij} the equilibrium length of the bond ij.

(iii) For the coordinate of torsion:

$$\Delta t = \sum_N \Delta\tau_{ijkl} / N , \qquad (5)$$

where $\Delta\tau$ denotes the change in the torsional angle between the bonds ji and kl belonging to the two rotating groups (See Fig. 2); τ_{ijkl} is thus the dihedral angle between planes ijk

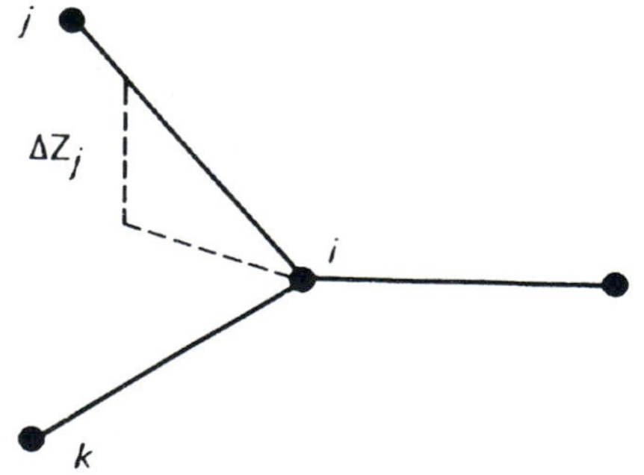

Fig. 1

and jkl. N is the number of torsional angles formed by the possible combinations of the bonds adjacent to the torsional axis (e.g. N is 2 for NH_2-OH and 4 for NH_2-NH_2).

The designation of angle coordinates by $\Delta\phi$ has the disadvantage that the dimensions of the force constants are different from those of bond-stretching force constants, whereas it has the advantages that the redundancy is more easily treated with these coordinates and moreover that an arbitrary choice of the value of r, when the adjacent bonds are inequivalent, can be avoided.*

The definition of the out-of-plane bending coordinates described above has the advantage that the bending displacement is defined uniquely, independently of the choice of the j numbering of the surrounding atoms.

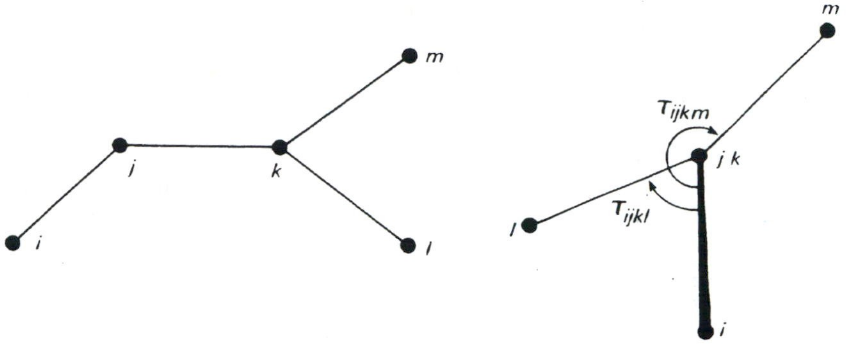

Figure 2

8. *The signs of the coordinates in the molecular force field of any type should be defined clearly, because they are indispensable for the assignment of the signs of the interaction force constants with other coordinates.*

(i) With the atoms arranged as in Figure 2, viewed along the bond jk with the atom j nearer to the observer than atom k, the sign of the torsional coordinate $\Delta\tau_{ijkl}$ is taken to be positive if the torsional angle τ_{ijkl} increases, where τ_{ijkl} is defined as the angle from the projection of bond ij to the projection of bond kl, traced in the clockwise sense. Thus $\Delta\tau_{ijkl} = \Delta\tau_{lkji}$.

(ii) The sign of the out-of-plane bending coordinate is defined by the direction of the vector Δz_j, which is taken from the plane ikl to the atom j, as shown in Fig. 1: thus

$$\Delta\theta_{i-jkl} = -\Delta\theta_{i-kjl} \ .$$

(iii) The sign of the CH_2 rocking, wagging or twisting coordinate of the XCH_2Y-part of a large molecule is taken to be positive if, viewing the XCH_2Y-group from the direction perpendicular to the HCH plane, with the atom X nearer to the observer, as shown in Fig. 3, the HCH group rotates in the clockwise direction (for rocking), the H atoms move away from the X atom to increase the XCH angles (for wagging), or the H atom on the right comes nearer and the H atom on the left moves away (for twisting). In other words, viewing from above, with the X atom nearer to the observer, the HCH group rotates in the clockwise direction for the twisting coordinate. Thus it should be noted that

$$S^{\text{rock}}_{XCH_2Y} = -S^{\text{rock}}_{YCH_2X} \ , \qquad S^{\text{wag}}_{XCH_2Y} = -S^{\text{wag}}_{YCH_2X} \ , \qquad S^{\text{twist}}_{XCH_2Y} = S^{\text{twist}}_{YCH_2X}$$

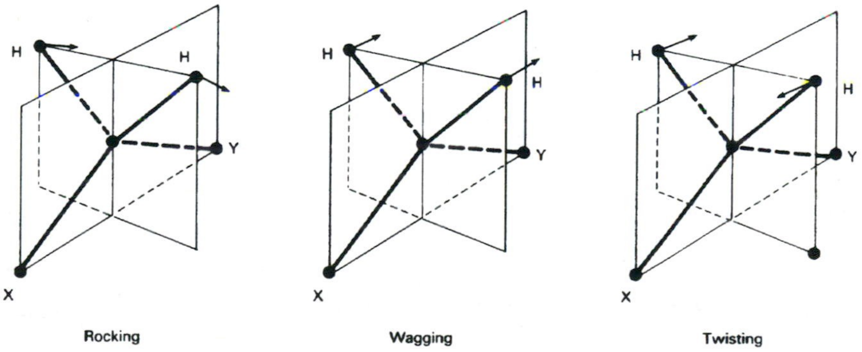

Figure 3

(iv) The following definitions of the symmetry coordinates of the CH_3 degenerate stretching (*ds*), degenerate deformation (*dd*), and degenerate rocking (*dr*) modes are recommended:

---

*In molecules of type XY_n, where the Y atoms are symmetrically equivalent and all are attached to the X atom, so that all the X-Y distances are equal (e.g., CO_2, H_2O, NH_3, CH_4 and SF_6), the force constants can be obtained from the frequencies even when we have no knowledge of the bond length, if we use $r\Delta\phi$ for the bending coordinate instead of $\Delta\phi$. In such cases $r\Delta\phi$ may be used as an exception to recommendation no. 7.

In the CH_3XY molecule shown in Fig. 4, the H atom located at the trans position relative to Y is numbered as 1. The H atoms 2 and 3 are numbered in the clockwise direction, viewing the atoms along the bond XC with the atom X nearer to the observer than C. Two components of the *ds*, *dd*, and *dr* symmetry coordinates are then expressed as:

$$S_{ds} = \Delta(2r_1 - r_2 - r_3)/\sqrt{6}$$
$$S'_{ds} = \Delta(r_2 - r_3)/\sqrt{2}$$
$$S_{dd} = \Delta(2\alpha_{23} - \alpha_{31} - \alpha_{12})/\sqrt{6}$$
$$S'_{dd} = \Delta(\alpha_{31} - \alpha_{12})/\sqrt{2}$$
$$S_{dr} = \Delta(2\beta_1 - \beta_2 - \beta_3)/\sqrt{6}$$
$$S'_{dr} = \Delta(\beta_2 - \beta_3)/\sqrt{2}$$

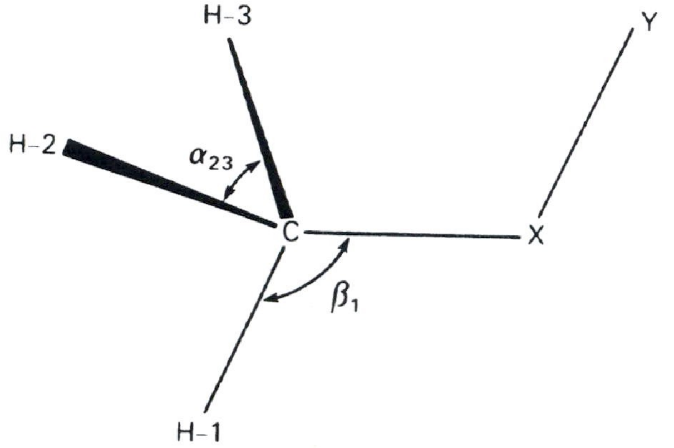

Fig. 4

APPENDIX

1. For the specification of force fields of molecules it is inevitable that the definitions and symbols of anharmonic force constants be considered, in addition to the quadratic force constants described above. The expansion of the anharmonic potential function can be carried out in an unrestricted Taylor series or in a restricted sum in which each cross term appears only once. The problem has been considered by several authors and is left for further exploration. The definitions in the foregoing recommendations for harmonic force constants are closely connected to the unrestricted expansion of the anharmonic potential function in the form of the Taylor series.

2. An attempt to make the dimensions of all the force constants identical by the use of dimensionless coordinates $\Delta r/r$ has been pursued with promising results. Considering, however, that there still are a number of points to be examined, this suggestion should be more thoroughly explored before it is recommended as an IUPAC standard.

3. An illustrative example:

The following example will illustrate the relation between the IVFF and the GFF through redundancy. In the in-plane vibrations of the H_2CO molecule the IVFF for angle deformation is expressed by (See Fig. 5)

$$V = \tfrac{1}{2}k_{11}\,(\Delta\phi_1)^2 + \tfrac{1}{2}k_{22}[(\Delta\phi_2)^2 + (\Delta\phi_3)^2] + k_{12}\,(\Delta\phi_1\Delta\phi_2 + \Delta\phi_1\Delta\phi_3) + k_{23}\Delta\phi_2\Delta\phi_3 \qquad (6)$$

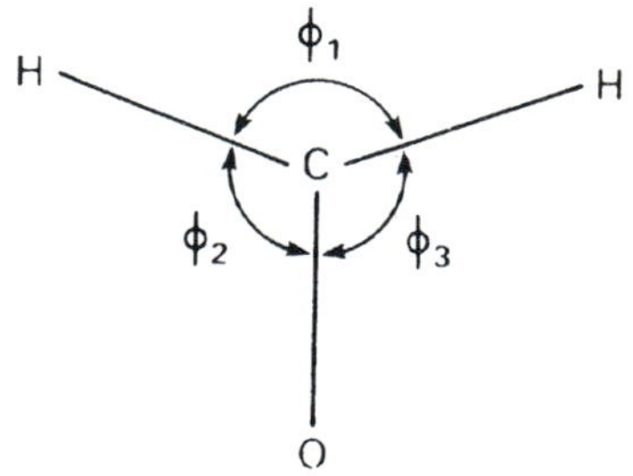

Fig. 5

On the other hand, by the use of symmetry coordinates

$$S_1 = (2\Delta\phi_1 - \Delta\phi_2 - \Delta\phi_3)/\sqrt{6}$$
$$S_2 = (\Delta\phi_2 - \Delta\phi_3)/\sqrt{2} \qquad (7)$$

and a redundant coordinate

$$S_3 = (\Delta\phi_1 + \Delta\phi_2 + \Delta\phi_3)/\sqrt{3} \tag{8}$$

the potential function Eq. (5) is transformed to

$$V = \frac{1}{2}f_{11}S_1{}^2 + \frac{1}{2}f_{22}S_2{}^2 + \frac{1}{2}f_{33}S_3{}^2 + f_{13}S_1S_3 \tag{9}$$

where

$$\begin{aligned} f_{11} &= (2k_{11} + k_{22} - 4k_{12} + k_{23}) \;/\; 3 \\ f_{22} &= k_{22} - k_{23} \\ f_{33} &= (k_{11} + 2k_{22} + 4k_{12} + 2k_{23}) \;/\; 3 \\ f_{13} &= (k_{11} - k_{22} + k_{12} - k_{23}) \left(\frac{\sqrt{2}}{3}\right) . \end{aligned} \tag{10}$$

Since S_3 is zero by the redundancy, the third and the fourth terms of Eq. (9) drop out automatically from the potential function and the remaining part becomes precisely the GFF of the molecule. It must be emphasized that the force constants f_{33} and f_{13} cannot be obtained by experiment. In other words, it is impossible to evaluate the force field expressed by Eq. (6), unless some assumptions are made on f_{33} and f_{13}. A similar situation is encountered whenever redundancy is involved in the coordinates.

Section 11

Automated Products

The rapid increase in the use of computers has led to automated databases, among other things. Most automated databases are intended for users needing data to solve a scientific or engineering problem and users with no interest in the data per se. The structure and format of the database are designed to make it easy to use, and the data have been selected and put into the desired format by the database preparer.

However, a different type of data banks contain raw measured data. Some data banks are publishing instructions to authors for preparing their data in the structure and format of the data bank so the data can be incorporated directly into the data bank without additional manipulation. The existence of such a protocol not only makes it easier for cooperating laboratories to exchange data by automated means, but also to use data as input to other programs. The protocol should also make it easier for journals to automate their production processes eventually, if they have not already done so. To the extent that the protocol requires that information about the character of the sample and measurement techniques and so on are covered, these instructions meet the aim of this book; however, the primary purpose of the protocol presently, is to expedite data exchange.

The crystallographers have a highly developed set of standards called "Standard Crystallographic File Structure 84" (*89*). The stated purposes are those previously mentioned and to serve as a standard for the journals of the International Union of Crystallography when they are ready to accept papers in machine-readable form. The existence of computer programs to obtain reduced cell parameters and to determine crystal symmetry as listed in Section 10 can also be mentioned again.

The materials properties field is also involved in preparing standardized formats for automated databases and networking. The work is being performed by the American Society for Testing and Materials (ASTM) Committee E–49 on the Computerization of Material Property Data, ASTM Committee G–1 on Corrosion of Metals, and by the National Association of Corrosion Engineers (NACE) Committee T–3 on Corrosion Science and Technology with cooperation from several other organizations. The *Proceedings of the 1st International Symposium on Computerization and Networking of Materials Property Databases* (*90*) contains a convenient survey of this work.

Two additional examples from the field of thermodynamics can be mentioned, although they are not yet as highly developed as the previous two. The International Council of Scientific Unions, Committee on Data for Science and Technology (CODATA) Task Group on Chemical Thermodynamic Data is in the process of developing reaction catalogs, a format for experimental data, and related bibliographic information that can be used to provide automated exchange of such data and to provide input to computer programs to determine optimum values of enthalpies or Gibbs energies for networks of related thermochemical measurements (*91*).

Finally, the *International Data Series: Selected Data on Mixtures, Series A* (*92*) is a journal for the publication of original thermodynamic data and data evaluations for nonreacting binary organic mixtures. It utilizes a highly stylized specific format. The editors have developed an International Data Series (IDS) data acquisitions diskette (INDES). Authors proposing to publish in the journal are faced with a series of queries that must be answered to proceed through the diskette. By following the instructions, the authors will have provided all of the required information about their measurements, and their data will have been converted into a standard file that can be processed further into a paper for the journal and into an automated database.

Appendix I: Abbreviated List of Quantities, Units and Symbols in Physical Chemistry

COMMISSION ON PHYSICOCHEMICAL SYMBOLS, TERMINOLOGY AND UNITS

Prepared for publication by K. H. Homann

This list is intended as a ready reference to the symbols most frequently used by authors, teachers and students in chemistry and related disciplines. It is based on the more comprehensive IUPAC document "Quantities, Units and Symbols in Physical Chemistry" referenced below.

1. Base SI Units and Physical Quantities

A *physical quantity* is the product of a *numerical value* (a pure number) and a *unit*.

Physical quantities are organized in a dimensional system built upon seven base quantities. The International System of Units (SI) is based on the seven base units having the same dimensions as the associated physical quantities. Their names and symbols are as follows:

Base Physical Quantity	*Symbol for Quantity*	*Name of SI Unit*	*Symbol for SI Unit*
length	l	metre	m
mass	m	kilogram	kg
time	t	second	s
electric current	I	ampere	A
thermodynamic temperature	T	kelvin	K
amount of substance	n	mole	mol
luminous intensity	I_v	candela	cd

The symbol for a physical quantity is a single letter of the Latin or Greek alphabet printed in italic (sloping) type. It may be modified by subscripts and/or superscripts of specified meaning, or further characterized in particular cases through annotations in parentheses put directly behind the symbol. The symbol for a unit is printed in roman (upright) type. Neither symbol should be followed by a full stop (period).

The physical quantity "amount of substance" is proportional to the number of elementary entities – specified by a chemical formula – of which the substance is composed. The proportionality factor is the reciprocal of the Avogadro constant L (6.022×10^{23} mol^{-1}). The "amount of substance" should no longer be called "number of moles".

Examples for relations between "amount of substance" and other physical quantities:

2 moles of N_2 contain 12.044×10^{23} molecules of N_2, amount of N_2 = number of N_2 molecules/L;

1.5 moles of Hg_2Cl_2 have a mass of 708.13 g;

1 mole of photons with frequency of 10^{14} Hz has an energy of 39.90 kJ;

1 mole of electrons, e^-, contains 6.022×10^{23} electrons, has a mass of 5.468×10^{-7} kg, and a charge of -96.49 kC.

(The numerical values are approximate.)

2. SI Prefixes

Prefixes to form the names and symbols of the decimal multiples and submultiples of SI units*

Multiple	*Prefix*	*Symbol*
10	deca	da
10^2	hecto	h
10^3	kilo	k
10^6	mega	M
10^9	giga	G
10^{12}	tera	T
10^{15}	peta	P
10^{18}	exa	E

Submultiple	*Prefix*	*Symbol*
10^{-1}	deci	d
10^{-2}	centi	c
10^{-3}	milli	m
10^{-6}	micro	µ
10^{-9}	nano	n
10^{-12}	pico	p
10^{-15}	femto	f
10^{-18}	atto	a

3. Examples of SI Derived Units with Special Names and Symbols

Physical Quantity (for symbols see Sect. 4)	*Name of SI Unit*	*Symbol for SI Unit*	*Expression in Terms of SI Base Units*
frequency	hertz	Hz	s^{-1}
force	newton	N	m kg s^{-2}
pressure, stress	pascal	Pa	m^{-1} kg s^{-2} (= N m^{-2})
energy, work, heat	joule	J	m^2 kg s^{-2} (= N m = Pa m^3)
power	watt	W	m^2 kg s^{-3} (= J s^{-1})
electric charge	coulomb	C	s A
electric potential	volt	V	m^2 kg s^{-3} A^{-1} (= J C^{-1})
electric capacitance	farad	F	m^{-2} kg^{-1} s^4 A^2 (= C V^{-1})
electric resistance	ohm	Ω	m^2 kg s^{-3} A^{-2} (= V A^{-1})
electric conductance	siemens	S	m^{-2} kg^{-1} s^3 A^2 (= Ω^{-1})
magnetic flux	weber	Wb	m^2 kg s^{-2} A^{-1} (= V s)
magnetic flux density	tesla	T	kg s^{-2} A^{-1} (= V s m^{-2})
inductance	henry	H	m^2 kg s^{-2} A^{-2} (= V A^{-1} s)
Celsius temperature**	degree Celsius	°C	K
plane angle	radian	rad	1 (rad and sr may be included or omitted in expressions for the derived units)
solid angle	steradian	sr	1

* Decimal multiples and submultiples of the unit of mass are formed by attaching prefixes to gram, examples: mg, not µkg; Mg, not kkg.

** The Celsius temperature is defined by $\theta/°C = T/K - 273.15$

4. Recommended Symbols for Commonly Used Physical Quantities

Several physical quantities have more than one entry in the symbol column for different reasons: (1) Both symbols are in use (e.g. p, P for pressure and q, Q for heat); (2) Different symbols are used for the same physical quantity in different physical systems (for example, electron spin quantum number s for a single electron and S for the sum over several electrons); (3) Alternative symbols are recommended to avoid conflict in the notation for quantities which otherwise would have the same symbols (e.g. E_a, to distinguish the energy of activation from another energy E in the same context). The unit 1 in the SI unit symbol column signifies a dimensionless quantity.

An extensive physical quantity, that is a quantity proportional to the mass or amount of substance of a system, is often symbolized by a capital letter. The corresponding *specific quantity* (quantity divided by mass) may be represented by the corresponding lower-case letter (example: volume V, specific volume $v = V/m$). A subscript m to the symbol of an extensive quantity denotes a *molar quantity* (quantity divided by amount of substance), example: molar volume $V_m = V/n$. It is permissible to omit the subscript m for *molar* when the text makes it obvious that a molar quantity is intended.

Vectors are printed in bold-faced italic type. They can alternatively be indicated by an arrow above the symbol.

4.1 Space and Time

Physical Quantity	*Symbol*	*SI Unit*
Cartesian space coordinates	x, y, z	m
position vector	$\boldsymbol{r}$	m
length	l	m
height	h	m
thickness, distance	d, δ	m
radius	r	m
diameter	d	m
path, length of arc	s	m
area	A, A_s, S	m^2
volume	V	m^3
plane angle	$\alpha, \beta, \gamma, \theta, \phi$	1, rad
solid angle	ω, Ω	1, sr
time	t	s
frequency	ν, f	Hz
circular frequency ($=2\pi\nu$)	ω	s^{-1}, rad s^{-1}
characteristic time interval, relaxation time, time constant	τ, T	s
velocity	$\boldsymbol{v}, \boldsymbol{u}, \boldsymbol{w}, \boldsymbol{c}$	m s^{-1}
acceleration	$\boldsymbol{a}$	m s^{-2}

4.2 Mechanics

mass	m	kg
reduced mass	μ	kg
(mass) density	ρ	kg m^{-3}
relative density	d	1
specific volume	v	m^3 kg^{-1}
moment of inertia	I	kg m^2
momentum	$\boldsymbol{p}$	kg m s^{-1}
angular momentum	$\boldsymbol{L}$	kg m^2 s^{-1} rad (=J s)
force	$\boldsymbol{F}$	N (=kg m s^{-2})
weight	G, W	N
pressure	p, P	Pa
surface tension	γ, σ	N m^{-1}, J m^{-2}
energy	E	J
potential energy	E_p, V, Φ	J
kinetic energy	E_k, T, K	J
work	w, W	J
Hamilton function	H	J
power	P	W

4.3 General Chemistry

Physical Quantity	*Symbol*	*SI Unit*
number of entities	N	1
amount of substance	n	mol
molar mass	M	kg mol^{-1}
relative molar mass, (molecular weight)	M_r	1
relative atomic mass, (atomic weight)	A_r	1
molar volume	V_m	m^3 mol^{-1}
mass fraction	w	1
volume fraction	ϕ	1
mole fraction	x, y	1
partial pressure of substance B	p_B	Pa
number concentration	C, n	m^{-3}
amount (of substance) concentration of B, or concentration of B	c_B, [B]	mol m^{-3}
molality	m	mol kg^{-1}
mass concentration	ρ, γ	kg m^{-3}
surface concentration	Γ	mol m^{-2}
stoichiometric coefficient	ν	1
extent of reaction	ξ	mol

4.4 Chemical Kinetics

rate of conversion	$\dot{\xi}$	mol s^{-1}
rate of concentration change of substance B (through chemical reaction)	v_B, r_B	mol m^{-3} s^{-1}
rate of reaction	v	mol m^{-3} s^{-1}
overall order of reaction	n	1
rate constant, rate coefficient	k	$(m^3\ mol^{-1})^{n-1}$ s^{-1}
half life	$t_{\frac{1}{2}}$	s
pre-exponential factor	A	$(m^3\ mol^{-1})^{n-1}$ s^{-1}
energy of activation	E, E_a	J mol^{-1}
Gibbs energy of activation	$\Delta^{\ddagger}G$	J mol^{-1}
reaction cross section	σ_r	m^2
collision frequency (of a particle A)	Z_A	s^{-1}
collision frequency factor	z_{AB}	m^3 mol^{-1} s^{-1}
quantum yield, photochemical yield	ϕ	1

4.5 Atoms, Molecules and Spectroscopy

Physical Quantity	*Symbol*	*SI Unit*
nucleon number, mass number	A	1
proton number, atomic number	Z	1
neutron number	N	1
decay constant	λ	s^{-1}
ionization energy	E_i, I	J
dissociation energy	E_d, D	J
electron affinity	E_{ea}	J
electron work function	Φ	J
quantum numbers:		
principal	n	1
electron orbital	l, L	1
—component	m_l, M_L	1
electron spin	s, S	1
—component	m_s, M_S	1
total angular momentum	j, J	1
—component	m_j, M_J	1
nuclear spin	I	1
—component	M_I	1
vibrational	v	1
magnetic moment of a particle	$\boldsymbol{\mu}, \boldsymbol{m}$	$J\,T^{-1}$, $A\,m^2$
magnetogyric ratio	γ	$C\,kg^{-1}$
g-factor	g	1
Larmor circular frequency	ω_L	s^{-1}
quadrupole moment	Q, Θ, eQ	$C\,m^2$
wavelength	λ	m
wavenumber in vacuum	$\tilde{\nu}$	m^{-1}
total term	T	m^{-1}
electronic term	T_e	m^{-1}
vibrational term	G	m^{-1}
rotational term	F	m^{-1}
rotational constants	A, B, C	m^{-1}

4.6 Electricity and Magnetism

Physical Quantity	*Symbol*	*SI Unit*
electric charge	Q	C
charge density	ρ	$C\,m^{-3}$
electric current	I	A
electric current density	j	$A\,m^{-2}$
electric potential	ϕ, V	V
electric potential difference, voltage	$U, \Delta\phi, \Delta V$	V
electric field strength	$\boldsymbol{E}$	$V\,m^{-1}$
electric displacement	$\boldsymbol{D}$	$C\,m^{-2}$
capacitance	C	F
permittivity	ε	$F\,m^{-1}$
relative permittivity	ε_r	1
dielectric polarization	$\boldsymbol{P}$	$C\,m^{-2}$
electric susceptibility	χ_e	1
polarization (of a particle)	α	$m^2\,C\,V^{-1}$
electric dipole moment	$\boldsymbol{p}, \boldsymbol{p}_e$	C m
magnetic flux	Φ	Wb
magnetic flux density	$\boldsymbol{B}$	T
magnetic field strength	$\boldsymbol{H}$	$A\,m^{-1}$
permeability	μ	$H\,m^{-1}$, $N\,A^{-2}$
relative permeability	μ_r	1
magnetization	$\boldsymbol{M}$	$A\,m^{-1}$
magnetic susceptibility	χ	1
molar magnetic susceptibility	χ_m	$m^3\,mol^{-1}$
resistance	R	Ω
conductance	G	S
resistivity	ρ	Ω m
conductivity	κ	$S\,m^{-1}$
self-inductance	L	H

4.7 Thermodynamics and Statistics

Physical Quantity	*Symbol*	*SI Unit*
heat	q, Q	J
work	w, W	J
thermodynamic temperature	T	K
Celsius temperature	θ, t	°C
internal energy	U	J
enthalpy	H	J
standard reaction enthalpy	$\Delta_r H^{\ominus}$	$J\,mol^{-1}$
entropy	S	$J\,K^{-1}$
Gibbs energy	G	J
Helmholtz energy	A	J
heat capacity	C	$J\,K^{-1}$
ratio C_p/C_v	γ, κ	1
Joule-Thomson coefficient	μ, μ_{JT}	$K\,Pa^{-1}$
isothermal compressibility	κ	Pa^{-1}
cubic expansion coefficient	α	K^{-1}
chemical potential	μ	$J\,mol^{-1}$
affinity of a reaction	$A, \mathcal{A}$	$J\,mol^{-1}$
fugacity	$f, \tilde{p}$	Pa
(relative) activity	a	1
activity coefficient:		
—mixtures	f	1
—solutes	γ	1
osmotic coefficient	ϕ	1
osmotic pressure	Π	Pa
equilibrium constant	$K^{\ominus}, K$	1
—on a concentration basis	K_c	$(mol\,m^{-3})^{\Sigma\nu}$
—on a pressure basis	K_p	$Pa^{\Sigma\nu}$
—on a molality basis	K_m	$(mol\,kg^{-1})^{\Sigma\nu}$
density of (energy) states	$\rho(E)$	J^{-1}
statistical weight	g	1
partition function:		
—particle	q, z	1
—canonical ensemble	Q, Z	1
—microcanonical ensemble	Ω	1
symmetry number	σ, s	1
characteristic temperature	Θ	K

4.8 Radiation

Physical Quantity	*Symbol*	*SI Unit*
radiant energy	Q, W, Q_e	J
radiant intensity	I, I_e	$W\,sr^{-1}$, W
emissivity, emittance	ε	1
absorptance	α	1
reflectance	ρ, R	1
transmittance	τ	1
absorption coefficient:		
—linear (decadic)	a	m^{-1}
—molar (decadic)	ε	$m^2\,mol^{-1}$
refractive index	n	1
molar refraction	R_m	$m^3\,mol^{-1}$
angle of optical rotation	α	1, rad

4.9 Electrochemistry

Physical Quantity	*Symbol*	*SI Unit*
charge number of an ion	z	1
ionic strength	I_m (I_c)	mol kg^{-1}, (mol m^{-3})
electromotive force, electrode potential	E	V
electrochemical potential	$\tilde{\mu}$	J mol^{-1}
overpotential	η	V
pH	pH	1
charge number of a cell reaction	n, z	1
electrokinetic potential	ζ	V
molar conductivity (of an electrolyte)	Λ	S m^2 mol^{-1}
molar conductivity (of an ion), ionic conductivity	λ	S m^2 mol^{-1}
electric mobility	u	m^2 V^{-1} s^{-1}
transport number	t	1

4.10 Transport Properties

Physical Quantity	*Symbol*	*SI Unit*
flux of a quantity X	J_X, J	(varies)
mass flow rate	$q_m, \dot{m}$	kg s^{-1}
volume flow rate	$q_V, \dot{V}$	m^3 s^{-1}
heat flow rate	Φ	W
thermal conductivity	κ, k, λ	W m^{-1} K^{-1}
coefficient of heat transfer	h	W m^{-2} K^{-1}
thermal diffusivity	a	m^2 s^{-1}
diffusion coefficient	D	m^2 s^{-1}
thermal diffusion coefficient	D_T	m^2 s^{-1}
viscosity	η, μ	Pa s
kinematic viscosity	ν	m^2 s^{-1}

5. Units Outside the SI

5.1 Units Used with the SI

Physical Quantity	*Unit*	*Symbol for Unit*	*Value in SI Units*	
time	minute	min	60	s
time	hour	h	3600	s
time	day	d	86 400	s
plane angle	degree	°	(π/180)	rad
volume	litre	l, L	10^{-3}	m^3
mass	tonne	t	10^3	kg
length	ångström	Å	10^{-10}	m
pressure	bar	bar	10^5	Pa
energy	electronvolt*	eV	1.602 18	× 10^{-19} J
mass	unified atomic mass unit*	u	1.660 54	× 10^{-27} kg

(* defined in terms of best values of certain physical constants)

5.2 Other Units

These units were used in older literature. They are given here for the purpose of identification and conversion to SI units.

force	dyne	dyn	10^{-5}	N
pressure	standard atmosphere	atm	101325	Pa
	torr(mmHg)	Torr	133.322	Pa
energy	erg	erg	10^{-7}	J
	thermo-chemical calorie	cal$_{th}$	4.184	J
magnetic flux density	gauss	G	10^{-4}	T
electric dipole moment	debye	D	3.335 64	× 10^{-30} C m
viscosity	poise	P	10^{-1}	N s m^{-2}
kinematic viscosity	stokes	St	10^{-4}	m^2 s^{-1}

6. Values of Some Fundamental Constants

permeability of vacuum*	μ_0	4π × 10^{-7}	N A^{-2}
speed of light in vacuum*	c_0	299 792 458	m s^{-1}
permittivity of vacuum*	ε_0	8.854 187 816 × 10^{-12}	F m^{-1}
elementary charge	e	1.602 177 33 (49) × 10^{-19}	C
Planck constant	h	6.626 0755 (40) × 10^{-34}	J s
Avogadro constant	L, N_A	6.022 1367 (36) × 10^{23}	mol^{-1}
rest mass of electron	m_e	9.109 3897 (54) × 10^{-31}	kg
rest mass of proton	m_p	1.672 6231 (10) × 10^{-27}	kg
Faraday constant	F	9.648 5309 (29) × 10^4	C mol^{-1}
Hartree energy	E_h	4.359 7482 (26) × 10^{-18}	J
Bohr radius	a_0	5.291 772 49 (24) × 10^{-11}	m
Bohr magneton	μ_B	9.274 0154 (31) × 10^{-24}	J T^{-1}
nuclear magneton	μ_N	5.050 7866 (17) × 10^{-27}	J T^{-1}
Rydberg constant	R_∞	10 973 731.534 (13)	m^{-1}
gas constant	R	8.314 510 (70)	J K^{-1} mol^{-1}
Boltzmann constant	k, k_B	1.380 658 (12) × 10^{-23}	J K^{-1}
gravitational constant	G	6.672 59 (85) × 10^{-11}	m^3 kg^{-1} s^{-2}
standard acceleration due to gravity*	g_n	9.806 65	m s^{-2}
triple point of water*	$T_{tp}(H_2O)$	273.16	K
zero of Celsius scale*	T(0°C)	273.15	K
molar volume of ideal gas (at 1 bar and 273.15 K)	V_0	22.711 08 (19)	L mol^{-1}

(* These values are exact)

7. References

Quantities, Units and Symbols in Physical Chemistry, IUPAC Commission on Physicochemical Symbols, Terminology and Units, to be published in 1987 by Blackwell Scientific Publications. This is a revised version of the 1979 edition published by Pergamon Press, and in Pure and Applied Chemistry, Vol. 51, p. 1–41 (1979).

ISO Standards Handbook 2, Units of Measurements, 1982, ISO Central Secretariat, Case Postale 56, Genève 20, Switzerland.

Symbols, Units and Nomenclature in Physics, Documents UIP 20, Physica, Vol. 93A, 1–60 (1978).

Fundamental Physical Constants: Report of the CODATA Task Group on Fundamental Constants, CODATA Bulletin No. 63 (1986).

References from the Green Book

8.1 PRIMARY SOURCES

1 IUPAC-Physical Chemistry Division
Manual of Symbols and Terminology for Physicochemical Quantities and Units
(a) 1st ed. *Pure Appl. Chem.* **21** (1970) 1–38.
(b) 2nd ed. Butterworths, London 1975.
(c) 3rd ed. *Pure Appl. Chem.* **51** (1979) 1–36.
(d) Appendix I—Definitions of Activities and Related Quantities, *Pure Appl. Chem.* **51** (1979) 37–41.
(e) Appendix II—Definitions, Terminology and Symbols in Colloid and Surface Chemistry, Part I, *Pure Appl. Chem.* **31** (1972) 577–638.
(f) Section 1.13: Selected Definitions, Terminology and Symbols for Rheological Properties, *Pure Appl. Chem.* **51** (1979) 1213–1218.
(g) Section 1.14: Light Scattering, *Pure Appl. Chem.* **55** (1983) 931–941.
(h) Part II: Heterogeneous Catalysis, *Pure Appl. Chem.* **46** (1976) 71–90.
(i) Appendix III—Electrochemical Nomenclature, *Pure Appl. Chem.* **37** (1974) 499–516.
(j) Appendix IV—Notation for States and Processes, Significance of the Word 'Standard' in Chemical Thermodynamics, and Remarks on Commonly Tabulated Forms of Thermodynamic Functions, *Pure Appl. Chem.* **54** (1982) 1239–1250.
(k) Appendix V—Symbolism and Terminology in Chemical Kinetics, *Pure Appl. Chem.* **53** (1981) 753–771.

2 Bureau International des Poids et Mesures, Le Système International d'Unités (SI), 5th French and English Edition, BIPM, Sèvres 1985.

3 IUPAP-SUN, Symbols, Units and Nomenclature in Physics, Document U.I.P. 20, *Physica* **93A** (1978) 1–60.

4 International Standards ISO, International Organization for Standardization, Geneva.
(a) ISO 31/0–1981, General Principles Concerning Quantities, Units and Symbols.
(b) ISO 31/1–1978, Quantities and Units of Space and Time.
(c) ISO 31/2–1978, Quantities and Units of Periodic and Related Phenomena.
(d) ISO 31/3–1978, Quantities and Units of Mechanics.
(e) ISO 31/4–1978, Quantities and Units of Heat.
(f) ISO 31/5–1979, Quantities and Units of Electricity and Magnetism.
(g) ISO 31/6–1980, Quantities and Units of Light and Related Electromagnetic Radiations.
(h) ISO 31/7–1978, Quantities and Units of Acoustics.
(i) ISO 31/8–1980, Quantities and Units of Physical Chemistry and Molecular Physics.
(j) ISO 31/9–1980, Quantities and Units of Atomic and Nuclear Physics.
(k) ISO 31/10–1980, Quantities and Units of Nuclear Reactions and Ionizing Radiations.
(l) ISO 31/11–1978, Mathematical Signs and Symbols for Use in Physical Sciences and Technology.
(m) ISO 31/12–1982, Dimensionless Parameters.
(n) ISO 31/13–1981, Quantities and Units of Solid State Physics.

5 ISO 1000–1981, SI Units and Recommendations for the Use of Their Multiples and of Certain Other Units.

All the standards listed here (4-5) are jointly reproduced in the ISO Standards Handbook 2, *Units of Measurement*, ISO, Geneva 1982.

6 ISO 2955–1983, Information Processing—Representations of SI and Other Units for Use in Systems with Limited Character Sets.

8.2 IUPAC REFERENCES

7 Use of Abbreviations in the Chemical Literature, *Pure Appl. Chem.* **52** (1980) 2229–2232.

8 Expression of Results in Quantum Chemistry, *Pure Appl. Chem.* **50** (1978) 75–79.

9 Recommendations for Presentation of Infrared Absorption Spectra in Data Collections: A—Condensed Phases, *Pure Appl. Chem.* **50** (1978) 231–236.

10 Presentation of Raman Spectra in Data Collections, *Pure Appl. Chem.* **53** (1981) 1879–1885.

11 Recommendations for the Presentation on NMR Data for Publication in Chemical Journals, *Pure Appl. Chem.* **29** (1972) 625–628.

12 Presentation of NMR Data for Publication in Chemical Journals: B—Conventions Relating to Spectra from Nuclei other than Protons, *Pure Appl. Chem.* **45** (1976) 217–219.

13 Nomenclature and Spectral Presentation in Electron Spectroscopy Resulting from Excitation by Photons, *Pure Appl. Chem.* **45** (1976) 221–224.

14 Nomenclature and Conventions for Reporting Mössbauer Spectroscopic Data, *Pure Appl. Chem.* **45** (1976) 211–216.

15 Recommendations for Symbolism and Nomenclature for Mass Spectroscopy. *Pure Appl. Chem.* **50** (1978) 65–73.

16 Definition and Symbolism of Molecular Force Constants, *Pure Appl. Chem.* **50** (1978) 1707–1713.

17 Names, Symbols, Definitions and Units of Quantities in Optical Spectroscopy. *Pure Appl. Chem.* **57** (1985) 105–120.

18 Nomenclature, Symbols, Units and their Usage in Spectrochemical Analysis. I: General Atomic Emission Spectroscopy, *Pure Appl. Chem.* **30** (1972) 651–679.

19 Nomenclature, Symbols, Units and their Usage in Spectrochemical Analysis. VI: Molecular Luminescence Spectroscopy, *Pure Appl. Chem.* **56** (1984) 231–245.

20 Recommended Standards for Reporting Photochemical Data, *Pure Appl. Chem.* **56** (1984) 939–944.

21 Nomenclature of Inorganic Chemistry, Butterworths, London 1971.

22 Nomenclature of Organic Chemistry, Pergamon, Oxford 1979.

23 An Annotated Bibliography on Accuracy in Measurement, *Pure Appl. Chem.* **55** (1983) 907–930.

24 Assignment and Presentation of Uncertainties of the Numerical Results of Thermodynamic Measurements, *Pure Appl. Chem.* **53** (1981) 1805–1825.

25 Glossary of Terms Used in Physical Organic Chemistry, *Pure Appl. Chem.* **55** (1983) 1281—1371.

26 The Absolute Electrode Potential: An Explanatory Note, *Pure Appl. Chem.* **58** (1986) 955–966.

27 Electrode Reaction Orders, Transfer Coefficients and Rate Constants: Amplification of Definitions and Recommendations for Publication of Parameters, *Pure Appl. Chem.* **52** (1980) 233–240.

28 Nomenclature for Transport Phenomena in Electrolytic Systems, *Pure Appl. Chem.* **53** (1981) 1827–1840.

29 Bard, A.J., Parsons, R. and Jordan, J. (eds) *Standard Potentials in Aqueous Solutions*, Marcel Dekker Inc., New York 1985.

30 Definition of pH Scales, Standard Reference Values, Measurement of pH and Related Terminology, *Pure Appl. Chem.* **57** (1985) 531–542.

30a Reporting Physisorption Data for Gas/Solid Systems, *Pure Appl. Chem.* **57** (1985) 603–619.
Reporting Experimental Pressure-Area Data with Film Balances, *Pure Appl. Chem.* **57** (1985) 621–632.
Reporting Data on Adsorption from Solution at the Solid/Solution Interface, *Pure Appl. Chem.* **58** (1986) 967–984.

31 Atomic Weights of the Elements 1979, *Pure Appl. Chem.* **52** (1980) 2349–2384.

32 Element by Element Review of Their Atomic Weights, *Pure Appl. Chem.* **56** (1984) 695–768.

33 Atomic Weights of the Elements 1985, *Pure Appl. Chem.* **58** (1986) 1677–1692.

34 Isotopic Compositions of the Elements 1983, *Pure Appl. Chem.* **56** (1984) 675–694.

35 *Nomenclature of Inorganic Chemistry*. Chapters 1–3: Elements, Atoms, and Groups of Atoms (in preparation).

8.3 ADDITIONAL REFERENCES

36 Jenkins, F.A. Notation for the Spectra of Diatomic Molecules, *J. Opt. Soc. Amer.* **43** (1953) 425–426.

37 Mulliken, R.S. Report on Notation for the Spectra of Polyatomic Molecules, *J. Chem. Phys.* **23** (1955) 1997–2011. (Erratum *J. Chem. Phys.* **24** (1956) 1118.)

38 Herzberg, G. *Molecular Spectra and Molecular Structure.*
I. Spectra of Diatomic Molecules, Van Nostrand, Princeton 1950.
II. Infrared and Raman Spectra of Polyatomic Molecules, Van Nostrand, Princeton 1946.
III. Electronic Spectra and Electronic Structure of Polyatomic Molecules, Van Nostrand, Princeton 1966.

39 Watson, J.K.G. Aspects of Quartic and Sextic Centrifugal Effects on Rotational Energy Levels. In: Durig, J.R. (ed), *Vibrational Spectra and Structure*, Vol. 6, Elsevier, Amsterdam 1977, pp.1–89.

40 Callomon, J.H., Hirota, E., Kuchitsu, K., Lafferty, W.J., Maki, A.G. and Pote, C.S. Structure Data of Free Polyatomic Molecules. In: Hellwege, K.-H. and Hellwege A.M., (eds), *Landolt-Börnstein*, New Series, II/7, Springer-Verlag, Berlin 1976.

41 Bunker, P.R. *Molecular Symmetry and Spectroscopy*, Academic Press, New York 1979.

42 Brown, J.M., Hougen, J.T., Huber, K.-P., Johns, J.W.C., Kopp, I., Lefebvre-Brion, H., Merer, A.J., Ramsay, D.A., Rostas, J. and Zare, R.N., The Labeling of Parity Doublet Levels in Linear Molecules, *J. Mol. Spectrosc.* **55** (1975) 500–503.

43 Brand, J.C.D., Callomon, J.H., Innes, K.K., Jortner, J., Leach, S., Levy, D.H., Merer, A.J., Mills, I.M., Moore, C.B., Parmenter, C.S., Ramsay, D.A., Narahari Rao, K., Schlag, E.W., Watson, J.K.G. and Zare, R.N. The Vibrational Numbering of Bands in the Spectra of Polyatomic Molecules, *J. Mol. Spectrosc.* **99** (1983) 482–483.

44 Hahn, Th. (ed) *International Tables for Crystallography*. Vol. A, 2nd ed.: *Space-Group Symmetry*, Reidel Publishing Co., Dordrecht 1987.

45 Domalski, E.S. Selected Values of Heats of Combustion and Heats of Formation of Organic Compounds, *J. Phys. Chem. Ref. Data* **1** (1972) 221–277.

46 Freeman, R.D. Conversion of Standard (1 atm) Thermodynamic Data to the New Standard State Pressure, 1 bar (10^5 Pa):
Bull. Chem. Thermodyn. **25** (1982) 523–530;
J. Chem. Eng. Data **29** (1984) 105–111;
J. Chem. Educ. **62** (1985) 681–686.

47 Wagman, D.D., Evans, W.H., Parker, V.B., Schumm, R.H., Halow, I., Bailey, S.M., Churney, K.L. and Nuttall, R.L. The NBS Tables of Chemical Thermodynamic Properties, *J. Phys. Chem. Ref. Data* **11** Suppl. 2 (1982) 1–392.

48 Chase, M.W., Davies, C.A., Downey, J.R., Frurip, D.J., McDonald, R.A. and Syverud, A.N. JANAF Thermochemical Tables, 3rd ed. *J. Phys. Chem. Ref. Data* **14** Suppl. 1 (1985).

49 Glushko, V.P. (ed) *Termodinamicheskie svoistva individualnykh veshchestv*. Vol. 1–4, Nauka, Moscow 1978–85.

50 CODATA Task Group on Data for Chemical Kinetics, The Presentation of Chemical Kinetics Data in the Primary Literature, *CODATA Bull.* **13** (1974) 1–7.

51 Cohen, E.R. and Taylor, B.N. The 1986 Adjustment of the Fundamental Physical Constants, *CODATA Bull.* **63** (1986) 1–49.

52 Particle Data Group, Review of Particle Properties, *Phys. Lett.* **170B** (1986) 1–350.

53 Wapstra, A.H. and Audi, G. The 1983 Atomic Mass Evaluation. I. Atomic Mass Table, *Nucl. Phys.* **A432** (1985) 1–54.

54 Lederer, M. and Shirley, V.S. *Table of Isotopes*, 7th ed., Wiley Interscience, New York 1978.

References to This Book

1. Ho, C. Y.; Powell, R. W.; Liley, P. E. *Thermal Conductivity of the Elements: A Comprehensive Review;* American Chemical Society; American Institute of Physics; National Institute of Standards and Technology: Washington, DC, 1974; Supplement Volume 3, Number 1 to *J. Phys. Chem. Ref. Data,* 786 pages.

2. Swietoslawski, W.; Keffer, L. *First Report from the Standing Commission for Thermochemistry;* International Union of Pure and Applied Chemistry General Secretariat: Paris, 1934.

3. "Resolution on Publication of Calorimetric and Thermodynamic Data"; International Union of Pure and Applied Chemistry, Physical Chemistry Section, Commission on Chemical Thermodynamics; *Pure Appl. Chem.* **1961,** *2*(1-2), 339–342.

   Also published in (a) *Phys. Today* **1961,** *14*(2), 47–50. (b) *Science* **1960,** *132,* 1658.

4. *Quantities, Units and Symbols in Physical Chemistry;* International Union on Pure and Applied Chemistry, Physical Chemistry Division, Commission on Physiochemical Symbols, Terminology, and Units, prepared for publication by Mills, I.; Cvitas, T.; Homann, K.; Kallay, N.; Kuchitsu, K.; Blackwell: Oxford, 1988; 134 pages.

   Note also *Manual of Symbols and Terminology for Physiochemical Quantities and Units;* International Union for Pure and Applied Chemistry, Physical Chemistry Division; (a) First edition: *Pure Appl. Chem.* **1970,** *21,* 1–38. (b) Second edition; *Manual of Symbols and Terminology for Physiochemical Quantities and Units;* Butterworths: London, 1975. (c) Third edition: *Pure Appl. Chem.* **1979,** *51,* 1–36. (d) "Appendix I—Definitions of Activities and Related Quantities"; *Pure Appl. Chem.* **1979,** *51,* 37–41. (e) "Appendix II—Definitions, Terminology and Symbols in Colloid and Surface Chemistry, Part I"; *Pure Appl. Chem.* **1972,** *31,* 577–638. (f) "Section 1.13: Selected Definitions, Terminology and Symbols for Rheological Properties"; *Pure Appl. Chem.* **1979,** *51,* 1213–1218. (g) "Section 1.14: Light Scattering"; *Pure Appl. Chem.* **1983,** *55,* 931–941. (h) "Part II: Heterogeneous Catalysis"; *Pure Appl. Chem.* **1976,** *46,* 71–90. (i) "Appendix III—Electrochemical Nomenclature"; *Pure Appl. Chem.* **1974,** *37,* 499–576. (j) "Appendix IV—Notation for States and Processes, Significance of the Word 'Standard' in Chemical Thermodynamics, Remarks on Commonly Tabulated Forms of Thermodynamic Functions"; *Pure Appl. Chem.* **1982,** *54,* 1239–1250. (k) "Appendix V—Symbolism and Terminology in Chemical Kinetics"; *Pure Appl. Chem.* **1981,** *53,* 753–771.

5. "Guide for the Presentation in the Primary Literature of Numerical Data Derived from Experiments"; International Council of Scientific Unions, Committee on Data for Science and Technology, CODATA Task Group on Publication of Data in the Primary Literature: Paris, September 1973, *CODATA Bull.,* No. 9, 1973, 6 pages.

   Note also (a) "Guidelines for the Reporting of Numerical Data in Experimental Procedures"; Garvin, D. *J. Res. Nat. Bur. Stand., Sect. A* **1972,** *76A,* 67–70.

6. "Guide for the Presentation in the Primary Literature of Physical Property Correlations and Estimation Procedures"; International Council of Scientific Unions, Committee on Data for Science and Technology, CODATA Task Group on Data for the Chemical Industry: Paris, 1978; *CODATA Bull.,* No. 30, 1978, 6 pages.

7. "A Guide to Procedures for the Publication of Thermodynamic Data"; International Union of Pure and Applied Chemistry, Physical Chemistry Division, Commission on Thermodynamics and Thermochemistry, prepared for publication by Kolesov, V. P.; McGlashan, M. L.; Rouquerol, J.; Seki, S.; Vanderzee, C. E.; Westrum, E. F., Jr. *Pure Appl. Chem.* **1972,** *29,* 399–408.

Also published in (a) *At. Energy Rev.* **1972,** *9,* 869. (b) *J. Chem. Thermodyn.* **1972,** *4,* 511–520. (c) *Indian J. Chem.* **1972,** *10,* 51. (d) *Indian J. Phys.* **1972,** *12,* 51. (e) *CODATA Newsletter* **1972,** *8,* 4. (f) *J. Chem. Eng. Data* **1973,** *18*(1), 3. (g) *J. Chim. Phys.* **1972,** *69*(10), 1407 (in French). (h) *Bull. Soc. Chim. Fr.* **March 1973,** Special number (in French). (i) The Society of Calorimetry and Thermal Analysis Special Publication, November 1971 (in Japanese). (j) *Zh. Fiz. Khim.* **1973.** *47,* 2459–2465 (in Russian). *Note also* (k) Reference 2. (l) Reference 3. (m) *Recommendations Concerning the Publication of the Results of Calorimetric Measurements;* Academy of Sciences of the USSR, Institute of General and Inorganic Chemistry, Scientific Council on Chemical Thermodynamics, prepared for publication by Sikolov, V. A.; Kolesov, V. P.; Vorob'ev, A. F. *Russ. J. Phys. Chem.* **1965,** *39,* 693–694 (English Translation). (n) *Zh. Fiz. Khim.* **1965,** *39,* 1298–1299.

8. "Guidelines for Reporting Experimental Data on Vapor–Liquid Equilibrium of Mixtures at Low and Moderate Pressures"; International Council of Scientific Unions, Committee on Data for Science and Technology, CODATA Task Group on Critically Evaluated Phase Equilibrium Data, *CODATA Bull.* **1989,** *21*(4), 69–78.

9. "Assignment and Presentation of Uncertainties of the Numerical Results of Thermodynamic Measurements (Provisional)"; International Union of Pure and Applied Chemistry, Physical Chemistry Division, Commission on Thermodynamics, Subcommittee on Assignment and Presentation of Uncertainties of Thermodynamic Data, *J. Chem. Thermodyn.* **1981,** *13,* 603–622.

Also published in (a) *Pure Appl. Chem.* **1981,** *53,* 1805–1825.

10. "Recommendations for Measurement and Presentation of Biochemical Equilibrium Data"; International Union of Pure and Applied Chemistry, International Union of Pure and Applied Biophysics, International Union of Biochemistry, Interunion Commission on Biothermodynamics, *J. Biol. Chem.* **1976,** *251,* 6879–6885.

Also published in (a) *Quant. Rev. Biophys.* **1976,** *9,* 439. (b) *CODATA Bull. Number 20, 1976. (c) Handbook of Biochemistry and Molecular Biology;* Fasman, G. D., Ed.; CRC Press: Cleveland, OH, 1976; Volume 1, Third Edition, Physical and Chemical Data Section, pp 93–106. (d) *Eur. J. Biochem.* **1977,** *72,* 1. (e) *Biochim. Biophys. Acta* **1976,** *461,* 1. (f) *Biofizika* **1978,** *23,* 739. (g) *Netsu Sokutei* **1977,** *4,* 172; **1978,** *5,* 77. (h) *Biochemical Nomenclature and Related Documents;* The Biochemical Society: London, 1978; pp 45–51.

11. "Recommendations for the Presentation of Thermodynamic and Related Data in Biology (Recommendations 1985)"; International Union of Pure and Applied Chemistry, IUPAB, IUB, Interunion Commission in Biothermodynamics, prepared for publication by Wads, I. *Eur. J. Biochem.* **1985,** *153,* 429–434

Also published in (a) *Pure Appl. Chem.* **1986,** *58* 1405–1410.

12. "Calorimetric Measurements on Cellular Systems: Recommendations for Measurements and Presentation of Results. (Provisional)"; International Union of Pure and Applied Chemists, IUPAB, IUP, Interunion Commission on Biothermodynamics, prepared for publication by Belaich, J. P.; Beezer, A. E.; Prosen, E.; Wads, I. *Pure Appl. Chem.* **1982,** *54,* 671–679.

Also published in CODATA Bull.; Pergamon Press: Oxford, 1981; Number 44.

13. Angus, S. "Guide for the Preparation of Thermodynamic Tables and Correlations of the Fluid State"; *CODATA Bull.;* Pergamon Press: Oxford, 1983; Number 51.

14. *Solubility Data Series Guidelines for Compilers, Evaluators and Editors;* International Union of Pure and Applied Chemistry, Analytical Chemistry Division, Commission on Solubility Data, Barton, A. F. M., Ed.; Pergamon: Oxford, 1984; 53 pages.

15. "Symbolism and Terminology in Chemical Kinetics (Provisional)"; International Union of Pure and Applied Chemistry, Physical Chemistry Division, Subcommittee on Chemical Kinetics, prepared for publication by Laidler, K. J. *Pure Appl. Chem.* **1981,** *53,* 753–771.

16. "The Presentation of Chemical Kinetic Data in the Primary Literature"; International Council of Scientific Unions, Committee on Data for Science and Technology, CODATA Task Group on Data for Chemical Kinetics, *CODATA Bull.;* CODATA Secretariat: Paris, 1974; Number 13, 7 pages.

17. "Manual of Symbols and Terminology for Physiochemical Quantities and Units: Appendix II. Definitions, Terminology and Symbols in Colloid and Surface Chemistry, Part II: Heterogeneous Catalysis"; International Union of Pure and Applied Chemistry, Physical Chemistry Division, Commission on Colloid and Surface Chemistry, prepared for publication by Burwell, R. L. *Pure Appl. Chem.* **1976,** *46,* 71–90.

18. *Enzyme Nomenclature (Recommendations 1984);* International Union of Biochemistry, Nomenclature Committee, Academic: New York, 1984.

19. Hanley, H. J. M.; Klein, M.; Liley, P. E.; Saxena, S. C.; Sengers, J. V.; Thodos, G.; White, H. J., Jr. "Recommendations for Data Compilations and for the Reporting of Measurements of the Thermal Conductivity of Gases"; *J. Heat Transfer* **November 1971,** 479–480.

20. "Manual of Symbols and Terminology for Physiochemical Quantities and Units: Appendix III, Electrochemical Nomenclature (Recommendations Approved 1973)"; International Union of Pure and Applied Chemistry, Division of Physical Chemistry, Commission on Electrochemistry, prepared for publication by Parsons, R. *Pure Appl. Chem.* **1974,** *37,* 499–516.

21. "Recommendations for Publishing Manuscripts on Ion-Selective Electrodes. (Recommendations 1981)"; International Union of Pure and Applied Chemistry, Analytical Chemistry Division, Commission on Analytical Nomenclature, prepared by Guilbault, G. G. *Pure Appl. Chem.* **1981,** *53,* 1907–1912.

    Note also (a) "Recommendations for Nomenclature of Ion-Selective Electrodes (Recommendations 1975)'; International Union of Pure and Applied Chemistry, Analytical Chemistry Division, Commission on Analytical Nomenclature, *Pure App. Chem.* **1976,** *48,* 127–132.

22. "Definition of pH Scales, Standard Reference Values, Measurement of pH and Related Terminology (Recommendations 1984)"; International Union of Pure and Applied Chemistry, Analytical Chemistry Division, Commission on Electroanalytical Chemistry, Physical Chemistry Division, Commission of Electrochemistry, prepared for publication by Covington, A. K.; Bates, R. G.; Durst, R. A. *Pure Appl. Chem.* **1985,** *57,* 531–542.

23. "The Absolute Electrode Potential: An Explanatory Note (Recommendations 1986)"; International Union of Pure and Applied Chemistry, Physical Chemistry Division, Commission on Electrochemistry, prepared for publication by Trasatti, S. *Pure Appl. Chem.* **1986,** *58,* 955–966.

24. "Recommendations on Reporting Electrode Potentials in Nonaqueous Solvents (Recommendations 1983)"; International Union of Pure and Applied Chemistry, Physical Chemistry Division, Commission on Electrochemistry, prepared for publication by Gritzner, G.; Kuta, J. *Pure Appl. Chem.* **1984,** *56,* 461–466.

Note also (a) "Recommendations on Reporting Electrode Potentials in Nonaqueous Solvents (Provisional)"; International Union of Pure and Applied Chemistry, Physical Chemistry Division, Commission on Electrochemistry, prepared for publication by Gritzner, G.; Kuta, J. *Pure Appl. Chem.* **1982,** *54,* 1527–1532.

25. Moody, G. J.; Thomas, J. D. R. "Development and Publication of Work with Selective Ion-Sensitive Electrodes"; *Talanta* **1972,** *19,* 623–639.

26. "Electrode Reaction Orders, Transfer Coefficients and Rate Constants Amplification of Definitions and Recommendations for Publication of Parameters (Recommendations 1979)"; International Union of Pure and Applied Chemistry, Physical Chemistry Division, Commission on Electrochemistry, prepared for publication by Parsons, R. *Pure Appl. Chem.* **1979,** *52,* 233–240.

27. "Nomenclature for Transport Phenomena in Electrolytes Systems (Recommendations 1979)"; International Union of Pure and Applied Chemistry, Physical Chemistry Division, Commission on Electrochemistry, prepared for publication by Ibl, N. *Pure Appl. Chem.* **1980,** *53,* 1827–1840.

28. "Manual of Symbols and Terminology for Physiochemical Quantities and Units: Appendix II. Definitions, Terminology and Symbols in Colloid and Surface Chemistry, Part I". International Union of Pure and Applied Chemistry, Division of Physical Chemistry, Commission on Colloid and Surface Chemistry, prepared for publication by Everett, D. H. *Pure Appl. Chem.* **1972,** *31,* 577–638.

29. "Reporting Physisorption Data for Gas/Solid Systems: with Special Reference to the Determination of Surface Area and Porosity (Recommendations 1984)"; International Union of Pure and Applied Chemistry, Physical Chemistry Division, Commission on Colloid and Surface Chemistry Including Catalysis, prepared for publication by Sing, K. S. W.; Everett, D. H.; Haul, R. A. W.; Moscou, L.; Pierrotti, R. A.; Rouquerol, J.; Siemieniewska, T. *Pure Appl. Chem.* **1985,** *57,* 603–619.

Note also (a) Reporting Physisorption Data for Gas/Solid Systems: with Special Reference to the Determination of Surface Area and Porosity (Provisional)"; International Union of Pure and Applied Chemistry, Physical Chemistry Division, Commission on Colloid and Surface Chemistry Including Catalysis, prepared for publication by Sing, K. S. W. *Pure Appl. Chem.* **1982,** *54,* 2201–2218.

30. "Reporting Data on Absorption from Solution at the Solid/Solution Interface (Recommendations 1986)"; International Union of Pure and Applied Chemistry, Physical Chemistry Division, Commission on Colloid and Surface Chemistry Including Catalysis, prepared for publication by Everett, D. H. *Pure Appl. Chem.* **1986,** *58,* 967–984.

31. "Reporting Experimental Pressure-Area Data with Film Balances (Recommendations 1984)"; International Union of Pure and Applied Chemistry, Physical Chemistry Division, Commission on Colloid and Surface Chemistry Including Catalysis, prepared for publication by Ter-Minassian-Saraga, L. *Pure Appl. Chem.* **1985,** *57,* 621–632.

Note also (a) "Reporting Experimental Pressure-Area Data with Film Balances (Provisional)"; International Union of Pure and Applied Chemistry, Physical Chemistry Division, Commission on Colloid and Surface Chemistry Including Catalysis, prepared by Ter-Minassian-Saraga, L. *Pure Appl. Chem.* **1982,** *54,* 2189–2200.

32. "Reporting Experimental Data Dealing with Critical Micellization Concentrations (C.M.C.'s) of Aqueous Surfactant Systems"; International Union of Pure and Applied Chemistry, Analytical Chemistry Division, Commission on Colloid and Surface Chemistry, prepared for publication by Mysels, K. J.; Mukerjee, P. *Pure Appl. Chem.* **1979,** *51,* 1083–1089.

33. "Recommended Standards for Reporting Photochemical Data. (Recommendations 1983)"; International Union of Pure and Applied Chemistry, Organic Chemistry Division, Commission on Photochemistry, prepared for publication by Lamola, A. A.; Wrighton, M. S. *Pure Appl. Chem.* **1984,** *56,* 939–944.

    Note also (a) "Recommended Standards for Reporting Photochemical Data (Provisional)"; International Union of Pure and Applied Chemistry, Organic Chemistry Division, Commission on Photochemistry, prepared for publication by Lamola, A. A.; Wrighton, M. S. *Pure Appl. Chem.* **1982,** *54,* 1251–1256.

34. "Recommendations for the Presentation of the Results of Chemical Analysis"; International Union of Pure and Applied Chemistry, Analytical Chemistry Division, Commission on Analytical Nomenclature, *Pure Appl. Chem.* **1969,** *18,* 439–442.

35. "Recommendations for the Usage of Selective, Selectivity and Related Terms in Analytical Chemistry"; International Union of Pure and Applied Chemistry, Analytical Chemistry Division, Commission on Analytical Reactions and Reagents, prepared for publication by Den Boef, G.; Hulanicki, A. *Pure Appl. Chem.* **1983,** *55,* 553–556.

36. Wilson, A. L. "The Performance Characteristics of Analytical Methods"; (a) *Talanta* **1970,** *17,* 21–29. (b) *Talanta* **1970,** *17,* 31–44. (c) *Talanta* **1973,** *20,* 725–732. (d) *Talanta* **1974,** *21,* 1109–1121.

37. Chalmers, R. A. "Writing, Reviewing and Editing in Analytical Chemistry"; *Rev. Anal. Chem.* **1970,** *1,* 217.

38. "Recommendations for Publication of Papers on Precipitation Methods of Gravimetric Analysis"; International Union of Pure and Applied Chemistry, Analytical Chemistry Division, Commission on Analytical Nomenclature, prepared for publication by Kirkbright, G. F. *Pure Appl. Chem.* **1981,** *53,* 2303–2306.

    Note also (a) Erdey, L.; Polos, L.; Chalmers, R. A. "Development and Publication of New Gravimetric Methods of Analysis"; *Talanta* **1970,** *17,* 1143–1155.

39. Rossotti, H. S. "Design and Publication of Work on Stability Constants"; *Talanta* **1974,** *21,* 809–829.

40. Mark, H. B., Jr. "Development and Publication of New Methods in Kinetic Analysis"; *Talanta* **1973,** *20,* 257–266.

41. "Nomenclature for Thermal Analysis IV (Recommendations 1985)"; International Union of Pure and Applied Chemistry, Analytical Chemistry Division, prepared for publication by Mackenzie, R. C. *Pure Appl. Chem.* **1985,** *57,* 1737–1740.

    Note also (a) "Nomenclature for Thermal Analysis II and III (Recommendations 1979)"; International Union of Pure and Applied Chemistry, Analytical Chemistry Division, prepared for publication by Guilbault, G. G. *Pure Appl. Chem.* **1980,** *52,* 2385–2391. (b) Mackenzie, R. C. "Nomenclature for Thermal Analysis"; *Talanta* **1969,** *16,* 1227–1230.

42. Newkirk, A. E.; Simons, E. L. "Suggestions for Reporting Dynamic Thermogravimetric Data, (Letter to the Editor)"; *Talanta* **1963,** *10,* 1199.

43. McAdie, H. D. "Recommendations of the Committee on Standardization of the International Conference on Thermal Analysis for Presenting Data on Differential Thermal Analysis and Thermogravimetry (Letter to the Editor)"; *Anal. Chem.* **1967,** *39,* 543.

44. "Recommendations on Nomenclature for Chromatography (Rules Approved 1973)"; International Union of Pure and Applied Chemistry, Analytical Chemistry Division, Commission on Analytical Nomenclature, prepared for publication by Ambrose, D.; Boyer, E.; Samuelson, O. *Pure Appl. Chem.* **1974,** *37,* 447–462.

Note also (a) "Recommendations on Ion Exchange Nomenclature"; International Union of Pure and Applied Chemistry, Analytical Chemistry Division, Commission on Analytical Nomenclature, prepared for publication by Samuelson, O.; Boyer, E.; Helfferich, F. G. *Pure Appl. Chem.* **1972,** *29,* 619–624. (b) "Recommended Nomenclature for Liquid-Liquid Distribution"; International Union of Pure and Applied Chemistry, Analytical Chemistry Division, Commission on Analytical Nomenclature, prepared for publication by Irving, H. M. N. H. *Pure Appl. Chem.* **1970,** *21,* 111–113. (c) "Proposed Recommended Practice for Gas Chromatography Terms and Relationship"; American Society for Testing and Materials Committee E 19, Gas Chromatography, *J. Gas Chromatogr.* **1968,** *6,* 1–4. (d) "Recommendations on Nomenclature and Presentation of Data in Gas Chromatography"; International Union of Pure and Applied Chemistry, Division of Analytical Chemistry, *Pure Appl. Chem.* **1964,** *8,* 553–562. (e) "Preliminary Recommendations on Nomenclature and Presentation of Data in Gas Chromatography"; International Union of Pure and Applied Chemistry, Analytical Chemistry Division, *Pure Appl. Chem.* **1960,** *1,* 177–185.

45. "Recommendations for Publication of Papers on a New Analytical Method Based on Ion Exchange or Ion-Exchange Chromatography"; International Union of Pure and Applied Chemistry, Analytical Chemistry Division, Commission on Analytical Reactions and Reagents, prepared for publication by Inczedy, J. *Pure Appl. Chem.* **1980,** *52,* 2553–2562.

46. "Recommended Terms, Symbols, and Definitions for Electroanalytical Chemistry (Recommendations 1985)"; International Union of Pure and Applied Chemistry, Analytical Chemistry Division, Commission on Electroanalytical Chemistry, prepared for publication by Meites, L.; Zuman, P.; Nurnberg, H. W. *Pure Appl. Chem.* **1985,** *57,* 1491–1505.

Note also (a) "Recommended Terms, Symbols and Definitions for Electroanalytical Chemistry (Provisional)"; International Union of Pure and Applied Chemistry, Analytical Chemistry Division, Commission on Electroanalytical Chemistry, prepared for publication by Meites, L. *Pure Appl. Chem.* **1979,** *51,* 1159–1174.

47. "Conditional Diffusion Coefficients of Ions and Molecules in Solution, an Appraisal of the Conditions and Methods of Measurement"; International Union of Pure and Applied Chemistry, Analytical Chemistry Division, Commission on Electroanalytical Chemistry, prepared for publication by Bishop, E.; Galus, Z. *Pure Appl. Chem.* **1982,** *54,* 1575–1582.

48. "General Aspects of Trace Analytical Methods—IV. Recommendations for Nomenclature, Standard Procedures and Reporting of Experimental Data for Surface Analysis Techniques (Provisional)"; International Union of Pure and Applied Chemistry, Analytical Chemistry Division, Commission on Microchemical Techniques and Trace Analysis, prepared for publication by Morrison, G. H.; Cheng, K. L.; Grassenbauer, M. *Pure Appl. Chem.* **1979,** *51,* 2243–2250.

49. "Nomenclature and Spectral Presentation in Electron Spectroscopy Resulting from Excitation by Photons (Recommendations 1975)"; International Union of Pure

and Applied Chemistry, Physical Chemistry Division, Commission on Molecular Structure and Spectroscopy, *Pure Appl. Chem.* **1976,** *45,* 221–224.

50. "Recommendations for Publication of Papers on Methods of Molecular Absorption Spectrophotometry in Solution Between 200 and 800 nm"; International Union of Pure and Applied Chemistry, Analytical Chemistry Division, Commission on Analytical Nomenclature, prepared for publication by Kirkbright, G. F. *Pure Appl. Chem.* **1978,** *50,* 237–242.

    Note also (a) Kirkbright, G. F. "Development and Publication of New Spectrophotometric Methods of Analysis"; *Talanta* **1966,** *13,* 1–14.

51. "Nomenclature, Symbols, Units and Their Usage in Spectrochemical Analysis"; International Union of Pure and Applied Chemistry, Division of Analytical Chemistry, Commission on Spectrochemical and Other Optical Procedures for Analysis"; (a) "General Atomic Emission Spectroscopy"; *Pure Appl. Chem.* **1972,** *30,* 653–679. (b) "Data Interpretation (Rules Approved 1975)"; *Pure Appl. Chem.* **1976,** *45* 99–103. (c) "Analytical Flame Spectroscopy and Associated Non-Flame Procedures (Rules Approved 1975)"; *Pure Appl. Chem.* **1976,** *45,* 105–123. (d) "X-ray Emission Spectroscopy (Recommendations 1979)"; prepared for publication by Jenkins, R. *Pure Appl. Chem.* **1982,** *52,* 2541–2552. (e) "Radiation Sources (Recommendations 1985)"; prepared for publication by Butler, L. R. P.; Laqua, K.; Strasheim, A. *Pure Appl. Chem.* **1985,** *57,* 1453–1490. (f) "Molecular Luminescence Spectroscopy (Recommendations 1983)"; prepared for publication by Melhuish, W. H. *Pure Appl. Chem.* **1984,** *56,* 231–245. (g) "Molecular Absorption Spectroscopy, Ultraviolet and Visible (UV/VIS) (Recommendations 1988)"; prepared for publication by Laqua, K.; Melhuish, W. H.; Zander, M. *Pure Appl. Chem.* **1988,** *60,* 1449–1460. (h) "Preparation of Materials for Analytical Atomic Spectroscopy and Other Related Techniques"; prepared for publication by Ure, A. M.; Butler, L. R. P.; Scott, R. O.; Jenkins, R. *Pure Appl. Chem.* **1988,** *60,* 1461–1472.

    Note also (a) Item e above) "Radiation Sources (Provisional)"; prepared for publication by Butler, L. R. P.; Laqua K. *Pure Appl. Chem.* **1981,** *53,* 1913–1950. (b) Item f above "Molecular Luminescence Spectroscopy (Provisional)"; prepared for publication by Melhuish, W. A. *Pure Appl. Chem.* **1981,** *53,* 1953–1966.

52. "Names, Symbols, Definitions and Units of Quantities in Optical Spectroscopy (Recommendations 1984)"; International Union of Pure and Applied Chemistry, Physical Chemistry Division, Commission on Molecular Structure and Spectroscopy, and Clinical Chemistry Division, Commission on Quantities and Units, prepared for publication by Sheppard, N.; Willis, H. A.; Rigg, J. C. *Pure Appl. Chem.* **1985,** *57,* 105–120.

53. *International Tables for Crystallography. Volume A, Second Edition: Space-Group Symmetry;* Hahn, Th., Ed.; Reidel: Dordrecht, 1987.

54. Kennard, O.; Speakman, J. C.; Donnay, J. D. H. "Primary Crystallographic Data (Recommendations of the International Union of Crystallography Commission on Crystallographic Data)"; *Acta Crystallogr.* **1967,** *22,* 445–449.

55. Ibers, J. A. "Information and Suggestions on Presentation of the Results of Crystal Structure Studies"; *Inorg. Chim. Acta* **1969,** *3*(1), 9–12.

56. Himes, V. L.; Mighell, A. D. *NBS.LATTICE: A Program To Analyze Lattice Relationships;* U.S. Department of Commerce, National Institute of Standards and Technology: Gaithersburg, MD, 1985; NBS Technical Note 1214, 74 pages.

57. Himes, V. L.; Mighell, A. D. "A Matrix Approach to Symmetry", *Acta Crystallog.* **1987,** *A43,* 375–384.

58. deWolff, P. M.; Belov, N. V.; Bertaut, E. F.; Buerger, M. J.; Donnay, J. D. H.; Fischer, W.; Hahn, Th.; Koptsik, V. A.; Mackay, A. L.; Wondratschek, H.; Wilson, A. J. C.; Abrahams, S. C. "Nomenclature for Crystal Families, Bravais-Lattice Types and Arithmetic Classes: Report of the International Union of Crystallography Ad-Hoc Committee on the Nomenclature of Symmetry"; *Acta Crystallogr.* **1985,** *A41,* 278–280.

59. Guinier, A.; Bokij, G. B.; Boll-Dornberger, K.; Cowley, J. M.; Durovic, S.; Jagodzinski, H.; Krishna, P.; deWolff, P. M.; Zvyagin, B. B.; Cox, D. W.; Goodman, P.; Hahn, Th.; Kuchitsu, K.; Abrahams, S. C. "Nomenclature of Polytype Structures: Report of the International Union of Crystallography Ad-Hoc Committee on the Nomenclature of Disordered, Modulated and Polytype Structures"; *Acta Crystallogr.* **1984,** *A40,* 399–404.

60. deWolff, P. M.; Billiet, Y.; Donnay, J. D. H.; Fischer, W.; Galuibin, R. B.; Glazer, A. M.; Senechal, M.; Shoemaker, D.P.; Wondratschek, H.; Hahn, Th.; Wilson, A. J. C.; Abrahams, S. C. "Definition of Symmetry Elements in Space Groups and Point Groups: Report of the International Union of Crystallography Ad-Hoc Committee on the Nomenclature of Symmetry"; *Acta Crystallogr.* **1989,** *A45,* 494–498.

61. Schwartzenbach, D.; Abrahams, S. C.; Flack, H. D.; Gonschorek, W.; Hahn, Th.; Huml, K.; Marsh, R. E.; Prince, E.; Robertson, B. E.; Rollett, J. S.; Wilson, A. J. C. "Statistical Descriptors in Crystallography: Report of the International Union of Crystallography, Subcommittee on Statistical Descriptors"; *Acta Crystallogr.* **1989,** *A45,* 63–75.

62. Kennard, O.; Hanawalt, J. D.; Wilson, A. J. C.; deWolff, P. M.; Frank-Kamenetsky, V. A. "Powder Data"; International Union of Crystallography, Commission on Crystallographic Data, *J. Appl. Crystallogr.* **1971,** *4,* 81–86.

63. Young, R. A.; Prince, E.; Sparks, R. A. "Suggested Guidelines for the Publication of Rietveld Analyses and Pattern Decomposition Studies (Letter to the Editor)"; *J. Appl. Crystallogr.* **1982,** *15,* 357–359.

64. Calvert, L. D.; Flippen-Anderson, J. L.; Hubbard, C. R.; Johnson, Q. C.; Lenhart, P. G.; Nichols, M. C.; Parrish, W.; Smith, D. K.; Smith, G. S.; Snyder, R. L.; Young, R. A. "The Standard Data Form for Powder Diffraction Data"; Subcommittee of the American Crystallographic Association, excerpt of final report presented at the Symposium on Accuracy in Powder Diffraction, Washington, DC, June 1979, International Centre for Diffraction Data, Swarthmore, PA, 23 pages.

65. Bartell, L. S.; Kuchitsu, K.; Seip, H. M. "Guide for the Publication of Experimental Gas-Phase Electron Diffraction Data and Derived Structural Results in the Primary Literature"; International Union of Crystallography, Commission on Electron Diffraction, *Acta Crystallogr.* **1976,** *A32,* 1013–1018.

66. Dillard, J. G.; Heller, S. R.; McLafferty, F. W.; Milne, G. W. A.; Venkataraghavan, R. "Critical Evaluation of Class II and Class III Electron Impact Mass Spectra. Operating Parameters and Reporting Mass Spectra (Letter to the Editor)"; *Org. Mass Spectrom.* **1981,** *16*(1), 48–49.

67. "Recommendations for Symbolism and Nomenclature for Mass Spectrometry", International Union of Pure and Applied Chemistry, Physical Chemistry Division, Commission on Molecular Structure and Spectroscopy, prepared for publication by Beynon, J. H. *Pure Appl. Chem.* **1978,** *50,* 65–73.

Note also (a) "Recommendations for Nomenclature of Mass Spectrometry (Rules Approved 1973)"; International Union of Pure and Applied Chemistry, Analytical

Chemistry Division, Commission on Analytical Nomenclature, *Pure Appl. Chem.* **1974,** *37,* 469–480.

68. "Specifications for Infrared Reference Spectra of Molecules in the Vapor Phase (Recommendations 1987)"; International Union of Pure and Applied Chemistry, Physical Chemistry Division, Commission on Molecular Structure and Spectroscopy in conjunction with the Coblentz Society Spectral Evaluation Committee, Vapor Phase Subcommittee, prepared for publication by Grasselli, J. G. *Pure Appl. Chem.* **1987,** *59,* 673–681.

   Note also (a) "Specifications for Infrared Reference Spectra of Molecules in the Vapor Phase (Provisional)"; Abstract given in *Chem. Int.* **1985,** *7*(3), 25.

69. "Recommendations for the Presentation of Infrared Absorption Spectra in Data Collections—A. Condensed Phases"; International Union of Pure and Applied Chemistry, Physical Chemistry Division, Commission on Molecular Structure and Spectroscopy, prepared for publication by Becker, E. D. *Pure Appl. Chem.* **1978,** *50,* 231–236.

70. "The Coblentz Society Specifications for Evaluation of Research Quality Analytical Infrared Spectra (Class II)"; *Anal. Chem.* **1975,** *47*(11), 945A–952A.

   Note also (a) *Coblentz Society Newsletter,* Number 41, 1969. (b) "Specifications for Evaluation of Infrared Reference Spectra"; The Coblentz Society Board of Managers, *Anal. Chem.* **1966,** *38*(9), 27A–38A.

71. Craver, C. D.; Grasselli, J. G.; Smith, A. L. "Criteria for Infrared Spectra Submitted to Journals (Letter to the Editor)"; *Anal. Chem.* **1975,** *47,* 2065.

72. "Presentation of Molecular Parameter Values for Infrared and Raman Intensity Measurements (Recommendations 1988)"; International Union of Pure and Applied Chemistry, Physical Chemistry Division, Commission on Molecular Structure and Spectroscopy, prepared for publication by Gussoni, M.; Zerbi, G.; Person, W. *Pure Appl. Chem.* **1988,** *60,* 1385–1388.

   Note also (a)"Recommendations for the Presentation of Molecular Parameter Values for Infrared and Raman Intensity Measurements (Provisional)"; Abstract given in *Chem. Int.* **1986,** *8,* 18.

73. "Presentation of Raman Spectra in Data Collections (Recommendations 1981)"; International Union of Pure and Applied Chemistry, Physical Chemistry Division, Commission on Molecular Structure and Spectroscopy, prepared for publication by Becker, E. D.; Durig, J. R.; Harris, W. C.; Rosasco, G. J. *Pure Appl. Chem.* **1981,** *53,* 1879–1885.

   Note also (a) "Recommendations for the Presentation of Raman Spectral Data"; National Research Council, National Academy of Sciences, Numerical Data Advisory Board, Ad Hoc Panel on Raman Spectral Data, *Appl. Spectrosc.* **1976,** *30,* 20–22. (b) "Recommendations for the Presentation of Raman Spectra for Cataloguing and Documentation in Permanent Data Collections"; International Union of Pure and Applied Chemistry, Physical Chemistry Division, Commission on Molecular Structure and Spectroscopy; *Pure Appl. Chem.* **1973,** *36,* 277–278.

74. "Recommendations for the Presentation of NMR Data for Publication in Chemical Journals"; International Union of Pure and Applied Chemistry, Physical Chemistry Division, Commission on Molecular Structure and Spectroscopy; *Pure Appl. Chem.* **1972,** *29,* 627–628.

75. "Standard Practice for Data Presentation Relating to High-Resolution Nuclear Magnetic Resonance (NMR) Spectroscopy," *ASTM Designation E 386–90 Annual Book of ASTM Standards;* American Society for Testing and Materials: Philadelphia, PA, 1990; pp 1–9.

Note also "Standard Definitions of Terms, Symbols, Conventions, and References Relating to High-Resolution Nuclear Magnetic Resonance (NMR) Spectroscopy"; *ASTM Designation: E 386–76. Annual Book of ASTM Standards;* American Society for Testing and Materials: Philadelphia, PA, 1976; p 1–7.

76. "Presentation of NMR Data for Publication in Chemical Journals—B. Conventions Relating to Spectra from Nuclei Other Than Protons (Recommendations 1975)"; International Union of Pure and Applied Chemistry, Physical Chemistry Division, Commission on Molecular Structure and Spectroscopy; *Pure Appl. Chem.* **1976,** *45,* 217–219.

77. "Recommendations for EPR/ESR Nomenclature and Conventions for Presenting Experimental Data in Publications (Recommendations 1989)"; International Union of Pure and Applied Chemistry, Physical Chemistry Division, Commission on Molecular Structure and Spectroscopy, prepared for publication by Kon, H. *Pure Appl. Chem.* **1989,** *61,* 2195–2200.

    Note also (a) "Recommendations for EPR/ESR Nomenclature and Conventions for Presenting Experimental Data in Publications (Provisional)"; Abstract given in *Chem. Int.* **1987,** *9,* 71.

78. "Nomenclature and Conventions for Reporting Mssbauer Spectroscopic Data (Recommendations 1975)"; International Union of Pure and Applied Chemistry, Physical Chemistry Division, Commission on Molecular Structure and Spectroscopy; *Pure Appl. Chem.* **1976,** *45,* 211–216.

79. Jenkins, F. A. "Report of Subcommittee f (Notation for the Spectra of Diatomic Molecules)"; *J. Opt. Soc. Am.* **1953,** *43,* 425–426.

    Note also (a) Mulliken, R. S. "Report on Notation for Spectra of Diatomic Molecules"; *Phys. Rev.* **1930,** *36,* 611–629.

80. "Report on Notation for the Spectra of Polyatomic Molecules, Report of the Joint Commission for Spectroscopy of the International Astronomical Union and International Union of Pure and Applied Chemistry"; *J. Chem. Phys.* **1955,** *23,* 1997–2011.

81. Brown, J. M.; Hougen, J. T.; Huber, K.-P.; Johns, J. W. C.; Ramsay, D. A.; Rostas, J.; Zare, R. N. "The Labelling of Parity Doublet Levels in Linear Molecules (Letter to the Editor)"; *J. Mol. Spectra* **1975,** *55,* 500–503.

82. Brand, J. C. D.; Callomon, J. H.; Innes, K. K.; Jortner, J.; Leach, S.; Levy, D. H.; Merer, A. J.; Mills, I. M.; Moore, C. B.; Parmenter, C. S.; Ramsay, D. A.; Narahari Rao, K.; Schlag, E. W.; Watson, J. K. G.; Zare, R. N. "The Vibrational Numbering of Bands in the Spectra of Polyatomic Molecules (Letter to the Editor)"; *J. Mol. Spectrsc.* **1983,** *99,* 482–483.

83. "Definitions and Symbolism of Molecular Force Constants (Recommendations 1978)"; International Union of Pure and Applied Chemistry, Physical Chemistry Division, Commission on Molecular Structure and Spectroscopy, prepared for publication by Morino, Y.; Shimanouchi, T. *Pure Appl. Chem.* **1978,** *50,* 1707–1713.

84. Moore, C. E. *Atomic Energy Levels—Volume 1;* U.S. Department of Commerce, National Institute of Standards and Technolgy: Gaithersburg, MD, 1971; Sections 3 and 5 NSRDS–NBS 35.

85. Martin, W. C.; Zalubas, R.; Hagan, L. *Atomic Energy Levels—The Rare-Earth Elements;* U.S. Department of Commerce, National Institute of Standards and Technolgy: Gaithersburg, MD, 1978; Section 2 NSRDS–NBS 60.

86. Wiese, W. L.; Smith, M. W.; Miles, B. M. *Atomic Transition Probabilities, Volume II, Sodium Through Calcium: A Critical Data Compilation;* U.S. Department of Commerce, National Institute of Standards and Technolgy: Gaithersburg, MD, 1969; Section C. NSRDS–NBS 22, pp i–xv.

87. Wiese, W. L.; Fuhr, J. R. "On the Accuracy of Atomic Transition Probabilities"; in *Lecture Notes in Physics;* Springer Verlag: New York, 1989; Volume 356, pp 7–18.

88. Wiese, W. L. "On the Critical Assessment of Atomic Spectroscopy Data"; in *Report of Workshop on the Assessment of Atomic Data; Queen University: Belfast, Ireland, 1987; 8 pages.*

89. *"The Standard Crystallographic File Structure"; Brown, I. D. Acta Crystallogr.* **1983,** *A39,* 216–224.

Note also (a) The current version, SCFS–84, December 1984, is available from Dr. I. D. Brown, Institute for Materials Research, McMaster University, Hamilton, Ontario, Canada L85 4M1. (b) Stalick, J. K.; Mighell, A. D. *Crystal Data: Version 1.0 Database Specifications;* U.S. Department of Commerce, National Institute of Standards and Technology: Gaithersburg, MD, 1986; NBS Technical Note 1229, 66 pages.

90. *Computerization and Networking of Materials Databases: Proceedings of the 1st International Symposium on Computerization and Networking of Materials Property Databases;* Glazman, J. S.; Rumble, J. R., Jr., Eds.; American Society for Testing and Materials: Philadelphia, PA, 1988; ASTM STP 1017, 355 pages.

Note particularly (a) Anderson, D. B.; Laverty, G. J. "Corrosion Data for Materials Performance Characterization"; pp 317–321. (b) Kaufman, J. G. *Standards for Computerized Material Property Data—ASTM Committee E 49; pp 7–22.*

91. *CODATA Thermodynamic Tables: Selections for Some Compounds of Calcium and Related Mixtures: A Prototype Set of Tables;* Garvin, D.; Parker, V. B.; White, H. J., Jr., Eds.; Hemisphere: Washington, DC, 1987; 355 pages.

92. Interested readers are referred to Dr. Henry Kehiaian, Universite de Paris VII CNRS, Institute de Topologie et de Dynamique des Systemes, 1 Rue Guy de la Brosse, 75005 Paris, France.

INDEX

A

B

T

Copy editing: Margaret J. Brown and Janet S. Dodd
Production: Margaret J. Brown
Acquisition: Robin Giroux and Cheryl Shanks
Indexing: Deborah H. Steiner

Printed and bound by Courier On-Demand